Wind Power:
The Industry Grows Up

WIND POWER

THE INDUSTRY GROWS UP

Rebecca L. Busby

PennWell®

PennWell Corporation
1421 South Sheridan Road
Tulsa, Oklahoma 74112-6600 USA

800.752.9764
+1.918.831.9421
sales@pennwell.com
www.pennwellbooks.com
www.pennwell.com

Marketing: Jane Green
National Account Executive: Barbara McGee

Director: Mary McGee
Managing Editor: Stephen Hill
Production Manager: Sheila Brock
Production Editor: Tony Quinn
Book Designer: Susan E. Ormston
Cover Designer: Meloni Janzen

Library of Congress Cataloging-in-Publication Data

Busby, Rebecca L.
Wind power : the industry grows up / Rebecca L. Busby.
p. cm.
Includes bibliographical references and index.
ISBN 978-1-59370-244-1
1. Wind power. 2. Wind energy conversion systems. I. Title.
TJ820.B87 2012
621.31'2136--dc23

2011047129

Printed in the United States of America

1 2 3 4 5 16 15 14 13 12

To Woody and Pinky—we miss you.

Contents

Illustrations

Figures

Tables

Preface

When I started this book in 2010, the world's industrial economies were stagnant and their governments hamstrung by debt—not a promising scenario for wind power development. Things haven't gotten much better since then. But the wind industry has adapted to changes in the marketplace, like exploding growth in China and offshore northern Europe.

Just one example is Spanish turbine maker Gamesa, which controlled more than a third of the Chinese market as recently as 2005, according to the *New York Times*. Gamesa assembled locally sourced components and trained domestic suppliers. By 2010, Chinese manufacturers controlled 90% of their market—but Gamesa was selling more than twice as many turbines, even with just a sliver of its original share.

In other words, the wind business has grown up, and the industry has become more resilient. Whether it can weather more financial storms, like Europe's and America's austerity policies, remains to be seen.

This book focuses on utility-scale turbines and wind farms. Wind energy can make a meaningful contribution to global power supply with large projects as well as distributed power, which I've always supported. I'm agnostic as to how we'll get our energy in the long run. We'll probably need all we can get from a variety of sources, not the least of which could be better energy efficiency.

As a footnote, I should mention that currency conversions in the book were calculated based on average rates between September 2010 and September 2011. The most common conversion is between euros and dollars (about 0.73 euros/$, or 1.4 $/euro).

Readers, I appreciate you. Thank you!

Acknowledgments

This book is founded on the hard work of hundreds of people, from the wind engineers who publish technical articles, to the journalists who translate them for the rest of us, and to the entrepreneurs who are hanging on to their dreams and keeping their wind businesses alive.

I'm very grateful to the wind energy professionals who took time out to answer questions when they were busy doing other things—Ed Duggan, Jack Kline, and Jim Webb. Thanks also to those who reviewed the book, including Jon McGowan, Herman Trabish, Malcolm Brown, and Dave Wright.

I also very much appreciate the people who provided artwork for the book. These include Niels Gylling Mortensen at Risø DTU, Alex Woodward at Natural Power, Todd Stone at 3TIER, Naomi Pierce at Second Wind, and Ben Bell at GL Garrad Hassan, who also shared his knowledge of wind farm financing, development, and operation.

I could not have written this book without relying on the wind industry reporters working for Greentech Media, *Renewable Energy World*, and many other news services. These journalists all keep a sharp eye on the global wind business, but among them I must single out Herman Trabish, whose work exasperates me by always reaching one step ahead of mine.

I also thank my family, friends, dogs, cats, and goats for putting up with more than a year of me being extra busy. Books devour time because writers collect way more information than they can hope to digest. But that's what writers always want—more information.

My husband Dan Handman remains my best friend and most critical reviewer. He's been in and out of the wind business since the 1970s, and I couldn't have written this book without his knowledge as my safety net. I would thank him for his help, but I'm still smarting from his mark-ups on torque, lift, and thrust in chapter 3.

Now as I set this manuscript aside, I'm reminded of some advice from a long-ago "Shoe" comic: The only way to stop editing something is to stop reading it. Okay, I'm done.

1

Overview: How Wind Power Works

Introduction

The idea of wind power is to capture *energy* from the wind and put it to work, doing something practical. *Wind energy*, like most other types of *renewable energy*, gets its power from the sun. Even our so-called *fossil fuels*—coal, oil, and natural gas—derive their energy from the sun indirectly, since they were formed from the remains of ancient plants that needed sunlight to grow.

Most of the sun's energy reaches us as heat or light, but about 1%–2% of it becomes wind energy. That's about 50–100 times more energy than all the plants in the world capture from the sun.

How the sun produces wind

Wind is created when solar radiation heats up the earth's land and water at different rates and at different times. For example, the air above land gets hotter faster than the air above water. Warm air is lighter in weight compared to cold air, which is denser. So the warm air on land rises into the sky, as much as six miles up, then begins to spread toward the north and south. Cooler air over the water rushes in to take its place, and that's what we call wind.

The same thing can happen when air over high mountains warms up quickly in the morning, and cooler air from nearby valleys flows up toward the mountaintops. Differences in air temperature can also create wind through canyons. These are all types of local winds.

On a larger scale, the air in countries around the earth's equator, because they're closer to the sun, gets hotter than the air in other areas farther to the north or south. When this warm tropical air rises, cooler air moves in. Another global factor in creating wind is that one side of the earth is dark at night while the other side is lit up by the sun, so the two sides are getting heated at different times.

Because the earth is spinning on its axis in space, the wind changes its direction while in motion. Instead of blowing straight north or south from warm areas around the equator to colder ones, the globe's rotation creates *trade winds*, which move from east to west.

Then farther away from the equator, the prevailing winds shift direction, moving from west to east. This band of wind, which lies in both the Northern and Southern hemispheres from about 30 to 60 degrees in latitude, is called the *westerlies*. Even farther away, near the North and South Poles of the earth, the wind direction changes again, creating bands of polar *easterlies*.

These global wind circulation patterns are a major force affecting the earth's climate, since they are involved in the formation of hurricanes, typhoons, and other severe weather. But they also play a role in identifying the best sites for developing wind power projects. Chapters 5 and 6 discuss wind resource assessment and wind farm siting.

Capturing and using the wind's energy

Air weighs very little, but it does have mass. Wind is air in motion, and the moving mass of the air contains *kinetic* energy (the energy of motion). The challenge of putting the wind's kinetic energy to work lies in converting it efficiently into more useful forms of energy, such as *mechanical power* or *electricity*. The motion of the wind can be captured by sails on a boat or by fans, which begin to spin when the wind blows into them. This rotating motion turns a *driveshaft* that can be connected to other machinery to produce mechanical power.

Wind energy systems that produce mechanical power are usually found in rural or remote places, like farms and undeveloped countries. Mechanical wind power typically is used to pump water up from the ground for livestock to drink or for irrigating crops. However, wind power can also be used mechanically for grinding grain in a millstone, sawing wood in a lumberyard, or pushing a sailboat.

In areas that are more densely populated, electric power is a more useful form of energy than mechanical power. When the driveshaft of a wind energy system is connected to a *generator*, the wind's mechanical power is converted into electricity. In modern wind power systems, the "fan" is a set of *blades* that look like airplane propellers but actually behave more like airplane wings.

The spinning blades drive the shaft of the generator, which contains magnets inside a coil of wire. When the driveshaft spins the magnets, electrical power is generated. Chapter 2 discusses blades, generators, and other components of wind energy systems.

The history of using wind power

Human beings have exploited the wind's energy since the dawn of ancient civilization. Thousands of years ago, the Egyptians invented the world's early sailboats, some time before 1000 BC. These boats, called feluccas, are still common on the Nile River today. Table 1–1 summarizes the history of wind power.

Table 1–1. History of wind power

5000–1000 BC	Egyptians, other early civilizations develop sailboats
200 BC	Small, simple windmills operate in China (not documented)
500–900 AD	Land-based vertical-axis windmills are used in Persia (Iran) for pumping water and grinding grain
~1000	Vikings cross the Atlantic from Scandinavia to North America in wooden ships with square sails
1390	Windmill design is refined by the Dutch people, who adapt it to drain water from low-lying areas
1400s–1800s	Explorers, settlers cross seas and oceans on ships using sails powered by wind energy
1600–1700	Settlers build windmills for grinding grain, pumping water, and sawing wood
1854	Halladay develops a windmill that pivots to catch the wind and controls its speed in high winds
1854–1900	Halladay windmills become widespread on America's farms and ranches for pumping water
1888	Electricity is first generated by a "large" windmill (12 kW), built by Charles F. Brush in Cleveland, OH
1890–1900	Windmills in Denmark built by Poul la Cour operate at higher speeds and take greater advantage of aerodynamic principles to generate electricity
1900–1918	Windmills generating 25 kW each spread throughout Denmark, but are soon eclipsed by larger fossil-fueled power plants
1931	The first vertical-axis wind turbine design is patented by French aeronautical engineer Georges Darrieus
1930s–40s	Rural electrification spreads through America, idling many small windmills
1930s–70s	Large experimental wind turbines (100 kW–1.25 MW) operate throughout Europe and the United States, but are not practical for widespread use
1973	Wind power and other renewable energy sources enter the spotlight when Middle Eastern oil producers cut back exports to oil-consuming nations
1970s–80s	Government programs in Europe and America spur research and development of multi-megawatt wind turbine technology
1990s	Europe takes the lead in wind power construction while U.S. developments languish
1991	The world's first offshore wind farm is built, Vindeby, located off the coast of Denmark in the Baltic Sea
2009	Wind power meets about 20% of Denmark's electricity demand
2010	SeaTitan, Britannia, and other wind turbines are being designed for 10–15 MW of capacity

People soon learned to build larger ships capable of crossing the Mediterranean Sea and, eventually, exploring new continents across the oceans. The Vikings first reached North America from Scandinavia about 1,000 years ago in wooden ships with square sails. All of the world's famous explorers and settlers—Columbus, Magellan, the Pilgrims—traveled in ships powered by wind energy.

People began using wind power on land more than 2,000 years ago, when the Chinese built small, simple windmills, although this has not been documented. About 1,300 years ago, around the 7th or 8th century, windmills were being used in Persia, now known as Iran.

The purpose of these early land-based windmills was the same then as it is now—to save labor by mechanically pumping water or grinding grain. Within a couple of hundred years, windmills were used extensively throughout the Middle East. Windmills are especially vital in dry climates, where no fast-running rivers are around to drive a pump or a millstone.

About 600 years ago, the Dutch people began to improve these primitive windmills, adapting them to the task of draining water from the land, much of which lies at or below sea level in the Netherlands.

By the late 1800s, windmills had spread throughout new settlements in America's Great Plains, where remote farms and ranches depended on them to pump water from wells. This type of windmill was based on a machine designed by Daniel Halladay in 1854 that was "self-governed," meaning that it pivots to face the wind and it's able to control its speed in high winds.

The invention of the steam engine and the spread of industrialization throughout Europe and America led to a gradual decline in the prevalence of windmills used for pumping water. Inventors switched their focus to electricity generation, and Charles Brush built the first large windmill for that purpose in 1888.

During the 1890s in Denmark, Poul la Cour developed more practical windmills that operated at higher speeds and took greater advantage of *aerodynamic* principles to produce electricity. In power generation applications, windmills eventually became known as *wind turbines*.

During the early to mid-1900s, wind turbines evolved from small machines that used fans or sails into larger, more powerful generators using modified airplane propellers or wings. Several experimental machines were built, generating a few hundred *kilowatts* to just over a *megawatt*. (A megawatt, MW, is equal to 1,000 kilowatts, kW, or 1 million *watts*.) But it was not until the 1970s that wind turbine research and technology development really took off, in response to rising energy prices and geopolitical tension.

Modern-day interest in harnessing wind energy has fluctuated with the price of coal, natural gas, and most importantly, oil. Throughout most of the 20th century, these fossil fuels were cheaper and easier to burn for power generation, compared with the complexity of wind power systems. Since around 1980, however, global oil prices have become more unpredictable, and the world's industrialized nations have experienced price shocks and interruptions in supply.

The conviction that our traditional fossil fuels will eventually run out—even though no one knows exactly when—has revived the search for economical ways to exploit wind power and other forms of renewable energy.

Wind energy conversion is a fascinating field to study, if only because its past has been so checkered and its exact future is so uncertain. Unlike the aerospace industry, the computer industry, and almost any other successful industry you can name, wind energy—the leading mechanically-based renewable energy for much of man's history—has never made anyone rich for long. But unlike many of these other industries, it has been around for thousands of years. It's a technology that has been reinvented numerous times. We are left with the promise and the drive to succeed despite daunting (and sometimes puzzling) obstacles.

—*Darrell M. Dodge,* Illustrated History of Wind Power Development

---

Today's Wind Energy Systems

Sizes

Wind turbines—the propeller-type machines that are most commonly used worldwide for power generation—come in all sizes. The smallest wind turbine blades are only about half a meter long (less than 2 feet), which yields enough power to charge a *battery* on a boat, for example. Other *micro* turbine systems, generating a few hundred watts or less, can also serve for lighting, pond aeration, and remote communication equipment.

In contrast, the largest wind turbines, called *utility-scale*, measure more than 100 meters (300 feet) in diameter and generate more than a megawatt of electricity. These big turbines are usually built on *wind farms* (also called *wind parks*), where dozens or even hundreds of machines provide power to electric utilities. Chapters 6 and 9 discuss wind farm development and operation on land and offshore.

In between these two extremes, a single *small-scale* wind turbine (less than 100 kW) can serve a variety of applications, including water pumping, telecommunications dishes, and power generation for homes, small businesses, schools, and farms. Small wind turbines can produce energy at relatively low wind speeds, but they still need to be mounted on tall poles, especially in wooded areas. Chapter 4 discusses small wind power systems.

Sometimes, small machines work in tandem with other power sources such as diesel-fueled generators or solar panels. These so-called *hybrid systems*, which often include batteries for storing electricity, are typically used in remote locations that have no access to the electric utility's *grid*.

Large and mid-range turbines (hundreds of kilowatts) are sometimes installed not on a wind farm, but in towns with good wind energy potential, often near a seacoast or in the mountains. A single *community-scale* wind turbine can contribute substantially to the town's power demand, or it can deliver electricity for sale to the utility if that's more profitable. Some towns band together in a cooperative to build community-scale wind farms that supply several megawatts of electricity.

Designs

Wind turbines also come in many different designs, but the most common is the *horizontal-axis* machine using three propeller-type blades. However, wind power can also be produced by *vertical-axis* turbines, which look sort of like an eggbeater. Chapter 4 discusses vertical-axis machines and other alternative designs.

Wind speed is a crucial factor in figuring out how much electricity a wind turbine will generate. The wind blows at higher speeds the higher up in the air you go, so wind turbines are mounted on tall *towers* to capture the maximum amount of wind.

A small difference in wind speed can make a big difference in how much electricity a turbine produces. When you double the wind speed, the turbine's power output increases by a factor of eight. That's because the mathematical relationship between the wind's potential power and its speed is not linear.

Instead, the amount of energy available in the wind is proportional to the cube of its speed. So a 10% increase in wind speed translates to a 33% increase in wind energy available for capture. Chapter 3 discusses wind energy and power.

Wind energy systems are made up of several major components in addition to the blades and generator. Most wind turbines have a *gearbox*, although many companies are developing and selling *direct-drive* machines that eliminate the gearbox to reduce maintenance costs. The gearbox and generator are enclosed in the *nacelle*, a pod-like structure that sits on top of the tower. Chapter 2 discusses wind turbine components.

Benefits

Clearly, the biggest advantage of wind power is that the wind is free. Wind turbines need no fuel to generate electricity. Also, the wind will always blow on earth as we know it, making it a renewable resource. And wind energy is available everywhere, so any country can identify its windiest locations and generate its own homegrown power.

Another benefit is that wind power has less of an impact on our environment than fossil-fueled electricity generation. It doesn't pollute the air, land, or water as coal-fired power plants do, and it's cleaner than natural gas–fueled power too. Nuclear plants emit no air pollution, but they produce toxic waste.

From an economic standpoint, a *kilowatt-hour* (kWh) of wind power is relatively cheap compared to solar *photovoltaic* (PV) technology. In some countries, the price of wind power already competes with that of conventional electricity.

Also, the best places for capturing wind energy are often remote, rural areas where land is plentiful, or even offshore, where wind turbines can be anchored in the ocean's shallow waters. Wind power could also provide new "green" jobs, an important goal for many government policy makers.

Challenges

Still, wind power faces many daunting challenges. Building a wind farm requires a much greater capital investment than a fossil-fueled power plant does. During the financial crisis around 2008, the money needed to develop wind farms dried up substantially in some countries, although staunch government support continued in others.

Another problem is the flip side of wind farm sites being remote—although plenty of land might be available for building turbines, the electricity they generate must travel over great distances to reach its customers, who are concentrated in big cities. Traversing this distance usually requires new electric transmission lines, which are also expensive to build, and some of the power is lost along the way.

Although most environmental activists support wind power, many are concerned with the impact of wind farms and their associated land development. The wind turbines themselves use relatively little land area, but wind farms need access roads for maintenance trucks and cranes. Also, other uses for the land might be more valuable than power generation, especially in densely populated, industrialized countries.

Many people object to the sight of wind turbines marching across a mountain ridge, spoiling the view. Similar objections have prevented some offshore wind sites from being developed, especially where ocean vistas are highly prized. Individuals respond in different ways to the appearance of wind turbines and other energy equipment (figs. 1–1 and 1–2).

Fig. 1–1. Post Oak wind farm, Texas (U.S.). *Image courtesy of LM Wind Power.*

Noise is also a concern, and complaints persist at wind farm sites in North America and Europe. Radar systems used by airports for air traffic control can be disrupted by wind turbines if they are located nearby.

Wildlife defenders are alarmed by the number of birds and bats that are killed in collisions with wind turbine blades. Activists on both sides of the issue cite statistics to support their case, and the wind industry has made some progress on resolving these problems.

Many energy experts argue about whether wind farms are reliable. This controversy sometimes confuses *reliability* with other technical factors. A related question on the horizon is how much power can be supplied by wind within a given utility distribution network or in a larger, regional power system encompassing several utility grids.

Subsequent chapters address these challenges in more detail. Chapters 6 and 9 discuss wind farm financing; chapter 7, wind energy integration into existing grids; and chapter 12, issues such as visual impact, turbine noise, and effects on wildlife.

Fig. 1–2. Middelgrunden wind farm, offshore Copenhagen (Denmark). *Image courtesy of LM Wind Power.*

References

American Wind Energy Association, http://www.awea.org.

Canadian Wind Energy Association, http://www.canwea.ca.

Dodge, Darrell M. Illustrated History of Wind Power Development. http://www.telosnet.com/wind.

Guided Tour (English). Danish Wind Energy Association. http://guidedtour.windpower.org/en/.

Morris, Neil. 2007. *Wind Power.* North Mankato, MN: Smart Apple Media.

Park, Jack. 1981. *The Wind Power Book.* Palo Alto, CA: Cheshire Books. (ISBN 0-917352-05-X).

U.S. Department of Energy, Energy Efficiency and Renewable Energy. Energy Basics, Wind Energy Technologies. http://www.eere.energy.gov/basics/renewable_energy/wind.html.

——— Wind and Water Program. http://www1.eere.energy.gov/windandhydro.

WindFacts (Wind Energy—the Facts), http://www.wind-energy-the-facts.org/.

2

Components: What's a Wind Turbine Made Of?

Introduction

Utility-scale wind turbines can have more than 8,000 parts, but for the purpose of understanding how a turbine works, we can narrow these components down to three major sections:

- Rotor (blades and hub)
- Drivetrain (gearbox and generator, connected to the rotor by a shaft)
- Control system

Figure 2–1 shows the key components of a utility-scale wind turbine, which are discussed below. Small machines incorporate most of the same components, except for the gearbox.

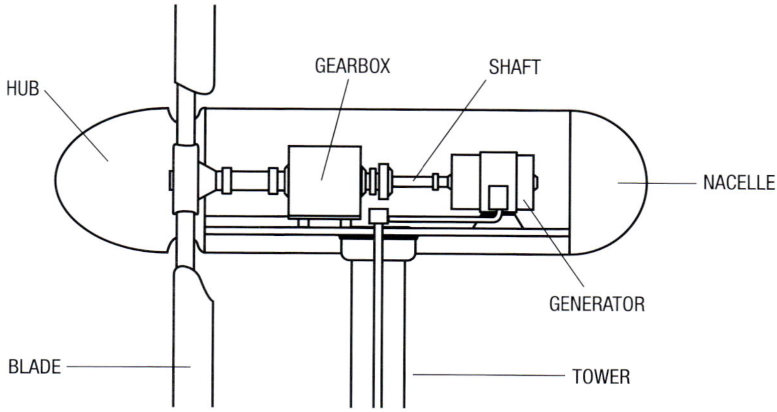

Fig. 2–1. Schematic of major wind turbine components. *Image courtesy of Renewable Energy Policy Project.*

Rotor

The term *rotor* usually refers to the whole blade assembly, consisting of the blades and their *hub*, where the blade roots are attached. *Rotor diameter* describes the entire diameter—that is, two blade lengths plus the diameter of the hub. The *swept area* is the circle defined by the blades' revolution.

The wind turbine blade is one of the machine's most critical components because it converts the kinetic energy of the wind's motion into mechanical power by turning a driveshaft. The turbine's propeller-type blades are shaped like an *airfoil*, similar to an airplane wing, to create *lift* from the moving air. Chapter 3 discusses blade design.

Most horizontal-axis wind turbines have three blades that are operated *upwind*, with the turbine facing into the wind. Several other types of wind turbines have also been designed and built, including vertical-axis machines (see chapter 4).

The blades and their hub, along with the rest of the wind power system, are mounted on tall towers, usually made of a huge, tubular steel column that can be tapered to a narrower diameter at the top. Lattice steel towers with an open framework are also still used but have become less popular. Lattice towers can't be locked up, and they afford little protection for maintenance engineers during the winter. They also provide places for birds to perch and nest, which likely increases bird deaths (see chapter 12).

Because of friction created by obstacles on the ground, the wind blows faster higher up in the air. In open country without obstruction, wind speed increases about 10% every time the height above the ground is doubled, boosting the amount of wind energy available. So turbine towers have become taller to allow the blades to work at the highest possible wind speeds.

Also, blades have become longer to increase their swept area and capture the maximum amount of wind energy. Longer blades require even taller towers to clear the ground. Towers for today's utility-scale turbines are typically 80–100 meters tall (260–330 feet), and some reach 130 meters (427 feet).

Tall towers can be especially advantageous in areas with a high *vertical wind shear*, where wind speed increases with height even more dramatically than normal. However, the taller the towers are, the stronger they must be, whether on land or offshore, and the heavier their foundations must be. The tower (not including its foundation) is the wind power system's heaviest component by far and also its third-most expensive, after the gearbox and blades (table 2–1).

The largest turbine blade developed to date, destined for service offshore, is 73.5 meters (241 feet) long from root to tip. That translates to an entire rotor diameter of about 150 meters (almost 500 feet). Figure 2–2 shows a 61.5-meter (200-foot) blade being trucked cross-country.

Table 2–1. Relative cost and weight of wind turbine components. *Data courtesy of the U.S. National Renewable Energy Laboratory.*

Component	% of total	
	Cost	Weight
Rotor		
Blades	14.66	8.15
Hub	4.15	5.94
Pitch mechanism and bearings	3.66	2.11
Spinner and nose cone	0.39	0.46
Subtotal rotor	**22.85**	**16.66**
Drivetrain and nacelle		
Low-speed shaft	2.03	1.78
Bearings	1.16	0.40
Gearbox	14.75	6.03
Mechanical brake, high-speed coupling, etc.	0.29	n/a
Generator	9.45	3.24
Variable speed electronics	11.48	n/a
Yaw drive and bearing	1.93	1.10
Main frame	8.97	11.64
Electrical connections	5.79	n/a
Hydraulic, cooling system	1.74	0.07
Nacelle cover	2.03	1.38
Subtotal drivetrain and nacelle	**59.59**	**25.64**
Controls, safety system, condition monitoring	**3.38**	**n/a**
Tower	**14.18**	**57.69**

Most utility-scale turbines for land-based wind farms have blades 30–50 meters long (about 100–150 feet) and rotor diameters of 60–100 meters (roughly 200–300 feet)—wide enough to sweep a football field. The opening at the blade's root is big enough to drive a car into. In today's turbines, the tip of the blade at the top of its revolution can be 140 meters up in the air (460 feet)—about the height of a 35-story office building.

Blades and rotor diameters for offshore machines are usually longer because the blades can be transported from coastal factories by ship or barge to their offshore destination, rather than traveling over land by railroad or truck.

Industrial haulers are overcoming these limits by developing special trucks that can handle very long blades, as shown in figure 2–2. Another constraint on the development of ever-larger machines is obtaining cranes big enough to lift the components to the top of the tower. Chapter 3 discusses turbine size constraints.

Fig. 2–2. Wind turbine blade being transported. *Image courtesy of LM Wind Power.*

Most wind turbine blades are white, so they blend in with clouds. Large blades are usually manufactured in three pieces—the top skin, the bottom skin, and the spar, which is the support that runs through the middle inside the blade. The blades are molded out of fiberglass—glass fibers held together by a plastic resin. The type of resin is typically polyester, a synthetic *thermoset resin* that can be molded or extruded.

Turbine blades must be stiff enough that they can't be pushed backwards into the tower by strong winds, an event called a *tower strike*, which is usually catastrophic. Yet blades must be flexible enough to bend during gusts of wind without breaking off (see chapter 3). To help prevent tower strikes, the rotor hub and blades are projected out in front of the tower by a considerable distance, and sometimes rotors are tilted slightly upward.

Some manufacturers reinforce fiberglass blades with carbon fibers to make them even stiffer and stronger, but still lightweight and flexible. However, carbon fibers cost 10 times more than fiberglass, so they must be used judiciously. *Carbon-fiber composite* materials are gradually becoming more common as their cost comes down. Researchers are also evaluating polyurethane resins reinforced with carbon nanotubes.

The rotor hub where the blades are attached is a heavy metal casting that connects the blades to the driveshaft. The hub contains large bearings that allow the blades to change their angle as the wind changes speed (see discussion of pitch control below).

The hub is located in front of the nacelle, a pod-like structure that encloses the drivetrain. Some nacelles are large enough for a helicopter to land on. The nacelle also houses electronic equipment such as controls and electrical cables. In some older turbines, maintenance engineers must climb out of the nacelle to access the hub for repairs, a job that requires training and strict safety procedures.

A *yaw* system sits at the top of the tower underneath or inside the nacelle. The yaw system uses a wind-direction sensor and a motor to keep the turbine's rotor pointed into the wind. The yaw system's huge bearings have to be extremely rugged, since they experience the highest loads, or stresses, in the whole wind turbine system. The entire assembly on top—blades, hub, drivetrain, and nacelle—aims toward the wind by turning on the yaw bearings.

The wind turbine's *pitch control system* angles the blades, rotating them around their long axis. In large, modern wind turbines, each blade has its own pitch-control motor and small gearbox, allowing the blades to be individually pitched.

On some turbines, the blade's pitch changes constantly as the rotor spins, because the best angle might be different when the blade is pointing up than when it's pointing down. About 60% of new turbines also incorporate batteries or *ultracapacitors* for pitch control backup power, in case the turbine loses its connection to the utility grid. All of the pitch control system's parts are contained in the rotor hub.

Pitch control is crucial to operating a wind turbine efficiently and safely. It allows the blades to capture the maximum amount of wind energy under optimal conditions. If wind speeds rise, the pitch control system begins to *feather* the blades, letting them "spill" a little wind and continue to operate safely.

If wind speeds become uncontrollably high, beyond what the machine can handle, the pitch control system *furls* the blades completely, turning them to slice horizontally into the wind (parallel with it) and allowing the turbine to stop running.

In some older machines, the blades are bolted into place, so they have a fixed pitch instead of active pitch control. But the blades are aerodynamically designed to *stall*, which prevents the rotor from overspeeding when winds get too strong.

Although every wind turbine is designed to operate best within certain ranges of wind speeds, in general large turbines "cut in" and start producing power at a wind speed of about 3–4 m/s (8–9 mph) and reach their maximum output around 12–14 m/s (27–31 mph).

Above these speeds, the pitch controls begin to limit power output to prevent overloading the generator and drivetrain. When winds exceed the turbine's "cut-out" speed, typically up to 27 m/s (60 mph), the pitch controls slow down and stop the turbine's rotation.

Drivetrain

The wind turbine's *drivetrain* includes the main driveshaft, gearbox, and generator, all of which are enclosed by the nacelle. Maintenance engineers are able to work on these components inside the nacelle, although sometimes it can be a tight fit.

The driveshaft is spun by the rotor blades, transferring mechanical power to the gearbox, which speeds up the rotation and transfers it to the generator, which produces electricity. The generator usually incorporates power electronics (see "Generator" below). The drivetrain and nacelle account for about 60% of the capital cost of a wind turbine (see table 2–1).

Gearbox

Wind turbine driveshafts typically rotate at 20 revolutions per minute (rpm) or less. The gearbox speeds up the slow rotation of the blades into a faster rotation that is better suited to driving the generator, usually 1,200-1,800 rpm. Figure 2–3 shows a wind turbine gearbox on a *dynamometer* test stand.

Fig. 2–3. Wind turbine gearbox on dynamometer. *Image courtesy of Burke Porter Machinery Co.*

Wind turbines encounter highly variable loads that are hard to predict, so wind power is one of the most demanding applications for gearboxes and, in particular, for gearbox bearings and gear teeth. Some turbines use *planetary gearboxes*, which consist of outer "planet" gears revolving around a central "sun" gear. Other types include helical gearboxes.

Manufacturers have developed bigger and stronger gears and bearings, stretching the limits of current technology. But gearboxes still require frequent maintenance, and this component has experienced high failure rates due to bearing issues and excessive loads. Gearboxes have been called the most costly failure in wind turbines. In offshore turbines, replacing a gearbox can cost $1.4 million (more than 1 million euros).

But no single manufacturer's products are vulnerable; rather, the problem is common to all gearboxes—it's the nature of the beast. Due to fleet-wide gearbox maintenance issues and related failures, wind power system designers have made it standard practice to perform extensive dynamometer testing on new gearboxes, and the component's reliability has dramatically increased over the past decade or so.

Government agencies in Europe and America are working to understand gearbox problems. The Reliawind Project, a consortium of European Union countries, is studying wind turbine reliability, including the role of gearboxes in underperformance, versus what was expected.

The U.S. National Renewable Laboratory's Gearbox Reliability Collaborative is examining gearbox loads (overloading and uneven load distribution) and modeling the effect of gusty wind on gearboxes (see chapter 10). The International Electrotechnical Commission and the International Organization for Standardization formed a joint committee to update the gearbox reliability standards for wind turbines.

In addition to the gearbox's reliability, its size and weight are important factors. That's because the more compact and lightweight any component can be, the easier it is to fit into the nacelle, and the less load it will put onto the tower. Gearboxes for utility-scale wind turbines typically weigh about 50–70 tons. However, by speeding up the driveshaft's rotation, they allow the use of higher speed generators that are smaller and lighter weight (see "Generator" below).

Most large-scale wind turbines use a caliper disk brake system. The brakes can be powered by fluid (hydraulics) or air (pneumatics), or they can operate mechanically in small wind turbines. Heavy-duty brakes are used on some large turbines to stop the driveshaft rotation. In most machines, however, the pitch control system furls the turbine down into idle, and the brakes are designed just to "park" the machine, holding it at rest for maintenance or other reasons.

Generator

Once the wind's kinetic energy has been captured by the blades and converted to mechanical power by the rotating driveshaft, it's the generator that takes the next step in energy transformation, converting the shaft's mechanical power (*shaftpower*) to electricity. The term "generator" can be confusing, because the wind turbine itself is often called a generator, but it's really this key component that produces the electrical power.

Generator design. Like the gearbox, the generator is enclosed in the nacelle at the top of the turbine tower. The generator contains magnets inside a coil of wire called the *windings*. When the driveshaft spins the magnets, *electrons* in the windings are forced to flow through an external electrical circuit, or closed loop. Electrons in motion are what we know as *electricity*. The windings create a magnetic field when the electric *current* passes through them.

The higher the rotational speed of the generator, the smaller and lighter it is. A generator slow enough to accept the turbine blades' rotation of 20 rpm or less, without it being sped up by the gearbox, would have to be much larger and heavier than a higher speed machine. Generator speed, size, and weight are common trade-offs in designing a turbine system (see chapter 3).

In contrast to gearboxes, generators contain only one or two bearings, but they are subject to heavy loads, and alignment of the generator's driveshaft with that of the gearbox is critical to the reliable performance of the wind power system. Generators' windings also are a potential source of failure.

In general, generator bearings wear out faster in bigger, heavier machines. However, direct-drive turbines that have large generators with bearings designed for low speed can be more reliable than wind power systems using gearboxes (see "Direct-drive wind turbines" below).

Types of generators. Many different types of generators can be used in large-scale wind turbines, but they fall into two classes: *induction generators* and *permanent-magnet generators*. A third class, *synchronous generators*, were rarely used in the past because their power output must be perfectly matched ("synchronized") to the utility grid's *frequency*, which in general is 60 *hertz* (Hz) in the Americas and 50 Hz in the rest of the world. Recently, however, synchronous generators have been applied successfully in direct-drive wind turbines (see below).

Traditionally, large-scale wind turbines have used induction generators running at a fixed speed of 1,200–1,800 rpm, producing *alternating-current* (AC) electricity at 50–60 Hz. In general, induction machines are simpler and more rugged than other types. The fixed speed of an induction generator actually has a bit of flexibility, or "slip," within a range of 100 rpm or so, letting it adjust to occasional gusts of wind.

Over the years, induction machines have been modified to make them much more useful in large-scale wind power applications. For example, *doubly fed* generators (also called *wound-rotor* machines) are equipped with *power electronics* that enable them to operate at variable rotational speeds.

Variable-speed operation allows the turbine to capture wind energy more efficiently over a greater range of wind speeds, thus increasing the amount of power it delivers. Theoretically, a variable-speed turbine can maximize its output regardless of wind speed, within a reasonable range.

Besides enabling variable-speed operation, power electronics also synchronize the generator's electricity to match conditions acceptable to the utility grid (50–60 Hz AC). This type of power electronics is called a *power converter* (fig. 2–4).

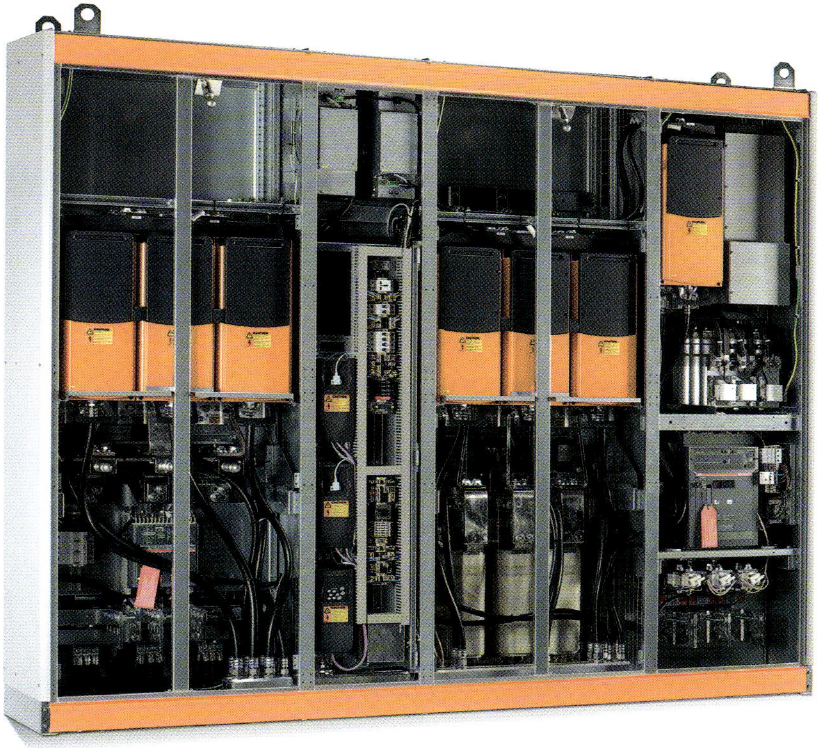

Fig. 2–4. Power converter. *Image courtesy of The Switch.*

Permanent-magnet generators are becoming more popular in today's utility-scale wind turbines, with most of the world's top manufacturers offering turbines using these generators. In permanent-magnet generators, the magnets on the rotor are permanently magnetized (thus the name), rather than being magnetized ("excited") by an electric current, as in conventional generators (fig. 2–5).

Fig. 2–5. Permanent magnet generator, outer rotor. *Image courtesy of The Switch.*

Permanent magnets in today's machines are made of rare-earth metals, which produce stronger magnetic fields than conventional iron or ceramic magnets. However, the price of these metals is a concern, since China controls the market for them (see chapter 11).

Although permanent-magnet generators are bigger and more expensive than induction machines, they have lower power losses and better reliability and efficiency, even when running at low loads. Also, like other types of generators, they incorporate power electronics for variable-speed operation and power conversion to utility requirements.

Direct-drive wind turbines

Permanent-magnet generators are being used in many direct-drive wind turbines, especially in offshore applications. Direct-drive, "gearless" systems use no gearbox to speed up the rotor, so they require a lower speed generator—meaning that it is larger, heavier, and more expensive.

But the trade-off is that gearbox maintenance and failures are avoided, which can result in a more reliable wind turbine overall. Without a gearbox, nearly all of the troublesome moving parts are eliminated. This is especially valuable offshore, where it's harder to get to the turbine and make repairs. Also, the generator's slow speed translates to fewer rotations and longer life. And with no gearbox inside the nacelle, maintenance engineers have more room to work and better access to the rotor hub.

Enercon (Germany) pioneered direct-drive technology in wind turbines. Enercon's machines use synchronous (annular) generators without permanent magnets. The E series turbines range up to 7.5 MW in capacity. Several other wind turbine manufacturers have introduced direct-drive wind turbine models using permanent-magnet generators. Virtually all small-scale, horizontal-axis wind energy systems also use direct-drive turbines and permanent-magnet generators (see chapter 4).

There is a lot of buzz in the industry about direct-drive turbines, but Acciona is waiting for that technology to mature, due to the unknowns. If it proves to be as reliable as hoped, it will take a lot of weight out of the turbine, and weight is cost.

—Joe Baker, CEO, Acciona Windpower (in Wind Today*)*

Control System

Most turbines on wind farms don't get frequent attention from a person on-site. Maintenance engineers are typically at the site every day, but they are working on dozens or hundreds of turbines, a few at a time. This means that the turbine's *control system* is critical for telling wind farm operators when something is wrong (see chapter 6).

As wind turbines have become taller, their structures have become more complex and dynamic. Wind power systems are subject to a lot of highly variable loads and stresses, many of which are repetitive or cyclic in nature, inducing fatigue (see chapter 3).

Ultimately, the control system is responsible for monitoring turbine conditions, recording the entire system's performance, alerting operators when maintenance is needed, detecting *faults*, and slowing down or stopping the rotor when hazardous conditions arise. Even a well-designed turbine can perform poorly or self-destruct if it's not properly controlled.

The turbine control system keeps a constant eye on the machine to monitor rotor speed, blade pitch, power output, nacelle position, and many other operating parameters. Dozens of sensors continuously feed information to the controller about wind direction and speed, temperatures and pressures at various points in the system, water and oil flows, vibration, and power characteristics (*voltage*, current, etc.).

Each wind turbine component interfaces with the main controller. For example, when the machine is starting up, the controller decides which pitch angle is best for the blades and sends that message to the pitch control system in the rotor's hub. The controller might also automatically turn on a fan if the generator windings are getting too hot, or it could turn on an oil pump to lubricate the gearbox.

Wind turbine controls usually incorporate a *SCADA system*, which stands for system control and data acquisition. New wind farms are required in some countries to have SCADA systems that can communicate in real time with the operator of the utility grid or regional power system.

SCADA systems collect all of the data being generated by the network of sensors and write a history of these readings to a computer, where the SCADA system stores the data and packages it. SCADA information is typically organized into 10-minute averages of each variable, otherwise the amount of data would be too vast to communicate and process. Chapter 6 discusses SCADA data and its uses.

In utility-scale wind turbines, control components are distributed throughout the structure, like a human nervous system, so that the wires connecting the sensors and their actuating devices don't get too long. Some parts of the control system are located in the tower base, the pitch controls are in the hub, and another set of controls is inside the nacelle. All of the turbine's sensors and controllers are integrated via communications channels. In older, smaller machines, the control system can be enclosed in a box on the ground near the turbine.

Data from the SCADA system is collected by the turbine's controllers and sent to a master SCADA computer for incorporation into a database. The master SCADA might be on site at the wind farm or far away at a control center that monitors several sites.

In addition to controlling turbine operation, the control system activates switchgear to connect the generator to the utility grid and to disconnect from it. This is an important function because the grid network is very sensitive to disturbances from power sources that don't match grid conditions perfectly. If the turbine sends the wrong kind of power to the grid, catastrophic accidents can result.

If a machine shuts down because of a fault and doesn't automatically restart, it needs human intervention. Sometimes, though, this intervention can make a bad situation worse. For example, if a machine experiences repeated faults in the pitch control system, a person can override the warning many times to keep the machine running. But eventually, the pitch system could fail, allowing the turbine to go into overspeed and self-destruct.

Other Equipment

At wind farm sites, other equipment is involved in delivering the turbines' electricity to the utility grid for transmission and distribution to homes, businesses, and industry. This equipment, which is external to the wind turbine itself, includes transformers that convert voltage, power lines that collect electricity from each turbine, and often one or more utility substations. Chapter 6 discusses these parts of a wind farm.

References

Ancona, Dan, and McVeigh, Jim. 2001 (August 29). Wind Turbine—Materials and Manufacturing Fact Sheet. Princeton Energy Resources International, LLC. http://www.perihq.com/documents/WindTurbine-MaterialsandManufacturing_FactSheet.pdf.

Burke Porter Machinery, http://www.bepco.com/.

Centurion Energy LLC, http://centurionenergy.net.

Del Franco, Mark. 2010. Permanent Magnet Generators Join Turbine Conversation. *North American Windpower.* 7(8): 48–50.

Eggleston, David M., and Stoddard, Forrest S. 1987. *Wind Turbine Engineering Design.* New York: Van Nostrand Reinhold Co. Inc. (ISBN 0-442-22195-9).

Enercon, http://www.enercon.de/en-en/Windenergieanlagen.htm.

Fingersh, L., Hand, M., and Laxson, A. 2006 (December). Wind Turbine Design Cost and Scaling Model. Report No. NREL/TP-500-40566, http://www.nrel.gov/wind/pdfs/40566.pdf.

Kalinoski, Gail. 2010 (November 10). Underperformance Issues Deserve Fresh Examination. *North American Windpower.* 7(10): 1, 41–43.

Laughlin, James. Drivetrain Innovation Raises Wind Turbine Efficiency, Offers Gearbox Improvements. Power-Gen Worldwide. http://www.powergenworldwide.com/index/display/articledisplay/258472/articles/power-engineering/volume-110/issue-6/field-notes/drive-train-innovation-raises-wind-turbine-efficiency-offers-gearbox-improvements.html.

Liebreich, Michael, and Young, William. 2005 (July 19). Offshore Wind: Europe's EUR 90 Billion Funding Requirement. New Energy Finance Ltd. http://bnef.com/Download/NEF_HTML_Press/Offshore-wind-funding.pdf.

Loos, Marcio R., et al. 2011 (August). World's First Carbon Nanotube Reinforced Polyurethane Wind Blades. Case Western Reserve University. http://polymers.case.edu/stories/World's%20First%20PU%20CNT%20Blades.html.

Maxwell Technologies, Inc., http://www.maxwell.com.

Northern Power Systems, http://www.northernpower.com.

Pacella, Rena Marie. 2010 (March 26). How It Works: The Next-Gen Wind Turbine. *Popular Science.* http://www.popsci.com/technology/article/2010-03/next-gen-wind-turbine.

Remington, K. 2010 (August 7). 4-MW direct drive headed offshore. *Windpower Engineering.* http://www.windpowerengineering.com/featured/business-news-projects/4-mw-direct-drive-headed-offshore.

Siemens, http://www.energy.siemens.com/hq/en/power-generation/renewables/wind-power.

Sterzinger, George, and Svrcek, Matt. 2004 (September). Wind Turbine Development: Location of Manufacturing Activity. Renewable Energy Policy Project. http://www.repp.org/articles/static/1/binaries/WindLocator.pdf.

Thresher, R., Robinson, M., and Veers, P. 2008 (August). Wind Energy Technology: Current Status and R&D Future. NREL/CP-500-43374, http://www.nrel.gov/wind/pdfs/43374.pdf.

Trabish, Herman K. 2010 (May 4). Wind Turbines, the Next Generation. Greentech Media. http://www.greentechmedia.com/articles/read/wind-tubines-the-next-generation.

———. 2011 (January 14). Wind Turbines Get Sensitive. Greentech Media. http://www.greentechmedia.com/articles/print/wind-turbines-get-sensitive.

———. 2011 (January 27). When Solar Power Plants and Wind Farms Talk to the Grid. Greentech Media. http://www.greentechmedia.com/articles/read/when-solar-power-plants-and-wind-farms-talk-to-the-grid.

3

Wind Power System Design: Combining Strength and Flexibility

Introduction

A smart engineer once said that a wind turbine should be designed like a tree. The roots and trunk should be the strongest, thickest, heaviest parts. That's the foundation and tower of a wind power system. The tips of the rotor blades, like leaves that twist in the wind, should be the lightest, thinnest, most flexible parts. In between are the branches and limbs of the system, such as the drivetrain and nacelle, which need intermediate amounts of strength and flexibility.

Thanks to ideas like this, wind turbine technology has made spectacular advances in the past 40 years or so. Blades made of sailcloth or sheet metal have progressed through wood to fiberglass and carbon-fiber composite materials. *Direct-current* alternators gave way to grid-synchronized induction generators, then to variable-speed machines with power electronics, and more recently to direct-drive turbines with no gearbox.

From mechanical cams and linkages for pitching blades, wind turbine designs moved to high-speed digital controls, motor drives, and batteries or ultracapacitors for pitch control and backup power. Today's airfoil designs are tested in wind tunnels, and blade materials are no longer as sensitive to surface roughness and dirt. Computer modeling and software make modern machines more robust than ever, yet less expensive than even a decade ago.

Overview

How much bigger can wind turbines get?

The average turbine is about seven times larger than in the 1990s, but produces 14 times as much electricity, thanks to design improvements. Since 1980, average wind turbine *power ratings* have grown almost linearly, from less than 50 kW to more than 6 MW for today's offshore machines, and turbines that will produce more than 10 MW are being developed (fig. 3–1).

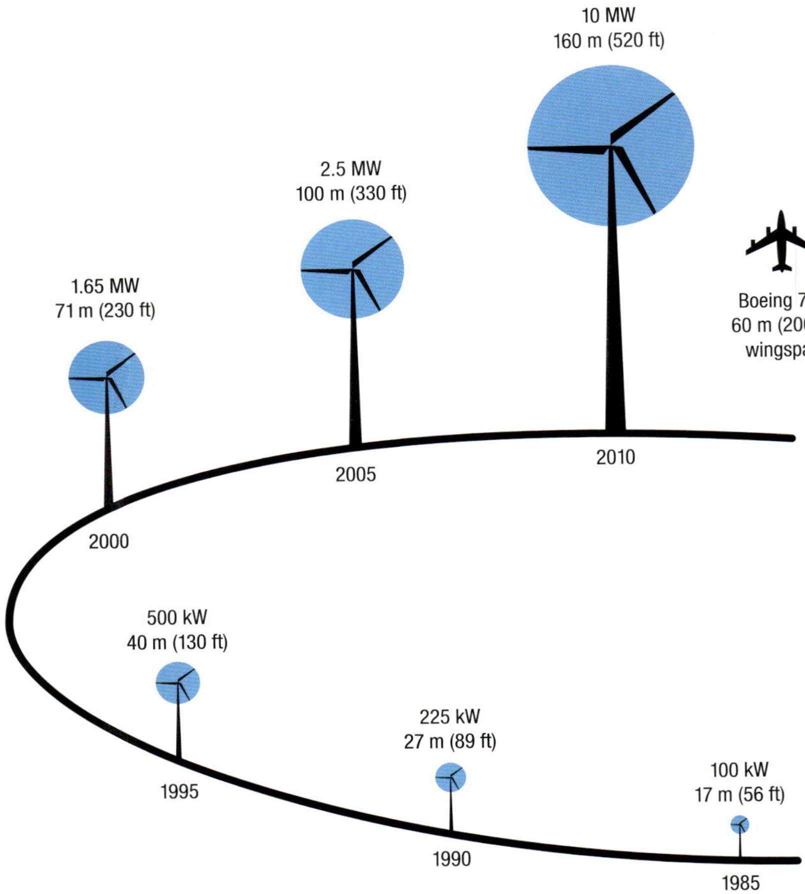

Fig. 3–1. Growth in turbine size and output

The reason for designing ever-larger turbines is to minimize the cost per kilowatt-hour of energy delivered, which goes down as turbine size and height go up. However, at some point, the cost of a larger turbine will begin to grow faster than the revenue derived from its greater output.

Design engineers call this the *square-cube law*. It states that the amount of material needed for a wind power system—and thus its weight and its cost—increases with size faster than the energy output does. Theoretically, this makes scaling up in size a losing economic game in the long run.

For example, you can double your blade length to quadruple the rotor's power (the "square"), but the weight and cost of materials needed goes up by a factor of eight (the "cube"). That's because a bigger rotor diameter, or swept area, requires a bigger hub, nacelle, tower, etc.—that is, a bigger everything else, which weighs more and costs more.

Clever designers have been able to evade this rule of thumb by using materials more efficiently, to trim weight and cost, and by taking advantage of new materials and better manufacturing methods. For example, longer turbine blades that boost rotor diameter and energy output do not weigh as much as you would expect from the square-cube law. Instead of increasing by an exponent of 3, the cube, their weight has grown by an exponent of only 2.3.

Research and development worldwide could allow designers to continue to scale turbines up in size and innovate around these weight and cost limits. Chapter 9 discusses tomorrow's super-size turbines.

But as mentioned in chapter 2, the logistics of transporting turbine blades by railroad or truck constrains their size. Blade roots and tower bases with diameters greater than 4.3 meters (14 feet) destined for use on land face hurdles during transportation in terms of cost and logistics. Transport costs can account for up to 10% of a turbine's delivered cost and, overall, as much as 10% of a wind farm's capital cost.

Special vehicles have been developed exclusively for transporting wind turbines. Also, wind power system manufacturers produce towers in sections for easier, cheaper transport and on-site assembly. Still, even a tower section can extend 25 meters (82 feet) and weigh 50–80 tons.

Another factor limiting the practical size of wind power systems is how to lift them onto the tower. Crane requirements for wind turbine construction are quite stringent because of the huge weight of the nacelle and its contents, the height of the lift, and the boom extension needed. Even when cranes are big and strong enough for the job, the cranes themselves are difficult to transport, and they require large, highly skilled crews, making them an expensive proposition overall.

Towers and foundations

Because of the great variety and magnitude of stresses placed on a wind power system, the tower and foundation must be designed to handle these loads safely. Conventional foundations, such as those for a building, simply transfer the dead weight of the structure (*vertical load*) to the ground. However, wind turbine foundations experience a significant *horizontal load* as well, like a tall building swaying in the wind.

International standards are generally applied in tower and foundation design. Also, wind farm developers test the soil at their prospective sites to see how well it will hold a turbine foundation. The tower and foundation are custom-designed for local site conditions (see chapter 6).

Today's large, onshore wind turbines almost all have tubular steel towers and concrete foundations. The major design variable is height, which must be selected to optimize energy capture within a reasonable cost. Most wind turbine towers are at least 60 meters tall (200 feet), but an 80–100 meter height (260–330 feet) is more prevalent for new construction.

Advanced towers are elevating onshore wind turbines even more, to hub heights up to 130 meters (427 feet), as noted in chapter 2. One new type of tapered tower called the Space Frame comes in smaller, lighter pieces, so it can be installed by a lighter-duty crane. It was scheduled for commercial introduction in 2012.

Another type of tower that was certified in 2011, the Atlas, uses a pre-cast concrete base (sort of like an above-ground foundation), allowing for hub heights greater than 110 meters (360 feet). The base tapers from a broad diameter of 15–18 meters on the ground (49–59 feet) and extends 31 meters up the tower (108 feet), where a conventional steel pole is installed. This design allows the foundation underground to be shallower and lighter.

Generators

Wind power applications place an unusual set of requirements on electrical generating equipment, which designers must account for in selecting and specifying the generator and, if used, the gearbox. Most electrical drive machines are designed to use electricity to turn a driveshaft and produce mechanical power, rather than using the mechanical power of a rotating driveshaft to produce electricity, as a wind turbine does.

Many applications that generate electricity from driveshaft rotation usually operate at a standard power level (*rated power*), which remains constant. In contrast, wind turbines must generate electricity at all power levels, and they spend a lot of time running at low power.

Also unlike most electrical drive machines, wind power systems must operate at the highest possible efficiencies in the low-power/low-wind region, as well as at mid-range and rated power, to squeeze every kilowatt-hour out of the available wind energy.

But wind power systems don't need to operate quite as efficiently in above-rated winds, when the rotor must "spill" a little wind to bring its output down to the rated level. So the wind power system can afford a little inefficiency at high power, but requires maximum efficiency during periods of low wind—the exact opposite of almost all other electrical applications, which need greater efficiency at higher loads.

As mentioned in chapter 2, the higher the rotational speed of the generator, the smaller and lighter it is. Generator speed, size, and weight are common trade-offs in designing a turbine system. Many turbine manufacturers are moving toward direct-drive designs, which require a lower speed generator that is larger, heavier, and more expensive—but the design eliminates the gearbox, which is a wind power system's number-one source of maintenance expenses and failures.

Control systems

Wind power control systems integrate the signals from dozens of sensors throughout the machine and make critical safety decisions, as described in chapter 2. Designers must consider how best to regulate the rotor speed to obtain peak efficiency in fluctuating winds.

This requires a controller that can continuously adjust blade pitch angle, rotor speed, and generator loading to maximize power while minimizing *transient loads*. These loads, which result from brief gusts of wind, include transient *torque* on the drivetrain and transient *thrust* on the whole machine (see "Torque" and "Bending" below).

Also as mentioned, large wind turbines today operate at variable speeds by using a power converter (power electronics). This technology also enables turbines to provide benefits to the grid such as protecting it by "riding through" a temporary fault in utility power, controlling the turbine's output voltage, and supplying *reactive power*, which is associated with energy losses in AC electrical systems.

Designing Today's Wind Turbines

Turbine design engineers are most concerned with how much power they expect to get out of the machine and how much that power will cost. One of the key design factors is the shape of the airfoil—the turbine blade and its aerodynamic performance. Blade design is the most crucial task for the engineer, because the wind power system's principal loads originate in the rotor.

But a turbine's output depends on many other factors, and design engineers must account for the great variety of stresses and loads that the entire system will encounter. Wind turbines are complex, vulnerable structures that are subject to vibration and are exposed to a wide range of environmental conditions—yet they must be as safe and reliable as airplanes.

Compared to airplane designers, wind turbine designers are less constrained by weight and more constrained by cost. The cost-effectiveness of a wind turbine relies ultimately on how practical it is—whether it can be installed, operated, and maintained safely and economically.

Let's take a look first at how designers predict the amount of power that their wind turbines will deliver, then how they handle the loads and stresses that affect the turbine's performance, reliability, and safety.

The design of a wind turbine is not a trivial task. Because of the variability of the power source, it is a little like designing a bridge that will be used by a succession of vehicles having random weights. The bridge can be built as strong as you like, but sooner or later a vehicle too heavy may come along and destroy it. Lest this thought make us too gloomy, however, remember that some windmills built in the early 1600s are still operational. Needless to say, a design life of 350 years isn't bad!

—*David M. Eggleston and Forrest S. (Woody) Stoddard,* Wind Turbine Engineering Design

---

Energy and power

The relationship between energy and power is fairly straightforward, but it gets more complex when distinguishing between wind energy and wind power. Basically, power is the rate at which energy is captured, so energy translates to power delivered over a period of time:

$$\text{energy} = \text{power} \times \text{time} \qquad \textbf{Eq. 3–1}$$

$$\text{power} = \text{energy} \,/\, \text{time} \qquad \textbf{Eq. 3–2}$$

Energy and power are derived from the wind by exploiting the forces it exerts on the turbine blades, causing them to rotate. Power is measured in watts or multiples, such as kilowatts or megawatts, while energy is measured in watt-hours or multiples, such as kilowatt-hours (kWh) or *megawatt-hours (MWh)*.

In calculating the amount of wind power a turbine will yield, *air density* is an important design factor. The force of the air blowing on the blades—that is, the kinetic energy that's available—depends on how dense, or heavy, the air is.

Engineers use the Greek letter *rho* (ρ) to designate air density in their calculations. Elevation above sea level plays a role in determining air density, as does climate to some extent.

The most important factors in predicting wind power are the wind speed at that instant, V, and the swept area, A (the circle defined by the rotating blades). Time is designated by the letter t in seconds. The energy and power available from the wind are proportional to the cube of the instantaneous wind speed, with air density as an important variable:

$$\text{available energy} = 1/2 \times \rho \times A \times V^3 \times t \qquad \textbf{Eq. 3–3}$$

$$\text{available power} = 1/2 \times \rho \times A \times V^3 \qquad \textbf{Eq. 3–4}$$

The design lesson here is that you can't control the air density, other than by choosing a site with heavy, powerful wind. This is a factor involved in planning a wind farm. You also can't control the wind speed, other than by analyzing specific sites and selecting the best ones. Wind speed is a factor—the single most important one—in assessing the potential wind resource at a given location.

But in wind turbine design, you can control the swept area, A, by selecting your blade length. The longer the blade, the larger the rotor's swept area, and the more power the wind turbine will deliver.

More importantly, you can control the shape of the turbine blade, or airfoil. That's where the ultimate efficiency of the wind turbine comes into play. Airfoil design is sometimes called the "secret sauce" of the turbine—the single design factor that's most influential on power output, other conditions being equal (see "Blade shape" below).

Rotor efficiency

A wind turbine rotor's efficiency is defined by its *coefficient of power*, designated as c_p. This factor (a ratio without dimensions) expresses the relationship between the total power available in the wind and the amount of power that the rotor actually produces or delivers:

$$\text{rotor efficiency } (c_p) = \text{power delivered / available power} \qquad \textbf{Eq. 3–5}$$

So to calculate how much power you'll get out of a rotor, you factor this coefficient into the equation for available power (eq. 3–4):

$$\text{power} = 1/2 \times \rho \times A \times V^3 \times c_p \qquad \textbf{Eq. 3–6}$$

No wind turbine rotor can capture all of the wind's energy and convert 100% of it to power. In theory, rotors are subject to the *Betz limit*, named after German physicist Albert Betz. This limit states that no wind turbine can convert more than about 59% of the wind's kinetic energy into mechanical power by turning a rotor.

If a wind turbine rotor converts 70% (0.7) of the theoretical Betz limit to mechanical shaftpower, its power coefficient is 0.41, meaning 41% conversion efficiency:

$$c_p = 0.7 \times 0.59 = 0.41 \quad \text{Eq. 3–7}$$

This is actually a pretty good coefficient of power. Well-designed turbine rotors generally fall into the 35%–45% range.

However, note that these calculations don't account for the process of converting the rotor's mechanical power to electricity. The overall conversion efficiency of the wind power system depends on the efficiency of the gearbox, generator, and power converter as well as that of the rotor. At each stage where power moves from one component to the next, the system loses some energy. Increasing the efficiency of each drivetrain component represents an opportunity to reduce the overall cost of energy from a wind turbine.

Power ratings and curves

Power curves show how much power the rotor produces at different wind speeds (fig. 3–2). Turbine manufacturers specify a rated power for which the electrical system has been designed. Power curves for today's wind turbines are very accurate, with many rules governing how they are developed and how turbines are tested by an independent third party for electricity output at different wind speeds.

Wind farm developers rely on power curves to project the energy output of their projects. When a wind farm is being developed, engineers typically measure the wind speed, direction, and other factors at a site for at least a year to define these characteristics as accurately as possible (see chapters 5 and 6). Then they analyze manufacturers' power curves for different wind turbines to see which ones might be the best match for their site.

The turbine design can be modified to become more suitable—for example, manufacturers can supply the same machine with shorter blades for a site with high wind speeds, or with longer blades for a site with low wind speeds.

The optimal turbine for one location might not be the optimal turbine for another spot with different wind speeds and other wind characteristics. The ultimate goal is the lowest cost of energy at the site where the turbine will be operating. The machine with the highest efficiency might not be the best one. Instead, a less efficient turbine might generate power at a lower cost, depending on site-specific wind conditions.

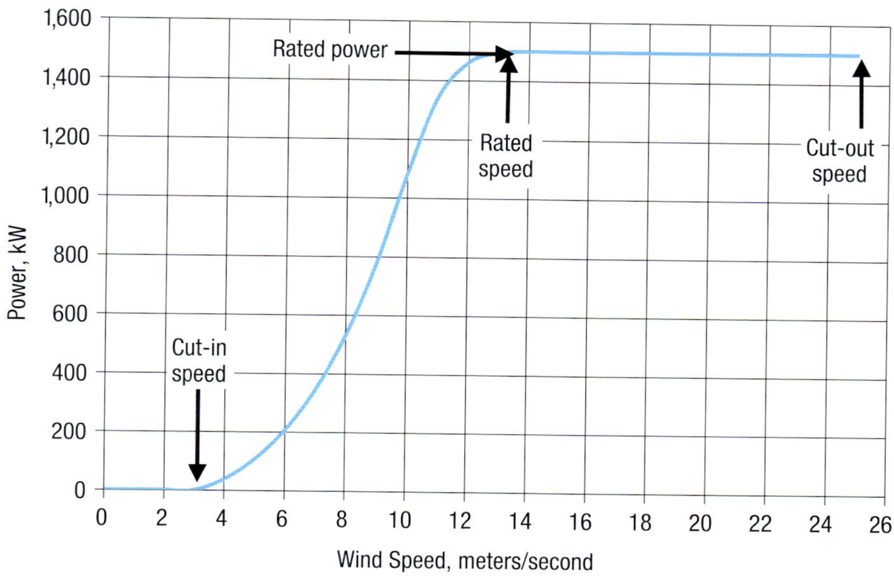

Fig. 3–2. Typical rotor power curve for a 1.5-MW wind turbine

Blade shape

Wind turbine blade designs are subjected to rigorous evaluation using the latest computer analysis tools so that blade weight can be minimized without sacrificing performance and safety. Design engineers pick one or more blade shapes, or airfoils, then they run models to predict the airfoils' behavior, and then they repeat these steps over and over to optimize the design. Airfoil design was pioneered by mathematician Abraham Robinson, who developed nonstandard analysis techniques in the 1960s.

> *The triumph of the modern era is that we can now explore wind turbine designs on paper and computer* before *physically constructing a full—or even reduced—scale model.*
>
> *—David M. Eggleston and Forrest S. (Woody) Stoddard, Wind Turbine Engineering Design*

Still, the process of designing a wind turbine blade is somewhat of a "black art" when it comes to tweaking the design. The most talented, experienced engineers use detailed models, but they also have an instinctive feel for what the blade's going to do with the wind and how the rest of the system will cope with that.

The aerodynamic performance of wind turbines has improved dramatically over the past 40 years. Today's rotors can capture about 80% of the theoretically potential amount of energy in the incoming air, called the *free stream*. This has been achieved by designing custom airfoils, often a new one for each individual turbine design.

Each section of the blade might have a slightly different airfoil shape that is blended into the next section. Computer software that was originally created for aerospace applications has been adapted for blade design.

These special airfoils attempt to optimize the turbine's efficiency at low wind speeds and limit aerodynamic loads in high winds, to protect the system from overspeeding. Blades have also been designed to be less susceptible to the accumulation of dirt and insects on their leading edge, though insects are less of a problem at greater tower heights.

Lift and drag. A wind turbine blade works like an airplane wing. When air blows past a rotor blade, its airfoil shape creates regions of uneven air pressure. The lower pressure air forms on the downwind side of the blade, as the wind begins to flow behind the airfoil. This pulls the blade toward the lower pressure region, causing the rotor to turn. This is called lift (fig. 3–3). Another way to think of it is, the lift force is created perpendicular to the wind, while a small *drag* force acts in opposition to the lift force.

The force of the lift is actually much stronger than the drag force—usually about 10 to 50 times as strong. Design engineers use a *lift coefficient*, c_l, to model the complex interactions of lift with airfoil shape, blade pitch angle, airflow conditions, and other variables.

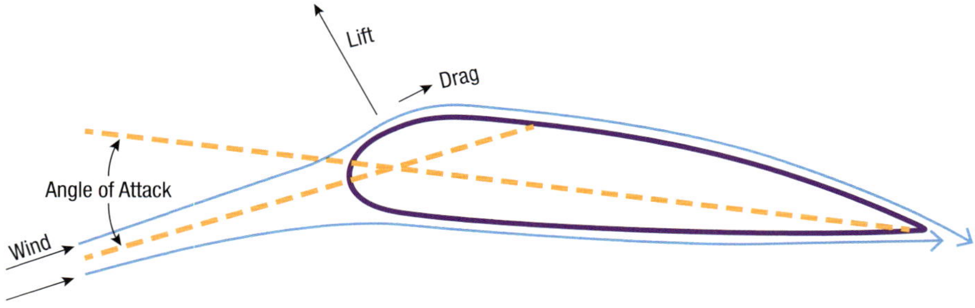

Fig. 3–3. Wind energy's aerodynamic lift force

Torque. The aerodynamic lift captured by the blades imparts a torque, or twisting motion, on the turbine rotor's shaft. Torque is defined by the following equations:

$$\text{power} = \text{torque} \times \text{rotational speed} \qquad \textbf{Eq. 3–8}$$

$$\text{torque} = \text{power} \,/\, \text{rotational speed} \qquad \textbf{Eq. 3–9}$$

Torque is measured by force and distance, in units such as *newton-meters* or *foot-pounds*. In a wind power system, torque flows from the rotor to the gearbox to the generator (or directly from the rotor to the generator in gearless, direct-drive systems). Turbine designers shape their airfoils specifically to increase lift and torque.

Solidity. *Solidity* is defined as the total surface area of a turbine's blades divided by the rotor's total swept area. If the blade surface area is large relative to the swept area, the turbine has a high solidity and typically would operate at low speeds and high torque. Turbines with low solidity (narrower blades) run at higher speeds and lower torque.

Tip speed ratio

In large, modern three-blade turbines, the speed of the blade at its tip is typically about four to seven times faster than the free-stream wind speed. Design engineers use the ratio of tip speed to wind speed to maximize the efficiency and power output of a turbine.

If the rotor turns too slowly, a lot of wind energy is lost in between blade revolutions. If it turns too fast, the rotor begins to act as a solid wall against the wind, repelling it rather than capturing it.

Tip speed ratio (TSR, a dimensionless factor), is calculated using the blade's length, the speed of the rotor's revolution, and the free-stream wind speed. The rotor's speed is designated by the Greek letter *omega* (Ω), and the blade length is the *radius*, R, so the tip speed is ΩR. The free-stream wind speed is defined as its velocity with nothing in its way, thus designated V_0, or V-*naught*. Tip speed ratio is then defined as:

$$TSR = \Omega R \,/\, V_0 \qquad \textbf{Eq. 3–10}$$

At tip speed ratios above 1, the rotor begins to capture aerodynamic lift. At ratios below 1, the blades are being pushed more by the drag force than by lift. Chapter 4 discusses drag-based machines.

Turbine designers develop graphs that describe the rotor's behavior by plotting its power coefficient, c_p, against the tip speed ratio (fig. 3–4). These graphs, which predict the rotor's performance at different rotating speeds, blade lengths, and wind speeds, are used by design engineers working on other parts of the wind power system.

These graphs often show different curves for several blade pitch angles, because the ideal angle varies at different tip speed ratios (fig. 3–5). The blade pitch angle is usually designated by the Greek letter *beta* (β).

The optimal tip speed ratio varies with blade design, and tip speed is limited by factors such as noise, which increases with tip speed. But higher tip speeds translate to lighter turbines and lower torque on the drivetrain. In general, the greater the number of blades on the rotor, the lower the optimal tip speed ratio. Primitive windmills designed to provide mechanical power for pumping water or grinding grain had multiple blades and very low tip speed ratios.

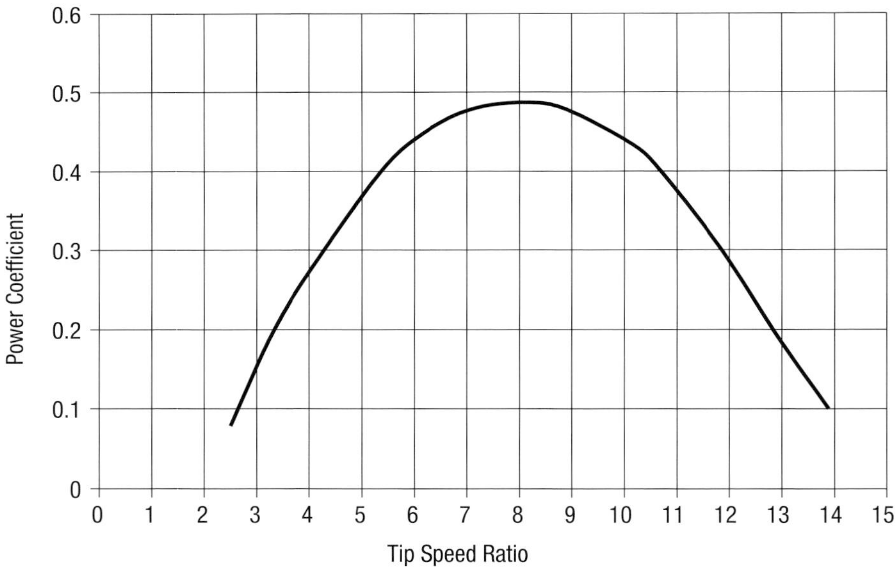

Fig. 3–4. Power coefficient vs. tip speed ratio

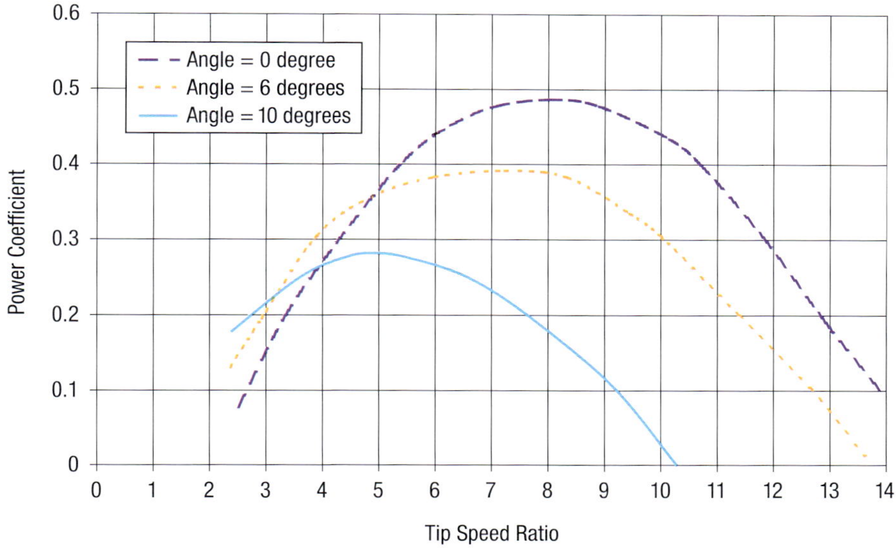

Fig. 3–5. Power coefficient vs. tip speed ratio for different pitch angles

Turbulence and wind shear

Turbulence is the opposite of a steady wind—it's strong, unpredictable gusts of wind or other sudden, frequent variations in wind speed and direction. Anyone who's been on an airplane during rough weather knows what turbulence feels like and how much power it can exert.

Turbulence is an important factor in designing a structurally sound, safe wind turbine—because a single very strong gust can take down a machine. Turbulence also causes fatigue loads (see "Cyclic loads and fatigue" below). The effective level of turbulence, referred to as turbulence intensity, is calculated according to standards defined by the International Electrotechnical Commission.

> *Before I die, I hope someone will clarify quantum physics for me. After I die, I hope God will explain turbulence to me.*
>
> —*Albert Einstein*

Typically, turbulence weakens at higher elevations above the ground, just as wind speed increases with height, so today's tall turbines are less vulnerable. Still, turbulence is a factor in assessing the local wind resource, and the *topography* of the surrounding land can contribute to wind turbulence (see chapter 5).

While turbine blades have become longer, and towers taller, to take advantage of the stronger, steadier winds at greater heights, bigger machines can also exploit *wind shear*. In areas with a high vertical wind shear, the wind speed increases with height even more dramatically than normal. However, longer blades and taller towers require stronger materials and heavier foundations, so it's a classic design trade-off.

Wind shear can also take other forms. Vertical wind shear can occur in reverse, with higher wind speeds at the bottom of the tower than at the top. Also, wind shear can be horizontal rather than vertical (*horizontal wind shear*), where the wind speed varies horizontally. For example, the wind can be stronger on the right-hand side of the turbine than on the left, or vice versa.

And wind shear can create some strange, very localized conditions, where blades encounter one wind direction on the left side of the rotor's revolution and the opposite wind direction on the right side, or opposite wind directions at the top and bottom of the revolution. Sometimes, the wind can blow from the direction opposite to that of normally prevailing winds. So a massive rush of air can come in, blowing into the "wrong" side of the turbine.

Although design engineers can't predict these conditions, wind turbines must be able to withstand such unusual events, within reasonable limits. No designer, however, can develop a machine that could survive (without shutting down) what one wind expert has called "the hammer of God"—weather that destroys turbines no matter how well designed and built they are.

Very few mathematicians have spent time hanging on at the top of a 100 ft (30.5 m) wind turbine in a 50 mph (22.4 m/s) wind to gain experience in the wind turbine craft. Although [we] do not recommend this experience, it is apparent that wind turbine design requires a judicious blend of analytical engineering and creative experimentation, not to mention a profound respect for the power of the wind.

—David M. Eggleston and Forrest S. (Woody) Stoddard, Wind Turbine Engineering Design

---

Natural frequencies

Wind turbine designers have to account for each of the wind power system component's *natural frequency*, or resonance. This is a specific point, measured in hertz, at which the component becomes "excited," creating vibrations that can cause it to self-destruct like a crystal glass when a singer hits that certain note.

Each wind power system component has a natural frequency that can become excited. For example, the rotor's frequency of revolution (20 rpm is equal to about 0.33 Hz) can cause the blades or the tower to vibrate if the rotor frequency approaches either of their natural frequencies. This vibration can collapse the tower or damage the blades by causing them to bend into abnormal shapes.

Large, heavy components such as the gearbox and generator have much higher natural frequencies, so they are not as much of a concern to design engineers. Natural frequencies can be more of a problem in vertical-axis wind turbines (see chapter 4).

Cyclic loads and fatigue

At the same time that the wind is spinning the blades and turning the turbine's driveshaft, the wind's energy creates forces that act upon the turbine system in complex ways. Like any rotating machine, a wind turbine experiences fatigue. Every revolution of its components—rotor, gearbox, generator, etc.—represents one cycle of fatigue on the blades, rotating shaft, gears, and bearings.

Each cycle causes a certain tiny amount of damage, which gradually reduces the life of the component. When enough cycles have been endured, a fatigue crack can develop. Additional cycles will propagate the crack until the component fails.

Throughout their life, wind turbine rotors endure these *cyclic loads*, which come and go repeatedly at different points during the rotor's operation. For example, stresses and loads at the top of the blade's revolution are much different from stresses at the bottom. Similarly, the loads are opposite when the blade is on the right side versus the left.

In contrast, stresses and loads that are constant, rather than cyclic, are more bearable, creating less wear and tear on the machine.

Bending. Two key cyclic loads that affect the spinning blade are *edgewise* and *flapwise bending* forces, both of which can cause vibration and fatigue, leading to blade failure. Edgewise forces occur through the leading and trailing edges of the blade, while flapwise forces are exerted perpendicular to the blade, thrusting it backward.

Edgewise forces can actually be helpful—they include the aerodynamic lift that causes the turbine's rotor to spin (see "Blade shape" above). Although edgewise forces can create vibration, they are not usually a problem for a properly designed wind turbine.

In contrast, flapwise thrust can be a powerfully catastrophic force, resulting in a tower strike, a tower collapse, or even an uprooted foundation. Flapwise bending is a cyclic load because it's more severe at the top of the blade's revolution, where wind speeds are usually higher. Engineers use the *thrust coefficient*, c_t, to calculate or predict thrust:

$$\text{thrust} = 1/2 \times \rho \times V^2 \times A \times c_t \qquad \textbf{Eq. 3–11}$$

Wind turbine designers need to account for flap-wise thrust by making the blades strong enough to resist it, yet flexible enough not to break. The thrust load on the blade varies considerably along its length. Today's longest blades can flex as much as 4–6 meters near their tips (12–18 feet) during strong gusts of wind.

To control flapwise bending, researchers at Risø DTU, Denmark's National Laboratory for Sustainable Energy, have performed tests in a wind tunnel on a prototype trailing-edge flap made of molded rubber. Sandia National Laboratories also has tested blades designed to change pitch when flapwise bending forces become too strong. Chapter 10 discusses these and other advanced technologies.

Other cyclic loads. Gravitational stresses are another example of a cyclic load. When the blades are pointing up or down, gravity doesn't affect them much, though it does squeeze or stretch them a bit (compression and tension, respectively).

But when the blades are pointing to either side, gravity causes the tips to bend downward. This gravitational force acts on opposite sides of the blade when it's in the left position versus the right, creating a cyclic stress. The bigger the wind power system, the greater is the gravity component of its loads and stresses, because every part of the machine is larger and heavier.

Torque can also cause cyclic stress on the turbine's driveshaft. That's one reason for using three blades rather than two or four, because the total torque generated by the blades is less cyclic—that is, more constant and bearable—in a three-bladed machine.

Gyroscopic loads occur when the turbine yaws, or pivots, to face into the wind. Gyroscopic loads are also cyclic, in that they change with blade position during the yawing motion.

Other common loads. Several other loads act upon wind power systems, but they are common to nearly all machinery and not unique to wind turbines, so they are well known and can be handled predictably by experienced design engineers. These include friction and inertia.

Similarly, ice accumulation on the blades is a concern, but one that is well understood, given experience with airplane wings icing up. Today's wind turbines are designed for safe operation at temperatures of –20° to 40°C (–4° to 104°F).

Site-specific remedies are undertaken for cold temperatures, such as water-repellent (*hydrophobic*) coatings to prevent ice formation. One de-icing technology uses electrothermal pulses, while other systems are based on ultrasonic signal detection. When ice causes the signal to weaken to a preset level, the system automatically begins to either heat the blades or stop the turbine. In some cases, simply painting the blades black is enough to keep them from icing up.

Additional information

Appendix A contains more details on wind power system design:

- Air density
- Energy and power
- Wind energy and wind power
- Maximum rotor efficiency
- Forces on an airfoil
- Shaftpower torque

For additional details, two excellent books on the subject are *Wind Energy Explained: Theory, Design and Application* (2nd edition), by James F. Manwell, Jon G. McGowan, and Anthony L. Rogers, and *Wind Turbine Engineering Design*, by David M. Eggleston and Forrest S. (Woody) Stoddard. Also, chapter 4 provides more information on the design of small wind power systems.

References

Atlas CTB. http://www.atlasctb.com/.

Eggleston, David M., and Stoddard, Forrest S. 1987. *Wind Turbine Engineering Design*. New York: Van Nostrand Reinhold Co. Inc. (ISBN 0-442-22195-9).

Gosman, Taylor. 2010 (May 4). Cutting Transport Costs Down to Size. *Windpower Monthly*. http://www.windpowermonthly.com/news/EmailThisArticle/1000918/Cutting-transport-costs-down-size.

Manwell, James F., McGowan, Jon G., and Rogers, Anthony L. 2010 (February 23). *Wind Energy Explained: Theory, Design and Application*. 2nd edition. Hoboken, NJ: Wiley. (ISBN-10: 0470015004; ISBN-13: 978-0470015001).

Risø DTU. 2010 (February 17). Controllable Rubber Trailing Edge Flap May Ease Stress on Wind Turbine Blades. *Renewable Energy World*. http://www.renewableenergyworld.com/rea/news/article/2010/02/controllable-rubber-trailing-edge-flap-may-ease-stress-on-wind-turbine-blades.

Robinson, Abraham. 1974. *Non-Standard Analysis*. Princeton, NJ: Princeton University Press. (ISBN: 9780691044903). http://press.princeton.edu/titles/5812.html.

Schmidt, Michael. Wind Turbine Design Optimization. Strategic Energy Institute, Georgia Institute of Technology. http://www.clemson.edu/scies/wind/Poster-Schmidt.pdf.

Thresher, R., Robinson, M., and Veers, P. 2008 (August). Wind Energy Technology: Current Status and R&D Future. NREL/CP-500-43374. http://www.nrel.gov/wind/pdfs/43374.pdf.

U.S. Department of Energy, Energy Efficiency and Renewable Energy. Energy Basics, Wind Energy Technologies. http://www.eere.energy.gov/basics/renewable_energy/wind.html.

Wind Tower Systems. http://www.windtowersystems.com/.

4

Small Wind Power: Traditional Systems and Alternative Designs

Market for Small Wind Power

The market for small-scale wind power has expanded recently throughout the world. *Off-grid* or *stand-alone* systems, which are not connected to utility grids, have been popular globally, from the United States to China, for many decades.

But recent market growth has also encompassed small wind power projects that do interconnect with the utility. These *grid-connected* systems are a form of *distributed energy* supply. Distributed power—which is generated at or near the customer's site, rather than from a central utility plant—includes other renewable sources such as solar power, *fuel cells* (an electrochemical technology similar to a battery), and old-fashioned engines burning a variety of fuels.

The term *small-scale* usually refers to wind turbines with capacities of less than 100 kW. This includes wind power systems that generate enough electricity for a single home, farm, or small business, typically about 1–10 kW. *Micro* usually refers to wind generators producing less than 1 kW for specialty applications, as noted in chapter 1.

Although the amount of power generated by small wind turbines is tiny compared to that from utility-scale wind farms, small wind systems are gaining popularity in Europe. Historically, the United Kingdom has been the second-largest market for small wind power (after the United States), accounting for 20%–25% of global demand.

Customers in the United Kingdom installed about 4,500 small wind turbines in 2009, reaching a total of 15,000 units (according to the latest public data available). Interest is also rising in Italy, especially for equipment that is manufactured domestically.

Globally, the small wind turbine market increased by about 10% in 2009, adding more than 21,000 units with a total capacity of 42.5 MW. Growth rates could nearly triple to 152 MW/yr by 2015, doubling the size of the market, according to Pike Research. Small wind turbines might soon reach a cumulative capacity of 1 gigawatt worldwide (1 GW, or 1,000 MW).

In 2010, the U.S. market grew 26% with 25.6 MW of new capacity installed, driven largely by homeowners and businesses taking advantage of financial incentives and streamlined permitting. About 8,000 new units were installed in America; this was 20% below 2009's number, reflecting a shift toward larger applications. For the first time, more than 90% of new small turbines were grid-connected.

China remains the largest market in the world for small wind turbines. About 50,000 units reportedly were sold in 2009, bringing China's total number of small wind systems to some 400,000.

About 250 companies in 27 countries around the world manufacture small wind turbines or have plans to start producing them. Just 10 years ago, no more than 50 such companies existed. Companies that supply small wind power systems tend to come and go, but the largest ones have been around for years or decades. Figure 4–1 and table 4–1 show where these manufacturers are based, and table 4–2 shows the top five in terms of capacity sold in 2009.

Fewer than 10 U.S. manufacturers accounted for up to half the world market in 2009. In 2010, however, six non-U.S. companies each sold more than a megawatt of small turbines, giving American manufacturers more competition.

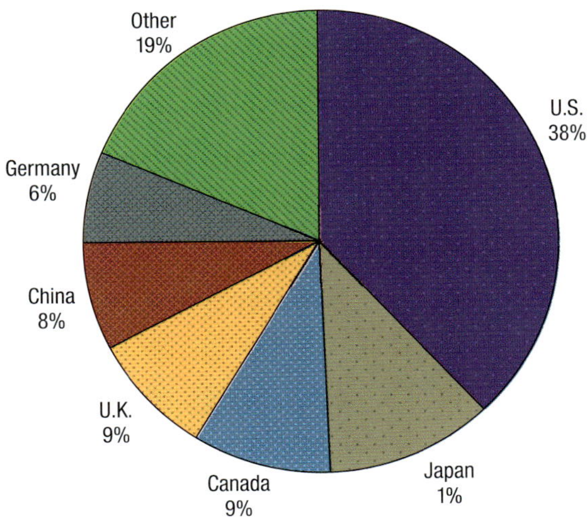

Fig. 4–1. Global distribution of small wind turbine manufacturers

Table 4–1. List of small wind turbine manufacturers by country. *Data courtesy of AWEA.*

Country	No. of Manufacturers	% of Total
U.S.	95	37.5
Japan	29	11.5
Canada	24	9.5
UK	22	8.7
China	19	7.5
Germany	16	6.3
Netherlands	6	2.4
Spain	5	2.0
Sweden	5	2.0
South Africa	4	1.6
France	3	1.2
India	3	1.2
Taiwan	3	1.2
Finland	2	0.8
Israel	2	0.8
Italy	2	0.8
Russia	2	0.8
Argentina	1	0.4
Australia	1	0.4
Austria	1	0.4
Denmark	1	0.4
Iran	1	0.4
Kenya	1	0.4
Poland	1	0.4
New Zealand	1	0.4
Switzerland	1	0.4
Tanzania	1	0.4
Total	**252**	**99.8**

Table 4–2. Five largest global manufacturers of small wind turbines

Company	Location	Worldwide sales, MW	Product capacities, kW
Southwest Windpower	U.S. (AZ)	11.7	0.2, 0.9, 1.0, 2.4, 3.0
Northern Power Systems	U.S. (VT)	9.2	100
Proven Energy	UK (Scotland)	3.7	2.5, 5.2, 12.1
Wind Energy Solutions	Netherlands	3.7	80, 250
Bergey WindPower	U.S. (OK)	2.1	1.0, 10

Small Wind Power Systems

Many small-scale wind power systems use three-blade, horizontal-axis turbines that look a lot like their much bigger counterparts. But many variations of this model, as well as many alternative designs, have been invented, built, and tested—including small wind turbines that look nothing like a propeller on a tower, and more like a modern sculpture or even something from science fiction or outer space.

Why so many choices?

Why do small wind power systems come in so many different shapes, while most large wind-farm turbines look alike? It's because everything costs so much less at such a small scale, so people can afford to experiment on a risky design. Look at it this way—a 1-MW turbine is 1,000 times bigger than a 1-kW turbine. So the inventor or developer can play around a bit, build a small machine, and refine the design for much less money.

Plus, all the components are relatively inexpensive, so if the turbine needs to be stronger, its structure can be "beefed up" for just a little extra money without redesigning the whole machine. Adding strength to a small wind power system is cheaper because the loads and stresses on the system are much lower, relative to the cost of dealing with them.

For example, when you add a thousand pounds or kilos to the drivetrain of a big machine, it has a huge ripple effect throughout the system, requiring designers to reexamine its performance and safety from top to bottom.

But you can add a pound or half a kilo to a small machine, and it's much easier to compensate for the extra weight safely and cheaply, because you're not operating at the very edge of the design limits of your materials and components, as you are on a big machine.

Think of a visit to the hardware store—do you need a ¼–inch bolt or a ⅜-inch bolt? The difference is in pennies. But when you're scaling a bolt from

6 inches to 8 inches on a big machine, it might cost you a couple of thousand dollars or euros per turbine, and even a little extra weight has to be accounted for in the design.

Horizontal-axis wind turbines

Although most utility-scale wind turbines have three blades, small-scale horizontal-axis machines can have two blades, four blades, or more. Certain cyclic stresses can become worse when the turbine has an even number of blades, allowing one blade to be pointed straight up at the same time that the opposite blade is pointed straight down, or one pointing right and the other left.

A small two-blade turbine can be less expensive than the same size machine with three or more blades, but it needs a teetering hub to cope with these unbalanced loads. The teetering hub also helps the turbine handle high wind shear, and it does work well under the right conditions. But in general, two-blade turbines are harder to design correctly, and three-blade machines run more smoothly, which usually means that they'll last longer.

Similarly, though most large turbines are upwind machines, small wind power systems can also be designed for *downwind* operation, with the rotor behind the nacelle and tower. Downwind turbines pivot freely as the wind direction changes, so they don't need a yaw system to mechanically pivot the rotor and nacelle. The machine can be less complex and easier to maintain. Another advantage is that when the blades bend backwards under high winds, they won't hit the tower.

In Japan, downwind turbines are used at some mountainous sites, where it can be hard to harvest wind power from conventional upwind machines. Japan has about 25 downwind turbines and plans to build dozens more.

However, downwind machines experience greater cyclic loads that can lead to fatigue and premature failure. For example, at one point during each revolution, the blade has to pass behind the tower. This blocks the wind and creates a cyclic load. Also, downwind machines typically are noisier.

Vertical-axis wind turbines

As its name implies, this wind turbine design has blades that rotate on a vertical axis rather than horizontal—in other words, the turbine's driveshaft is perpendicular to the ground like a flagpole. Sometimes the vertical driveshaft is called a tower, even though the turbine is mounted on the ground. Although many vertical-axis wind turbines have been built, these designs are not in widespread use on big wind farms.

The vertical-axis turbine has many advantages, but its disadvantages have so far outweighed them except in a handful of applications (see "Pros and cons of vertical-axis turbines" below). Engineers continue to work on new vertical-shaft turbine designs that might broaden this technology's usefulness. Researchers are also working on other alternative designs such as structures that funnel the wind to increase its speed before it enters the turbine (see chapter 10).

Vertical-axis wind turbines (often dubbed VAWTs) come in two flavors—machines based on aerodynamic lift, and those based on drag. As described in chapter 3, lift comes from a region of low-pressure air behind the blade, which causes the blade to turn. Lift is created perpendicular to the wind, while a small drag force acts in opposition to the lift force. Vertical-shaft turbines can exploit drag to deliver high torque—strong mechanical rotation of the driveshaft.

In vertical-axis applications, wind turbines that use lift as their primary driving force are mainly *Darrieus* machines, which are named for their inventor, an aeronautical engineer in France who patented the design in the 1920s. These turbines, with their curved blades, are often described as looking like an eggbeater. Darrieus wind turbines are the most common type of vertical-axis machines for generating electricity in all size ranges.

Darrieus-type turbines can also have straight blades, rather than curved. These machines are called *giromills*, and a variation of the giromill is the *cycloturbine*. Vertical-axis turbines can even be tipped on their side, an option called the *transverse-axis* design. All of these turbines are based on the same concept.

Most vertical-axis machines that are driven by lift capture more of the wind's energy than those based on drag. The classic, two-blade Darrieus wind turbine captures maximum lift only twice per revolution, but delivers a huge amount of torque (fig. 4–2).

Another type of vertical-axis wind turbine is the *Savonius* machine, which produces mechanical power from the wind's drag force. The blades on Savonius turbines and other drag-type machines operate sort of like a canoe paddle, where you propel the canoe by dragging the paddle through the water. Drag-based turbines have a maximum rotational speed equal to the wind speed (a tip speed ratio of 1).

When viewed from above, the Savonius rotor is shaped like the letter S. Savonius wind turbines rotate relatively slowly and yield a high torque. They can also start up at very low wind speeds, so they are sometimes used in conjunction with Darrieus turbines as "starter motors" (fig. 4–3).

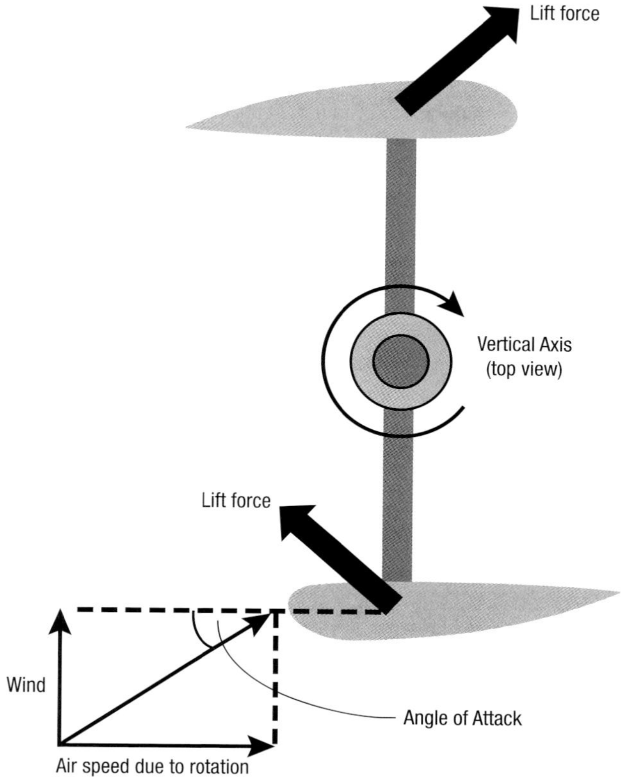

Fig. 4–2. How a Darrieus wind turbine works

The low-speed mechanical shaftpower from Savonius turbines is useful for many purposes such as pumping water and grinding grain, but it's not the best kind of power for generating electricity (its rotating speed is too slow). However, their blades can be built out of just about anything—buckets, paddles, sails, or cut-apart oil drums, for instance.

Pros and cons of vertical-axis turbines. Rather than being perched on towers, vertical-axis wind turbines are built with their base close to the ground, which makes it easier to access parts for maintenance. This also allows them to be built in places where tall towers are prohibited.

A more important advantage is that the wind direction is irrelevant to the operation of a vertical-axis machine, so the rotor doesn't need a yaw mechanism to pivot it into the wind. Also, some vertical-shaft turbines can start up at low wind speeds.

Fig. 4–3. Combination Darrieus/Savonius vertical-axis wind turbines. *Image courtesy of Fred Hsu.*

Moreover, proponents claim that vertical-axis wind turbines are simple to install and maintain, and they have a more pleasant appearance than propeller-type horizontal-axis machines, though obviously this is a subjective factor.

One potential advantage of small-scale vertical-axis turbines is that they might be sited in densely populated cities with many obstacles that create turbulent wind, such as buildings and signs. Although software is on the market for siting turbines in cities, data are scarce on the real-world performance of vertical-axis turbines in urban settings. To test these applications, Boston's Museum of Science is operating a rooftop wind turbine lab with five small machines, including one vertical-axis turbine.

The biggest disadvantage of vertical-axis wind turbines is that they are less efficient overall than horizontal-axis types (fig. 4–4). On a vertical shaft, each turbine blade is in the best position to capture lift and generate torque at only one point, or position, during its revolution. Thus, a two-blade vertical-axis rotor gets the most lift only twice per revolution. The rest of the time, it's not catching the most energy available from the wind.

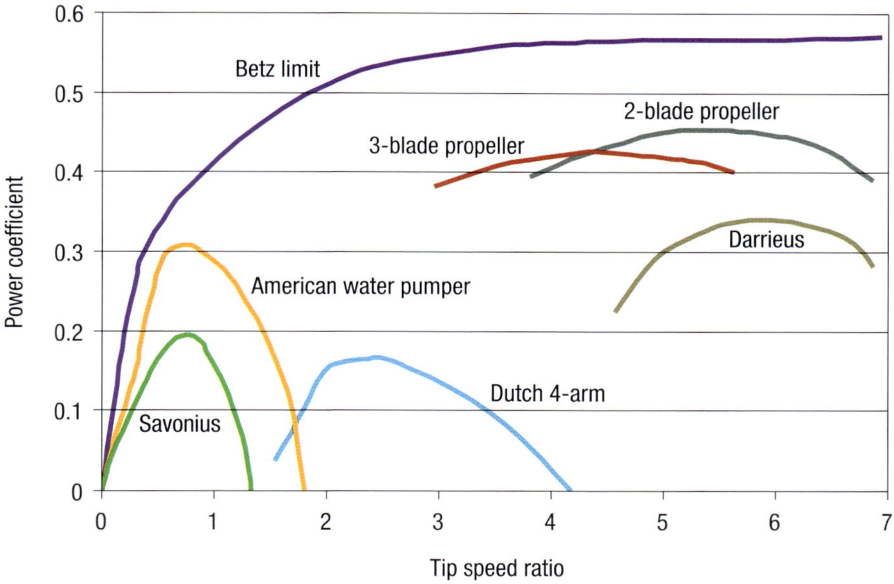

Fig. 4–4. Power coefficient vs. tip speed ratio for different wind turbine designs

Also, vertical-axis wind power systems have some natural frequencies that must be avoided while the machine is running. Because each blade gets maximum lift only once per cycle, the whole structure endures pulsating torque, where stresses and loads climax every time each blade reaches that point. With a classic two-blade Darrieus machine, the pulses occur twice during every spin of the rotor. This is sort of like twanging a guitar string repeatedly or banging on a pipe to make it resonate.

At certain rotational speeds, the blades' pulsations can match the natural frequency of other system components. This causes vibration that fatigues the equipment and leads to premature failure. The dynamics of this situation are challenging, to say the least.

Also, because vertical-axis turbines are difficult to mount on tall towers, their capacity is limited by wind speeds near the ground, which are generally lower, translating to less wind power available for capture. In addition, some large Darrieus turbines require guy wires or cables to keep them balanced, increasing the system's *footprint* (the area covered by the equipment). Guy wires also impose a large thrust load on the main turbine bearings.

Although vertical-axis machines perform well in turbulent winds, turbulence can still introduce vibration, which makes noise and can wear out the turbine bearings prematurely. Also, some Darrieus machines need an external source of power to start their rotation in low winds, though Savonius turbines do not. Savonius machines are often used in applications where high reliability is required, such as *anemometers* for measuring wind speed. They also deal better with turbulence than the Darrieus type.

A saga of scams. People who are new to the wind industry sometimes think that vertical-axis turbines are a novel idea that hasn't been exploited, or at least not sufficiently. But experienced wind engineers are all too familiar with the vertical design concept. Among wind experts, the general opinion is that vertical-axis machines just aren't as good as horizontal-axis turbines in terms of efficiency and power output (as shown above in fig. 4–4). Some experts also question the reliability or survivability of vertically oriented turbines.

Vertical-axis technologies make up only about 1% of the market for small wind systems, although new products are frequently introduced. Unfortunately, vertical wind turbine products suffer from a long history of abuse by unscrupulous salespeople who make exaggerated, sometimes fraudulent claims. One experienced engineer calls them "shysters" who prey upon naive buyers.

In addition, well-intentioned inventors who might truly believe in their idea have sucked others in. The Internet makes it easy for anyone to promote their products online, so consumers can become fascinated by vertical-axis wind turbines, which are all over YouTube, for example. If you didn't know any better, the technology might seem like an energy revolution that's waiting to happen.

Traditional horizontal-axis machines have not been immune to overselling either, and many customers have been disappointed in their machine's performance. Some manufacturers might even have exploited government rebates for small wind systems by inflating their estimates of how much wind energy the machines deliver. A California incentive program was suspended due to questionable applications for rebates.

What [California's] rebates are for is to get renewable energy that works. There are installers who will go to a customer and tell them 'Look, you don't have to put down any money upfront; we'll take care of all the expenses, just sign here.' What the installer is doing is making money from the state . . . by installing a very inexpensive machine that hasn't been proven or tested and in most cases is worthless.

—Tal Mamo, Talco Electronics (in Greentech Media)

But the wind industry has begun to get a handle on this problem, since it could undermine consumers' confidence in any type of wind energy. Some turbine manufacturers are having their vertical-axis products (as well as small horizontal-axis turbines) analyzed independently to verify their performance.

In the United Kingdom, turbine certification is required to qualify for special rates for small-scale wind power. Two turbines made by Scotland's Proven Energy have been certified through the UK's Microgeneration Certification Scheme.

Also, countries around the world are cooperating through the International Energy Agency in an effort to develop a labeling protocol for small wind turbines (Task 27 under a collaborative research program). Underwriters' Laboratories (UL), a global agency that certifies safety, covers small wind power systems that generate electricity.

The U.S. National Renewable Energy Laboratory (NREL) began testing small wind turbines in 2007 using standards adopted from the International Electrotechnical Commission. These tests include power performance and quality, acoustics, and safety. Previously, NREL and its predecessors had been testing small machines for more than 30 years.

As part of the new program, NREL tested a three-blade, 1-kW vertical-axis turbine (giromill type) manufactured by Windspire Energy, as well as small horizontal-axis turbines made by Abundant Renewable Energy, Entegrity Wind System, Gaia, and Ventera.

Since then, the American Wind Energy Association (AWEA) formed a Small Wind Certification Council that certifies turbines' rated power, annual energy output, and sound level. The council issues labels confirming that the product meets or exceeds performance and safety standards for small wind turbines, which include durability. AWEA coordinates with the UK's certification program.

More than 20 companies have applied to the council for at least 28 turbines to be certified—a welcome sign of transparency in the small wind energy market. Two of the world's largest manufacturers of small horizontal-axis machines, Southwest and Bergey, are awaiting turbine certification. The list also includes four vertical-axis machines:

- Windspire's 1.2-kW giromill
- Urban Green Energy, 1-kW and 4-kW (a smaller product is shown in fig. 4–5.)
- UrWind's 1.7-kW, also a three-blade giromill

Rigorous testing will determine whether vertical-axis and other, more conventional small wind turbines will perform as their customers expect.

Some manufacturers of small wind turbines post their products' power curves on their company Web sites, but often this information is missing or hard to find. Numerous power curves have been collected for reference by Better Generation. These include curves for several vertical-axis machines and for six classes of horizontal-axis turbines, based on their rated kW output.

"What drives the continued development of mechanical devices like wind turbines in the face of…widespread lack of support? In the case of wind turbine technology, I suspect that part of the reason for persistence of this vision is how accessible wind turbines are to [human] understanding. They are personal in a way that almost no other form of power generation is. This 'personal' scale has been both the blessing and the curse of wind power development. The field tends to attract people who are committed, creative, and passionate. It also attracts a few people who are a little too much of all of those things, to the point that sometimes the grounding of reality is lost."

—*Darrell M. Dodge,* Illustrated History of Wind Power Development

---

Fig. 4–5. Vertical-axis wind turbine (600-watt). *Image courtesy of Urban Green Energy.*

Small Wind Power Applications

Stand-alone versus grid-connected

Traditionally, small wind energy systems have been stand-alone (off-grid) machines serving people without access to utility power, such as homeowners in remote areas or farmers who need to pump water for livestock or crops.

More recently, people who want to be energy-independent have embraced wind power and other renewable energy sources. Stand-alone systems have also brought mechanical power and electricity to people in remote areas of undeveloped countries. Most of these turbines generate less than 10 kW and incorporate batteries or other devices to store electricity.

Stand-alone wind power equipment is often integrated with another source of energy in a hybrid system. Wind turbines can work in tandem with solar PV panels or with engines that generate power from diesel fuel, gasoline, propane, or natural gas if available. Hybrid systems offer an extra measure of reliability and flexibility for isolated customers.

Historically, wind power systems that interconnect with the utility grid have been larger, with capacities up to 50 or even 100 kW. As noted earlier, however, grid-connected systems are growing in popularity worldwide as utility customers, both residential and commercial, try to reduce their electric bills. Small wind turbines have flooded the marketplace, from online vendors such as Amazon to big-box retail stores.

One selling point for grid-connected systems is that the utility still provides power if your wind turbine doesn't generate enough electricity to keep all your appliances and electronics running. However, if the utility has a power outage, grid-connected wind power systems also shut down for safety reasons.

Another advantage is, if your wind turbine produces more electricity than you need, utilities generally will compensate you for the extra power—in other words, your electric meter can spin backward. This is called *net metering*, and it's become common in industrial countries. Some utilities match the retail rate when accepting excess power from homeowners and businesses, while others compensate customers at lower rates.

Small wind power applications have expanded to include new uses such as aerating municipal wastewater ponds to control odor and sludge buildup. Businesses and towns are beginning to consider wind turbines that supply electricity for lighting streets and parking lots, often in tandem with solar power and batteries. Turbine manufacturers have developed small hybrid systems for such applications.

Hybrid wind/solar systems can also be ideal for powering remote telecommunications equipment such as cellphone towers, which might have used diesel-fueled generators in the past. Diesel-powered systems are less efficient when running at such typically low loads, and they cost a lot for fuel and maintenance. Small wind/solar hybrids might pay for themselves in certain applications by reducing the amount of time that the diesel generator needs to run.

Technology options and requirements

Most small-scale wind power systems are three-blade, horizontal-axis, propeller-type machines, though other versions are on the market with two, four, or more blades. Small propeller machines have a long tail to steer them into the wind, just like old-fashioned windmills did, and they incorporate automatic devices to keep the rotor from spinning out of control in very high winds. Vertical-axis systems are also making inroads into the small turbine business, as noted above.

Nearly all of today's small wind power systems incorporate a permanent-magnet generator with a power converter. No gearbox is needed to speed up the driveshaft's rotation. In other words, they are direct-drive machines—the same technology that's recently become more prevalent in multi-megawatt turbines (see chapter 2).

In general, small wind turbines can operate effectively and deliver power economically at relatively low wind speeds—a minimum of 4–5 m/s (9–11 mph), depending on how they're being used (mechanical power, battery charging, or electricity generation). However, even in residential applications, the site-specific wind speed is a critical factor in figuring out how a small wind power system will perform (see below).

Also, even small machines must be mounted on a tower, especially in wooded areas. Ideally, the tower should be at least 10 meters (30 feet) taller than any surrounding objects such as trees and houses, which create a region of very turbulent wind up to twice their own height. Plus, the area around the tower must be clear of any tall obstacles within a 100-meter (300-foot) radius.

Economics of small wind power

Until recently, the up-front cost of wind power systems has deterred many customers from taking advantage of the technology. Homeowners, for instance, might stay in the same house for only five or six years, whereas a wind turbine will probably take longer than that to pay for itself in energy cost savings. This is called the *payback period*, the time required for investors to recoup their initial costs.

But governments throughout the industrial world have stepped in to offer financial incentives for ordinary people to invest in wind power and other renewable resources. Policy makers have begun to recognize the potential benefits of a more diverse, less centralized electricity supply (see chapter 11).

In deciding whether a small wind power system is practical and economical, homeowners and businesses have a wealth of information at their disposal—perhaps too much information. The best place to look for preliminary information is your government—whether national or federal, state or provincial—and your local electric utility.

Usually, government agencies can provide information on the practicality of a wind power system for homes and businesses, including whatever financial incentives might be offered to offset the cost of its installation and operation. For example, the U.S. federal government offers a tax credit equal to 30% of a small wind turbine's total installed cost, effective until 2017.

Europe's Small Wind Industry Implementation Strategy (SWIIS) project, funded by the European Commission, stimulates industry growth by providing information and analyzing barriers. The group is compiling data on successful small turbine installations and will develop a comprehensive manual for policy makers and the public. In America, the Distributed Wind Energy Association addresses barriers to small wind power such as burdensome permitting, zoning, and regulatory requirements.

Many free tools are available online to help people decide whether a wind turbine will work for them. Table 4–3 lists just a few sources of information on small wind power systems.

Many of these available tools are simplified calculators that let you plug in your location, electricity consumption, and other basic information, then the tool "guesstimates" the size of a wind power system to meet your needs—its capacity, its output per year, etc. This is basically a screening exercise, not a way to make a final decision.

Table 4–3. Small wind power system information and screening tools

Tool	Source	URL
FAQs on Small Wind Systems	U.S. DOE	http://www1.eere.energy.gov/wind/small_wind_system_faqs.html
Is Wind Right for Me?	Southwest Windpower	http://www.windenergy.com/assessment
Small Wind Energy	CanWEA	http://www.smallwindenergy.ca/en/SmallWind.html
Small Wind for Homeowners, Ranchers, and Small Businesses	Wind Powering America (U.S. DOE)	http://www.windpoweringamerica.gov/small_wind.asp

In general, for a wind turbine to be economical, the customer must be located in an area where the following apply:

- Electricity prices are relatively high.
- The utility does not impose unreasonable barriers to connecting with the grid.
- Local building codes or zoning laws do not prohibit wind turbines or towers.

Net metering with the utility and financial incentives from various levels of government are other economic factors to consider, as well as favorable zoning by-laws that are meant to encourage small wind power.

But the most important variable is wind speed. Most existing maps showing wind resources worldwide are not detailed enough to predict the wind speed at locations specific enough for siting small systems, although some U.S. maps are increasing in detail and resolution (see chapter 5). Moreover, wind conditions can vary even within the boundaries of a small property, so selecting a site free of obstacles for testing and installation is also paramount.

Good wind conditions—primarily high speed, but also steady (nonturbulent) winds—are critical to ensuring that a turbine will operate safely and provide the amount of power expected by the homeowner or business. Turbine vendors should perform a professional, thorough site analysis before selling and installing a machine.

Alternatively, prospective customers might want to put up an anemometer on a tower to measure the wind speed on their property for about a year before making a final decision to invest in a wind power system. Anemometers can cost anywhere from $150 on up (110 euros), not including mounting it on the tower and collecting data. Complete systems that measure not just the speed of the wind but also its direction are also on the market.

Appendix B contains more information on drag-type machines, small rotor sizing, and small wind power system costs.

References

American Wind Energy Association. 2010. AWEA Small Wind Turbine Global Market Study. http://www.awea.org/documents/2010_AWEA_Small_Wind_Turbine_Global_Market_Study.pdf.

Bergey WindPower, http://www.bergey.com/.

Better Generation. Comparing Wind Turbine Power Curves. http://www.bettergeneration.com/wind-turbines/comparing-wind-turbine-power-curves.html.

———. Power Predictor. http://www2.powerpredictor.com.

Distributed Wind Energy Association. http://www.distributedwind.org/.

Distributed Wind Site Analysis Tool, http://dsat.cadmusgroup.com/.

The Economist. 2011 (June 2). Japan's Winds of Change. http://www.economist.com/node/18750586?story_id=18750586.

Eggleston, Eric. What Are Vertical-Axis Wind Turbines (VAWTs)? American Wind Energy Association. http://www.awea.org/faq/vawt.html.

Gay, Paul L. Why Two Bladed Rotors Make Sense. Aerostar Wind. http://www.aerostarwind.com/Why%20Two%20Bladed%20Rotors%20Make%20Sense.html.

Gipe, Paul. 2009 (April). *Wind Energy Basics Revised: A Guide to Home- and Community-Scale Wind Energy Systems*. White River Junction, VT: Chelsea Green Publishing. (ISBN 978-1-06358-030-4). http://www.chelseagreen.com/bookstore/item/wind_energy_basics_second_edition:paperback.

GlobalData. 2009 (September). The U.S. Small Wind Market Set to Become BIG. http://www.reportsandreports.com/reports/15111-the-us-small-wind-market-set-to-become-big.html.

Lacey, Stephen. 2010 (July 21). Will Vertical Axis Wind Ever Scale Up? *Renewable Energy World*. http://www.renewableenergyworld.com/rea/news/podcast/2010/07/will-vertical-axis-wind-ever-scale-up.

Meteodyn, http://www.meteodyn.com/en/town-planning.html.

Northern Power Systems, http://www.northernpower.com/.

Proven Energy, http://www.provenenergy.co.uk/.

REN21 Renewable Energy Policy Network for the 21st Century. 2010 (July 15). Renewables 2010 Global Status Report. http://www.ren21.net/globalstatusreport/g2010.asp.

Ritchie, Ed. 2010 (September–October). Tipping Point. *Distributed Energy*. 8(5): 38-42. http://www.distributedenergy.com/september--october-2010/tipping-point-windturbines.aspx.

Small Wind Certification Council, http://www.smallwindcertification.org/.

Small Wind Implementation Industry Strategy (SWIIS), http://www.smallwindindustry.org/index.php?id=114.

Southwest Windpower, http://www.windenergy.com/.

Trabish, Herman K. 2011 (March 15). Have Small Wind Manufacturers Exploited Loopholes in California Rebates? Greentech Media. http://www.greentechmedia.com/articles/read/have-small-wind-manufacturers-exploited-loopholes-in-california-rebates.

———. 2011 (September 19). Is Distributed Wind the Answer? (Part Three). Greentech Media. http://www.greentechmedia.com/articles/read/is-distributed-wind-the-answer-part-three/.

———. 2012 (January 31). One Solar-Wind Hybrid Goes Down, Another Goes Up. Greentech Media. http://www.greentechmedia.com/articles/read/one-solar-wind-hybrid-goes-down-another-goes-up/.

Types of Wind Turbines. Centurion Energy LLC. http://centurionenergy.net/types-of-wind-turbines.

Urban Green Energy, http://www.urbangreenenergy.com.

U.S. Department of Energy, Energy Efficiency and Renewable Energy. Energy Basics, Wind Energy Technologies. http://www.eere.energy.gov/basics/renewable_energy/wind.html.

U.S. Department of Energy, Energy Efficiency and Renewable Energy, Wind and Water Program. Frequently Asked Questions on Small Wind Systems. www1.eere.energy.gov/windandhydro/small_wind_system_faqs.html.

Webb, Chris. 2011 (March 22). Small Wind: Market Update. Renewable Energy World. http://www.renewableenergyworld.com/rea/news/article/2011/03/small-wind-market-update?cmpid=WNL-Wednesday-March23-2011.

Wind Energy Solutions, http://www.windenergysolutions.nl/.

5

Assessment: Mapping and Measuring the World's Wind

Introduction

Most of the sun's energy reaches us as heat or light, but about 1%–2% of it becomes wind energy, as noted in chapter 1. That's about 50–100 times more energy than all the plants in the world capture from the sun.

Researchers have estimated that globally, wind power from land areas with wind speeds of at least 6.9 m/s (15 mph)—considered the minimum for economical power generation—could produce about 72 *terawatts* (TW) of electricity. That's 72 million megawatts, compared to today's global wind power capacity of about 200,000 megawatts (200 gigawatts).

Of course, this theoretical potential is impossible to tap fully in the real world. Wind energy is a variable resource, and its contribution to our power supply depends on when it's available. But if even 20% of global wind power could be captured and put to work, it could easily supply the world with all the electricity it needs. And that's not counting the vast amount of wind resources offshore.

Estimating Wind Power Potential

The global wind resource estimate above was based on measurements taken mostly at 10 meters (33 feet) above the ground and extrapolated to 80 meters (260 feet). Since then, researchers have made progress collecting data and estimating wind speeds at greater heights, first at 50 meters (160 feet) and more recently at 80–90 meters (260–300 feet). These estimates, which are more representative of the rotor hub heights on modern wind turbines, make today's wind resource assessments more accurate than ever.

Measuring the amount of wind power available and characterizing it precisely are critical steps in projecting the performance of a wind turbine at a particular site (see "Assessment Tools and Procedures" below). A small difference in wind speed can make a big difference in how much electricity a turbine produces.

When the wind speed doubles, a turbine's power output increases by a factor of eight. That's because the wind's available energy is proportional to the cube of its speed, $(m/s)^3$, as noted in chapters 1 and 3. So a 10% increase in wind speed translates to a 33% increase in kinetic wind energy available for capture.

In general, wind speeds of 3–4 m/s (7–9 mph) are strong enough for small, off-grid wind turbines used for mechanical power or charging a battery. For grid-connected wind power systems generating electricity for homes and small businesses, about 5 m/s (11 mph) should be fast enough.

But for utility-scale wind farms, developers look for strong, steady winds of at least 6–7 m/s (13–16 mph), preferably blowing much of the time at about 12–14 m/s (27–31 mph), the region where large turbines generate their maximum output. To expand wind power beyond these best sites, turbine manufacturers sell machines that are designed to operate effectively at lower wind speeds (see chapter 11).

Scientists have developed systems to classify wind resources based on wind speed and the *wind power density*. The power density, which is measured in watts per square meter (W/m^2), indicates how much potential wind energy is available over a given area of land for a turbine to convert to power. One system assigns wind resources to seven classes, which are used as a measure of the wind resource potential (table 5–1).

Table 5–1. Classes of wind

Class	Height, 10 m (33 ft)		Height, 50 m (164 ft)	
	Wind power density, W/m^2	Wind speed, m/s (mph)	Wind power density, W/m^2	Wind speed, m/s (mph)
1	<100	<4.4 (9.8)	<200	<5.6 (12.5)
2	100-150	4.4-5.1 (9.8-11.5)	200-300	5.6-6.4 (12.5-14.3)
3	150-200	5.1-5.6 (11.5-12.5)	300-400	6.4-7.0 (14.3-15.7)
4	200-250	5.6-6.0 (12.5-13.4)	400-500	7.0-7.5 (15.7-16.8)
5	250-300	6.0-6.4 (13.4-14.3)	500-600	7.5-8.0 (16.8-17.9)
6	300-400	6.4-7.0 (14.3-15.7)	600-800	8.0-8.8 (17.9-19.7)
7	>400	>7.0 (>15.7)	>800	>8.7 (19.7)

Notes: Power density and wind speed are annual averages.

Wind speeds were extrapolated vertically based on a power law that describes typical vertical wind shear. However, wind shear does not always behave as predicted.

Wind speed is the mean based on the Rayleigh distribution (see appendix D).

Wind speed is for standard sea-level conditions. To maintain the same power density at higher elevations, wind speed has to increase by 3% per 1,000 m (5% per 5,000 ft).

Areas designated class 4 or greater are preferred for utility-scale wind farms, as noted above. However, stronger winds can sometimes be found in certain parts of lower class areas. For example, exposed hilltops can be a prime location for high wind speeds, but they might not appear on a wind resource map because their area is too small to show up.

In other words, the wind data's level of detail and quality affect how accurate the wind class designation is. Local geography can be a deciding factor in siting a wind farm, since it affects not only wind speed but also turbulence and other wind characteristics (see "Assessment Tools and Procedures" below).

Other, similar classification systems based on wind speed are used internationally, sometimes with power density estimates as well. Also, a well-known wind speed and turbulence standard, established by the International Electrochemical Commission, is used to classify turbines (see chapter 6, table 6–3).

Mapping the Global Wind Resource

The potential of wind energy in Europe and North America has been well defined on a broad geographical scale, as well as at particular locations where large wind farms have been developed. Researchers continually work to refine their maps of wind energy resources and to extend work to other countries where mapping has been less comprehensive.

Figures 5–1 through 5–6 show wind resource maps for Europe and North America, as well as a global map of wind speeds, which shows that most countries worldwide have at least some level of wind resources to exploit.

Europe

Much of the European mapping work has been done using WAsP (Wind Atlas Analysis and Application Program), a methodology that was used to produce the Wind Atlases of the World. This online atlas contains maps for more than 30 countries and wind data for about 80 more countries. The atlas is managed by Risø DTU, Denmark's National Laboratory for Sustainable Energy.

The European Union (EU) also is funding the Norsewind project to develop high-resolution, hub-height atlases of offshore wind resources in three northern European seas—the North, Irish, and Baltic. Norsewind also will provide data acquired from offshore measurements, test advanced techniques for remote data sensing (see "Advanced assessment tools" below), and develop a short-term forecasting technology.

The European Wind Energy Association (EWEA) estimates that the EU's "raw" wind energy potential is huge, equivalent to almost 20 times the total energy demand anticipated for the EU in 2020, if environmental, social, and economic considerations are set aside. However, EWEA takes a conservative approach to projecting the EU's actual wind capacity for 2020, which it estimates at 230,000–265,000 MW (230–265 GW). By comparison, Europe had an installed base of 86.3 GW as of 2011, most of it in EU countries.

EWEA also projects wind power generation, measured in *terawatt-hours* (TWh). This unit simply converts wind capacity (power) to electricity produced (energy). Similar measurements are kilowatt-hour (familiar on home electric bills), megawatt-hour, and *gigawatt-hour* (kWh, MWh, and *GWh*)—they're all units of energy, defined as power produced over a period of time (see chapter 3).

A terawatt-hour equals 1,000 gigawatt-hours. According to EWEA, by 2020 wind power in Europe could be producing anywhere from 12,200–70,000 TWh/yr, meeting 3%–20% of the EU's total energy demand.

United States

The U.S. National Renewable Energy Laboratory (NREL), through its Wind Powering America program, has mapped the nation's wind resource, including individual maps for each state. Map data are also presented in tables that are accessible online. Table 5–2 summarizes the national wind resource by state.

According to NREL's figures, the wind power resource in the United States is almost 10,500 GW—10.5 million MW—in terms of potential installed capacity on land, estimated at a height of 80 meters (260 feet). By comparison, a total of only 40.2 GW had actually been built as of 2011.

When higher wind speeds at 100 meters (330 feet) are factored in—not an unreasonable height for today's turbine towers—NREL's U.S. resource estimate jumps to 12,000 GW. This potential wind power capacity is about three times greater than the U.S. economy's electricity demand at its annual peak (typically on a hot summer day).

If installed, this wind turbine capacity could generate roughly 10 times the amount of electricity America needs each year: 37–45 million GWh, compared to annual consumption of about 4 million GWh.

NREL also estimated America's potential wind resource offshore (fig. 5–4 and table 5–3), even though the country lags far behind Europe in offshore wind development and many areas might be excluded. About 4,150 GW of wind power could be available in waters within 93 km of shore (50 nautical miles). This estimate was based on wind turbine rotor hub heights of 90 meters (300 feet).

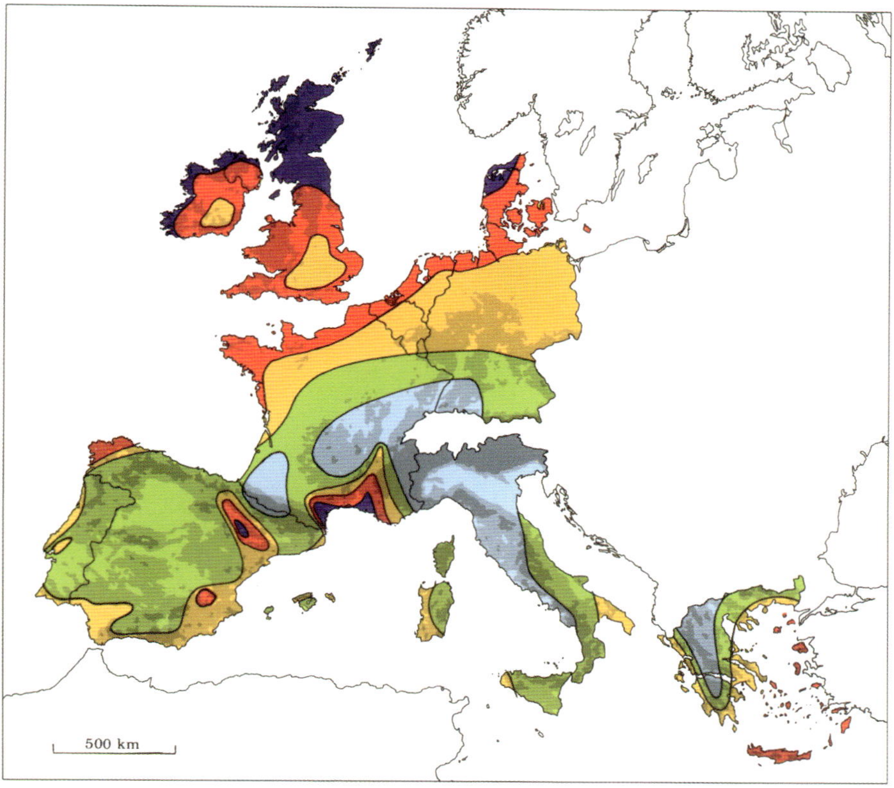

	Wind resources[1] at 50 metres above ground level for five different topographic conditions									
	Sheltered terrain[2]		Open plain[3]		At a sea coast[4]		Open sea[5]		Hills and ridges[6]	
	$m\,s^{-1}$	Wm^{-2}	$m\,s^{-1}$	Wm^{-2}	$m\,s^{-1}$	Wm^{-2}	$m\,s^{-1}$	Wm^{-2}	$m\,s^{-1}$	Wm^{-2}
(dark blue)	> 6.0	> 250	> 7.5	> 500	> 8.5	> 700	> 9.0	> 800	> 11.5	> 1800
(red)	5.0-6.0	150-250	6.5-7.5	300-500	7.0-8.5	400-700	8.0-9.0	600-800	10.0-11.5	1200-1800
(yellow)	4.5-5.0	100-150	5.5-6.5	200-300	6.0-7.0	250-400	7.0-8.0	400-600	8.5-10.0	700-1200
(green)	3.5-4.5	50-100	4.5-5.5	100-200	5.0-6.0	150-250	5.5-7.0	200-400	7.0- 8.5	400- 700
(light blue)	< 3.5	< 50	< 4.5	< 100	< 5.0	< 150	< 5.5	< 200	< 7.0	< 400

Fig. 5–1. Europe's onshore wind resources. *Image courtesy of European Wind Atlas (©Risø DTU).*

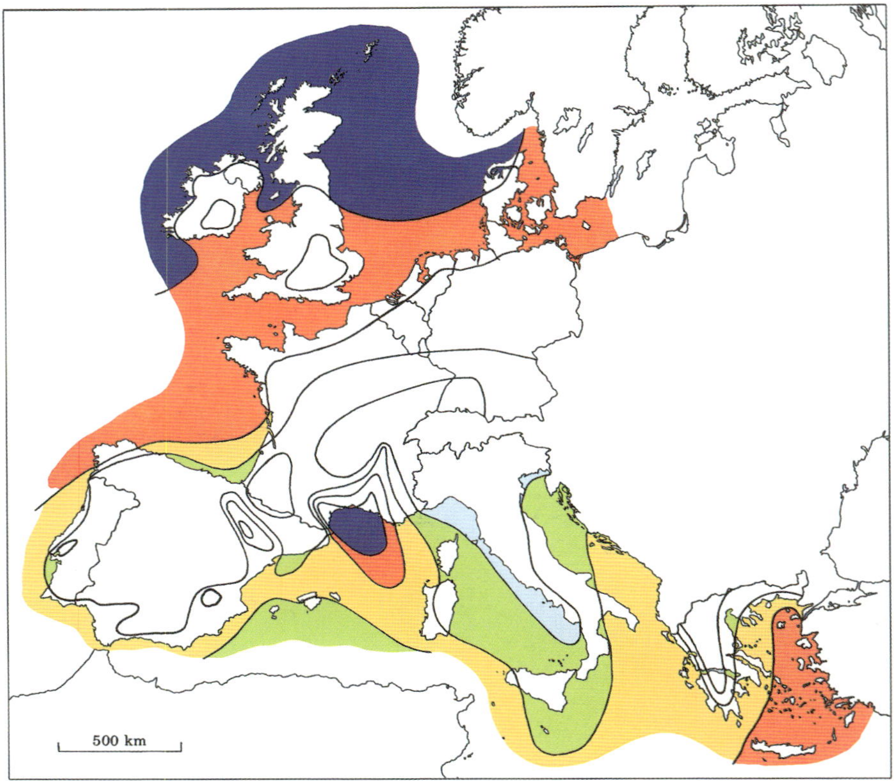

Wind resources over open sea (more than 10 km offshore) for five standard heights										
	10 m		25 m		50 m		100 m		200 m	
	m s^{-1}	Wm^{-2}	m s^{-1}	Wm^{-2}	m s^{-1}	Wm^{-2}	m s^{-1}	Wm^{-2}	m s^{-1}	Wm^{-2}
	> 8.0	> 600	> 8.5	> 700	> 9.0	> 800	> 10.0	> 1100	> 11.0	> 1500
	7.0-8.0	350-600	7.5-8.5	450-700	8.0-9.0	600-800	8.5-10.0	650-1100	9.5-11.0	900-1500
	6.0-7.0	250-300	6.5-7.5	300-450	7.0-8.0	400-600	7.5- 8.5	450- 650	8.0- 9.5	600- 900
	4.5-6.0	100-250	5.0-6.5	150-300	5.5-7.0	200-400	6.0- 7.5	250- 450	6.5- 8.0	300- 600
	< 4.5	< 100	< 5.0	< 150	< 5.5	< 200	< 6.0	< 250	< 6.5	< 300

Fig. 5–2. Europe's offshore wind resources. *Image courtesy of European Wind Atlas (©Risø DTU).*

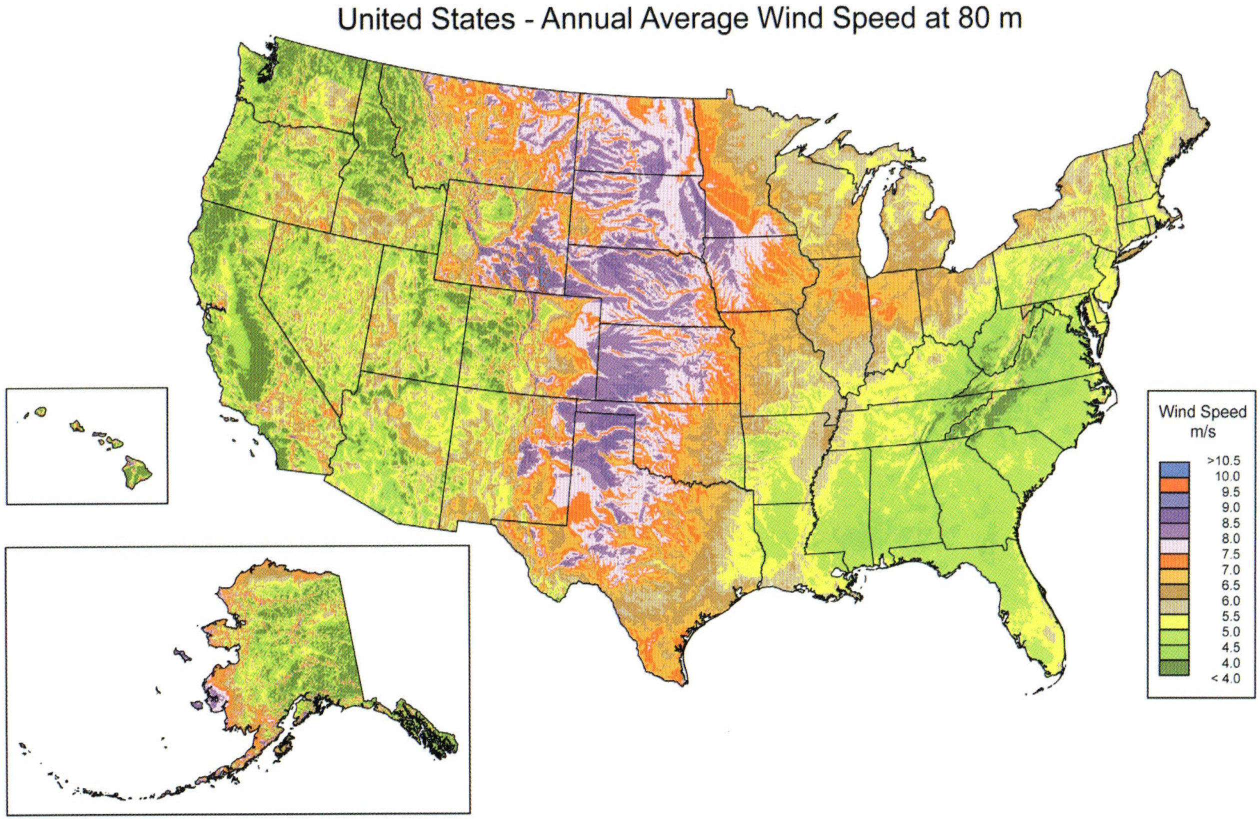

Fig. 5–3. U.S. onshore wind resources. *Image courtesy of Wind Powering America.*

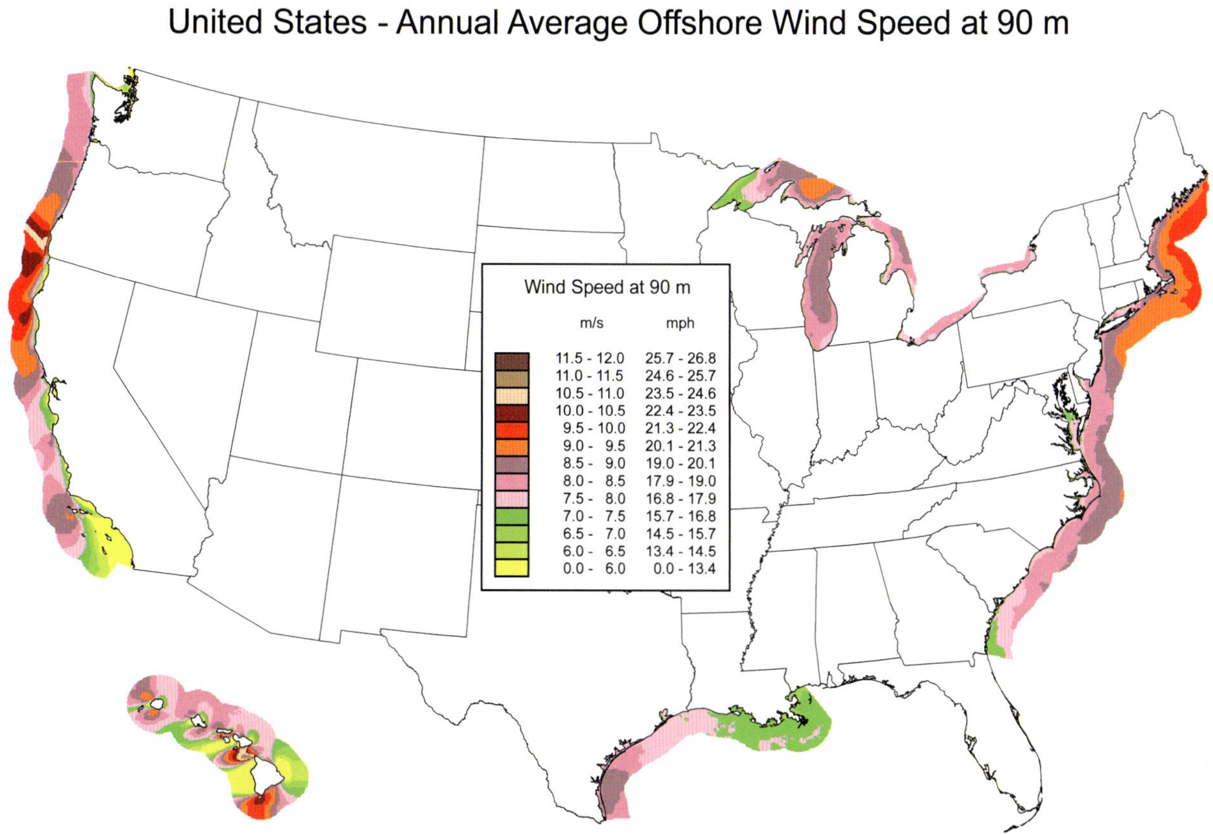

Fig. 5–4. U.S. offshore wind resources. *Image courtesy of the U.S. National Renewable Laboratory.*

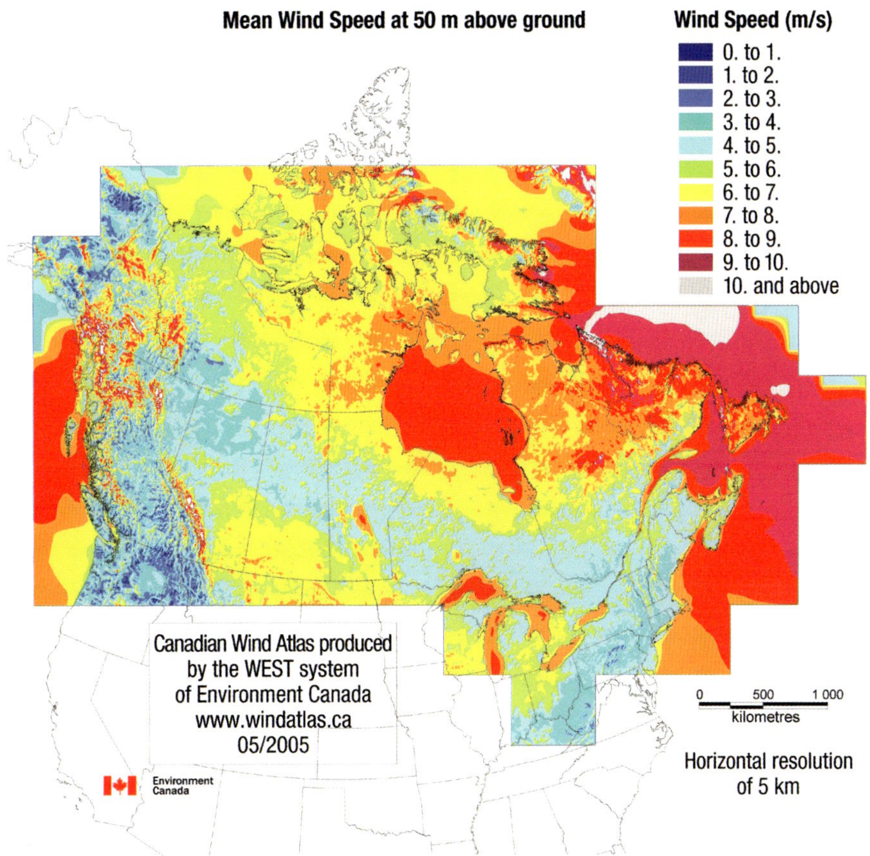

Fig. 5–5. Canada's wind resources. *Image courtesy of the Canadian Wind Energy Atlas.*

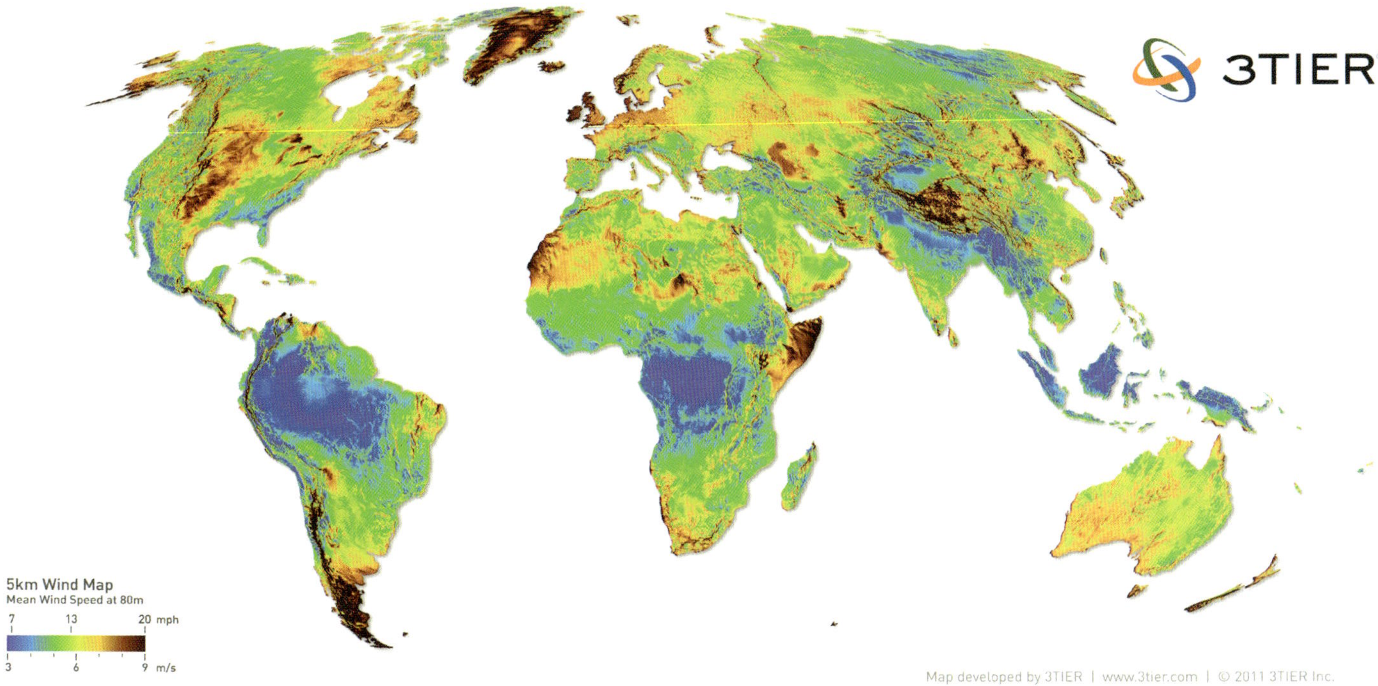

Fig. 5-6. Global mean wind speeds (at 80 meters/260 feet, the hub height of most utility-scale wind turbines). *Image courtesy of 3TIER, Inc.*

Appendix C includes tables of data on wind power potential by wind speed, water depth, and distance from shore for each of the 26 U.S. coastal states.

NREL has worked closely on many of these maps with AWS Truepower, a private wind assessment company. The latest wind resource maps can be overlaid with geographical features such as power lines and roads. Work is continuing to increase the horizontal resolution of the U.S. maps to 1 kilometer from the current 2.5 kilometers (to 0.6 mile from 1.6 miles).

AWS also reports regularly on U.S. and European wind speeds for the past quarter-year, mapping local anomalies and climatic fluctuations. This "hind cast" (as opposed to forecast) is useful to wind farm operators and investors.

In addition to U.S. wind resource maps, NREL has created similar maps for about 20 other countries worldwide, including 15 specific areas of China—the world's largest and fastest-growing market for wind power—and seven areas of Mexico. This work has led to the recognition of higher-class winds in areas where none were thought to exist.

NREL also provides access to maps available through the Asia Sustainable and Alternative Energy program (ASTAE) and the Solar and Wind Energy Resource Assessment (SWERA). ASTAE, a World Bank program, maintains the Wind Energy Resource Atlas of South Asia, which covers Cambodia, Laos, Thailand, and Vietnam.

SWERA, a United Nations project, provides information on wind and solar energy resources in 13 developing countries in Latin America, the Caribbean, Africa, and Asia. Partners on the project include NREL, Risø DTU, the International Energy Agency, and several other government and private organizations worldwide.

China

China possesses a huge land area and long coastline, implying a major source of wind energy. Measurements by the China Meteorological Administration, performed at 50 meters (160 feet), suggest that the country could develop about 3,500 GW of wind power onshore, according to the Global World Energy Council. This includes 2,380 GW of class 3 wind potential and 1,130 GW of class 4, plus 200 GW of class 3 potential offshore in shallow waters up to 25 meters deep (82 feet).

China plans on an installed wind power capacity of 30 GW by 2020, but due to its fast-growing market, the country could far surpass its goal, reaching some 150–230 GW (see chapter 11). External studies of China's wind resource estimate a total exploitable capacity of 700–1,200 GW, including offshore. Other assessments suggest even higher figures, up to 2,500 GW. One analysis pegs China's offshore resource alone at 740 GW.

Table 5–2 (a). U.S. onshore wind resources by state. *Data courtesy of NREL.*

State	Windy Land Area >= 30% Gross Capacity Factor at 80m: Total (km^2)	Excluded[2] (km^2)	Available (km^2)	Available % of State	% of Total Windy Land Excluded	Wind Energy Potential: Installed Capacity[3] (MW)	Annual Generation (GWh)
Alabama	80.4	56.7	23.6	0.01%	70.6%	118.2	333
Arizona	4,545.0	2,364.1	2,180.8	0.74%	52.0%	10,904.1	30,616
Arkansas	4,663.2	2,823.2	1,840.1	1.34%	60.5%	9,200.3	26,906
California	26,901.3	20,079.2	6,822.0	1.67%	74.6%	34,110.2	105,646
Colorado	95,830.4	18,386.5	77,443.9	28.73%	19.2%	387,219.5	1,288,490
Connecticut	31.4	26.1	5.3	0.04%	83.1%	26.5	73
Delaware	36.6	34.7	1.9	0.04%	94.8%	9.5	26
Florida	9.6	9.5	0.1	0.00%	99.2%	0.4	1
Georgia	281.3	255.3	26.0	0.02%	90.7%	130.1	380
Idaho	13,420.4	9,805.3	3,615.1	1.67%	73.1%	18,075.6	52,118
Illinois	70,763	20,787.1	49,976.4	34.25%	29.4%	249,882.1	763,529
Indiana	46,255.2	16,609.7	26,345.5	31.63%	35.9%	148,227.5	443,912
Iowa	134,900.1	20,757.3	114,142.8	78.32%	15.4%	570,714.2	2,026,340
Kansas	211,861.3	21,387.1	190,474.2	89.38%	10.1%	952,370.9	3,646,590
Kentucky	48.7	36.6	12.1	0.01%	75.1%	60.6	173
Louisiana	125.5	43.6	82.0	0.07%	34.7%	409.8	1,100
Maine	6,026.5	3,776.2	2,250.2	2.69%	62.7%	11,251.2	33,779
Maryland	567.7	271.1	296.6	1.18%	47.8%	1,482.9	4,269
Massachusetts	1,709.0	1,503.4	205.6	0.99%	88.0%	1,028.0	3,323
Michigan	19,761.3	7,952.9	11,808.5	7.85%	40.2%	59,042.3	169,221
Minnesota	121,884.7	24,030.6	97,854.1	44.83%	19.7%	489,270.6	1,679,480
Mississippi	0.0	0.0	0.0	0.00%	N/A	0.0	0
Missouri	69,676	14,805.8	54,871.0	30.9%	21.2%	274,355.1	810,619
Montana	232,768.6	43,967.7	188,800.9	49.60%	18.9%	944,004.4	3,228,620
Nebraska	199,627.8	16,028.0	183,599.7	91.64%	8.0%	917,998.7	3,540,370

Nevada	5,873.6	4,424.2	1,449.4	0.51%	75.3%	7,247.1	20,823
New Hampshire	1,663.8	1,236.8	427.1	1.78%	74.3%	2,135.4	6,706
New Jersey	280.8	254.5	26.4	0.14%	90.6%	131.8	373
New Mexico	111,445.8	13,029.1	98,416.7	31.25%	11.7%	492,083.3	1,644,970
New York	7,705.8	12,549.6	5,156.3	4.10%	70.9%	25,781.3	74,695
North Carolina	1,155.6	994.1	161.5	0.13%	86.0%	807.7	2,395
North Dakota	182,374.6	28,335.4	154,039.2	84.25%	15.5%	770,195.8	2,983,750
Ohio	17,189.9	6,205.9	10,983.9	10.28%	36.1%	54,919.7	151,881
Oklahoma	123,243.6	19,879.2	103,364.4	57.10%	16.1%	516,822.1	1,788,910
Oregon	17,109.8	11,689.7	5,420.1	2.16%	68.3%	27,100.3	80,855
Pennsylvania	2,123.5	1,462.1	661.4	0.56%	68.9%	3,307.2	9,673
Rhode Island	74.0	64.7	9.3	0.35%	87.4%	46.6	153
South Carolina	102.8	65.8	37.0	0.05%	64.0%	185.0	504
South Dakota	193,828.3	17,345.8	176,482.5	88.36%	8.9%	882,412.4	3,411,690
Tennessee	359.9	298.1	61.9	0.06%	82.8%	309.3	900
Texas	435,638.6	55,332.7	380,305.9	55.54%	12.7%	1,901,529.7	6,527,850
Utah	5,273.6	2,652.8	2,620.7	1.19%	50.3%	13,103.7	37,104
Vermont	2,569.6	1,979.8	589.7	2.39%	77.0%	2,948.7	9,163
Virginia	1,567.2	1,208.5	358.7	0.35%	77.1%	1,793.3	5,395
Washington	11,932.6	8,236.9	3,695.7	2.12%	69.0%	18,478.5	55,550
West Virginia	1,495.2	1,118.6	376.6	0.60%	74.8%	1,883.2	5,820
Wisconsin	30,228.8	9,477.3	20,751.4	14.29%	31.4%	103,757.1	300,136
Wyoming	146,166.2	35,751.7	110,414.5	43.58%	24.5%	552,072.6	1,944,340
US 48 Total	2,571,180	479,391	2,091,789	26.89%	18.6%	10,458,945	36,919,551

1 NREL's wind potential estimates were based on maps produced by AWS Truewind using the MesoMap® system.

2 Excluded lands include protected lands (national parks, wilderness, etc.), incompatible land use (urban, airports, wetland and water features), and other considerations.

3 Assumes 5 MW/km^2 of installed nameplate capacity

Table 5–2 (b). U.S. onshore wind resources by state. *Data courtesy of NREL.*

Wind Resource Exclusions
Criteria for Defining Available Windy Land (numbered in the order they are applied):

Environmental Criteria	Data/Comments
2) 100% exclusion of National Park Service and Fish and Wildlife Service managed lands	USGS Federal Lands shapefile, Dec 2005
3) 100% exclusion of federal lands designated as park, wilderness, wilderness study area, national monument, national battlefield, recreation area, national conservation area, wildlife refuge, wildlife area, wild and scenic river or inventoried roadless area.	USGS Federal Lands shapefile, Dec 2005; Inventoried Roadless Areas, 2004: BLM Areas of Critical Environmental Concern (2008)
4) 100% exclusion of state and private lands equivalent to criteria 2 and 3, where GIS data is available.	State/GAP land stewardship data management status 2, from Conservation Biology Institute Protected Lands database, 2004
7) 50% exclusion of remaining USDA Forest Service (FS) lands (incl. National Grasslands) except ridgecrests	USGS Federal Lands shapefile, Dec 2005
8) 50% exclusion of remaining Dept. of Defense lands except ridgecrests	Military Lands boundary files, internal dataset (2007)
9) 50% exclusion of state forest land, where GIS data is available	State/GAP land stewardship data management status 2, from Conservation Biology Institute Protected Lands database, 2004
Land Use Criteria	
5) 100% exclusion of airfields, urban, wetland and water areas	USGS North America Land Use Land Cover (LULC), version 2.0, 1993; ESRI airports and airfields (2006); U.S. Census Urbanized Areas (2000 and 2003)
10) 50% exclusion of non-ridgecrest forest	Ridge-crest areas defined using a terrain definition script, overlaid with USGS LULC data screened for the forest categories
Other Criteria	
1) Exclude areas of slope >20%	Derived from 90m national elevation dataset
6) 100% exclude 3 km surrounding criteria 2–5 (except water)	Merged datasets and buffer 3 km

Note: 50% exclusions are not cumulative. If an area is non-ridgecrest forest on FS land, it is just excluded at the 50% level one time.

Table 5–3. U.S. offshore wind resources by state. *Data courtesy of NREL.*

State	Wind Speed (m/s) at 90 m 7.0–7.5 km² *(GW)*	7.5–8.0 km² *(GW)*	8.0–8.5 km² *(GW)*	8.5–9.0 km² *(GW)*	9.0–9.5 km² *(GW)*	9.5–10.0 km² *(GW)*	>10.0 km² *(GW)*	Total >7.0 km² *(GW)*
California	11,439 *(57.2)*	24,864 *(124.3)*	23,059 *(115.3)*	22.852 *(114.3)*	13,185 *(65.9)*	15,231 *(76.2)*	6,926 *(34.6)*	117,555 *(587.8)*
Connecticut	530 *(2.7)*	702 *(3.5)*	40 *(0.2)*	0 *(0.0)*	0 *(0.0)*	0 *(0.0)*	0 *(0.0)*	1,272 *(6.4)*
Delaware	223 *(1.1)*	724 *(3.6)*	1,062 *(5.3)*	931 *(4.7)*	0 *(0.0)*	0 *(0.0)*	0 *(0.0)*	2,940 *(14.7)*
Georgia	3,820 *(19.1)*	7,741 *(38.7)*	523 *(2.6)*	0 *(0.0)*	0 *(0.0)*	0 *(0.0)*	0 *(0.0)*	12,085 *(60.4)*
Hawaii	18,873 *(94.4)*	42,398 *(211.5)*	33,042 *(165.2)*	13,913 *(69.6)*	7,779 *(38.9)*	6,720 *(33.6)*	4,852 *(24.3)*	127,477 *(637.4)*
Illinois	92 *(0.5)*	166 *(0.8)*	3,844 *(19.2)*	90 *(0.4)*	0 *(0.0)*	0 *(0.0)*	0 *(0.0)*	4,192 *(21.0)*
Indiana	82 *(0.4)*	216 *(1.1)*	286 *(1.4)*	0 *(0.0)*	0 *(0.0)*	0 *(0.0)*	0 *(0.0)*	584 *(2.9)*
Louisiana	48,043 *(240.2)*	15,032 *(75.2)*	0 *(0.0)*	0 *(0.0)*	0 *(0.0)*	0 *(0.0)*	0 *(0.0)*	63,075 *(315.4)*
Maine	906 *(4.5)*	1,142 *(5.7)*	1,976 *(9.9)*	3,331 *(16.7)*	8,429 *(42.1)*	15,485 *(77.4)*	42 *(0.2)*	31,311 *(156.6)*
Maryland	2,192 *(11.0)*	1,946 *(9.7)*	1,540 *(7.7)*	5,078 *(25.4)*	0 *(0.0)*	0 *(0.0)*	0 *(0.0)*	10,756 *(53.8)*
Massachusetts	202 *(11.0)*	526 *(2.6)*	1,639 *(8.2)*	3,606 *(18.0)*	20,351 *(101.8)*	13,674 *(68.4)*	0 *(0.0)*	39,997 *(200.0)*
Michigan	4,459 *(22.3)*	18,074 *(90.4)*	31,086 *(155.4)*	34,305 *(171.5)*	8,719 *(43.6)*	0 *(0.0)*	0 *(0.0)*	96,642 *(483.2)*
Minnesota	3,102 *(15.5)*	994 *(5.0)*	0 *(0.0)*	0 *(0.0)*	0 *(0.0)*	0 *(0.0)*	0 *(0.0)*	4,096 *(20.5)*
New Hampshire	19 *(0.1)*	46 *(0.2)*	171 *(0.9)*	336 *(1.7)*	102 *(0.5)*	0 *(0.0)*	0 *(0.0)*	672 *(3.4)*
New Jersey	528 *(2.6)*	1,508 *(7.5)*	4,965 *(24.8)*	12,934 *(64.7)*	0 *(0.0)*	0 *(0.0)*	0 *(0.0)*	19,935 *(99.7)*
New York	1,105 *(5.5)*	4,358 *(21.8)*	8,324 *(41.6)*	2,876 *(14.4)*	7,453 *(37.3)*	5,322 *(26.6)*	0 *(0.0)*	29,439 *(147.2)*
North Carolina	1,847 *(9.2)*	4,098 *(20.5)*	13,655 *(68.3)*	39,875 *(199.4)*	16 *(0.1)*	0 *(0.0)*	0 *(0.0)*	59,491 *(297.5)*
Ohio	341 *(1.7)*	3,067 *(15.3)*	5,829 *(29.1)*	0 *(0.0)*	0 *(0.0)*	0 *(0.0)*	0 *(0.0)*	9,237 *(46.2)*
Oregon	388 *(1.9)*	1,493 *(7.5)*	8,644 *(43.2)*	13,925 *(69.6)*	7,394 *(37.0)*	6,065 *(30.3)*	5,986 *(29.9)*	45,894 *(219.5)*
Pennsylvania	34 *(0.2)*	211 *(1.1)*	1,679 *(8.4)*	0 *(0.0)*	0 *(0.0)*	0 *(0.0)*	0 *(0.0)*	1,924 *(9.6)*
Rhode Island	224 *(1.1)*	126 *(0.6)*	283 *(1.4)*	671 *(3.4)*	1,461 *(7.3)*	2,360 *(11.8)*	0 *(0.0)*	5,126 *(25.6)*
South Carolina	1,457 *(7.3)*	8,202 *(41.0)*	10,384 *(51.9)*	6,007 *(30.0)*	0 *(0.0)*	0 *(0.0)*	0 *(0.0)*	26,049 *(130.2)*
Texas	2,019 *(10.1)*	24,823 *(124.1)*	16,556 *(82.8)*	12,273 *(61.4)*	0 *(0.0)*	0 *(0.0)*	0 *(0.0)*	55,671 *(278.4)*
Virginia	889 *(4.4)*	3,658 *(18.3)*	6,549 *(32.7)*	7,794 *(39.0)*	0 *(0.0)*	0 *(0.0)*	0 *(0.0)*	18,890 *(94.4)*
Washington	1,573 *(7.9)*	4,621 *(23.1)*	18,261 *(91.3)*	0 *(0.0)*	0 *(0.0)*	0 *(0.0)*	0 *(0.0)*	24,445 *(122.3)*
Wisconsin	3,715 *(18.6)*	3,405 *(17.0)*	7,761 *(38.8)*	8,417 *(42.1)*	0 *(0.0)*	0 *(0.0)*	0 *(0.0)*	23,298 *(116.5)*
Total	108,102 *(540.5)*	174,040 *(870.2)*	201,159 *(1,005.8)*	189,213 *(946.1)*	74,888 *(374.4)*	64,856 *(324.3)*	17,805 *(89.0)*	830,064 *(4,150.3)*

Other global wind resources

Canada. Like China, Canada possesses a huge land area (second only to Russia's) and the world's longest coastline. Canada's government has identified more than 100 GW of wind energy potential in Quebec Province alone, all of it within 25 km of transmission lines (16 miles). Canada's national goal is to get 20% of its electricity from wind power by 2025.

Canada's wind maps come from Environment Canada's Wind Energy Atlas, a product of the EOLE Wind Energy project. The online map shows 65 squares (tiles) on which numerical simulations were performed. The map can display mean wind speed or wind energy at 30, 50, or 80 meters (100, 160, or 260 feet), as well as topographical features, seasonal differences, and other variables.

Russia. Russia's wind resources have been mapped by Grigori Dmitriev of VetrEnergo and by Risø DTU and the Russian-Danish Institute for Energy Efficiency. NREL has also mapped Russia's Murmansk region, a peninsula in the Barents Sea within the Arctic Circle.

VetrEnergo estimates that Russia's potential wind power capacity could generate 6,218 TWh/yr, or if gross wind energy resources are considered, up to 80,000 TWh/yr. This would exceed Europe's estimated maximum wind power generation. Russia is the world's biggest country in terms of land area, and its coastline is one of the world's longest.

Australia. The continent of Australia also contains a large potential wind resource, which has been studied by the Commonwealth Scientific and Industrial Research Organization. Some of Australia's strongest winds are found in the New South Wales State, which has been mapped more extensively than other areas. Areas along the continent's southern and western coasts also have strong wind potential. Australia's government released a Renewable Energy Atlas in 2008 that contains a wind resource map.

India. The Centre for Wind Energy Technology recently increased its estimate of India's potential for wind power to 48.5 GW. However, the assessment, conducted at a 50-meter height (160-foot), included only nine of India's 28 states. Other agencies place the country's potential at 65–70 or even 100 GW.

Brazil. Brazil's wind resource was remeasured at 80–100 meters (260–330 feet), increasing the country's potential to 350 GW. Large unpopulated land areas and a long coastline could give Brazil a tremendous wind resource. The country is expected to lead wind power growth in Latin America, accounting for about 69% of the capacity expected to be installed in South America by 2025 (see chapter 11).

Egypt. Egypt cooperated with Denmark on a wind atlas that was expanded to cover all of Egypt in 2005. Wind speeds average 10 m/s (22 mph) in the Gulf of Suez—where at least 7.2 GW of wind power could be developed by 2022, according to the World Bank—and 7–8 m/s (16–18 mph) in large desert regions along the Nile River and in parts of Sinai. Egypt's electrical grid serves nearly 100% of the population and gives dispatching priority to renewable energy projects.

Conclusion

Clearly, the world has no shortage of sites for capturing wind energy. The Global Wind Energy Council calls wind a vast untapped resource capable of supplying the world's electricity needs many times over.

But wind power development in any particular country depends more on government policies than wind energy potential. Some countries with relatively poor wind resources such as Germany boast a high degree of wind power development, while others with a plentiful resource such as Argentina, Russia, and South Africa have barely started developing their wind power. The world's energy industry might worry about "peak oil," but it need not worry about "peak wind."

Assessment Tools and Procedures

Introduction

As mentioned above, characterizing the local wind resource precisely and accounting for the geography and topography of the surrounding terrain are critical to projecting wind turbine performance at a particular site. Wind speeds vary by the time of day or night (diurnal variations) and from month to month, as well as from year to year.

Also, wind can be a fickle source of energy, in terms of how much power is available at a certain location. Wind speed can vary dramatically with temperature and humidity, as well as with elevation and terrain. For example, one spot that's 50 meters (160 feet) lower than another might produce 10%–15% less energy.

Even though today's wind resource maps are more accurate than ever, they can serve only as screening tools for preliminary selection of potential wind farm sites. They cannot substitute for onsite measurement of wind speed and direction, which define turbulence, wind shear, and gust intensity—all important factors in wind farm siting (see chapter 6).

Once a good area has been identified from resource maps, the first step in the assessment process is prospecting, or exploring for wind on a leasable tract of land. In the old days, engineers would look for places where trees and bushes had been bent over by the constant force of strong, steady winds. Even today, with modern instruments and computer software, deformed vegetation is considered a good sign in deciding where to start measuring the wind.

Meteorological towers

The most common, least expensive way to assess a site's wind resource is to put up a *meteorological tower* (typically nicknamed a "met" tower or "met mast"). This tower is basically a pole on a hinged base (fig. 5–7).

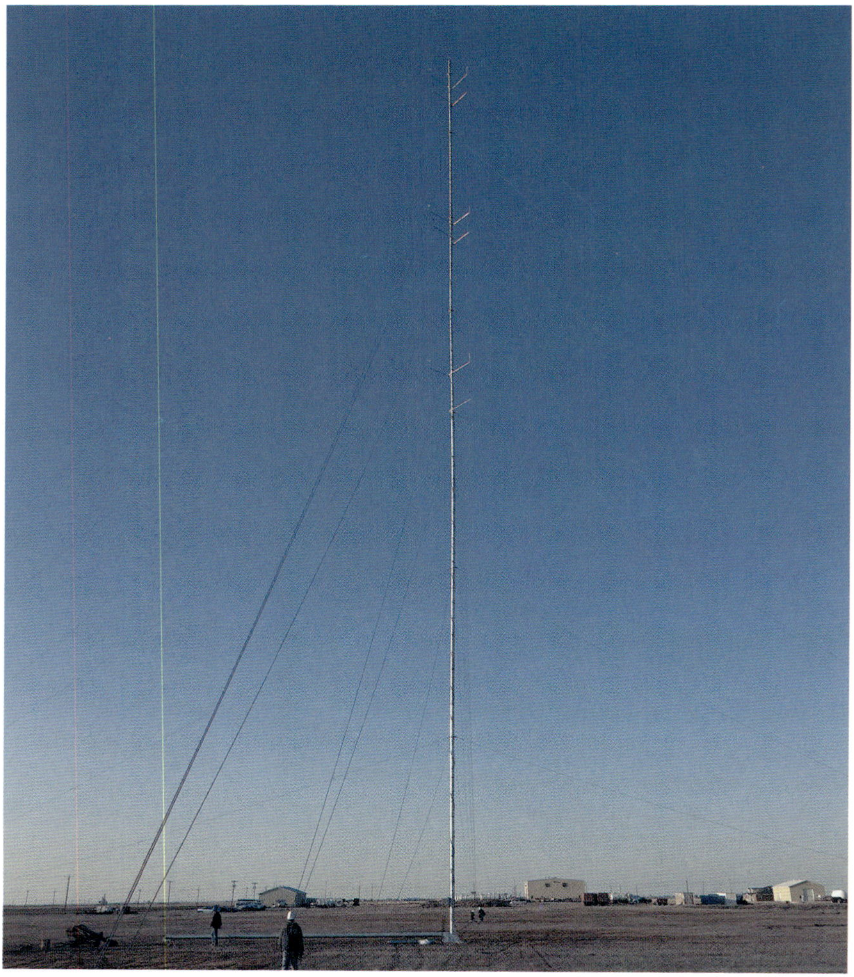

Fig. 5–7. Meteorological tower. *Image courtesy of Second Wind. Photo by Steve Butman.*

Equipment. Met towers are equipped with anemometers to measure wind speed and vanes to measure wind direction. Multiple anemometers and vanes are placed at several heights along the pole, mounted on booms that extend horizontally from the pole. The instruments, to get accurate results, have to be placed a certain distance away from the tower, which disturbs the air flowing past it. Most met towers also incorporate temperature and barometric-pressure sensors for additional data on weather conditions such as air density.

Usually, met towers are temporary, "tilt-up" structures that are held up by guy wires. They typically don't need a foundation, so they can be dismantled and moved around. However, some local governments require building permits or zoning variances even for temporary met towers.

Sometimes existing towers, such as those used to convey cellphone data, are adapted to wind measurement, although this type of "piggyback" operation is not as easy to do as it might seem. As with met towers, the wind instruments have to be mounted on long booms away from the tower.

Still, some companies specialize in converting or adapting cellphone towers to collect wind data. One advantage is avoiding possible opposition to a new tower from neighboring homes and businesses. Also, the use of existing towers can cost less than erecting a new, purpose-built tower. One U.S. company has a portfolio of about 12,000 existing towers, some of which are more than 80 meters tall (260 feet).

Met towers also incorporate a data logger at their base to continuously record and process a vast amount of wind data. The tower's instruments send low-voltage electrical signals to the data logger, which collects instantaneous data on wind speed and direction and processes the data into snapshots of 1 minute or 10 minutes. Each snapshot includes various statistics such as averages, maximums, minimums, and standard deviations.

The data logger stores the statistics, then sends the information from the met tower via cellular or satellite modems to a server. If the logger has no communications unit, an operator comes out to physically collect the data.

As mentioned, wind measurements should be taken for at least a year, but even then, the data might not reflect actual wind conditions over the long term. Before building a wind farm, the developer's engineers or consultants also check data from the closest weather station, often at an airport, where conditions have been recorded for many years or even decades. Correlation of this data with records from the met tower helps ensure that the wind speeds recorded at the location are representative of long-term conditions (see "Wind data analysis" below).

Siting. Ideally, the met tower should be located at the exact spot and rotor hub height of the wind turbines likely to be sited there. In practice, however, met towers usually have to be shorter than turbine towers. Today's turbine towers can be 100 meters tall or more (over 300 feet), with rotor tips reaching as high as 140 meters (460 feet). Most standard, tilt-up met towers are 60–80 meters tall (200–260 feet).

Taller met towers can be built, but they are more expensive because they require poured concrete foundations, additional permitting, and beacon lights. A met tower built to rotor hub height or higher can cost more than twice as much as a standard tower.

Also, met towers can't always fit into the same land area as the turbine. The met towers' guy wires take up more space on the ground than the base of a wind turbine does, and there must be enough room to lay the tower down when measurements are finished.

So the measurements collected from a met tower, in addition to being correlated with long-term data from weather stations, have to be extrapolated to simulate wind speed and direction at a higher distance above the ground and sometimes at a different turbine location nearby. Computer models that analyze the terrain around the met tower can predict the wind characteristics at a different turbine site and hub height.

Wind data analysis

The wind industry has spawned a healthy business in resource assessment products, not only new instrumentation (see "Advanced assessment tools" below) but also a variety of computer programs and third-party services that analyze raw wind data.

Wind distribution. Wind speed data are processed to determine the average wind speed and its *distribution*—that is, the amount of time that the wind blows at any given speed, measured either in hours or as a percentage of total time (fig. 5–8). This results in a *wind distribution curve* (sometimes called a velocity duration curve or a frequency distribution curve).

These curves are developed by *binning*, or categorizing, the wind speeds. Each small range of speeds (0 to 1 m/s or mph, 1–2 m/s or mph, etc.) is binned as a single category. Then the number of times that the wind speed falls within each bin is counted up—that is, the frequency of the wind occurring in a particular bin is calculated. This results in a bar graph from which a curve can be derived.

The shape of the curve is similar to a *Weibull distribution*, named after a Swedish engineer. This standardized statistical tool reflects actual hourly wind speeds fairly accurately at many locations. It can be used to predict a site's wind distribution if you don't have a full year of wind data or if other details are lacking.

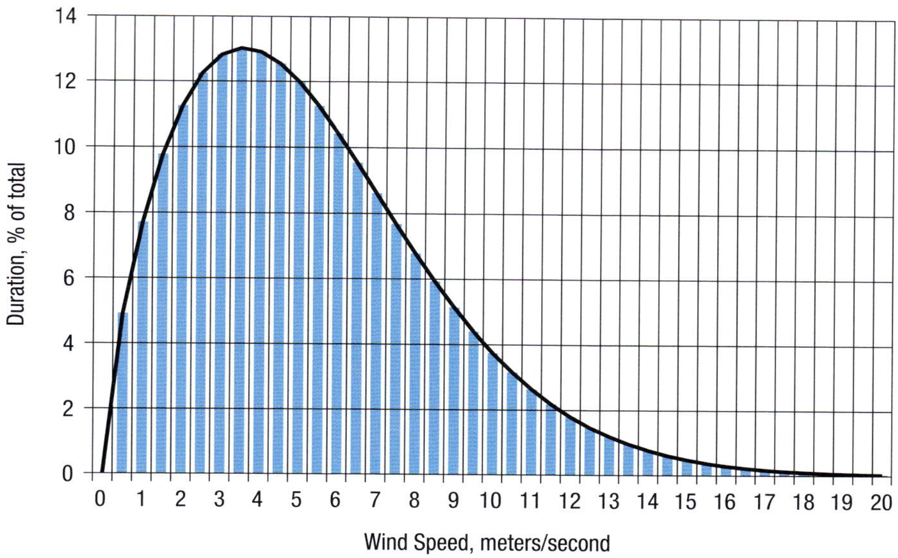

Fig. 5–8. Wind speed and duration

Another way to approximate the wind speed distribution, rather than measure it precisely, is the *Rayleigh distribution*, named after a British physicist. This tool also provides a reasonable description of wind speed and duration in some locations (see appendix D).

Wind distribution curves reveal that a lot of wind power comes in bursts during short periods of strong winds. That's because a wind turbine's output rises so dramatically with wind speed. Of course, each site's wind characteristics are unique, but the uneven nature of wind turbines' power production can make wind power tricky to integrate into utility grids (see chapter 7).

Wind direction and other characteristics. Once the wind speed distribution has been measured or approximated, many other factors must be considered as well, such as the following:

- Wind direction at any given moment
- Data on turbulent conditions (gusts and sudden, frequent changes in direction)
- Variations in wind speed at different heights above the ground (vertical wind shear) and at different horizontal positions (horizontal wind shear)
- Information on the site's terrain (surface roughness)

All of these factors affect wind turbine performance in the real world. Energy loss factors also must be accounted for, such as the amount of time a turbine might be shut down due to high wind speeds or for other reasons.

One useful tool for analyzing the wind resource is called the *wind rose* (fig. 5–9). This diagram shows not only the direction of the wind, but also the average wind speed from each direction, translating to the potential amount of wind power by direction.

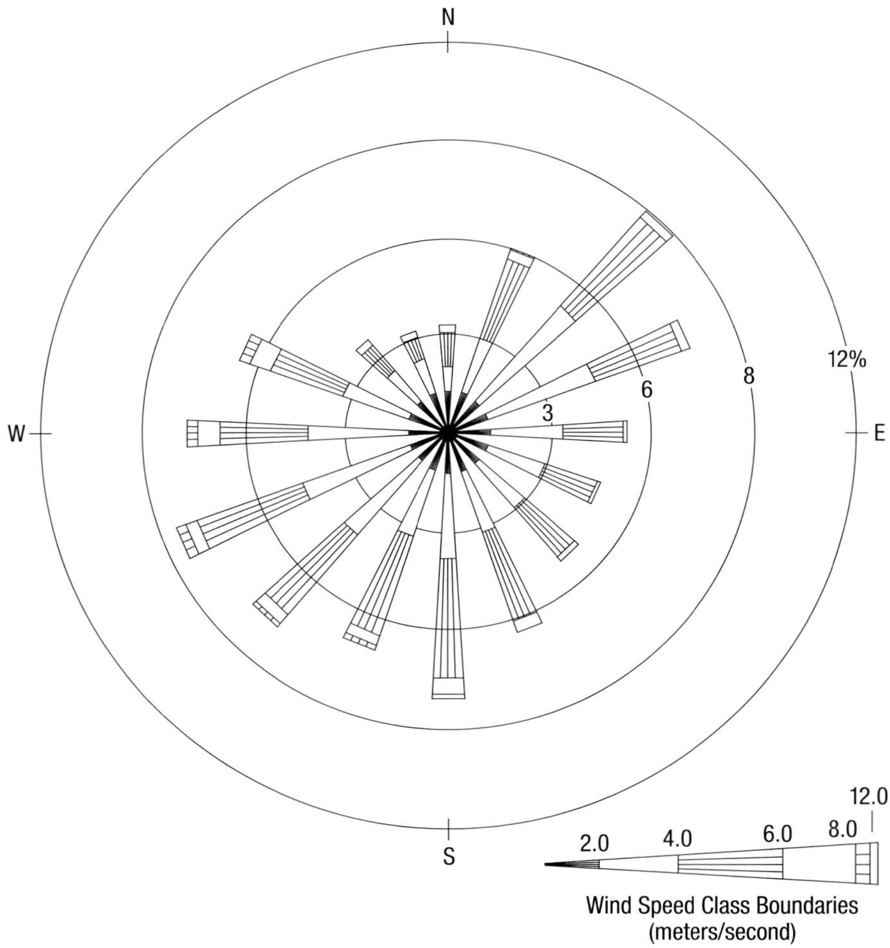

Fig. 5–9. Wind rose showing wind direction and average wind speed by direction. *Image from Arnett, Karapatakis, Mamaley (1993).*

In addition to wind speed and direction on the horizontal plane, *vertical wind speed* also should be measured if possible, especially at sites with complex terrain. This should not be confused with vertical wind shear—rather, vertical wind speed refers to wind actually blowing vertically, upward or downward, either at an angle or as a sudden downdraft or updraft.

Vertical wind speed is sometimes called wind's "third dimension," according to Jack Kline of RAM Associates. It can have a serious impact on turbine performance and life by reducing the turbine's power output and increasing wear and tear on the machinery. Special anemometers and new instrumentation are on the market for measuring vertical wind speed (see "Advanced assessment tools" below).

PERSPECTIVE: Meteorology, Assessment, and Siting
Jack Kline, RAM Associates

Jack is a certified consulting meteorologist who has been working in the wind industry for almost 30 years. He has a B.S. in meteorology from Florida State University and an M.S. in atmospheric sciences from Georgia Institute of Technology. RAM Associates, headed by Jack Kline, is a meteorological consulting firm specializing in wind resource assessment, turbine micrositing, and energy and wake loss calculations.

Q: What changes have you observed in wind turbine siting practices over the past several years?

A: The siting of turbines in wind farms now provides generally more open arrays—that is, wider spacing between turbines in a row and between rows of turbines. We have learned much about the degree to which turbine wake losses propagate, and this has been mitigated to a certain extent by more open arrays.

Q: What are some of the other factors involved in turbine micrositing, and what are the biggest limitations on putting each turbine in the best spot?

A: Wow, you could write a book on that topic alone! Many limitations on micro-siting are related to setbacks or constraints issues. Turbines must be sited some distance away from dwellings, roads, transmission lines, pipelines, wells, microwave beam paths, wetlands, culturally sensitive areas, and so forth. Sometimes these constraints prevent one from siting turbines on the best wind locations.

Then one must consider array losses, terrain formations, and the local wind regime, in terms of directionality, and how those factors interact to affect how turbines can be sited and the optimal location for the turbines. This is within the context of the considerations for the turbine type (size in rotor diameter and hub height) and the overall project size in terms of installed capacity. The larger the rated power of the turbine being considered, the fewer number of turbines will be required, and vice versa.

Then there are constructability issues. In more complex terrain, it's more difficult to prepare a crane pad for larger turbines, and it's harder to construct roads to get the various components to the site, particularly large rotor blades.

Q: What are the biggest pitfalls facing developers in designing a wind farm, and how can they be avoided?

A: Probably one of the biggest pitfalls would be for an inexperienced developer to rely implicitly on the results of wind modeling software that they don't fully understand. There are many shortcomings to most models, yet they can be seductive with their fancy graphics and such. People who lack the experience to know a bad result when they see one might take the result at face value.

For example, the "industry standard," WAsP [Wind Atlas Analysis and Application Program], is known to perform poorly in various terrain conditions, yet it's still the standard. Another type of model in vogue is the computational fluid dynamics [CFD] type, which can be difficult to understand due to the complexity of the models, so getting good results from them is challenging.

Users who are unaware of the potential biases and errors in any model, or don't know how to make reasonable compensation for suspect results, will be off the mark in their energy estimates.

Q: What are the most frequent reasons that wind farms underperform in real life?

A: Part of this problem might be due to the models used, but another important factor is that when people prospect an area, they often put up the met tower at the highest spot, so they get their measurements at the best location. Then when they use the software to analyze the measurements, it's inherently biased. It predicts higher energy estimates because it starts from a high wind spot.

Q: How can this affect the energy predictions for a wind farm and its chances of getting built?

A: An associate in the business told me about some consultants who were investigating the observed underperformance of wind projects, in an effort to understand where mistakes were made. It turned out, due to the competitive bidding process, projects that were predicted to have a higher output and lower cost of energy got picked and built over those with a lower output and higher cost of energy.

If mistakes were made on the upside of the assessment, the projects that got selected were based on overestimates of the wind resource, while the ones that got it right didn't look that good. This was not done purposely, even though assessment consultants might have felt pressured at times to make the project look as good as possible.

But that's not the case anymore, given the history of wind project underperformance. The resource assessment process has come back down to earth, and professionals are making a greater effort to stay objective, to avoid bias and overcome the desire to steer estimates toward the upside. Instead, assessments are more carefully restricted to what the data will support. There's almost been an overcompensation by some consultants to take every opportunity to look at the downside to the maximum extent.

Also, wind projects usually now undergo a third-party assessment. The developers have their consultant's report, then the banker's engineer comes in with their report. The investors will look at the two and take the lowest numbers, or might not even look at the developer's report, in the interest of being conservative.

Correlation with weather data. As mentioned above, wind data from weather stations are used as reference points to gain a longer-term picture of wind characteristics. Analysts use the "M-C-P" process (measure-correlate-predict) to estimate the energy output of a wind farm over its lifetime by correlating the data actually measured in the field with a long-term record from a nearby reference site.

U.S. sources for reference data include the National Center for Atmospheric Research and the National Centers for Environmental Prediction. National agencies throughout Europe provide similar data.

However, obtaining historical weather data can be difficult, especially in remote locations, so wind data analysts sometimes turn to vendors such as 3Tier and AWS Truepower. One of 3Tier's products is online access to global reference data at exact hub heights. The database incorporates numerical modeling techniques to improve the accuracy of its correlations.

Computer models. Most wind resource assessments these days rely on computers to account for the wind's numerous variables quickly and precisely. Many manufacturers of data loggers, as well as third-party services, sell compatible software for analyzing and managing the vast amount of data collected. As noted earlier, computer models are available to simulate wind conditions where actual site measurements are lacking, accounting for factors such as terrain and elevation.

Similar computer models are also used to predict future wind conditions and thus the performance of a proposed wind farm—a key requirement for developers to get financing. Banks, venture capitalists, and other investors demand a high level of confidence in the developer's estimates of a farm's energy output, based on wind forecasting. In fact, most investors require verification by an independent third party.

Thus, accurately assessing the wind resource, forecasting wind conditions at the site, and predicting turbine output have become crucial hurdles for wind farm developers to make a financial case for their projects (see chapter 6).

"You don't need a weather man to know which way the wind blows," Bob Dylan once sang. But if you're a wind farm developer, you absolutely do need someone to help you know not only which way the wind blows, but also where, when, how often, and for how long. That's where wind energy forecasting comes in.

—*Jeremy Shere,* Renewable Energy World

Advanced assessment tools

Wind industry researchers have developed new instruments for collecting and processing more wind data than ever, without a tower. These tools are based on technologies similar to radar that employ waves of sound (sonic) or beams of light (laser) to capture and record wind speeds and other data at greater heights than met towers can reach.

Advanced sonic- and laser-based systems have begun to supplement met tower data or even in some cases replace it, doing away with the met tower.

Sodar. Sonic detection and ranging technology—called *sodar*—works like radar but uses sound waves rather than radio waves. Most sodar systems issue a high-pitched acoustic pulse, sort of like the "cheep" of a bird, then listen for the sound to bounce back.

The sonic wave is scattered by atmospheric turbulence and echoes back to the sodar system. Changes in the frequency and intensity of this *backscatter* allow the system's instrumentation to define wind speed, direction, and turbulent conditions—thus the technology's usefulness in wind resource assessment. Figure 5–10 shows a sodar unit in the field.

Sodar systems can be installed at a prospective site much more quickly than met towers. One sodar manufacturer claims that a unit can be up and running within two hours. Also, sodar instruments are based on the ground, so they aren't visible to neighbors like towers are, and they usually don't need a permit or costly liability insurance.

More importantly, sodar technology can easily measure data at heights up to 200 meters (660 feet), far greater than those reached by physical met towers (fig. 5–11). However, a major drawback is that sodar units usually can't collect data during heavy rain or snow.

Fig. 5–10. Triton sodar wind assessment unit. *Image courtesy of Second Wind. Photo by Tim Clarke.*

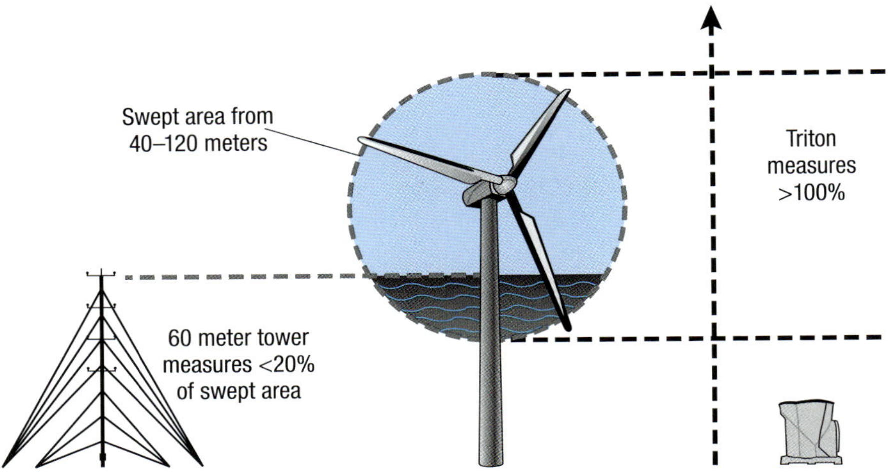

Fig. 5–11. Sodar measurement height vs. met tower. *Image courtesy of Second Wind.*

Until recently, conventional sodar systems generally provided only average wind speeds, not deviations such as sudden gusts of wind. However, modern sodar technology can capture extensive data on anomalous wind events such as shear and turbulence, as well as vertical wind speed.

Lidar. Another advanced method for assessing wind resources is *lidar*, or light detecting and ranging technology, which uses beams of light, in the form of laser pulses, instead of sonic or radio waves. Lidar instruments define wind speed and turbulence by measuring how long it takes for the laser pulse to be reflected back, or backscattered. Figure 5–12 shows a lidar unit in the field.

A laser beam is very highly focused (that is, it has a very short wavelength), allowing lidar systems to map wind conditions with very high resolution. Light waves are reflected back by tiny particles carried in the wind such as water droplets, dust, pollen, and salt crystals.

Like sodar systems, lidar equipment for wind assessment is based on the ground rather than perched on a tower, so lidar technology has similar advantages—easy to transport and quick to install (one manufacturer claims 25 minutes)—and a similar disadvantage, the inability to collect data during heavy precipitation.

Lidar technology uses a set of spinning mirrors to focus and emit a cone of laser beams that encompasses the area surrounding a prospective turbine site (fig. 5–13). Lidar assessment tools have an effective range up to 300 meters (almost 1,000 feet).

Fig. 5–12. ZephIR lidar wind assessment unit. *Image courtesy of Natural Power.*

A UK company, Natural Power, pioneered the use of lidar technology and now sells the ZephIR products, including a unit designed for mounting on a turbine to "see" the incoming wind before it encounters the blades and to adjust their pitch accordingly. This allows wind farm operators to monitor and optimize turbine performance (see chapter 6). The turbine-mounted technology was developed in cooperation with Risø DTU.

Other vendors of lidar equipment include NRG Systems and partner Leosphere, which sell the WindCube v2 technology; SgurrEnergy, which markets Galion systems, and Catch the Wind, which sells the Vindicator product line, including a turbine-mounted lidar unit and handheld lidar equipment capable of measuring data within 1,000 meters (3,300 feet).

For the most accurate readings, you need to get high. [Lidar and sodar] systems are more cost-effective ways than anemometer-equipped meteorological towers to measure the power a site can really be expected to produce at the loftier heights at which wind is now being harvested.

—Herman K. Trabish, Greentech Media

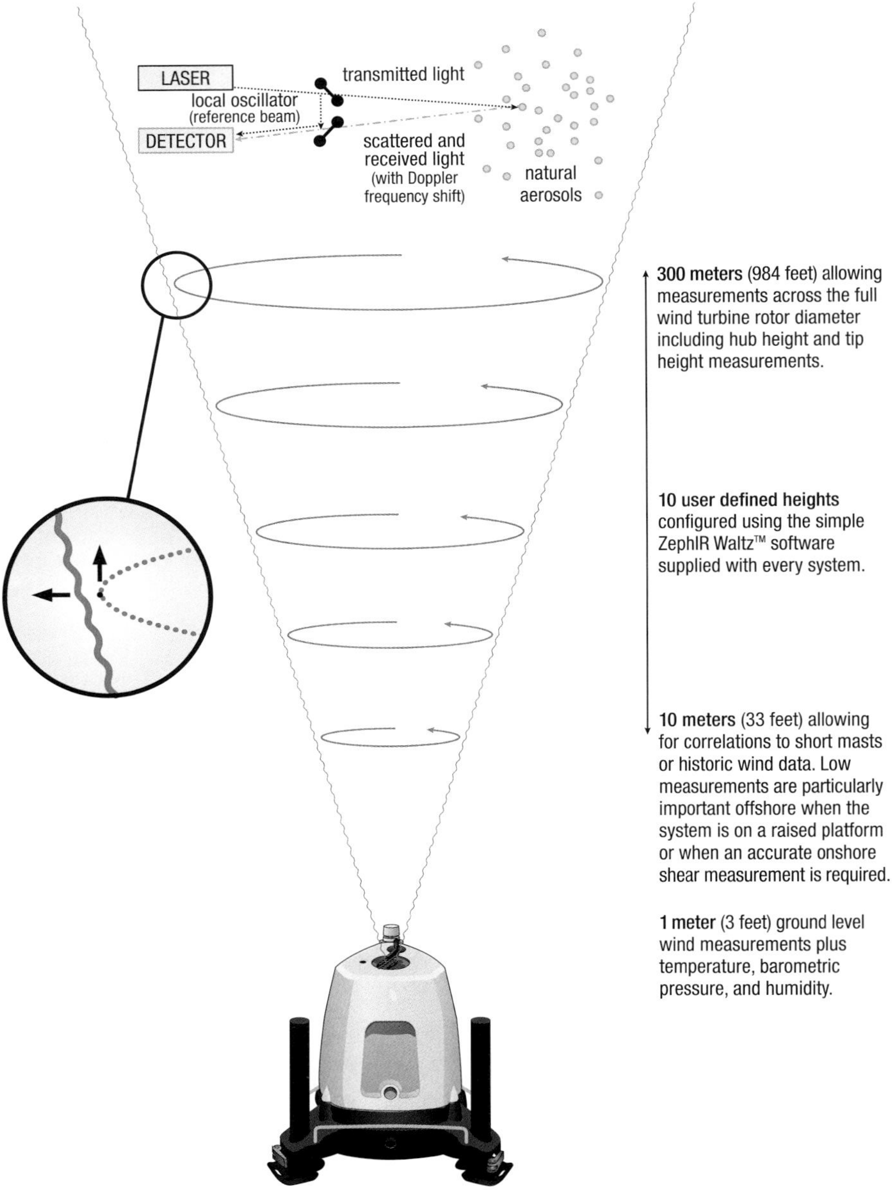

Fig. 5–13. How lidar technology works. *Image courtesy of Natural Power.*

Lidar and sodar tests and comparisons. Natural Power operates a lidar and sodar test site for field validation of data at heights up to 90 meters (300 feet). Lidar and sodar technologies are also being tested offshore, where much of the world's wind energy development is taking place (see chapter 9).

Natural Power's ZephIR lidar system has been tested at a weather station off British Columbia, Canada (fig. 5–14), and a floating SeaZephIR platform, powered by wind and solar energy, will be installed off New Jersey, U.S.

A Vindicator lidar unit made by Catch the Wind has performed well on a buoy, also off British Columbia, and NRG Systems calls offshore applications "ideal" for its WindCube lidar system.

As might be expected, however, both sodar and lidar are more expensive than met towers for assessing wind resources—but they can be worth it when highly accurate wind characterization is critical to predicting turbine performance and obtaining financing for wind farm development.

Sodar systems typically cost around \$50,000–\$60,000 installed (36,000–44,000 euros). Lidar systems are more expensive, at least \$100,000 (73,000 euros). Some sodar and lidar equipment can be rented or leased on a short-term basis instead of purchased.

The accuracy of data collected by sodar and lidar equipment appears comparable, and both types provide data on vertical wind speed, which met towers typically don't capture. Neither sodar nor lidar interferes with the flow of the wind while measuring it, in contrast to the obstruction created by a met tower.

However, both lidar and sodar data collection can be interrupted by heavy rain and snow or distorted by obstacles within their range. Other pluses and minuses are that lidar systems can be less rugged, while the acoustic chirps from a sodar unit can be disturbing in some locations. Lidar systems require a 100-watt power source, while sodar units are powered by solar panels onboard.

Ultrasonic sensors. As an alternative to conventional anemometers, ultrasonic sensors are becoming more common. Although they're more expensive, they work well in freezing weather, and they can be more accurate. Mechanical devices are affected by inertia, which inhibits their parts from beginning to move. In contrast, ultrasonic sensors measure changes in wind speed and direction immediately and in real time.

One final note on wind resource assessment is that it doesn't stop when a commercial wind farm has been sited and built. Defining the wind's character is an ongoing part of monitoring the turbines' operation and providing data to forecast how much electricity the wind farm will generate within the next hour, day, or week, for purposes of integrating wind power into the utility grid. Chapters 6 and 7 discuss wind farm operation and wind power integration.

Fig. 5–14. ZephIR lidar unit at meteorological station off British Columbia. *Image courtesy of Natural Power.*

References

American Wind Energy Association. Basic Principles of Wind Resource Evaluation. http://www.awea.org/faq/basicwr.html.

Archer, Cristina L., and Jacobson, Mark Z. 2005. Evaluation of Global Wind Power. *Journal of Geophysical Research—Atmospheres*, 110, D12110, doi:10.1029/2004JD005462. http://www.stanford.edu/group/efmh/winds/2004jd005462.pdf.

Asia Sustainable and Alternative Energy Program, http://www.worldbank.org/astae.

AWS Truepower, http://www.awstruepower.com/.

Blau, John. 2011 (May 18). Europe Leads Offshore Wind Market. Renewable Energy World. http://www.renewableenergyworld.com/rea/news/article/2011/05/wind?cmpid=WNL-Friday-May20-2011.

Canadian Wind Energy Association. WindVision 2025. http://www.canwea.ca/images/uploads/File/Windvision_summary_e.pdf.

Catch the Wind, Inc., http://www.catchthewindinc.com.

Center for Energy Efficiency & Renewable Energy, University of Massachusetts at Amherst. Wind Power: Resource Assessment. http://www.ceere.org/rerl/about_wind/RERL_Fact_Sheet_5_Resource_Assessment.pdf.

Commonwealth Scientific and Industrial Research Organization, http://www.csiro.au/.

Coppin, P. A., Ayotte, K. A., and Steggel, N. 2003 (October). Wind Resource Assessment in Australia—a Planner's Guide. Australian Commonwealth Scientific and Industrial Research Organization (CSIRO), Version 1.1. http://www.csiro.au/files/files/pis7.pdf.

Dmitriev, Grigori. 2001 (June). Wind Energy in Russia. VetrEnergo for Gaia Apatity and Inforse-Europe. http://www.inforse.org/europe/word_docs/ruswind2.doc.

EcoBusinessLinks. Wind Maps. http://www.ecobusinesslinks.com/wind_maps.htm.

Environment Canada. Wind Energy Atlas. http://www.windatlas.ca/en/index.php.

EOLE Wind Energy Project, http://collaboration.cmc.ec.gc.ca/science/rpn/modcom/eole/.

European Wind Energy Association. 2009 (February 13). Wind Energy—the Facts. http://www.ewea.org/fileadmin/ewea_documents/documents/publications/WETF/1565_ExSum_ENG.pdf.

Global Security. 1993. Chapter 3. Affected Environment. Figure 3–8, Wind rose for SRS, 1987 through 1991. http://www.globalsecurity.org/wmd/library/report/enviro/eis-0217/eis0217_3.html.

Global Wind Energy Council, www.gwec.org.

Global Wind Energy Council, Greenpeace International. 2010 (October). Global Wind Energy Outlook 2010. http://www.gwec.net/fileadmin/documents/Publications/GWEO%202010%20final.pdf.

International Wind Resource Maps. U.S. National Renewable Energy Laboratory. http://www.nrel.gov/wind/international_wind_resources.html.

Kline, Jack. 2011. Interviewed by Rebecca Busby. E-mail message to author. June 23.

Kropper, Steve. 2011 (January 26). Wind Measurement Towers: How High Should They Be? *Renewable Energy World.* http://www.renewableenergyworld.com/rea/news/article/2011/01/wind-measurement-towers-how-high-should-they-be.

Lacy, Virginia. 2010. The Answer Is Growing in the Wind. Rocky Mountain Institute. http://www.rmi.org/rmi/AnswerGrowinginWind.

Natural Power, http://www.naturalpower.com.

Norsewind, http://www.norsewind.eu/public/index.html.

Northwest Community Energy, http://nwcommunityenergy.org/wind/resource-assessment.

NRG Systems, http://www.nrgsystems.com.

Remington, K. 2010 (August 9). Ultrasonic Wind Sensors or Cup Anemometers? That Is the Question. *Windpower Engineering.* http://www.windpowerengineering.com/construction/simulation/ultrasonic-wind-sensors-or-cup-anemometers/.

Renewable Energy Atlas. Government of Australia. http://climatechange.gov.au/en/what-you-need-to-know/renewable-energy/~/media/publications/renewable-energy/atlas/atlas-help.ashx.

Rohn Products, http://www.rohnnet.com.

Russian Wind Atlas, http://www.wasp.dk/Download/DownloadFiles/General/Russian%20Wind%20Atlas.pdf.

Schwartz, M., Heimiller, D., Haymes, S., and Musial, W. 2010 (June). Assessment of Offshore Wind Energy Resources for the United States. NREL/TP-500-45889. http://www.nrel.gov/docs/fy10osti/45889.pdf.

Second Wind, Inc., http://www.secondwind.com/index.html.

Shere, Jeremy. 2011 (January 6). Knowing Which Way the Wind Blows. *Renewable Energy World.* http://www.renewableenergyworld.com/rea/blog/post/2011/01/knowing-which-way-the-wind-blows?cmpid=WindNL-Thursday-January13-2011.

Solar and Wind Energy Resource Assessment, http://swera.unep.net.

3Tier, Inc., http://www.3tier.com/en.

Trabish, Herman K. 2010 (July 8). Measuring the Wind and the Sun. Greentech Media. http://www.greentechmedia.com/articles/read/measuring-the-wind-and-the-sun.

———. 2010 (August 31). Advanced Wind Measurement Technologies Take Hold in the Industry. Greentech Media. http://www.greentechmedia.com/articles/read/advanced-wind-measurement-technologies-take-hold-in-the-industry.

———. 2010 (December 29). Wind's 2010 Top Ten: Bust and Building, Coming and Going. Greentech Media. http://www.greentechmedia.com/articles/print/winds-2010-top-ten-bust-and-building-coming-and-going/.

U.S. Department of Energy, Energy Efficiency and Renewable Energy, Wind Powering America. Wind Maps and Wind Resource Potential Estimates. http://www.windpoweringamerica.gov/wind_maps.asp.

WAsP—Wind Atlas Analysis and Application Program. http://en.openei.org/wiki/Wind_Atlas_Analysis_and_Application_Program_(WAsP).

Wind Atlases of the World, http://www.windatlas.dk/index.htm.

WindData.com, http://www.winddata.com.

6

Wind Farms: Developing and Operating Wind Power Plants

Introduction

Hardly a day goes by without an announcement in the energy business news that another wind farm will be built somewhere in the world. Although the wind market's growth slowed in many countries in 2010 after the financial crisis, developers are still seeking to take advantage of incentives offered by many governments to get more wind power and other renewable energy sources online.

Wind farm sizes

At a modern wind farm or park, dozens or even hundreds of large turbines—each more than 1 MW in capacity—are located in the same area and controlled by an operating company, whether on-site or remotely. These utility-scale wind farms generally range in size from 20 MW to 200 MW when built on land, but can be much larger (tables 6–1 and 6–2). The Shepherds Flat wind farm being built in Oregon could exceed 900 MW in capacity.

Table 6–1. Largest onshore U.S. wind farms. *Data courtesy of American Wind Energy Association.*

Name	State	Capacity, MW
Roscoe	Texas	782
Horse Hollow	Texas	736
Capricorn Ridge	Texas	663
Sweetwater	Texas	585
Buffalo Gap	Texas	523

*Many of these wind farms are being built in stages, with more capacity planned.

Table 6-2. World's largest onshore wind farms (excluding U.S.)

Name	Location	Capacity, MW
Jiuquan	China	1,150
Dabancheng	China	500
Fantanele	Romania	348
Whitelee	UK	322
Huitengliang	China	300
Lake Bonney	Australia	279
Vankusawade	India	259
Eurus	Mexico	251
Alto Minho	Portugal	240
Vento Minho	Portugal	240

*Many of these wind farms are being built in stages, with more capacity planned.

Offshore wind farms can also have capacities in the hundreds of megawatts, but their turbines are typically bigger than those onshore, generating up to about 6 MW each, and larger ones are in the works (see chapter 9).

At the smaller end of the wind farm size range, community-scale wind power can vary from a single large or mid-range turbine to a dozen or so, contributing power supply to a town or university campus or delivering electricity for sale to the utility (see "Community Wind" below).

A large wind farm might spread over hundreds of square kilometers, but most of the land area (more than 95%) can still be used by farmers for crops, livestock grazing, or other purposes. Each turbine needs to be separated by a certain distance from neighboring turbines to minimize interference of airflow among them (see "Final design" below).

Whenever possible, wind farms on land are built in remote areas with small populations, but with access to utility transmission lines that have enough capacity to accept additional power. Wind farm siting can be controversial when neighboring communities raise objections (see chapter 12).

Connecting with utilities

Wind farms connect with the electrical grid network operated by a utility, which transmits and delivers wind power along with conventional electricity to homes, businesses, and industrial customers. Wind power is often sent to customers hundreds or even thousands of kilometers away—such as utilities that might need wind energy to satisfy government requirements for purchasing renewable power.

Individual turbines are connected by medium-voltage power lines (34.5-kV) and other electrical and communications equipment, called a *collection system*. Large wind farms usually have at least one utility substation, which steps up the turbines' power to a higher voltage for long-distance transmission.

To an electric utility, a wind farm acts as a power plant, with a key difference. The amount of power generated is similar to that produced by power plants that burn natural gas or coal (though many coal-fired plants are substantially larger). However, the amount of electricity being produced by a wind power plant can vary quite a bit, so utilities have to plan more carefully when integrating it into their electrical systems.

Today's methods of forecasting wind power have become quite accurate for short-term planning, so power supply from wind farms has become more predictable. But the issue of wind's variability is still hotly debated (see chapter 7).

From the wind farm developer's perspective, proximity to transmission lines is a high priority in selecting a site, second only to the wind resource itself. If turbines will be sited in an area without access to the power grid, someone has to build transmission lines to connect the wind power supply—an expensive project that can also embroil a wind farm in lengthy delays.

Construction of some wind farms in America has been postponed because the sites lack access to the power transmission network. Sometimes, wind farms that are connected to the grid can be *curtailed*—forced to reduce their output or shut down temporarily—when the amount of wind power being generated exceeds the available transmission capacity.

Development Process

The process of developing a wind farm is complex, with a lot of steps that have to be coordinated simultaneously. Many of the tasks are concurrent rather than sequential (fig. 6–1), so the developer must wear several hats at once. (Wind farms at the small end of the size range can be built more quickly than shown below.)

The development process is sort of like managing the design and construction of a house, but on a much larger scale. A big wind farm might have more than 10 suppliers and service providers and more than 20 supply and service contracts. And much of the equipment comes from places all over the world.

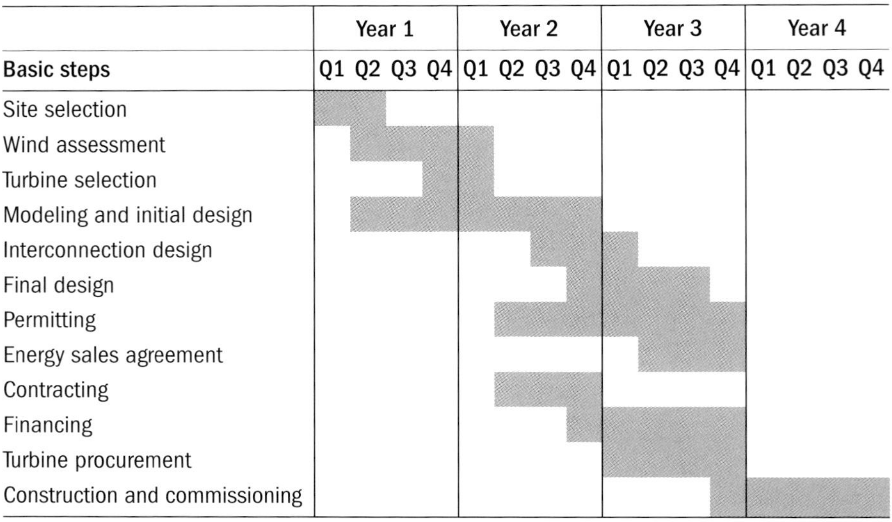

Fig. 6–1. Wind farm development process

The developer has to schedule everything so that the materials, equipment, and people needed for each part of the wind farm are there at the right time and place, without costly delays. For example, if the main power transformer doesn't arrive on time, the whole project is held up. At times, the logistics can become overwhelmingly difficult. Constant communication among all parties is key to keeping a project on track.

This review of wind farm development describes the steps in the process as they might apply to a typical wind farm of 20–200 MW on land. Land-based wind power projects bigger than that are usually developed in smaller stages or phases (often dubbed I, II, etc.). U.S. projects averaged 76 MW in 2010, while average sizes on land have traditionally been smaller in Europe, where community wind is more popular and practical. Appendix E presents two wind farm case studies, one U.S. and one European.

Selecting a site

Wind energy potential. Finding the best spot for a wind farm is the most important factor in developing a successful project, and wind speed is the most important factor in finding the best spot. In the perfect location, the wind would blow nearly constantly throughout the year at a high speed, without turbulence (strong, unpredictable gusts of wind or other sudden, frequent variations in wind speed and direction, discussed in chapter 3).

Lacking this conceptual ideal, developers of large wind farms look for strong, steady winds of at least 6–7 m/s (13–16 mph), preferably blowing much of the time at 12–14 m/s (27–31 mph), the region where big turbines generate their maximum output. No wind farm project will succeed without a good wind resource.

The first place to start is a factor called power density (measured in W/m^2), which indicates how much potential wind energy is available over a given area of land for a turbine to convert to power. Wind resources worldwide are classified according to wind speed and power density, as described in chapter 5 and defined in table 5–1.

But many physical objects on the land's surface can interfere with airflow and contribute to turbulent winds, which impair turbine performance. These obstacles include trees, buildings, and large topographical features such as hills or valleys. Also, the site's vegetation and terrain cause friction that slows down the wind close to the ground. This creates vertical wind shear (different wind speeds at different heights above the ground, as described in chapter 3).

On the other hand, the topography of the land can enhance a wind farm site's potential. For example, many wind turbines are installed along ridgelines in areas with canyons, hills, or mountains. The wind accelerates along the ridgeline, boosting the amount of energy that can be captured.

Land agreements and other issues. Before wind farm developers can put up meteorological towers or other equipment to analyze the wind characteristics at a prospective site, they must find out whether the land is available and acquire the rights to surface development, as well as the right to access the land (ingress and egress).

Developers must negotiate a lease with the landowner specifying such factors as lease term, expected project life, and payment, which is usually based on a percentage of project revenues, plus annual flat payments per turbine or per megawatt. Lease contracts also cover indemnification—how to prevent harm, loss, or damage and how to compensate the landowner for it, should it occur—and provisions for restoring the land at the end of the project life.

The developer must also begin working with the electric utility at this stage. The cost of building one or more utility substations at the wind farm has to be negotiated, in addition to figuring out how to provide access to the grid for long-distance transmission. The utility must determine that transmission capacity for the wind power will be available by performing a grid stability study (see "Interconnection design").

Also at this point, the developer takes a first look at the prospective site's environmental characteristics. Is there a nature preserve or local conservation area nearby? Is the land ecologically sensitive? Is the area treasured for any historical or archeological reason? And the neighboring communities must be considered as well. Are they likely to rise up against a wind project? How close are homes to the turbines, and how much noise might people hear?

All of these questions are addressed in greater detail later in the process. But smart developers wouldn't consider siting a wind farm in an environmentally vulnerable area, and investors probably wouldn't finance such a project anyway.

Wind assessment

Once the land agreements have been signed and the utility is on board provisionally, the developer puts up at least one meteorological tower to measure wind speed, direction, distribution, and other characteristics, as described in chapter 5. More "met" towers might be added as the site's wind becomes better understood and the developer gains confidence that the resource is promising enough to pursue.

Met towers installed during the assessment phase are sometimes left in place when the wind farm starts operating. They can continue to provide valuable information on how the turbines are performing, given how much wind energy is available. Met tower data also can be used in forecasting the wind farm's output on an hourly or daily basis for purposes of integrating its power with the utility grid (see chapter 7).

In addition to met towers—or in some cases, instead of them—developers are using sodar and lidar technologies to assess the wind resource more accurately and comprehensively. These systems generate more data on wind characteristics than met towers alone can (see chapter 5). Knowledge of the wind's turbulence and other complex variables are crucial to selecting the right turbines for the site.

The extra wind data provided by sodar and lidar technologies is not only helpful to the developer in designing the wind farm, but even more valuable in terms of raising investors' level of confidence in predictions of the wind farm's energy output. To get financing for huge, expensive wind farms, developers must bring mountains of wind data to the table.

During the wind resource assessment, developers begin to get a feel for how big the turbines should be and where they might be positioned. But at least a year's worth of wind measurements, and correlation with longer-term data from weather stations, are needed before design details can begin to emerge.

Turbine selection

It might seem early to begin selecting turbines for a wind farm site, but the developer might need a long lead time to procure the machines, depending on market conditions. At this point, computer modeling has begun (see below), and developers have an idea of what they're going to need in terms of turbine size (output capacity) to make the wind farm practical financially.

But first, developers perform what's called a site suitability assessment to determine how local conditions will affect the structural integrity, operation, and performance of the turbines being considered for the wind farm. This assessment covers factors such as wind loads (extreme wind speeds and fatigue loads), seismic loads, soil conditions, temperatures, and other weather conditions (lightning, icing, etc.).

The assessment also analyzes the ability of the utility grid to accept wind power. Wind turbines can be idled for many reasons, such as unscheduled maintenance or extreme weather, but a common cause of curtailment is transmission constraints. The developer's model must anticipate the future capacity limitations of the utility grid and the larger power system serving the region (see chapters 7 and 8).

Site suitability assessments are needed because actual site conditions never match perfectly to a turbine's design parameters on paper. Wind turbines are designed to a class—a specific set of wind speeds and turbulence (table 6–3), as defined by the International Electrotechnical Commission. The turbine's design certification verifies that the machine will perform as specified over a 20-year life—if it's operated in conditions corresponding to its *turbine design class*.

Typically, wind farm developers get performance curves and other specifications for at least two different turbines, then plug the turbines' design specifications and other data into their model. They might run the model several times to optimize the size and class of the turbines to be selected.

Table 6–3. Wind turbine design classes

IEC 61400-1 (Edition 3)	Class I	Class II	Class III	Class IV
Wind speed, m/s (mph)				
Average	10.0 (22.4)	8.5 (19.0)	7.5 (16.8)	6.0 (13.4)
Extreme 50-yr gust	70.0 (157)	59.5 (133)	52.5 (117)	42.0 (94.0)
Characteristic turbulence				
Class A	18%			
Class B	16%			
Wind shear exponent	0.20			

Notes: IEC = International Electrotechnical Commission

All wind speeds are at the turbine rotor's hub height.

Wind speeds are measured every 3 seconds; wind speed and standard deviation are recorded every 10 minutes.

Turbulence is the standard deviation of wind speed measured at 15 m/s (34 mph).

An important factor in turbine selection besides the power curve, which shows how much power the rotor produces at different wind speeds, is the machine's projected *availability*, defined as the percentage of time that a wind turbine is running or is ready to run. Availability is a measure of how often the turbine might have to shut down, based on the performance of existing same-model machines in the field (see "Performance factors" below).

Once all these factors have been defined, developers solicit bids from several big turbine manufacturers. Often, the manufacturer will make a deal for more than one wind farm at a time, if the developer is large enough to be involved in more than one project.

One final note on turbine selection: Developers often stick with certified machines that are commercially proven, as defined by the wind farm investors' consultants. This definition includes such factors as the number of same-model turbines in operation, their number of run-hours, and the fleet availability. However, the developer can sometimes get a better deal by choosing a new turbine model or design if the manufacturer offers a longer warranty, discounted price, or other incentives.

Modeling and initial design

The developer's computer model must answer a few basic questions: How much wind power might be generated, and how much money will the wind farm make by producing that power? It's basically a financial model—a roadmap to profitability.

Developers use the model to design the wind farm, analyze risks, and document predictions of its performance. Wind farm design is always a compromise between energy output, accessibility, ease of permitting, and financial practicality.

To take a first cut at wind farm design, the software incorporates analytical tools for using data from maps of wind speed, terrain, and other variables. The model predicts how these factors will affect turbine performance under specific site conditions. The land's topography is an important factor. While flat, open land might be preferable, the terrain at wind farm sites is often rugged.

The model answers questions such as, where will the roads be built, and how much gravel will they need? Wind farm roads must accommodate huge vehicles bringing in construction cranes, heavy electrical equipment such as transformers, and mammoth wind power system components—not only ever-longer blades, but tower segments that can weigh more than 50 tons.

Soil conditions must also be considered. For example, will the turbines' foundations be a standard design, or will a more expensive foundation be needed to ensure stability in the site's soil?

It seems that logistics can sometimes be left last in the design processes If there is a message for turbine manufacturers, it's that they should start thinking about modular turbines that can be shipped below the upper limits imposed by most [regulators]. Otherwise, they could face a boat-in-the-basement problem.

—*Alan Redding, ATS Wind Energy Services*

This initial design phase is when a developer might conduct a FEED study (front-end engineering and design) or hire a contractor to perform one. A separate FEED study answers all of the questions raised above and then some. But typically, FEED studies are more common for offshore wind farm projects, due to their greater complexity and risk.

Turbine output and energy revenues. The ultimate energy output of the wind farm is a key projection to be calculated using the model. The goal is to understand the wind farm's costs and figure out what the returns are projected to be, based on revenue from energy generated. As explained in chapter 3, energy is power produced over a period of time. A turbine's power capacity is measured in megawatts, while its energy output is measured in megawatt-hours.

The model evolves during the development process and will be used to finalize the design and to plan construction of the wind farm. Table 6–4 lists the factors considered in the model, such as the cost of turbines and other equipment. Table 6–5 shows the relative cost of the various components of a utility-scale wind power system over a 20-year life. The *balance of plant* is the turbine's support system and equipment such as the foundation and the electrical connections.

To predict wind farm revenues, the developer uses the model to plan for the costs of replacing blades and other major components such as the gearbox. Developers have to get a good handle on maintenance costs before they can arrive at an annual figure for the cost of operation. Maintenance requires some turbine downtime, which in turn affects the amount of power generated over a given period of time.

Table 6–4. Elements of wind farm model. *Data courtesy of GL Garrad Hassan.*

Energy output projections (MWh)
Cost of turbines, blades, towers, and electronics
Balance of plant costs
Foundation
Pad-mount transformer
Collection system
Cables
Erection
Substation
SCADA system
Taxes
Sales
Property
Income
Depreciation
Operating and maintenance costs
Finance costs
Tax credits/loan guarantees
Equity rate of return
Debt rate and term

This downtime is related to the turbine's availability, as described above. It also affects the wind farm's overall *capacity factor*, a measure of productivity. This is a percentage defined as the actual amount of energy produced by a wind turbine or farm over a given period of time divided by how much energy it should have produced during that time if running at maximum output:

$$\text{capacity factor} = \text{MWh produced} / (\text{rated MW} \times \text{hours}) \qquad \textbf{Eq. 6–1}$$

A wind farm's actual energy production can be lower than anticipated if wind speeds are slower than predicted or if losses are greater than planned, due to unscheduled maintenance, for example. Today's utility-scale wind turbines typically operate 65%–90% of the time, but often at less than full capacity (lower than their maximum power rating). So when actual output is accounted for, capacity factors of 25%–40% are common, and 33%–35% is about average.

For conventional power plants, capacity factor is almost entirely a matter of reliability. In contrast, a wind turbine's capacity factor is a matter of economical design. A turbine with very long blades and a small generator would run at full capacity whenever the wind blows and could achieve capacity factors of 60%–80%—but it would produce very little electricity, being limited by generator size in relation to rotor size. To maximize return on investment, wind farm operators use a larger generator to produce more electricity at a lower capacity factor.

Obviously, a wind farm's precise capacity factor is impossible to predict during design, but the model seeks to account for as many variables as possible, based on data from existing machines and manufacturer specifications. Both availability and capacity factor are measured when the wind farm is up and running (see "Performance monitoring" below).

Table 6–5. Cost of utility-scale wind power system components. *Data courtesy of the U.S. National Renewable Energy Laboratory.*

Wind Turbine	% of total cost
Rotor	
Blades	10.83
Hub	3.06
Pitch mechanism and bearings	2.71
Spinner and nose cone	0.28
Subtotal rotor	16.88
Drivetrain and nacelle	
Low-speed shaft	1.50
Bearings	0.85
Gearbox	10.90
Mechanical brake, high-speed coupling, etc.	0.21
Generator	6.98
Variable speed electronics	8.48
Yaw drive and bearing	1.42
Main frame	6.62
Electrical connections	4.27
Hydraulic, cooling system	1.28
Nacelle cover	1.50
Subtotal drivetrain and nacelle	44.01
Controls, safety system, condition monitoring	2.49
Tower	10.47
Subtotal turbine	**73.86**
Balance of Plant	
Foundations	3.28
Transportation	3.56
Roads, civil work	5.63
Assembly and installation	2.71
Electrical interface/connections	8.69
Engineering and permits	2.28
Subtotal balance of plant	**26.14**

Another variable in the model is the price the developer expects to get for however much wind power will be sold to the customer on a net basis (minus electricity consumed by the wind farm). Power prices vary by time of day, increasing the complexity of projecting wind farm revenues. All of these factors are intertwined in the model's software.

In calculating the net cost of power generated, developers commonly *levelize* expenses and incomes over the life of the turbines and the wind farm. This means taking an expense or income that occurs in a particular year and spreading it over the total years of the project life (converting it to an equivalent amount for each year, accounting for interest rates and other economic factors).

Of course, it's impossible to predict all these variables with unwavering certainty. Wind speeds, initial costs, machine performance, and maintenance costs can vary in the real world from what the model projects. But the more accurate the data used as input to the model, the more accurate its calculation of revenues will be.

One way to improve a project's chance of meeting revenue goals is to hedge against underperformance. Developers can take advantage of financial tools that calculate the risk of failing to deliver the expected energy output due to wind's variability. These products identify a specific price for hedging the risk to avert any impact on profitability, in essence functioning as insurance against sub-par production. Prospective investors are more likely to approve projects that have a built-in way to handle such risks before they come up.

Risk assessment. It's important to note that each stage in the development process costs money and involves a certain amount of risk that the project might not go forward, so why spend more money on it? Decisions must be made at every step to invest even more in a wind farm that might turn out to be impractical. The model's predictions are vital to making these decisions at each step.

The model analyzes risks related to government regulations, environmental concerns, economic issues, and many other factors. For example:

- What are the chances that the developer will encounter an environmental showstopper that halts the project?
- How will the wholesale power market affect the price to be paid for the wind power?
- Will a government financial incentive expire or be revoked before the wind farm can qualify for it?
- What if a large transmission system operator decides to curtail wind power without compensating operators?

Land-based wind farms, with their fairly solid history of development and operation, carry relatively little risk related to the technology (except for new turbine models), construction costs, or operating and maintenance expenses. All of these can be reasonably well defined. In contrast, offshore wind farms still are riskier to design, finance, build, and operate (see chapter 9).

Interconnection design

As noted above, the developer begins working with utilities early in the process to make sure that transmission lines will be accessible and will have enough capacity to export the wind power. Separate permits will be needed if a transmission line must be built to reach the grid. By now, the developer will have applied for a position in the line of projects waiting to connect with the transmission system, called the interconnection queue.

At this stage, the developer uses the model to design the wind farm's electrical collection system, while the utility specifies the substation and designs the grid interconnection, such as where to tie in to the utility's high-voltage line and where to meter the electricity.

The utility performs a grid stability study to address all of the interconnection issues. One important issue to be considered in the design is the transmission system's resilience or robustness—its ability to accept wind power whenever it's generated, in whatever amounts.

Another issue is grid protection. The utility's electrical network is very sensitive to disturbances from power sources that don't match grid conditions perfectly, as noted in chapter 2. If the wind farm sends the wrong kind of power to the grid, catastrophic accidents can result. All of the wind farm's electrical equipment must be compatible with the transmission system's requirements, or grid codes, as they're called in Europe.

In today's wind farms, turbines with power electronics are able to support the grid when it needs generating capacity to maintain voltage. This is called *low-voltage ride-through*. In the past, wind farms were required to disconnect whenever the transmission system experienced problems, even for a few milliseconds.

Ride-through enables wind power to remain stable and connected during voltage dips or other transient faults on the grid. China experienced three large grid failures in 2011 when voltage dipped and wind turbines disconnected from the transmission system. Chinese regulators urged that existing turbines without low-voltage ride-through be upgraded soon. Whole wind farms can be retrofitted with equipment that provides this feature and other grid-support functions for multiple turbines.

A related function of wind power systems, which also supports electrical grids, is to provide *dynamic reactive power compensation* (also called VAR support, because it's measured in volt-amperes reactive, or VAR). This function, which stabilizes voltage fluctuations on the power system, was required routinely in industrial countries before low-voltage ride-through also became common.

Final design

The final design, which evolves from the model's initial design, is a detailed plan that includes the electrical design, a turbine foundation design, a map of roads, sites for additional met towers, locations for substations and a workshop, and many other details.

When it's complete and ready to present to prospective investors, the final design consists of a massive number of documents, drawings, and blueprints. Appendix F lists the many factors involved in evaluating the economic feasibility of a wind project.

Turbine output. The final design specifies the operating parameters of the turbines that have been selected. The developer must make a trade-off between the financial need to generate as much energy as possible versus the technical need to operate the turbines realistically.

The temptation is to design the wind farm such that every turbine will be maxed out in terms of its power output. This means that the machines would continue to operate in high and/or turbulent winds, up to the edge of the design limit specified by the manufacturer.

But when turbines are run that way, they don't last as long or perform as reliably. Operating at maximum capacity all the time takes an exceptional amount of life off of the turbine. The developer might make more money in the long run by planning to run machines at lower power and to turn off some machines during turbulent winds or other adverse conditions. This can help ensure that the turbines will last longer and need less maintenance.

Turbine siting. At this point, the developer also *microsites* the turbines, or pinpoints each turbine's location on the wind farm map. This process ultimately affects the farm's power production because small differences in turbine position or elevation can dramatically increase the machine's output or, conversely, reduce it.

Another important factor in siting the turbines is a phenomenon called the *wake effect*. Every wind turbine extracts some energy from the wind, so the wind leaving the turbine has less potential energy left in it. In other words, the turbine casts a "wind shadow" called a wake in the downwind direction. The wake is a long trail of wind that is not only slowed down, but also quite turbulent. This is called the wake effect.

Proper spacing of wind turbines is critical to obtaining the maximum energy output of a wind farm. The machines must be far enough apart that they can avoid other turbines' wakes and capture enough wind power. Typically, however, some energy losses from the wake effect are unavoidable and must be accounted for in the developer's wind farm output calculations. Downstream turbines might lose as much as 20%–30% of their power compared to the machines in front.

For turbines that are side by side, perpendicular to the site's prevailing wind, the turbines ideally should be spaced at least three rotor diameters apart or up to five diameters, if enough land is available. Turbines that face into the prevailing wind, arranged front to back, should be at least five rotor diameters apart or up to ten diameters.

Figure 6–2 shows a sample wind farm layout with spacing of four rotor diameters side by side and seven diameters front to back. Figure 6–3 shows the same layout under adverse wind conditions, with selected turbines shut down.

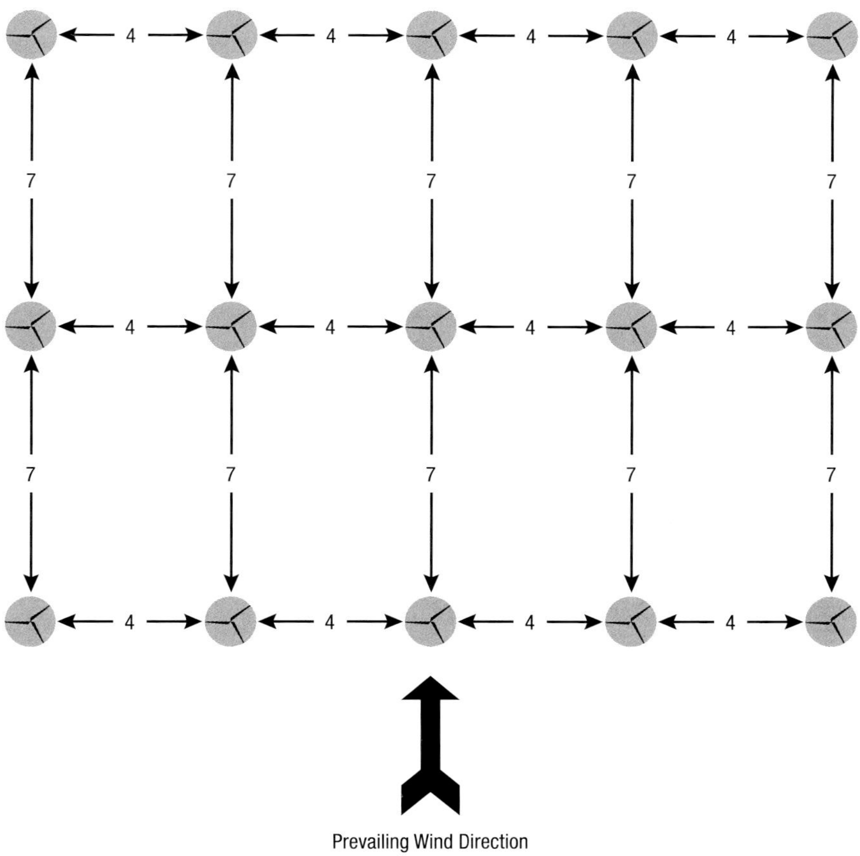

Fig. 6–2. Wind farm layout for minimizing wake effect

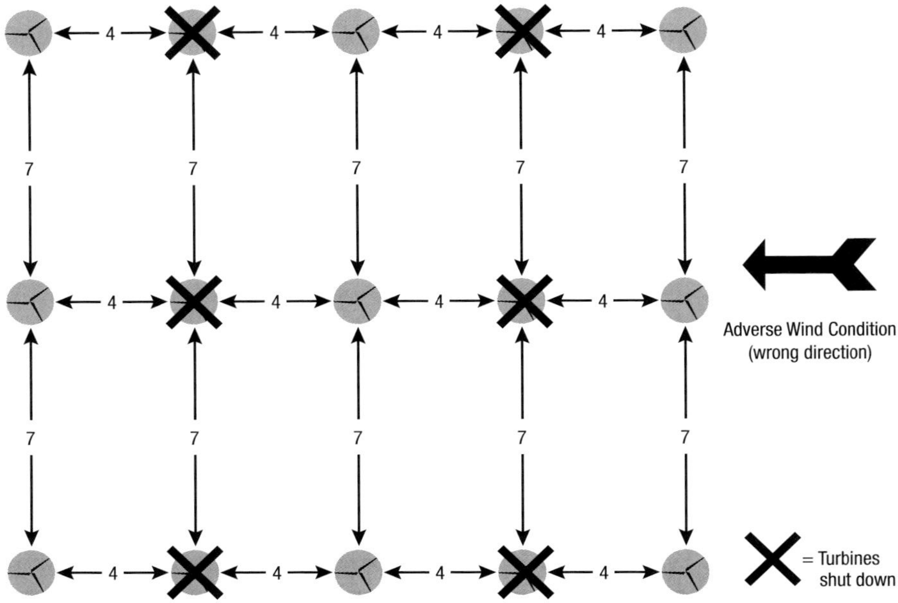

Fig. 6–3. Turbines selected for shutdown during adverse wind conditions

The wind farm design must balance energy output losses associated with the wake effect against land area constraints. Another limit to turbine spacing is the cost of connecting the machines to the grid. Minimizing the area of the electrical collection system (the lengths of cable needed) helps control construction costs. A smaller area also means shorter roads and lower costs. Researchers have recommended doubling the space between turbines to increase wind energy output, but real-world conditions tend to make this impractical.

Remember that rotor diameters for today's wind turbines are typically 60–100 meters (about 200–300 feet). So a spacing of five rotor diameters would require that each turbine be 300–500 meters away from its neighbor (1,000–1,600 feet)—that's up to half a kilometer apart, or almost a third of a mile.

Permitting

Like all large construction projects, wind farms raise issues among the public that are often contentious. Any type of big project concerns the people around it, who might worry about truck traffic and noise, for example, or in the case of transmission lines, the perceived threat of electromagnetic fields. As mentioned earlier, developers take an initial look at the wind farm site's environment and community to avoid any obvious showstoppers.

But wind farms face unique, additional constraints, primarily because they are so visible. The machines are huge, and they are often sited in prominent spots like mountains, hills, or ridgelines. As a result, laws and regulations that govern wind farm siting have become more complex. Wind farms require permits from several different government agencies at all levels of authority—local or municipal, state or province, and federal or national.

Permitting requirements. Before permits can be obtained, the developer must conduct a detailed environmental review that considers the wind farm's potential impact on endangered species, wetlands, and many other factors. Studies analyze the effect on birds and other wildlife, the aesthetics of the farm as envisioned when complete, and how much noise will be produced, how far away it can be heard, etc. A historical and archeological review is also conducted.

Several issues usually dominate the wind power permitting process, though they might differ depending on the state, province, or country where the wind farm is being considered for siting:

1. Noise
2. Birds, bats, other wildlife, and endangered species
3. Visual or aesthetic concerns
4. Compatibility with existing land use
5. Soil erosion, wetlands, and water quality
6. Interference with radar stations and air traffic control
7. Public health and safety
8. Cultural and historical significance
9. Community infrastructure and services
10. Air quality and climate change
11. Solid and hazardous waste

On some of these issues, wind power passes with flying colors, while others can be a sticking point in the development process (see chapter 12). Developers should provide stakeholders with hard data and offer to compromise when necessary. For example, generous setbacks from roads and property lines can help make wind farms less intrusive.

Community outreach. During the environmental review process, the smart developer meets and greets the neighbors of the proposed wind farm as soon as possible. The earlier that developers get involved with the community, the less likely people will later object to the wind farm plans.

Experts recommend that the developer set up local meetings to consult with the public. In addition to sales and marketing people, the developer's technical staff should be on hand to answer questions. The key is to appreciate input from every stakeholder who's involved, listen carefully to what they have to say, and respond to it thoughtfully.

Taking polls or canvassing the area can help the developer find out what people's concerns are. It's important to analyze the feedback correctly and accumulate quantitative data whenever possible. Developers might also want to sponsor local events, even after the wind farm is built.

To involve the public and other stakeholders effectively, developers should follow these best practices:

- Make consultation inclusive. All stakeholders deserve to have their views heard and addressed.
- Treat people equally. Judge ideas on their merit, not their source.
- Share responsibility. Let stakeholders know how their input has affected the developer's decisions.
- Consider using an independent facilitator for workshops and public meetings so that stakeholders feel they are being treated impartially.
- Build a coalition. Include established leaders of the community to help support the project.
- Don't disguise any uncertainties. Untruths and omissions can destroy people's trust.

Energy sales agreements

Many wind farm developers rely on *power purchase agreements* (PPAs) as a way to ensure a stable price for the wind power over the life of the project. The PPA, which is sometimes called a power off-take agreement, is a contract under which a utility or other company agrees to buy wind power from the wind farm developer/owner at a predefined rate or price, typically for a 20-year period. In other words, the PPA provides the project's cash flow.

Wind farm developers can also plan to sell *merchant power* into wholesale spot markets, without a contract from a buyer. Some wind farms are built and operated by companies that specialize in merchant power and assume most or all of the financial risk. Also, wind farms can sell energy to aggregators who buy relatively small amounts of power from several suppliers and resell it as a larger package.

Wholesale power markets are operated by regional, national, or multinational agencies, usually quasi-governmental and regulated. In the United States, these are called *independent system operators* (ISOs), which allocate space on transmission lines, maintain backup power reserves, and match electric supply and demand.

These agencies work closely with utility grid operators and *regional transmission organizations* (RTOs) to ensure that electricity flows smoothly where it's needed, without congestion or overloading of vulnerable transmission lines. In Europe, all of these functions are performed by *transmission system operators* (TSOs). Chapters 7 and 8 discuss global power system operations.

In general, prospective investors in a wind farm prefer a PPA over the merchant power option because prices in wholesale markets (spot or short-term contract) can be volatile from day to day and unpredictable over the longer term. Moreover, with merchant buyers, the wind farm operator has little or no prior commitment to take the turbines' output.

In contrast, a PPA reduces the price risk of the wind farm project and helps ensure that it will be practical financially. The contract's prices can be tied to inflation or linked to commodity prices in related markets. However, the credit-worthiness of the company signing the PPA should be taken into account, by judging its bond rating, for example.

PPAs, though most commonly contracted with utilities, can be signed with municipalities, schools, etc., or with any private or public company. In the U.S. market, utilities own or buy power from 64% of wind power capacity, with merchant-type projects accounting for less than 25% and other power marketers about 10%.

Contracting

Wind farm developers have to make a choice between two basic contracting strategies: whether to go turnkey, or whether to manage multiple contracts, which can exceed 20 on large projects. This is sort of like deciding whether to hire a general contractor on a construction project—the turnkey option—or act as your own general contractor, hiring and managing all the subcontractors—the multi-contract option.

One consideration is the wind farm developer's size and know-how. Smaller, less-experienced developers might do best with a turnkey contract, which has the advantage of being easier to coordinate and manage. A typical turnkey contract is called EPC—engineering, procurement, and construction. Many companies worldwide offer turnkey wind farm services that include management of the detailed design and the entire construction process.

Larger developers with more experience might make more money acting as their own general contractor and managing several suppliers and subcontracts. However, suppliers are risk-averse and expect the developer to assume most or all of the risk. Either way, the contracting strategy decision has to be made before financing is obtained, at which point all the contractors have to be lined up.

Turbine procurement

By now, the developer has selected the turbines that will deliver the most energy from the wind farm cost-effectively, based on their power curves, the wind resource characterization, and the model's output projections. The next step is for the developer (or turnkey project manager) to sign a procurement contract with the manufacturer that specifies the turbines' warranties, power curves, availability, noise levels, and other performance parameters.

Typically, this contract represents a firm order for supplying the wind turbines (blades, rotor, nacelle, controls, etc.) and commissioning the wind power systems. It might also include towers as well as transportation/delivery and installation/erection. Separate contracts are usually let for foundations, electrical collection systems, and other balance of plant.

Often, the developer also signs an agreement for the manufacturer to service and maintain the wind power system for a specified number of years, which can be extended if needed. Beyond that, technical assistance agreements are another option.

Financing

Securing money to finance a wind farm project is a long, complex process that occupies the developer almost from the start. It culminates in a financial "closing" (see below) where dozens of people might meet over a period of several days, discussing hundreds of documents and negotiating the final details.

The huge investment required to build a wind farm is one factor acting against wind power in favor of traditional natural gas–fueled plants (see chapter 12). Figure 6–4 shows a typical financing structure for a wind farm project.

Historically, prospective investors were willing to finance wind farms with little analysis as to how the project would perform in terms of energy output and revenues. But that all changed with the financial crisis in industrial nations, which resulted in tighter standards for business credit. Also, the underperformance of some wind farms, compared to original projections, has further increased investor requirements for greater scrutiny, or due diligence, in advance of financing a project.

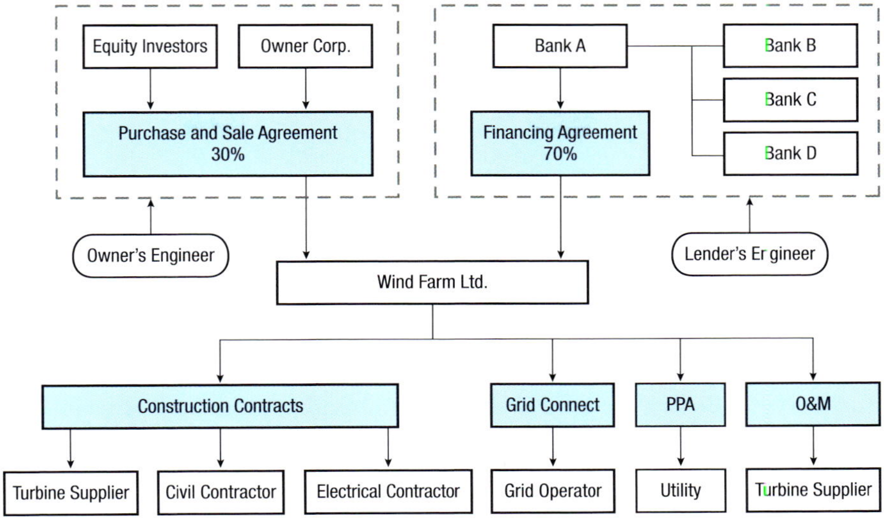

Fig. 6–4. Wind farm financing structure

Financing issues. Elements of the financing package include the source of equity, the source of debt (and the market interest rates), the term of debt, and the implied debt/equity ratio, as well as equipment and construction financing, titles/deeds and other assignable documents, and third-party due diligence. Transmission access, insurance, and safety planning are also important issues. Investors and insurance agents often require that wind farm developers select certified, well-proven turbine models.

As with previous tasks, the developer uses the model to juggle the various options and determine their potential outcomes. As the project takes shape and the model becomes more detailed, its projections become more accurate, but developers should allow for the inevitable escalation in costs that occurs with time.

Onshore wind farms usually rely on *project financing*, where the lender or underwriter focuses on the credit of the project rather than the credit of the borrower. The opportunity for repayment is confined to the wind farm's performance and its value as collateral.

Due diligence performed by an independent third party is required for most large wind farm projects these days. Several consultants worldwide specialize in monitoring and analyzing the cost and risk of a wind farm while the project is in development and then giving it their stamp of approval before the financial closing can take place.

These independent consultants can serve as either the owner's or the lender's engineer, sort of like real estate agents can represent the buyer or the seller, but not both. For some large projects, government regulators also hire their own independent evaluator.

Government incentives. Most wind farm developers seek to take advantage of various government incentives to fund their projects, in addition to traditional sources of debt and equity. These incentives include feed-in tariffs (described below), cash grants, loan guarantees, *production tax credits*, *investment tax credits*, and *renewable energy credits*, as well as *accelerated depreciation*, where capital costs are deducted within a few years, rather than over the entire lifetime of the project.

A production tax credit is based on the amount of energy generated by the wind farm over time. An investment tax credit allows a developer to take a one-time tax credit up front against the capital expense needed to build the wind farm. Another, more complex tax credit is tied to projects that invest in low-income communities.

Tax credits are more valuable to wind farm owners who have substantial tax obligations that might be offset by the credit. Some wind farm owners don't have enough tax liabilities to make use of credits, so governments have offered grants and loan guarantees to spur development. The U.S. investment tax credit can be converted to a cash grant, making it more popular than the production tax credit (see chapter 11).

Loans guaranteed by the government are an effective way to make investors more comfortable with a wind energy project. In 2010, the U.S. government guaranteed a $1.3 billion loan (0.95 billion euros) to a development group planning to build the Shepherds Flat wind farm mentioned earlier. The project already had transmission capacity in place or planned and a PPA with a utility, which will credit the wind power toward its state renewables mandate.

> *There has been a tendency by some commentators to lose sight of the fact that no successful wind energy project—in any country—has been conducted without some form of government intervention in the form of financial, technical, or regulatory support. This is really no different than any other kind of power production facility.*
>
> —*Darrell M. Dodge,* Illustrated History of Wind Power Development

---

Renewable energy credits are another type of incentive, sometimes sanctioned by government regulators. Basically, wind farm operators and other energy producers receive a credit for every unit of renewable energy they produce. The credits can also be called "green tags" and renewable energy or electricity certificates. These credits and certificates can be traded on the open market for cash, separately from the electricity itself, or they can be bartered.

Markets for renewable energy credits thrive wherever a certain percentage of energy demand or electricity supply is required to come from renewable sources, usually by a certain date in the future. This requirement is called a *renewable energy standard* or renewable portfolio standard. Governments use these mandates as a policy tool to stimulate development of wind power and other renewables and to attract investment in the clean energy sector (see chapter 11).

Many large utilities also are required by government regulations to generate a certain percentage of their electricity from renewables. If they don't have enough renewable capacity, they can purchase power from a wind farm and/or purchase the corresponding renewable energy credits.

Aside from such requirements, though, many energy consumers—especially businesses that want to present a green image—buy renewable energy credits voluntarily to show their commitment to clean energy.

Some companies have gone far beyond that, such as Google, which has invested millions of dollars in American wind farms and is supporting development of an offshore transmission network. Google claims that these are cost-effective business decisions, not just green for green's sake.

Feed-in tariffs. This tariff is simply a mechanism for wind farm owners to be paid for the energy they generate. *Feed-in tariffs* are one of the most widely used government policies in the world for accelerating development of wind power and other renewable energy sources (see chapter 11).

A feed-in tariff is a special rate for electricity that a wind farm (or other producer of renewable power) "feeds" into the grid. The rate, based on the cost of generation, can be set at a fixed price independent of market rates or can be pegged to prevailing electricity prices. Some feed-in tariffs pay higher rates in the earlier years of a project, then decline with time.

Feed-in tariffs drive market growth by giving wind farm developers and owners a long-term contract for their electricity (typically 15–20 years), a guaranteed connection to the grid, and a price high enough to recover a reasonable return on their investment. A specific tariff reduces financial risk by providing investors with rates that are clear and generally stable, making wind farms easier to develop and finance.

Financial closing. The closing is a major event in the birth of a wind farm, as noted above. One of the main requirements for the closing is a firm price for the wind power (unless the merchant option has been approved). Contracts must be negotiated to the point where price certainty is within 5%–10% of the final price. Key terms and conditions also must be confirmed to back the offered price.

The financial instruments at the closing can take many forms, such as fixed-income securities similar to mortgages, construction bridge loans, convertible notes, and ancillary credit mechanisms. Sometimes, a financial closing can cover more than one wind farm project with a single developer.

One final note on wind farm financing: Small developers sometimes put a deal together and "flip" the project to a large wind developer or owner, sort of like people renovate houses and flip them into the real estate market. This can happen at almost any point in the development process, from land agreements to closing or beyond. Sometimes the larger company buys the smaller developer as part of the deal.

PERSPECTIVE: Wind Farm Development and Operation
Ed Duggan, Executive Vice President, Oak Creek Energy Systems

Ed has 27 years of experience in the wind industry, including management of repowering Oak Creek's pioneer wind farm site in Tehachapi, land acquisition and wind resource assessment for the Alta Wind Energy Center, and working with the military to coordinate turbine siting with the operation of low-flying aircraft. Ed holds an engineering degree from the University of Massachusetts.

Oak Creek Energy began developing wind farms in 1982 and in 2006 signed the largest-ever renewable energy contract with a U.S. utility. The agreement calls for delivering 1.55 GW of wind power to southern California, with over 1 GW of this total to be installed and operational by 2012.

Q: What are your biggest headaches during the development process?

A: One of the biggest headaches during development is the permitting process—specifically, the continually changing environmental review process that has gone from being fairly reasonable in the 1990s to being onerous to outrageous since Y2K. This is true in certain states and on federal land. Prime examples of this are the review processes required by the California Environmental Quality Act [CEQA] and the federal National Environmental Policy Act [NEPA].

These processes are supposed to identify any significant environmental risks and then develop reasonable mitigation measures for the wind project to take to avoid such potential risks. However, in many cases, the review process is hijacked by environmental extremist or NIMBY groups that will do and/or say anything to protect their perceived interest. Like our politicians of today, many of these groups do not seem to be capable of compromise, and they often resort to use of the legal system to delay projects and stress them financially.

Q: How accurately can models estimate a wind farm's energy production and revenues before it's built?

A: Models have been evolving through the years to better predict wind farm energy production. As recently as three to five years ago, most of the models were typically over-predicting wind farm production by anywhere from 5% to 20% depending on the size of the project, the wind regime the project is installed in, and other factors.

However, as the models have become more sophisticated, they have continued to get better at predicting the deep wake effects and other impacts that may reduce the energy output of the wind project. Thus, such models have been closing the energy estimate gap in recent years.

Q: Can wind energy's cost continue to go down in upcoming power purchase agreements [PPAs]?

A: Wind energy costs are dropping as the energy market gets more competitive. This is happening through several means, including:

1) Design simplification to reduce the number and cost of parts in a wind turbine. An example of this would be using direct-drive generators that eliminate the need for expensive and failure-prone gearboxes.
2) Manufacturing process improvements to reduce the labor required to build a turbine. For example, the automation and/or use of industrial robots in critical component manufacturing for historically high-labor items such as blades is expected to reduce the cost of turbines.
3) More careful matching of turbines to their site allows for better energy extraction by having turbines with greater ratios of rotor swept area to generator nameplate.
4) Increased competition among manufacturers has resulted in reduced profit margins for them and better pricing for wind turbines. These manufacturers are now also offering better

extended service and warranty agreements, which also help drive down the cost of energy produced by these turbines.

5) Increased competition among engineering, procurement, and construction [EPC] contractors, due to a lack of other construction work in this current economy, has also resulted in reduced project construction costs.

Q: What's the outlook for financing in the 2012–2013 time frame?

A: It is somewhat up in the air due to expiration of the federal cash grant program for projects that have not started construction by the end of 2011 and the expiration of wind energy production tax credits in 2012. These programs have helped in recent years to spur the wind energy industry by leveling the playing field between wind and other subsidized energy sources. However, in light of budget deficit concerns, such programs are likely to come under heavy scrutiny when being considered for renewal.

Q: What might be the impact of California's 33% renewable portfolio standard [RPS] on wind power development in the west?

A: It should spur development, but wind is likely to be very difficult to develop in California due to CEQA and NEPA processes run amuck. Thus, many of the projects that serve California may be in Mexico as well as in Oregon, Washington, Nevada, and other nearby states.

Q: What do you like most about your job—development, operations, engineering, something else?

A: My favorite part of my job has been development. The process of locating a potential project area, conceiving a project, acquiring the land rights, measuring the wind, proving that a project is feasible, getting the project permitted, and then building it can be very gratifying. In the Tehachapi area, I can point to hundreds of 65-kilowatt to 150-kilowatt turbines that I picked the location of in the late 1980s that are still operating today, as well as over 300 more (much bigger) turbines, 1.5 to 3 megawatts, that were installed in the past year that should operate for the next 25 to 40 years.

Construction

The process of building a wind farm starts with excavation of the roads and turbine foundations. Civil and electrical works are ongoing while the reinforcing steel bars (rebar) are set in place, concrete is poured, foundations are assembled, and towers are erected.

Once that's done, the nacelle containing the drivetrain components is lifted onto the top of the tower first. Then the rotor is lifted and attached to the nacelle. The final steps are connecting with the utility grid and commissioning the project.

Of course, it's not all that simple. For example, a single wind turbine foundation can contain more than 40 tons of rebar and roughly 200–400 cubic meters of concrete (300–500 cubic yards), which is equal to 30–50 truckloads.

Tubular towers are usually erected in four pieces, but only two can be stacked at the same time. Otherwise, wind blowing over the open top of the tower could create natural-frequency vibrations capable of collapsing the structure (see chapter 3). So the second two tower sections are added when the nacelle is ready to seat in place, which closes off the top and prevents vibration.

For the rotor assembly, blades are inserted into the hub on the ground and "stabbed" into place. Alternatively, the hub can be hoisted first and the blades inserted at the tower top. When the rotor is lifted, engineers on top of the tower bolt the hub into the nacelle.

Either way, this can be a tricky business if the wind is blowing—which it should be, given that the site was selected for that very reason. Figures 6–5 and 6–6 show a turbine hub and blades being assembled and a rotor being lifted onto the tower.

Sometimes, the construction crew has to wait several days or even weeks for the wind to die down during the daytime before it's safe to lift the rotor and connect it to rest of the wind turbine. Safety should be paramount not only during construction, but through the life of the wind farm project (see "Safety").

Throughout the construction process, the project manager—whether it's the developer or a turnkey contractor—has to make sure that everyone's talking to each other. For example, whoever's running wire for the electrical infrastructure needs to know when a turbine has been relocated on the site map.

The simplest communications can become problematic. Imagine sending packages to a wind farm site, where there's no "address" per se. The delivery driver's instructions might say, look for a dirt road near the intersection of rural routes 18 and 27, then turn right and drive 1.7 miles until you see a trailer.

Fig. 6–5. Assembling a rotor at Mojagazas wind farm in Spain. *Image courtesy of LM Wind Power.*

Fig. 6–6. Rotor and blades being hoisted at Dabancheng wind farm in China. *Image courtesy of LM Wind Power.*

Before the turbines can begin operating, the utility has to bless the connection to the grid and turn on the juice. Although the developer has been working with the utility for months or years by now, this is still a critical step that must be conducted by the utility.

Once the wind farm is officially interconnected, the turbine supplier comes in to commission the machines. This involves inspecting the construction, turning on the turbines, and double-checking every feature of their operation, as well as testing the wind farm's electrical infrastructure. A single turbine can be commissioned in two or three days by experienced engineers.

At the end of its useful life, a wind farm can be repowered or decommissioned. In repowering, the existing turbines and foundations are replaced, usually with larger ones. Decommissioning involves dismantling and removing not only the turbines but also the underground and overhead power collection and transmission system.

In some cases, foundations can be left in the ground, but other times they must be excavated and carted off, an expensive project. Site restoration might be required by the wind farm operator's lease or permits.

Operation and Maintenance

The wind farm operator's chief concern is to generate energy at the lowest possible cost ($ or euros/MWh). To do this, the operator must get the most energy out of each turbine, without overworking it, and must prevent unscheduled maintenance—the main driver of the cost of operation. The key tool to accomplish these goals is to monitor the performance of the wind power system.

Performance monitoring

Wind farm operators monitor turbine performance by analyzing data collected from each turbine by its SCADA system. SCADA data are the key input for operators to maximize energy output, plan maintenance, and avoid unplanned maintenance.

SCADA data. Every turbine has a SCADA system that collects all of the data being generated by the control system's network of sensors, as described in chapter 2. A good SCADA system has a computer processor at each turbine that keeps collecting data even if communications are lost, then sends the data in when communications are back up.

Typically, the SCADA system organizes the information into 10-minute averages of turbine operating variables such as rotor speed, blade pitch, energy output, and nacelle position, as well as data from other wind power system components, such as temperatures and pressures at various points, water and oil flows, vibration, and power characteristics.

In addition to this information, the SCADA system tracks the turbine's fault logs, recording the exact time of a fault, how long it lasted, and what caused the turbine to "trip" (shut off). The SCADA system also collects data from the wind farm's met towers (wind speed and direction, air temperature and pressure, etc.) and tracks the amount of energy being metered at the electrical substation or other point of connection to the utility grid.

A 100-MW wind farm can create more than 2 *gigabytes* of operational data each year. An effective SCADA system is able to distill these data into informative reports for engineers on the wind farm's operation and maintenance (O&M) staff. The reports identify trends and show the results of corrective actions. Analytical SCADA software, designed for wind power applications, can enable operators to improve turbine availability and efficiency.

Although O&M people are on the front line, the wind farm operator should also have specialists who analyze certain sections of the SCADA reports. For example, an engineer specializing in wind turbine generators should be looking closely at the SCADA data on generator performance.

Performance factors. As discussed earlier, wind farm developers go to great pains to select the right turbines for their site. However, the machines have to be constantly monitored to check the difference between the amount of energy they are expected to produce and the amount they are actually producing. To do this, operators compare the predicted turbine power curves against power curves measured at the wind farm.

In a recent study of a North American wind farm's power curves, researchers found that turbines overperformed in steady wind, producing more energy than predicted, and underperformed in turbulent wind, which is understandable.

However, even at identical hub-height wind speeds, the difference in performance approached 20%, which was much greater than expected. Turbulence and wind shear evidently affect turbine performance more strongly than had been thought.

Another measure of performance is the availability of the turbine. As noted above, availability is the percent of time that a wind turbine is running or is ready to run—that is, a measure of how often the turbine shuts down for whatever reason.

Wind farm operators use the SCADA data to determine whether a turbine's downtime was caused by high wind speeds that forced it to stop, for example, or if it was compelled to shut down by the utility to protect the grid. If high winds were the cause, were they actually that high, or was the anemometer misreading wind speed for some reason? SCADA data can help clarify these issues.

The operator then compares actual availability with the figures reported by the turbine manufacturer for the existing fleet of same-model turbines. This comparison is used to assess damages if actual availability is sub-par or to award bonuses for extra availability. Data collected worldwide at more than 300 wind farms showed that the global average availability was 96.1%, with American wind farms lagging slightly at 94.6%.

Wind turbine manufacturers don't always calculate availability in exactly the same way, making it hard to compare performance among sites and turbines. Some manufacturers don't count scheduled maintenance as downtime, which can amount to 120 hours annually. Some focus on just the turbine's downtime, while others include shutdowns attributed to balance-of-plant equipment (such as the transformer and electrical collection system), events (such as lightning), or low wind speeds.

To remedy this situation, the International Electrotechnical Commission is developing a standard for measuring availability. However, manufacturers likely will continue to seek their own competitive edge. For example, one plans to offer guarantees of turbine availability based on the machine's energy output rather than its run-time.

Capacity factor is also monitored but is largely irrelevant for a single wind farm. This measure of productivity is more useful for comparing the performance of one wind farm versus another.

Minimizing the cost of maintenance

Unscheduled maintenance is the key driver of a wind farm's operating costs, as noted above. The effect of maintenance costs is evident in the formula for calculating wind power generation costs:

cost of energy = (capital costs + O&M costs)/lifetime energy capture **Eq. 6–2**

Figure 6–7 shows the life-cycle cost of each component over the 20-year operating lifetime of a wind power system.

O&M costs are affected by many variables, but data suggest that the age and the size of the wind farm are important factors. In the first few years of operation, U.S. projects that were installed more recently incurred, on average, lower O&M costs than older projects. Not surprisingly, O&M costs rise with age. Also, larger wind farms tend to experience lower O&M costs than smaller developments. However, data on these factors is limited and is spread over a considerable range.

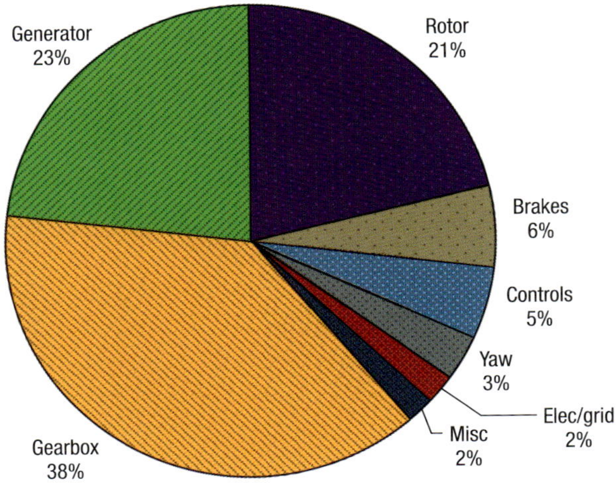

Fig. 6–7. Wind power system life-cycle cost by component. *Data courtesy of Upwind Solutions.*

Types of maintenance. Several strategies can be employed to minimize the economic impact of both unscheduled maintenance and planned, *preventive* maintenance or routine service. Unscheduled maintenance (also called *reactive*) means, basically, fix it when it breaks.

Preventive maintenance follows a schedule based on the equipment manufacturer's service manual. Figure 6–8 shows routine wind turbine blade inspection and servicing.

Other types of preventive maintenance include inspecting the drivetrain components with an optical device (*borescope*), finding hot spots in the equipment via infrared scanning (also called *thermographic imaging*), and analyzing the oil in the various components to detect evidence of metal degradation. The electrical collection system also undergoes periodic testing of the transformers, distribution cables, circuit breakers, protective relays, etc.

To reduce the cost of planned maintenance, operators should schedule service when the wind isn't blowing, to avoid lost revenues. This requires the operator to anticipate windows of opportunity for service, which is possible with today's more accurate weather forecasting methods (see chapter 7).

Also, maintenance procedures should be performed in "sweeps," with the crew doing the same job on as many turbines as possible. Consolidating maintenance into one plant-wide outage, and scheduling one crane, can save hundreds of thousands to millions of dollars or euros per year in lost revenue, depending on the size of the wind farm.

Fig. 6–8. Blade inspection and servicing. *Image courtesy of LM Wind Power.*

Most wind farm operators have hotlines to their maintenance technicians and engineers, as well as communications via smart phones, streaming video, and other digital technologies. People arriving at the site have instructions on where to go and exactly what to do. They can use hand-held devices to identify parts, check procedures, and access drawings.

Controlling unplanned maintenance is trickier, but can be done. Operators should have enough trained technicians on hand or on call to fix problems when they arise. A corollary of this is to keep enough spare parts nearby and cranes on call to perform the work.

Condition monitoring. Beyond these basics, wind farm operators can predict the need for maintenance before it happens unexpectedly by *condition monitoring*, or structural health monitoring, of the wind power system's key components. This is called *predictive* maintenance as opposed to preventive (time-based, on a schedule) or reactive maintenance (fix it when it breaks).

Research has shown that preventive maintenance costs 24% less than reactive maintenance, and predictive maintenance based on condition monitoring costs 47% less. As an extreme example, the cost of replacing a broken wind turbine blade can be 10–20 times greater than the cost of repairing it before it breaks. Replacing a gearbox ahead of berakdown costs about one-third as much as an emergency job.

Monitoring the condition of the drivetrain—driveshafts, gearbox, and generator—is fairly well understood in the wind industry, which has picked up the technology from other, more mature industries. One key technique is continuously monitoring and analyzing the frequency of vibration. For example, excessive vibration in the gearbox is a clear indicator of an impending failure.

Vibration in nacelle components is usually monitored by sensors connected to the turbine controller. Vibration monitors can also be added on to the equipment by third parties. The monitors can communicate wirelessly by satellite, cellphone, or radio with the wind farm's O&M people. Some wind farms use hybrid systems with more than one type of wireless communication.

International vibration standards for wind turbines were released in 2010 by the Association of German Engineers (Verein Deutscher Ingenieure, or VDI). The standards specify threshold values for vibration of drivetrain components. This will make it easier for wind farm operators to use vibration-analysis software for condition monitoring.

Digital techniques are also emerging to monitor component health. The wind power system's lubricating oil and mechanical drivetrain components can be remotely monitored using sensors embedded in the equipment that send data via the SCADA system to the operator's computer. Software then analyzes the data to understand the condition of the system. Ongoing real-time analysis can yield better insight than routine inspections.

Monitoring the health of the rotor and blades is more challenging. The vibration of a blade can be monitored, to warn when it might be coming loose. Blades can also be inspected as shown above (fig. 6–8), but this requires stopping the turbine, and it provides no continuous blade-health data.

Newer techniques to monitor rotor condition use fiber-optic sensors to measure load rather than conventional electrical strain gauges. Ideally, the fiber optics would be embedded in the blade when it's manufactured, but cables can be attached inside existing blades. Condition monitoring with fiber optics in the blades is still considered an expensive option. Each turbine blade might need as many as 20 sensors measuring loads and other variables such as temperature.

Load control. Another technique for minimizing unscheduled maintenance is load control. This means taking into account the excessive wear and tear on a wind power system that results from operating at maximum capacity all the time. Instead, operators can "dial in" a specific fatigue life by counting the cycles that the turbine is undergoing at full load and then throttling back just enough to extend the turbine's life.

This type of operation is sometimes called smart load control or smart rotor control. With condition monitoring and advanced controls, the operator can predict the remaining useful life of components and how much output the wind power system can deliver without suffering too much damage.

By deciding to run the turbines at less-than-maximum output, or to turn them off before they reach their design limit in turbulent or strong winds, wind farm operators sacrifice some energy revenues in the near term, as explained earlier.

But as a result, the operator can reap financial rewards later when the turbines last longer and need less maintenance. Think of driving a car 24/7 for a year at 100 km/h (60 mph)—you'd put more than 800,000 kilometers on the engine (500,000 miles). Similarly, a wind power system shouldn't be stressed to its design limits all the time. Smart load control helps the operator analyze the trade-offs.

Load control technology will be tested at a 50.6–MW wind farm to be built in Maine, funded in part with a U.S. federal loan guarantee. A system of sensors and software could allow the turbines to continue generating electricity during turbulent wind conditions, rather than shut down completely, if the additional wear and tear would cost less than the additional energy output gained.

Fault analysis. The availability of a turbine is by nature a historical measure, a report on past performance in terms of run-time. A better statistic for wind farm operators seeking to improve their turbines' run-time is called the *mean time between faults*. This is defined as the average number of generating hours between faults (unplanned events that can be reset automatically or manually).

Many wind turbines can experience anywhere from 100 to 300 faults each year. The machines trip, or shut down, whenever any operating condition exceeds its limits, such as high oil temperature in the gearbox or high vibration in the generator bearings. Besides reducing availability, cycling the turbine on and off in response to faults inflicts wear and tear on all of the wind power system's equipment.

Instead of just resetting the trip and restarting the machine, smart wind farm operators investigate the underlying reason for the fault and do something about it. This can not only improve the machine's availability, but also increase the lifetime of the equipment.

One type of fault—circuit breaker trips—can be responsible for up to 50% of a typical wind turbine's total faults. Analyzing the SCADA data to find out why this type of fault occurs is clearly a path to mitigating turbine downtime.

For example, the circuit breaker trips when it's overloaded with current, but that's not the root cause of the fault. SCADA data might reveal that the turbine is putting out too much power and overloading the circuit breaker, or the data might show that the circuit-breaker trip level is set too low. The wind farm operator would want to either reduce output from that turbine, or reset the breaker's trip level.

Another way to analyze fault data is by how much energy revenue is lost due to the fault. Fig. 6–9 shows the type of wind turbine faults in terms of lost revenue for one wind farm example. Energy loss analysis can help the wind farm operator decide whether to step up the preventive maintenance schedule on a component that is tripping the system frequently. For example, in this case more attention to the pitch control system would be warranted.

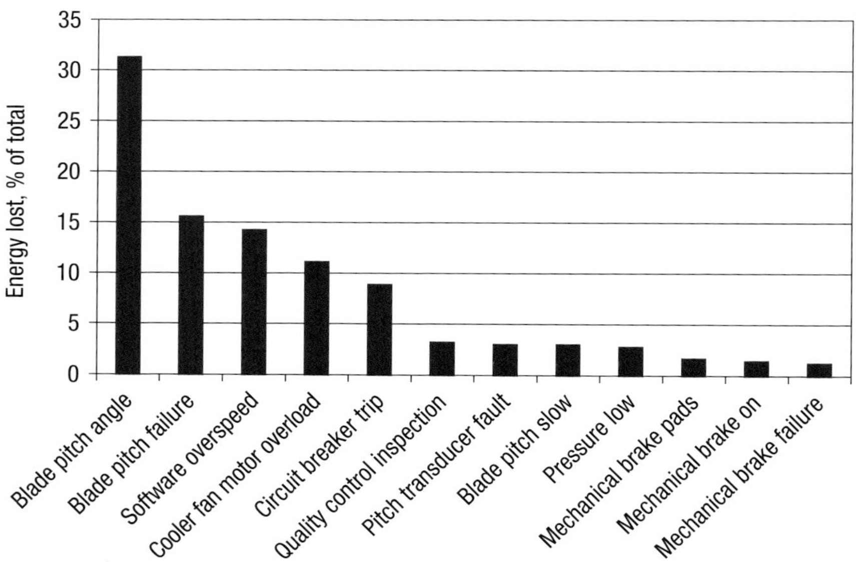

Fig. 6–9. Prioritizing service activity by identifying sources of lost revenue. *Data courtesy of GL Garrad Hassan.*

PERSPECTIVE: Wind Farm Management
Jim Webb, Site Supervisor

Jim is a supervisor at a large wind farm (greater than 250 MW) in West Virginia. He has worked in the wind industry for 30 years. Jim is a licensed electronics technician (U.S. Federal Communications Commission), a certified welder (American Welding Society), and a licensed aircraft mechanic (U.S. Federal Aviation Administration).

Q: What are the most common causes of faults tripping a turbine?

A: It depends entirely on the manufacturer of the turbine. Each one has specific problems that lead to repeat turbine trips. There are some generalities, however.

Pitch systems usually give problems, most often because of slow pitching, or the blades not moving together. There are two types of pitch systems in common use, hydraulic and electric. The electric type usually has a motor at each blade. Neither system seems to be more troublefree than the other, although the electric motors are much cleaner.

Power inverters on variable-speed turbines usually do not cause large numbers of faults, but the faults that do occur can take much longer to troubleshoot. Like computers, the error messages can be cryptic, and many different faults can have the same symptoms.

Q: How do you decide how to handle each fault?

A: The only practical way is to send out good troubleshooters and let them make the diagnosis at the turbine. Unless a problem is so widespread that everyone has it memorized, the SCADA alarms do not give enough information to diagnose it from there.

Well-trained techs equipped with a systematic troubleshooting method are the best way to handle a fault. Preferably, the troubleshooting method is written so that techs who aren't as good at troubleshooting have a guide. Troubleshooting trees, the ones that direct a tech to a specific action based on the results of the last test, have not worked well, at least for me.

It seems to work better to have a set routine of tests for each subsystem. No matter what the symptoms of that subsystem, the tech always does all the tests. Some tests pass, and others fail, and the tech can then look over the list to see which ones were which, and usually make a diagnosis.

This method does not tell the tech what the bad part is, but leaves it up to the tech to decide, based on which tests passed and which ones failed. I've had much better results with techs using that system than the troubleshooting tree.

Q: How do you minimize unscheduled maintenance costs?

A: Good routine maintenance, done right, and done on schedule. To accomplish this, you must have enough techs available, keep a well-stocked inventory of parts and equipment, and periodically review the inventory, because parts and equipment needs change as turbines age.

Also, you must analyze the SCADA data. A lot of useful information is hidden in SCADA data. Some failures are related to running hours, some are related to megawatt-hours generated. There are correlations with turbulence, wind direction, temperature, specific turbine sites, different manufacturers of the same sub-assemblies on different turbines, etc. The list is long.

Once the turbines are off warranty, it is possible for the owner to make minor changes to the turbines, usually by buying better replacement parts such as electric motors. That can help prevent unscheduled maintenance if the particular part has caused trouble in the past.

Q: If you could magically transform existing SCADA systems, what would you make them do?

A: I would make the SCADA systems and all the attached IT hardware easier for non-IT techs to troubleshoot. Although SCADA systems from a particular manufacturer are the same, the installation is always a custom one-off. The documentation is usually opaque, and most sites do not have an IT staff.

The HELP number, for either the SCADA system itself or all the IT hardware, always refers the caller to voice-mail. I would have someone always on call to answer the HELP number. I would have documentation that is readable, and that reflects the actual installation.

Imagine a large office complex using their most skilled air-conditioning repair tech to fix the network when it crashes—and the documentation was never updated with the changes that were made during construction.

Q: What do you like most about your job?

A: It is still a new business, and the rules are not fully written yet.

Q: What advice would you give to someone who's thinking about training for a technician job?

A: Learn to think logically and to troubleshoot. Good troubleshooters are necessary in this business, and they're quite valuable. It is also great fun to get paid to solve puzzles. Good troubleshooters are often used only for troubleshooting, while the other techs are doing preventive maintenance, which consists of doing the same thing every day forever.

Learn to plan ahead. There is nothing like spending an hour to get to the turbine, and another half hour to get to the top and haul up the equipment, only to discover something was forgotten. Even worse is to realize that the crane is at the turbine, and the entire job has to be delayed while someone scrounges parts or equipment. It can cost tens of thousands of dollars to deliver and set up a crane, and several more thousands of dollars a day for the crane and crew to be on-site. Long delays are hard to explain.

Understand that no matter what education or experience a tech has, there will be quite a lot to learn about each specific turbine. Everyone starts off behind the 8-ball.

Anticipate that there may be travel involved, and there may be on-call involved.

Optimizing turbine performance

Until recently, methods of measuring wind speed and direction at the top of a turbine were limited to wind vanes and anemometers on the nacelle. However, the turbine's rotating blades and the bulk of the nacelle disturb the air flow around the machine, limiting the accuracy of wind measurements. Moreover, the turbine's blade pitch and yaw control systems are not proactive—they can only react to fluctuations after the wind has hit the turbine. These shortcomings can affect the dynamic loads that stress the wind power system.

Lidar technology, which measures wind conditions using laser beams, has begun to evolve from a wind assessment tool (described in chapter 5) to a turbine optimization tool. A few companies are commercializing lidar systems that can be mounted on the nacelle or in the rotor hub's nose cone (spinner). Typically, these lidar units use fiber optics to direct the lasers rather than the mirrors that are used in ground-based lidar. Natural Power's lidar system, the Control ZephIR (fig. 6–10), was developed with Risø DTU, Denmark's National Laboratory for Sustainable Energy. Another turbine-mounted system, the Wind Iris, is being marketed by Avent Lidar Technology, a private joint venture between NRG Systems and Leosphere. Catch the Wind is also testing turbine-mounted lidar technology.

Fig. 6–10. Lidar unit mounted in turbine nose cone. *Image courtesy of Natural Power.*

A turbine-mounted "look-ahead" lidar unit scans the incoming wind 150–400 meters in front of the turbine (500–1,200 feet) and characterizes the wind before it reaches the rotor. The control system can pitch the blades to maximize energy capture, depending on how the wind is behaving at that instant and at that height. Also, the yaw system can pivot the turbine into proper alignment before the wind arrives.

Turbine-mounted lidar systems could also help reduce wear and tear on the turbine. When the lidar unit sees a gust of turbulent wind coming, pitching the blade to avoid the extreme load can lead to longer blade life. Drivetrain stress can also be reduced.

Conclusion

Regardless of which strategies are best for a specific wind farm site, one trend is clear. The O&M business is booming and will only grow faster and become more competitive. As existing wind farms age, the warranties on their components expire, prompting some operators to seek out an O&M contract.

Also, mechanical O&M procedures are being supplemented by digital techniques, as noted above. Although technicians still have to go out there with a wrench at some point, service companies offer software that mines through all of the data being collected, detects anomalies or impending problems, and automatically notifies the wind farm operator. This can help the operator pinpoint the best time to perform the O&M procedure or replace the part.

Safety

It's hard to overstate the importance of safety in operating and maintaining a wind farm. The industry provides little data on accidents or other emergencies, but safety training is well established. After all, technicians and engineers are often working at heights of over a hundred meters (hundreds of feet), in harsh weather, and sometimes in tight quarters inside the nacelle.

Five factors drive safety at wind farms:

1. Proper orientation for new employees and site visitors
2. A training program that complies with or exceeds government regulations for electrical safety, fire safety, protection from unintended machinery start-up, etc.
3. Reporting and investigation of accidents, incidents, and near misses
4. Site safety audits
5. Sharing lessons learned

Safety is about training and consciousness, says industry expert Paul Gipe. All those working in wind energy must be trained, including those with experience. Gipe has documented a growing fatality rate in the wind industry, starting with one death in 1975 and rising to 8 in the 1980s, 10 in the 1990s, and at least 25 since 2000. Gipe maintains a database recording the location, year, and other information on wind energy-related deaths.

Unfortunately, serious incidents like these are not unusual in any industry that involves working with heavy machinery high off the ground, but accident rates reported in some countries are no worse than those in safer jobs.

Industry observers agree that wind farm operators should share best safety practices to minimize accidents. Several wind turbine manufacturers have joined forces to develop a basic safety training program for the industry under the framework of the Global Wind Organisation. The program was expected to launch in 2011. Also, the American Wind Energy Association started a benchmarking program that will collect safety data confidentially from industry participants.

Today's wind turbine technicians and engineers benefit from an array of specialized equipment that has been developed to serve the wind O&M market and other industries where work commonly takes place at great heights. This equipment includes special ladders, ergonomic fall-arrest harnesses, and better helmets. Government regulations typically specify protective equipment for workers, but it's vitally important for wind farm operators to enforce rules requiring the correct use of ropes and harnesses when climbing towers or working at the top.

Another innovation is the aerial work platform (as shown in figure 6–8), which comes equipped with electrical, hydraulic, and pneumatic outlets for tools. These platforms are safer for workers than being hoisted in a basket by a crane or rappelling down a tower from the nacelle. They can also be less costly than very large cranes, which are expensive to deliver, set up, and take down.

Besides falls, turbine fires and electrical shocks are a hazard that can endanger technicians or engineers working in a nacelle. Lightning strikes are the most common cause of fires, and the risk increases with rising turbine heights.

Another cause of fires and electrical damage is an arc fault, or arc flash. Arc faults occur when two parts of a system become connected that normally shouldn't be, resulting in a flash of electrical current and an explosion. This happens most often during maintenance in the nacelle, when a tool might be left out, but can occur at ground level too.

Arc fault explosions can cause serious injuries and equipment damage. Incidents occur at rates of up to 200/yr in Germany, 400–500/yr in the United Kingdom, and 2,000/yr in the United States. Not all incidents are catastrophic, but arc faults cause an estimated 10 injuries annually in Germany and 4–5 deaths annually in the United Kingdom.

Another electrical danger is equipment such as a transformer being turned on unexpectedly, shocking and burning any worker nearby. It's up to the wind farm operator to implement safe practices for working with and around electrical equipment.

Fire safety codes are becoming stricter and more specific to wind turbines, and safety regulators can impose large fines for willful violations. Also, in addition to the cost of repairing or replacing damaged equipment, a wind farm operator can be held liable for damages and firefighting expenses if the fire spreads to the turbine's surroundings.

Safe wind farm operation also includes proper lighting on met towers, nacelle tops, and other structures that could present hazards to aircraft. Heights and types of lighting are specified by government regulators. Operators must prove to investors that their obstruction lighting meets requirements and is certified as such.

Federal safety regulators announced a crackdown on U.S. wind farms in 2011. Tougher enforcement could compel operators to modify their turbine supply agreements, O&M contracts, and maintenance procedures.

Community Wind

Community-scale wind power varies from a single large or mid-range turbine to a dozen or so, contributing power supply to a town or university campus or delivering electricity for sale to the utility. With community wind projects, local people often have a direct financial stake in the project, assuming a share in both the risks and the rewards of the investment.

Community wind farms can be developed by groups of farmers or other rural landowners, school districts, municipal utilities, rural electric cooperatives, or native tribes. These stakeholders have some control over how the wind farm project is managed, such as voting on decisions that matter to the community.

The community wind business model delivers economic benefits such as land lease payments, temporary construction jobs, long-term operation jobs, and perhaps most importantly, a bigger local tax base.

Another benefit of community wind projects versus larger scale developments is public acceptance. Some studies indicate that opposition to wind power virtually disappears when the community is invested in the project and has a say in how it's conducted.

Europe leads the way

Compared with America, Europe has embraced community wind to a much larger extent. Most wind power projects in Denmark (more than 80%) and a good share in Germany (up to 45%) were developed with local ownership. In contrast, community wind accounts for about 2% of installed wind power capacity in the United States, although the market is growing.

European countries employ different mechanisms to promote community wind, but they all require cooperation among the participants. In Denmark, for example, the government guarantees community wind power developers a grid connection and a fair price from utilities, which are required to buy the electricity. Revenues are shared among cooperative members. Danish law encourages mutual ownership of community wind projects by exempting owners from taxes on the portion of wind power that offsets the individual's electricity consumption.

In the Danish model, the developer pays for connection costs at the nearest feasible point, while the utility pays for grid upgrades or other improvements needed for the transmission system to handle the extra power. Danish cooperatives, because they can sell their wind power, are able to select larger, more cost-effective turbines than would be needed just to supply the cooperative's members.

Germany's North Frisia area, which lies mostly in the state of Schleswig-Holstein, offers another example of Europe's success on the community wind front. About 60 wind parks are operating there, with a total installed capacity of over 700 MW from more than 600 turbines. About 90% of these wind power plants are community farms.

While other European countries lag Denmark and Germany, community wind is spreading its wings throughout the European Union. In the United Kingdom, an independent company has developed 12 community projects.

In Italy and other nations with high electricity prices, towns are putting up wind turbines to generate revenue for local services such as schools and waste disposal, even though legal restrictions can prevent the town from reducing its electric rates by adding wind power supply.

Development and jobs

Developing a community turbine or small wind farm involves most of the same steps as a utility-scale farm, but on a smaller scale and with less complexity. For example, an existing transmission line might be able to easily handle an additional 5 or 10 MW of wind power from a community project, but not 100 MW from a utility-scale wind farm.

On the other hand, costs can be higher per megawatt, as would be expected. Construction costs don't scale down proportionately with the number of turbines to be installed. However, if the developer is managing several small projects, the turbine price might be bargained down by purchasing multiple units, and construction costs can be spread more widely. Alternatively, turbines that are unproven in the broader marketplace can sometimes be procured at a lower price with a stronger warranty.

In terms of employment, community wind farms create more jobs, both temporary and long-term, than absentee projects operated by companies without local roots, according to a study by the U.S. National Renewable Energy Laboratory. The greater the local ownership and return on investment in a community project, the greater the impact on job creation.

Community wind employs four to six people for one year per megawatt during construction (1.1–1.3 times higher than a commercial farm) and creates 0.3 to 0.6 jobs per megawatt for wind power plant operations (1.1–2.8 times higher). These are averages for U.S. projects.

Financing a community project can be problematic since lenders have become more conservative. Likely sources of financing include government agencies and turbine manufacturers, which have some understanding of community wind. Community planners can improve their chance of obtaining financing by working with experienced professionals, structuring the project properly, and offering personal and/or corporate repayment guarantees.

References

American Wind Energy Association. 2011 (May). 2010 U.S. Wind Industry Annual Market Report: Rankings. http://www.awea.org/learnabout/publications/factsheets/upload/2010-Annual-Market-Report-Rankings-Fact-Sheet-May-2011.pdf.

Angelou, Nikolas. 2010 (October 14). Predicting the Future Using Lidar. *Renewable Energy World*. http://www.renewableenergyworld.com/rea/news/article/2010/10/lidar-forecasts-boost-wind-harvests.

Avent Lidar Technology, http://www.aventlidartechnology.com/.

Bell, Benjamin. 2010 (January 12). Wind Project Development Value Chain. Maine Wind Seminar Series 2010.

Bell, Benjamin. 2009 (September 2). Operating Efficiently in a New Age of Renewable Power Generation. Presentation at the GreenPower Conferences: Optimizing Wind Power Performance. Houston.

Bonnon, Ian D. 2010 (September 21). Project Planning—Concept to Finance to Build. Presentation at the Offshore Wind Best Practice Summit. Boston.

Chambers, Ann. 2003. *Renewable Energy in Nontechnical Language*. Tulsa: PennWell. (ISBN 978-1-59370-005-8).

Chelsea Green. A Case Study in Community Wind: Denmark. http://www.chelseagreen.com/content/a-case-study-in-community-wind-denmark.

Community Windpower Ltd., http://www.communitywindpower.co.uk/index.asp.

Couture, Toby D., et al. 2010 (July). A Policymaker's Guide to Feed-in Tariff Policy Design. NREL/TP-6A2-44849. http://www.nrel.gov/docs/fy10osti/44849.pdf.

Crotty, Marty. 2010 (September). What Gets Measured Gets Done. Upwind Solutions, Engineering White Papers. http://www.engineeringwhitepapers.com/wp-content/uploads/2011/01/Day-1_Upwind-Sol_Marty_Crotty.pdf.

Deign, Jason. 2011 (March 11). Wind Industry Needs to Put Safety First. Wind Energy Update. http://social.windenergyupdate.com/industry-insight/wind-industry-needs-put-safety-first?utm_source=newsletter&utm_medium=email&utm_campaign=WEU.

Diaz, Chris. 2010 (November). Community Wind Projects Still Have Financing Options. http://www.seminolefinancialservices.com/wp-content/themes/seminole/images/NA_Windpower_Article.pdf.

Dodge, Darrell M. Illustrated History of Wind Power Development. http://www.telosnet.com/wind.

Duggan, Ed. 2011. Interviewed by Rebecca Busby. E-mail message to author. July 12.

Economist. 2010 (March 4). Smarting from the Wind. http://www.economist.com/node/15582251.

European Wind Energy Association. Wind Energy: The Facts. http://www.wind-energy-the-facts.org.

Farrell, John. 2010 (October 27). Hermann Scheer's Feed-In Tariff Legacy: 75% of World's Solar PV. http://energyselfreliantstates.org/content/hermann-scheers-feed-tariff-legacy-75-worlds-solar-pv.

Fingersh, L., Hand, M., and Laxson, A. 2006 (December). Wind Turbine Design Cost and Scaling Model. Report No. NREL/TP-500-40566, http://www.nrel.gov/wind/pdfs/40566.pdf.

Gipe, Paul. 2009. Wind Energy—The Breath of Life or the Kiss of Death: Contemporary Wind Mortality Rates. http://www.wind-works.org/articles/BreathLife.html.

———. Community Wind: The Third Way. Wind-Works.org. http://www.wind-works.org/articles/communitywindthethirdway.html.

Guided Tour (English). Danish Wind Energy Association. http://guidedtour.windpower.org/en/tour/wres/wake.htm.

Heller, Arnie. 2010 (April). Extracting More Power from the Wind. Lawrence Livermore National Laboratory, Science and Technology Review. https://str.llnl.gov/AprMay10/mirocha.html.

Hsu, Jeremy. 2010 (January). Wind Turbines Leave Clouds and Energy Inefficiency in Their Wake. *Popular Science*. http://www.popsci.com/technology/article/2010-01/wind-turbines-leave-clouds-and-energy-inefficiency-their-wake.

Jordan, Kim. 2011 (June 20). Developing a Wind Farm Support Campaign. *Renewable Energy World*. http://www.renewableenergyworld.com/rea/news/article/2011/06/developing-a-wind-farm-support-campaign?cmpid=WNL-Wednesday-June22-2011.

Kropper, Steve. 2011 (January 26). Wind Measurement Towers: How High Should They Be? *Renewable Energy World*. http://www.renewableenergyworld.com/rea/news/article/2011/01/wind-measurement-towers-how-high-should-they-be.

Lantz, E., and Tegen, S. 2009 (April). Economic Development Impacts of Community Wind Projects: A Review and Empirical Evaluation. NREL/CP-500-45555. http://www.nrel.gov/docs/fy09osti/45555.pdf.

Melnyk, Markian M.W., and Andersen, Robert M. 2009. *Offshore Power: Building Renewable Energy Projects in U.S. Waters*. Tulsa: PennWell. (ISBN 978-1-59370-211-3).

Moore, Mike. 2010 (August). Maintaining the Wind Fleet. *Power Engineering*. http://www.powergenworldwide.com/index/display/articledisplay/2077898726/articles/power-engineering/volume-114/Issue-8/Features/Maintaining-the-Wind-Fleet.html.

National Wind Coordinating Collaborative. 2002 (August). Permitting of Wind Energy Facilities. http://www.nationalwind.org/assets/publications/permitting2002.pdf.

Natural Power, http://www.naturalpower.com.

Pfahl, Greg. 2010 (July 10). ITC or PTC for Your Renewable Energy Project? *Renewable Energy World*. http://www.renewableenergyworld.com/rea/news/article/2010/07/itc-or-ptc-for-your-renewable-energy-project.

Remington, K. 2010 (August 12). Laser-Based Wind Sensor Lets Turbines Grab More Power. *Windpower Engineering*. http://www.windpowerengineering.com/design/electrical/sensors/laser-based-lets-turbines-grab-more-power/.

Robb, Drew. 2010 (December 1). Finding Fault: Improving Wind Farm Availability. *Power*. http://www.powermag.com/business/3218.html.

Schlez, Wolfgang, and Tindal, Andrew. Wind Farm Siting and Layout Design. World Wind Energy Association. http://www.wwindea.org/technology/ch02/en/2_4_2.html.

Stancich, Rikki. 2011 (May 30). Arc Flash Prevention Saves Money and Lives. Wind Energy Update. http://social.windenergyupdate.com/qa/arc-flash-prevention-saves-money-and-lives?utm_source=http%3a%2f%2fcommunicator.firstconf.com%2flz%2f&utm_medium=email&utm_campaign=31+may+weu+ebrief&utm_term=Wind+farm+security&utm_content=605714.

Starr, Scott. 2011 (February). Newer Safety Standards Include Precautions for Turbine Fires. *North American Windpower*. 8 (1): 64, 68, 70.

Susman, Jacob. The "Coming of Age" of Community Wind. Electric Energy Online. http://www.electricenergyonline.com/?page=show_article&mag=61&article=470.

Trabish, Herman K. 2011 (January 21). Flow, Good Vibrations and Other Wind Machine Profit Opportunities. Greentech Media. http://www.greentechmedia.com/articles/read/flow-good-vibrations-and-other-wind-machine-profit-opportunities.

———. 2011 (April 27). Psst—Hey Buddy, Wanna Build a Wind Farm?. Greentech Media. http://www.greentechmedia.com/articles/read/psst-hey-buddy-wanna-build-a-wind-farm/.

———. 2011 (May 23). Financial Tools and Tourism: Some Ideas for US Wind. Greentech Media. http://www.greentechmedia.com/articles/read/three-solutions-for-wind/.

———. 2011 (August 22). Community Wind: Small Now, But Growing. Greentech Media. http://www.greentechmedia.com/articles/read/comunity-wind-small-now-but-growing/.

———. 2010 (October 1). Operations and Maintenance: Keys to Wind Farm Profitability. Greentech Media. http://www.greentechmedia.com/articles/read/operations-and-maintenance-the-key-wind-farm-profitability.

———. 2012 (January 25). Wind Tower Reliability Means Being Proactive in O&M. Greentech Media. http://www.greentechmedia.com/articles/read/Wind-Tower-Reliability-Means-Being-Proactive-in-OM/.

U.S. Department of Energy, Energy Efficiency and Renewable Energy. 2010. Wind Power Today—2010. http://www1.eere.energy.gov/windandhydro/pdfs/47531.pdf.

——— 2005 (May). Wind Energy Myths. DOE/GO-102005-2137. http://www.nrel.gov/docs/fy05osti/37657.pdf.

Webb, Jim. 2011. Interviewed by Rebecca Busby. E-mail message to author. June 28.

Wharton, S., and Lundquist, J. K. 2010 (February 23). Atmospheric Stability Impacts on Power Curves of Tall Wind Turbines—an Analysis of a West Coast North American Wind Farm. LLNL-TR-424425. http://atoc.colorado.edu/~jlundqui/wharton_lundquist2010LLNLTR424425.pdf.

Williams, Andrew. 2011 (July 15). Hedging Against Wind Variability: A Risky Business? *Renewable Energy World*. http://www.renewableenergyworld.com/rea/news/article/2011/07/hedging-against-wind-variability-a-risky-business??cmpid=WNL-Wednesday-July20-2011.

Wilson, Ryan D. 2011 (January 20). Wind Farms Built from the Ground Up. *Clay Center Dispatch*. http://www.zwire.com/site/news.cfm?newsid=20454645&BRD=1160&PAG=461&dept_id=190958&rfi=6.

Wind Energy: The Facts. European Wind Energy Association, http://www.wind-energy-the-facts.org/en/part-i-technology/chapter-4-wind-farm-design/commissioning-operation-and-maintenance.html.

Wind Today. 2010 (July). GE Energy Services Increases Wind Farm Profitability by Reducing Gearbox Maintenance Costs with Condition-Based Maintenance. http://www.windtoday.net/info/articles.html?ID=96752.

Zipp, Kathleen. 2010 (October 3). A Safer Way to Reach High Places. *Windpower Engineering*. http://www.windpowerengineering.com/maintenance/safety/a-safer-way-to-reach-high-places.

7

Wind Power Integration: Striking a New Balance

Introduction

Intermittent . . . variable . . . unpredictable . . . erratic . . . All of these terms and more have branded wind power as an "unreliable" source of energy. Some of the criticism is based on misconceptions about how an electrical power system operates. Other objections are valid but sometimes exaggerated, such as the variable nature of wind power. Still other problems, like forecasting wind energy, are steadily being resolved.

Power System Operation

Frequently, people misunderstand the essence of the electrical power transmission system, which is extremely dynamic. The issue of integrating wind power is often treated in isolation, when the network as a whole should be considered.

The key challenge in managing a power system is that the amount of electricity going out must precisely match the amount coming in, and vice versa. This requires both long-term planning and split-second response on the part of the power system operator.

But the system must cope with planned and unplanned events whether it's carrying wind power or not. The power system's design, controls, and connections with other grids enable the operator to manage it effectively despite constantly changing conditions. The flexibility that has been built into modern power systems can handle tremendous amounts of variability from electrical loads.

Before wind farms became common, the variability of wind energy was virtually indistinguishable from natural fluctuations in consumer demand for power. People switch their appliances, electronic equipment, and heating and cooling units on and off all the time.

In addition to fluctuating demand, variability in the power system arises from transmission line and power plant outages, scheduling slip-ups, and dispatch errors, as well as from variable generating sources such as wind and solar power. Large, dynamic power markets with robust connections to neighboring markets help system operators cope with varying demand and supply.

Bulk power markets

Wholesale, or bulk, power markets are operated by regional, national, or multinational agencies, usually quasi-governmental and regulated. Power system operators allocate capacity on transmission lines, maintain generation reserves, and match electric supply and demand, as noted in chapter 6.

These agencies work closely with operators of individual utility grids and regional transmission systems to ensure that electricity flows smoothly where it's needed, without congestion or overloading of vulnerable transmission lines. This also involves anticipating when and where new generation sources or transmission capacity will be needed.

Operators of power systems in different countries, provinces, or states cooperate with each other over larger geographic areas, forming regional markets or *power pools*. Each grid can be operated more efficiently when resources are shared—that is, when surplus electricity in one area of the market can be sold and delivered to another grid, or when generation resources in one area can come online to supply a neighboring market. Regional pools let system operators see the big picture beyond their own grid network.

Setting prices. The power system operator holds auctions and takes bids from suppliers, then schedules and dispatches the various generating units that supply the system. In large, regional power markets, operators take bids on different time scales, in increments from five minutes ("real time") to a day ahead or more.

In the United States, wind power output is generally bid into short-term markets (15 minutes to an hour) rather than day-ahead. The Midwestern power system in 2011 created a new designation for renewables that allows wind power to participate in the real-time five-minute market as a "dispatchable intermittent" resource. Shorter time frames allow wind energy output to be predicted more accurately. In Europe, cross-border exchanges of electricity among countries are typically scheduled a day ahead, reducing the efficiency of the power market.

The European Wind Energy Association found that intraday scheduling (hourly or shorter intervals) could save \$1–\$3 billion per year (1–2 billion euros) in system operating costs. If wind power producers were allowed to forecast output closer to real time, they could provide more accurate information.

Power market auctions usually have a maximum allowable price cap. Bids are awarded starting with the lowest marginal price. Bids must reflect how much energy and operating reserves the market participant can provide, as well as information about generating unit constraints such as ramping rates (how quickly it can ramp up to full capacity and vice versa) and start-up costs and times.

Wind energy producers can afford to supply power at just about any market price because once a wind farm is up and running, its fuel costs are zero. Operation and maintenance costs per megawatt-hour range from \$10–\$27 (7–20 euros). If transmission capacity is available, wind power displaces more expensive output from conventional generation and can contribute to surplus generation, lowering prices overall. Occasionally, excess wind energy can result in negative market prices, where sellers give away electricity or even pay to deliver it.

Generating units are dispatched by the hour according to a cost ladder (or "stack"), starting at the bottom with the cheapest power, until enough capacity is online to meet demand. The price of electricity from the last unit to be dispatched, typically a gas-fired turbine system, sets the market clearing price for the hour. Wind power and other generators then receive that price regardless of their cost of energy production.

Emergency measures. If power supplies on offer are insufficient, the system operator might raise the price cap, solicit additional bidders, and/or buy power on the spot market, which is typically more expensive. Occasionally, a power system simply doesn't have enough supply coming in, so the operator starts a series of rolling blackouts.

This process shuts off electricity for small groups of customers for a short time, then turns it back on and moves the blackout to another group. Rolling blackouts are an emergency measure intended to prevent an uncontrolled, catastrophic failure of the grid. A less extreme alternative is brownouts, which lower the voltage on the system overall.

Balancing supply and demand

To anticipate demand, the power system operator gets input from electric utilities, which can predict how much power they'll need over the next 24 hours reasonably accurately, within 1%–2%. This allows the system operator to schedule just enough capacity to balance supply and demand.

However, if a utility's prediction is wrong, or if a power plant goes down unexpectedly, or if demand rises suddenly for some reason, the system operator must resolve the imbalance. Demand for electricity is considered a load on the system, so this is called *load balancing*. The system operator must turn up or turn down the output from certain generating units to accommodate changing loads (see "Power supply mix" below).

The power grid's voltage and frequency are system-wide indicators that signify imbalances. To maintain balance, *regulation service* fine-tunes the system's voltage and frequency, keeping them within a narrow, stable range over a period of split seconds to minutes. When demand is being met over a longer time scale, from 10 minutes to several hours, it's called *load following* or ramping.

Load balancing, regulation, voltage control, and load following are all types of *ancillary services* managed by system operators to support their basic service of reliable power delivery. Power traders have to buy ancillary services on the market as part of the delivery package. Figure 7–1 and table 7–1 summarize the time frames and services in power system operations and power markets.

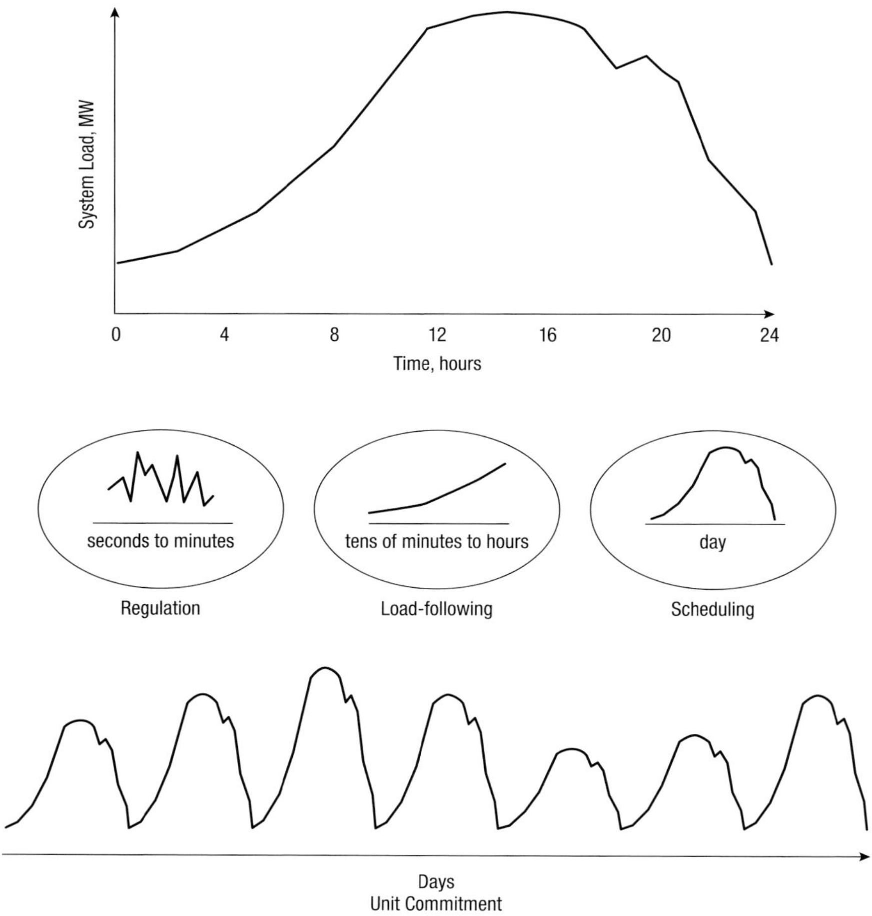

Fig. 7–1. Power system operational time scales

Table 7–1. Power market time horizons

Market	Time frame	System operator use	Generator use
Real time	5–10 min	· Ancillary services	Ancillary services
Intraday	1–24 hr	· Congestion management, economic dispatch, unit commitment	
Day-ahead	12–24 hr	· Day-ahead reserve setting	Wind farm maintenance planning
		· Congestion management, economic dispatch, unit commitment · Transmission line maintenance planning	Wind farm storage coordination
Short term	1–3 days (24–72 hr)	· Congestion management, economic dispatch, unit commitment · Transmission line maintenance planning	Wind farm maintenance planning
Longer term	3–7 days (72–168 hr)	· Congestion management, unit commitment · Transmission line maintenance planning	Conventional generator maintenance

Electricity is more a will o' the wisp than a commodity. It can't be "held" somewhere in a storage tank or reservoir. It is generated, and used, in a second. Transmission system operators are like conductors of the "orchestra" of power supply, bringing in one section of the orchestra and quieting another; or are like controllers of traffic at a busy and complex intersection.

—Global Wind Energy Council

As more wind energy is integrated into power systems, the impact must be considered in terms of time frames—seconds to minutes, minutes to hours, hours to days, and days to weeks. To ensure that enough transmission and generation capacity will be available, planners must address even longer time frames, sometimes up to 30 years.

To ensure reliability, the largest U.S. transmission system operator, PJM Interconnection, holds capacity auctions using a reliability pricing model that commits generation resources three years in advance. An auction held in 2011 for power to be delivered in 2014–2015 procured almost 700 MW of wind power.

Power supply mix

Power systems have a diverse portfolio of generating units. Large *baseload* power plants (around 500–1,000 MW), usually nuclear or coal-fired, provide the bulk of a system's electricity 24/7. They take time to start up, and they don't have much flexibility in their output, regardless of demand. *Hydroelectric power* and natural gas–fueled power plants are much more controllable, so they can be adjusted quickly to meet changing conditions on the grid.

Peakshaving plants deliver power when consumer demand peaks seasonally or daily, usually during the extremes of a summer's heat or winter's cold. These plants are very costly for utilities to maintain and run. Power system operators also keep extra generation capacity on hand that can quickly respond to fluctuating demand (see "Power reserves" below).

All of these power supplies are considered to be *dispatchable*, meaning that the system operator can decide when to turn them on or off and, to some extent, ramp them up or down. Large generating units that require hours or even days to come up to speed are scheduled well in advance, with the system operator assigning a *unit commitment* to that plant.

Lining up too much generation increases costs, but not having enough on hand could jeopardize reliability and also increase costs if power has to be purchased at high market prices or if expensive quick-start units have to be brought online.

An additional power system resource that is becoming more widespread is *demand response* (also called load response). This is a coordinated reduction in demand by a group of consumers, usually industrial and large commercial. Utilities and private companies solicit these customers in advance to agree to turn down certain electrical loads, or turn them off completely, when the grid is running short on supply. The demand reduction is often nicknamed a *negawatt* (negative watt).

Demand-response customers typically get a 10- or 15-minute warning, and their energy-use reduction might last an hour or less. Many of them have their own emergency generation equipment on standby. As compensation, the customers get a break on their electric bills and/or other type of payment.

Demand response contributes to the stability of a utility grid, enhancing the reliability of the power system as a whole. In some regional markets, demand response competes head-to-head with other generation resources. Projects were underway to test whether small-scale demand response can provide frequency regulation service within 4-second intervals on PJM's system.

When used for load-balancing purposes, demand response can also work in reverse, if the provider can absorb extra power and store it—in cold storage facilities, for example. Some foods can tolerate lower temperatures when excess power is used to ramp up the freezer compressors. One U.S. utility, Bonneville Power Administration, plans to use demand response to balance wind power as it ramps up, rather than down.

The UK government, expecting a huge expansion of wind power supply, is planning to strengthen existing utility demand-response programs and possibly let negawatts bid competitively in the UK power market.

U.S. federal regulators require that demand-response resources get the full market price in real-time and day-ahead markets—that is, the same payment as conventional generation resources. The rule applies to power systems that can use demand response for load balancing, provided that it's dispatched cost-effectively.

Figure 7–2 shows the sources and relative cost of power system supplies, including stored electricity (see chapter 8).

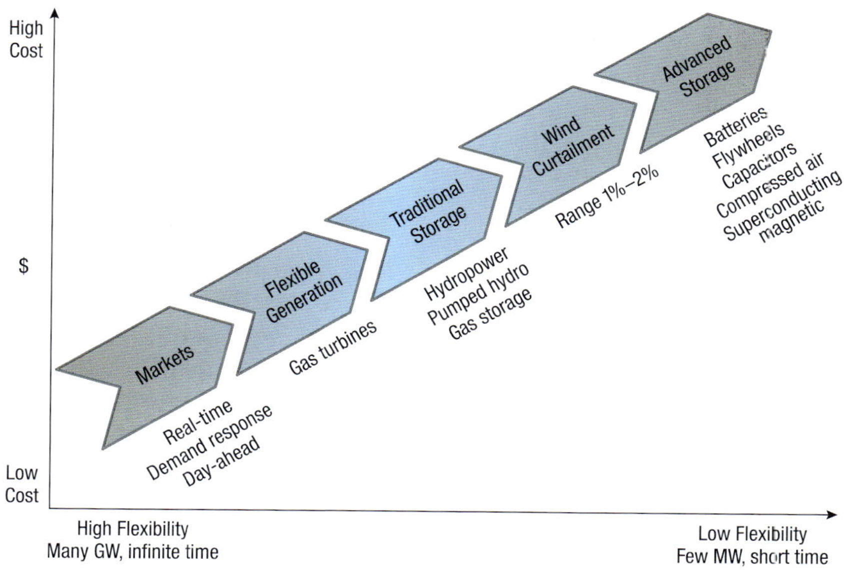

Fig. 7–2. Flexibility of power system supply

Integration Issues

Reliability

Measurements of reliability aren't specific to any single generating unit or resource. Reliability describes the power system as a whole. In case of emergencies, power is supplied by spare generating units throughout the system. Extra capacity is designed into the system, not only to accommodate fluctuating loads but also to manage a transmission line failure or an unexpected generation outage, whether at a conventional plant or a wind farm.

No power plant is 100% reliable. When a central utility station goes down, it does so nearly instantaneously. Wind power usually does not suddenly cut out of the power system in large, well-connected networks. Instead, it takes several minutes or, more likely, hours for the total wind energy output to die off. That's because instead of a single central plant, hundreds or thousands of units are running, spread over hundreds of kilometers or miles (see "Variability" below).

Power reserves

Generation reserves, often called backup power, take the form of *spinning reserves*, which are running online at all times, and *non-spinning reserves*, which can start up quickly when needed, within 10–30 minutes. Non-spinning reserves (also called supplemental reserves) are not used as often as the spinning kind.

Hydropower plants and natural gas–fueled turbines are the most common types of both spinning and non-spinning reserves, because they're quick to start up and flexible to control. Diesel engines can also be used where air quality regulations allow them. Demand response also qualifies as a spinning reserve on some power systems. For example, it has provided up to 16% of spinning capacity in PJM's network.

Spinning reserves (also called operating reserves or contingency reserves) are ready to be ramped up or down nearly instantly to balance power supply and demand. These units are the first line of defense for the power system operator. Spinning reserve generators have local automatic controls, governors, or other devices that constantly adjust the generation output to respond to the power system's voltage, frequency, and other variables.

Spinning reserves get some price premium for being constantly available. They're also expensive for utilities to have on hand. In some cases, they are idling and burning fuel even when they're not actually being used.

Utilities generally are required to maintain certain levels of spinning and non-spinning reserves—typically about 15% of the generating fleet's capacity. But the total amount of reserves depends on the individual utility—how much excess generating capacity it has, how quickly the generating units can be turned on, how much power might be available to purchase from other connected systems, etc.

Here's where a common misperception about wind energy comes into play. Many reports claim that for every unit of wind capacity, multiple units of traditional, dispatchable capacity must be kept in reserve as backup power. In reality, additional reserve requirements depend on the power system's supply mix, load characteristics (peaks and averages), and other factors (see "Impact on services and reserves").

The U.S. Department of Energy estimates that 5 MW of wind power capacity requires less than 0.9 MW of additional balancing reserves, and far less than that in many power systems, especially those with intrahour scheduling.

Wind farm owners in some power systems are required to pay for their own backup power, and natural gas–fueled plants are often installed to supplement large amounts of new wind energy coming into the system. Gas turbine manufacturers are busy promoting their products as solutions to enable wind power integration.

It's true that a wind farm cannot be considered a dispatchable power supply in that it can't be turned up on command—it doesn't have a control knob. However, today's wind turbines, with blade pitch control and power electronics, can limit their ramp rate and power output, and a wind farm can be gradually ramped back up after a curtailment or shutdown. Some advanced wind farms are equipped with automatic generation controls.

Wind power does get a *capacity credit* in some markets, which accounts for how much conventional generation capacity is displaced by wind energy. As noted above, the PJM transmission system procures some firm wind power supply three years in advance. The credit is based on average wind power production, or capacity factor, as defined in chapter 6. Essentially, the capacity credit estimates how much firm wind power will contribute to overall power system reliability.

However, wind is better treated as an energy resource, not a capacity resource. Its output can't be easily matched with demand, and wind energy is not usually coincident with peak loads. But regardless of how much capacity wind power supplies, it saves the fuel that would otherwise be burned by conventional generating plants and avoids the emissions that would be produced (see chapter 12).

Variability

Power generation and transmission systems throughout the developed world are capable of handling large, variable amounts of wind power within reasonable limits. System operators manage the grid as a whole to ensure that it can take the maximum output from wind farms during high winds yet maintain a reliable power supply to consumers during low/no wind periods.

Calling wind "intermittent" is misleading, because wind energy does not stop and start at intervals, which is the definition of intermittency. Some regulations consider any nondispatchable resource as intermittent. At the risk of quibbling with definitions, the term intermittent should not be confused with variable.

Still, the variable nature of wind energy does make wind power tricky to integrate into the power system, or to do so cost-effectively. Wind's natural variability can challenge the system operator's management of ancillary services and reserves. The grid must be resilient enough to absorb unexpectedly large amounts of wind power and robust enough to fill the gap with power from other sources when winds drop off.

A wind turbine's output rises dramatically with wind speed, so a lot of wind power can come in bursts during short periods of strong winds. These are called *ramp events*, which also include sudden falls in wind speed and power generation. (These events should not be confused with ramping rates that apply to all types of power generators.)

Also, wind energy production, at the mercy of the weather, can vary on any time scale, from seconds to years, or even decades if our global climate changes. So utilities and power system operators have to plan more carefully when integrating wind power into their electrical systems.

For example, power consumption falls dramatically during the night, when most residual demand is met by inflexible coal or nuclear plants. If the wind kicks up at several wind farms during the night, the system operator can have problems balancing supply and demand and might have to curtail the wind energy output.

Unlike other generators, which stop burning fuel when they shut down, turbines lose the potential wind energy for good, translating to a financial hit to wind farm operators. In the future, new loads that can use excess nighttime wind energy might alleviate this problem.

Because of imbalances and lack of transmission capacity, curtailment has increased at some U.S. wind farms, nearly tripling on the Midwest power system, from 292 GWh in 2009 to 824 GWh in 2010 (figs. 7–3 and 7–4). On the Texas grid, about 17% of all potential wind energy was curtailed in 2009, but aggressive transmission line build-out helped cut this figure below 8% in 2010.

Geographical distribution. The geographical distribution of wind energy is considered to be an important factor in mitigating wind's natural variability. Theoretically, while a single wind farm's output might swing from one hour to the next by 60% of its capacity, several wind farms producing hundreds of megawatts might vary hourly by less than 20%. If you consider larger areas encompassing even more wind farms, maximum variations of 10% of capacity might be more likely.

However, some industry analysts question whether this smoothing effect actually occurs. Data from the Bonneville Power Administration (BPA) for one week of system operation in the U.S. Northwest and from the United

Kingdom for a two-year period show extreme swings in wind power output, yet these areas are quite large geographically.

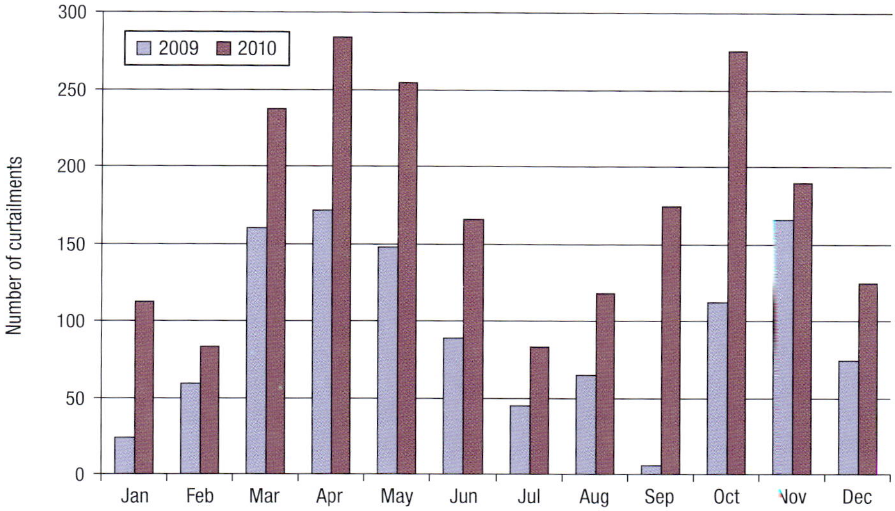

Fig. 7–3. Number of curtailments on U.S. Midwest power system. *Data courtesy of Midwest Independent System Operator (MISO).*

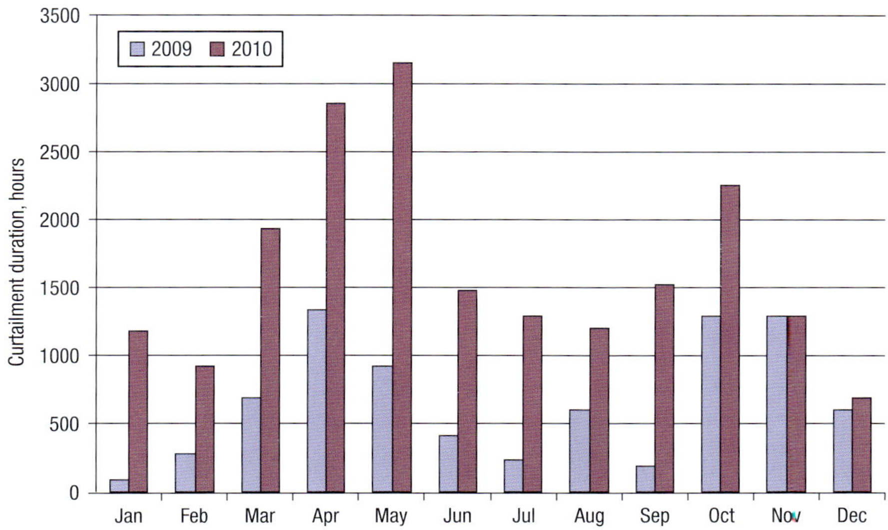

Fig. 7–4. Duration of curtailments on U.S. Midwest power system (total hours for turbines and wind farms feeding into the system). *Data courtesy of MISO.*

The United Kingdom spreads over roughly 250,000 sq km (almost 100,000 sq mi), while BPA's territory is about 800,000 sq km (300,000 sq mi). By comparison, northwestern Europe including the North Sea has an area of about 1.3 million sq km (500,000 sq mi).

BPA's power system has more than 3.5 GW of connected wind capacity, which could rise to 6 GW by 2014. The utility's charges for wind-balancing services total $1.29/kW per month for a 100-MW wind farm (0.94 euros), or $129,000 monthly (94,000 euros). As noted earlier, BPA is exploring demand response as a balancing option. The utility is also planning a new substation and a 61-kilometer (38-mile) transmission line that will carry 840 MW of wind power to an existing substation.

BPA's situation is complicated during spring and early summer, when melting snow increases hydropower production at the same time that strong winds maximize wind power production. The utility can't divert too much water around hydro turbines because of threats to salmon (though this is debated), and diversion wouldn't reduce hydropower enough to solve the problem anyway.

Wind farm curtailment is the BPA's last resort, but more than 30 GWh of wind power reportedly was curtailed in just one week during spring 2011, without compensating wind operators—a decision that federal regulators called discriminatory. An exceptionally wet season created enormous amounts of hydropower, driving not only wind energy out of the market but also fossil fuel generation.

In the broader Pacific Northwest region, oversupply events rose quickly during 2008–2010 but should level off, according to the Northwest Power and Conservation Council (fig. 7–5). Growth in demand for electricity is expected to catch up with increases in wind power and other renewable generation. The timing depends on escalation of state renewable standards in the region and the ability to export wind power to California via new transmission lines.

"I fear one arm of the federal government—the Bonneville Power Administration—really is not acting in the best interest of the administration's policy to try and integrate and build the wind industry. . . . Their role is going to require some stretching, and there will be a little pinch on the way, but we can't afford to lose this opportunity and put the burden on the wind industry itself. We've got to fix this integration issue."

—Representative Earl Blumenauer (Democrat–Oregon)

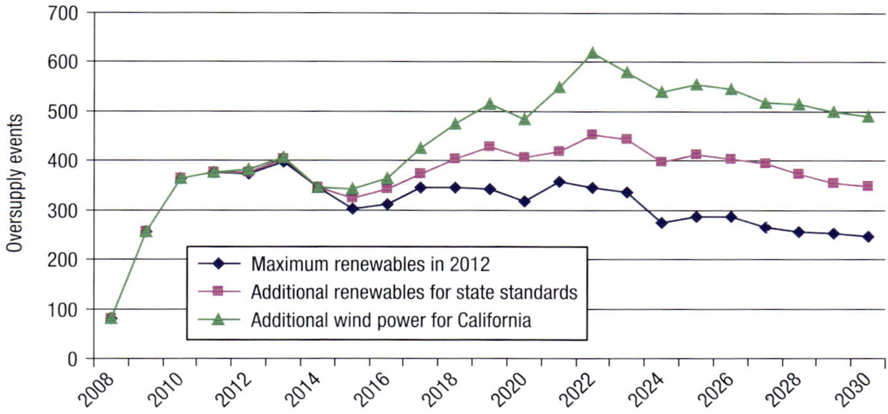

Fig. 7–5. Electricity oversupply in U.S. Pacific Northwest. *Data courtesy of the Northwest Power and Conservation Council.*

Another analysis, which modeled wind power production in Britain, Denmark, and Spain, concluded that the European grid would fail without adequate natural gas supplies. The study began by considering hours of low wind power output and the difference between installed capacity and actual production. It concluded that neither high-voltage connections nor storage and demand response would be capable of balancing power supplies.

Impact on services and reserves. In terms of the impact on ancillary services and spinning/non-spinning reserve requirements, wind power adds little variability to the power system on a minute-to-minute time scale. Large amounts of wind energy can be added with minimal effect on the use of spinning reserves and with relatively little impact on the need for voltage and frequency regulation service.

Regulation is the most expensive ancillary service, and its cost remains high even when electricity demand falls throughout the night. Evidence from the Horns Rev wind farm off Denmark and a smaller wind farm in Quebec shows that wind energy can actually add regulation service to the power system rather than requiring more of it. This is done by equipping the turbines with automatic generation controls.

On an hour-to-hour time scale, modest amounts of wind power hardly affect an entire power system's overall variability. However, greater amounts of wind energy would require adding non-spinning reserves to accommodate hourly changes in wind supply and to maintain the power system's load-following capability. This requirement doesn't add much to the cost of integrating wind power, however, because load-following service does not command much of a price premium in large power markets.

Wind's share of electricity supply

The size of an electricity market and the extent of the load-balancing area play very important roles in the ease and cost of integrating wind power. Larger power systems have more balancing resources, a more diverse generation portfolio, more interconnections for importing and exporting electricity, and a greater variety of financial mechanisms for reducing integration costs. A liquid, real-time market is the most economical way to balance wind power.

Global experience indicates that generally, wind energy can provide around 20% of the power supply within a given regional market or power pool without significant problems. The electrical system's existing controls and reserves, designed to deal with constantly changing demand and supply, can typically handle the additional variability from wind power at that level.

Some systems can cope with much higher amounts of wind power, while others would have trouble with less. It all depends on the flexibility of the particular power system, in terms of its ability to respond to fluctuating loads. Hydroelectric power and natural gas–fueled generating units tend to enhance a power system's flexibility.

Penetration levels, which refer to the percentage of the market supplied by wind power, can be measured in three ways:

- *Capacity penetration* is the ratio of wind farm capacity to the power system's peak load. In a system with a 1,000-MW peak load and a 300-MW wind farm, the capacity penetration is 30% (0.3).
- *Energy penetration* is the ratio of the amount of wind energy delivered to the total amount. If a power system delivers 1,000 MWh of energy and 200 MWh of that comes from wind, the energy penetration is 20% (0.2).
- *Instantaneous penetration* is the ratio of wind farm output to total power system load at a specific point in time or over a short period.

Researchers have analyzed the extra cost of increasing the penetration of wind power in utility grid networks. European studies show that up to the 20% level, existing reserves and mechanisms can correct unpredicted imbalances that might result from wind power variations. The extra balancing costs ranged from negligible up to $4 (3 euros) per megawatt-hour. These costs add less than 10% to the cost of wind energy generation, and the effect of wind power's component on electricity prices to the consumer is close to zero.

U.S. research concluded that wind power can meet up to 40% of a power system's peak load at additional costs per megawatt-hour below $10 (7 euros) and often below $5 (4 euros), without risking the system's reliability.

Larger load-balancing areas, better forecasting, and subhourly scheduling can ease integration and minimize costs, not only for wind power penetration but also for overall system operation. Figures 7–6 through 7–8 show savings and operational benefits for the western U.S. Table 7–2 summarizes wind integration costs from selected U.S. studies, where costs vary depending on the power generation mix and ancillary service requirements.

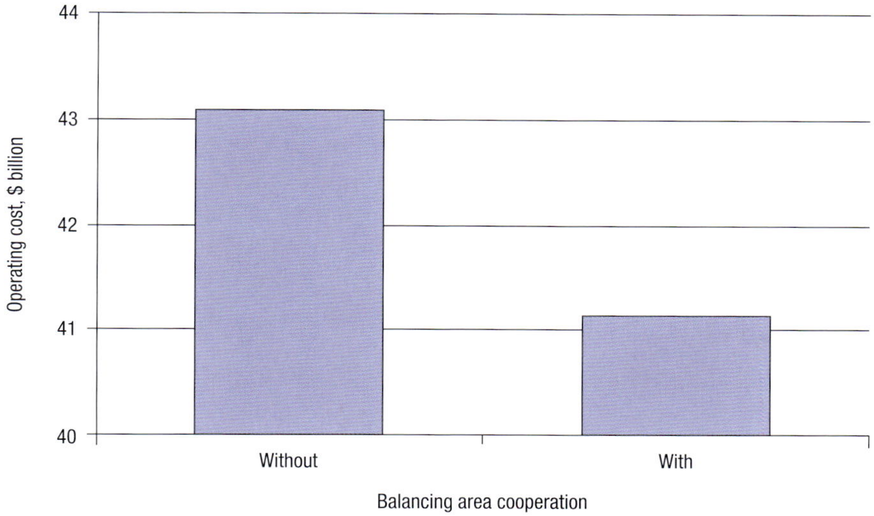

Fig. 7–6. Savings from larger load-balancing areas (five regions vs. 106 zones at 10% wind power penetration). *Data courtesy of the U.S. National Renewable Energy Laboratory (NREL).*

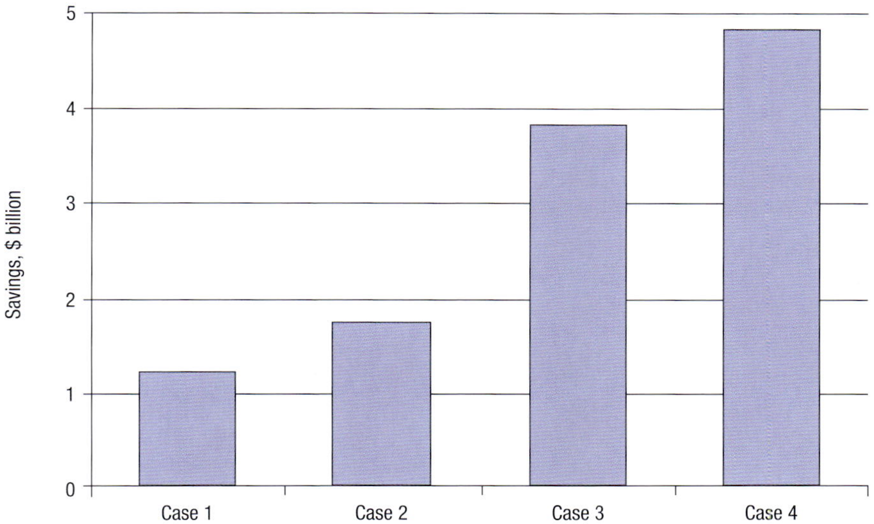

Fig. 7–7. Savings from advanced forecasting (state-of-the-art day-ahead forecast at wind power penetrations of 10%–30%). *Data courtesy of NREL.*

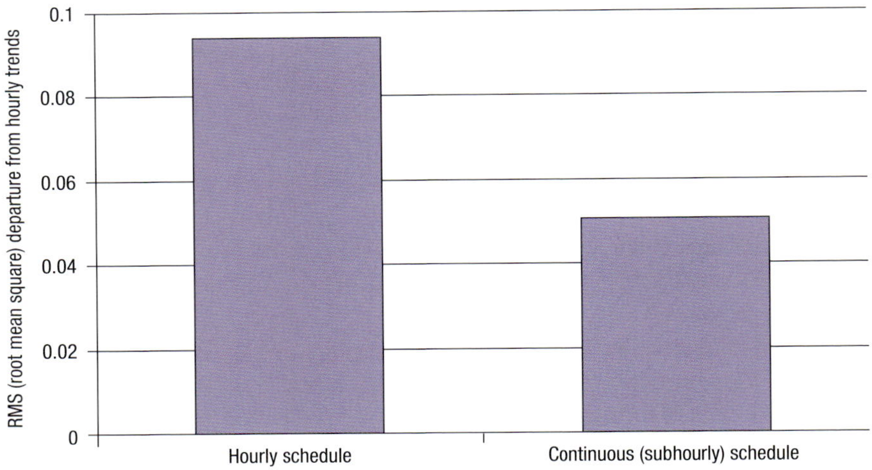

Fig. 7–8. Reduction in fast maneuvering of gas generators (at 30% wind power penetration). *Data courtesy of NREL.*

Table 7–2. Wind energy integration costs. *Data courtesy of NREL.*

Wind capacity penetration, %	Integration cost, $/MWh			
	Regulation	Load following	Unit commitment	Total
48	–	–	1.61	4.54
36	0.22	1.14	–	5.70
31	–	–	–	4.41
30	1.43	4.40	3.00	8.84
29	1.02	0.15	1.75	2.92
20	–	–	–	7.92
20	–	–	–	8.56
18	–	1.10	4.00	5.10
15	0.37	2.65	1.06	4.08
15	0.23	–	4.37	4.60
12	–	–	–	6.94
11	0	1.48	3.16	4.64
4	0.45	trace	trace	0.45
3.5	0	0.41	1.44	1.85

In a large, regional transmission system (100 GW), an extra 50 GW of wind could be integrated comfortably with an additional 1 GW of regulation service to balance the wind power, according to the Brattle Group. This is a fairly modest level, although smaller power systems with relatively higher penetrations of wind power would need to add more regulation service. The New York State's power system operator estimates that 20 MW of additional regulation service could support the adoption of 4 GW of wind power.

Many European countries and some U.S. states and cities have accepted high levels of wind power into their systems successfully. According to the European Wind Energy Association, Denmark got about 24% of its electricity from wind power in 2010, followed by Portugal at 14.8% and Spain at 14.4% (fig. 7–9). Ireland got just over 10%, and Germany approached that level with 9.3%. Smaller areas of Europe have accommodated much higher levels of integration.

During March 2011, wind energy was Spain's biggest source of electricity, meeting 21% of demand (fig. 7–10). The country's transmission operator, Red Eléctrica, created a separate power control center for renewable energy, allowing real-time monitoring and control of wind power and other renewable sources around the country.

In Portugal, the instantaneous penetration of wind energy in the power system reached 70% during the winter of 2009. No technical problems were reported, but Portugal's wind power capacity was less than 3.7 GW, indicating that grid management strategies should be developed before much more wind capacity can be installed.

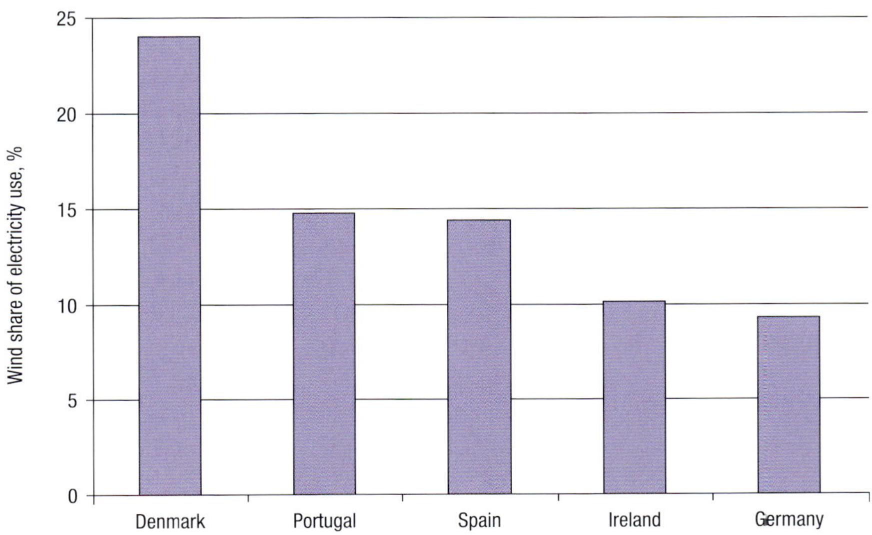

Fig. 7–9. Highest wind energy penetrations in Europe. *Data courtesy of the European Wind Energy Association.*

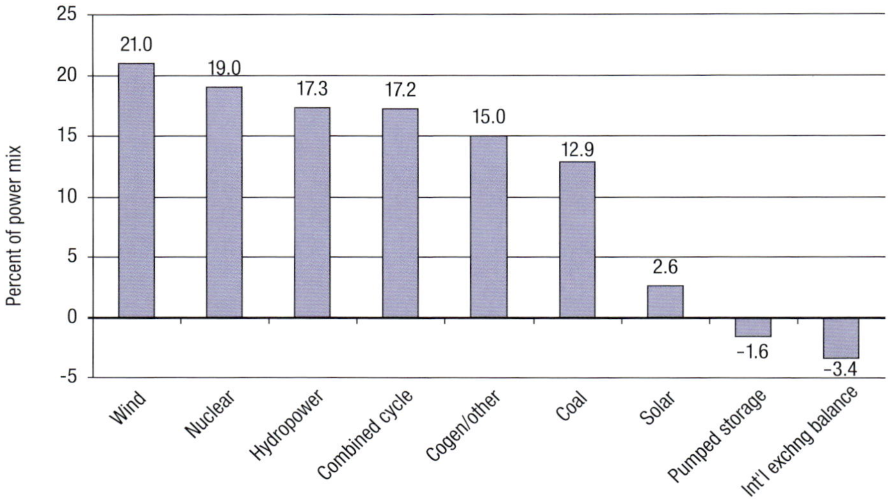

Fig. 7–10. Spain's power mix (March 2011). *Data courtesy of Red Eléctrica de España S.A.*

One study analyzed the cost of wind power supplying 45% of Ireland's power mix by 2020 and found that wholesale electricity costs could drop by 11.5%, resulting in consumer savings of $140 million (100 million euros). The analysis accounted for the cost of additional transmission lines. Wind energy was expected to reduce wholesale prices mainly by displacing coal- and gas-fired generation. If these fossil-fuel prices escalate, savings would increase.

In China, wind power contributed 21.1% to local power consumption in eastern Inner Mongolia Province, 8.7% in the western area of the province, and 4.6%–5.6% in two other provinces.

In America, the national goal is for wind energy to supply 20% of electricity demand by 2030. This would mean integrating some 300 GW of wind capacity into the nation's power systems, requiring an estimated 50 GW of new flexible generation, most of it likely fueled by natural gas.

The state of Iowa got more than 15% of its electricity from wind power in 2010, with North Dakota second at 12% and Minnesota at almost 10%. South Dakota has enough wind power capacity to generate about 23% of the state's electricity in an average year. By this metric, Iowa has 17%, North Dakota almost 14%, and Minnesota more than 12%.

Nationwide, America got 2.3% of its electricity from wind, up from 1.8% in 2009. The Los Angeles metro area received 20% of its electricity from renewable sources in 2010, of which 47% was wind power—more than 9% overall.

System operators in New York State and the New England region are confident that their networks can absorb substantial amounts of new wind power. California's system operator studied wind energy integration issues and concluded that the grid could handle 20% of its power supply from wind and other renewable resources without unmanageable problems or reliability impairment. The study found that wind energy and solar power can complement each other's output in terms of daily and nightly variations.

To manage greater amounts of renewable generation (California's mandate is 33% by 2020), the grid operator developed a new control center with advanced weather forecasting and automated electronic auctions. California's operator is also introducing software to make the power system more flexible by allowing generators to bid into a market for ramping services.

The California power system's nodal market, which replaced a three-zone market, sets real-time and day-ahead prices every five minutes, taking bids from all forms of generation at 3,000 *nodes* (central points of transmission). The center's visualization screen shows a real-time snapshot of the power system, focusing on management of supply and demand.

Isolated problems with wind power have attracted lots of publicity and condemnation from antiwind activists, while wind farm curtailments anger operators and squander potential energy. One oft-cited example of a wind energy problem occurred in Texas in winter 2008. About 1,200–1,400 MW of wind power capacity was lost unexpectedly, while other generating units were also out due to extremely cold weather. This prompted the system operator (Electric Reliability Council of Texas, or Ercot) to temporarily cut off large interruptible customers.

Although widespread rolling blackouts were avoided, the incident prompted much public debate about why it happened, why it wasn't predicted, and how to avoid it in the future. As it turned out, the drop in wind energy had been predicted, but the system operator wasn't aware of it because forecasting hadn't yet been integrated into their procedures.

Since then, Ercot has installed a hub-height network of real-time wind data sensors covering nearly all of the state's wind farms. In mid-2011 during a record-breaking heat wave, wind power outperformed expectations, at one point generating 131% of Ercot's planned amount.

Forecasting

Predicting a wind farm's energy output accurately is key to integrating ever-larger amounts of wind power into transmission systems at the lowest possible cost. Forecasting helps power system operators meet dynamic loads by juggling their generating resources, both conventional and variable, while delivering electricity reliably.

In an extensive survey of grid operators worldwide, 94% said that integrating a significant amount of wind will largely depend on the accuracy of wind-power forecasts. The survey, by Alstom Grid for the U.S. DOE, interviewed 33 operators in 18 countries who, together, are responsible for integrating 72% of the world's wind power.

Obviously, wind power forecasts—just like weather forecasts—are more accurate in the short term than farther out. Today's methods of forecasting short-term wind power have become quite reliable for a collection of wind farms covering a large geographical area—such as a regional power market or pool—and wind energy supply is becoming more predictable.

Seamless integration of wind farm output forecasts, power market operations, and utility control room operations is crucial to accommodate large amounts of wind power while maintaining the reliability of the grid network. Tools that present wind forecasts in a useful, meaningful way that supports decision making are essential for operators to integrate wind energy successfully, according to the international survey noted above.

System operator applications. Data from wind farm monitoring are used to forecast how much electricity the wind farm will generate within the next minute, hour, or day for purposes of integrating wind power into the utility grid.

Forecasting allows power system operators to anticipate the energy output of wind farms and adjust the electrical output of other generating plants on their system to balance the amount of wind power coming in. More accurate forecasts of wind production would let operators make better decisions about real-time operations, day-ahead markets, and the amount of power to be committed from other generators.

For example, shutting down and re-starting a baseload power plant is no trivial task. It's an expensive process that can take hours or even days, and a shutdown can jeopardize system reliability if it's timed wrong. Also, keeping a backup plant idling wastes fuel and emits pollution unnecessarily. When operators have accurate information about upcoming wind power supply, they can schedule the commitment of the other generators in their power system more efficiently and cost-effectively.

Although system operators already use short-term wind forecasts (hourly or subhourly), day-ahead wind forecasts are more commonly used in reliability planning but not in power market decisions. Longer term wind forecasts could help reduce costs and improve the efficiency of market operations and dispatching of generating units, even when the longer term forecasts are less accurate than short-term ones.

For example, if 5 GW of wind energy is forecast, a 20% error would translate to 4–6 GW, so scheduled backup power could be 1 GW too high or too low. But without any forecast at all, 6 GW of backup power would have to be scheduled to ensure grid reliability.

Even with more accurate wind power predictions, however, forecasting tools and methods must be integrated into the power system's operating procedures and supporting software, as noted above in the Texas case.

Wind farm operator applications. Continuous monitoring of wind conditions is an ongoing job at commercial wind farms. When finding a site, developers are concerned with long-term average wind characteristics, as described in chapter 6. But when the turbines are up and running, the wind farm operator focuses on specific, short-term forecasts of weather and energy output, which translates to revenues.

Forecasting also provides wind farm operators with a tool to schedule turbine maintenance and wind farm outages. Servicing the machines during times of low wind saves money by avoiding maintenance during windier periods when the turbines could be producing more energy.

Many wind farm operators hire a contractor to forecast site weather conditions and the output of the machines. One American company, AWS Truepower, sells a service that forecasts one-hour-ahead wind farm output to within 4% of the site's rated capacity. Prediction errors for day-ahead forecasts range from 14% to 22%.

Erroneous forecasts, however, can erode the operator's profits. Depending on how bids are placed into the power system, penalties might be charged when a wind farm produces less energy than promised, or if it generates more than expected, the operator might not be paid in full for the additional amount.

Numerical weather prediction models. Computer models for forecasting weather and wind energy output are an increasingly common tool used by utilities and power system operators to schedule generating units. Advanced weather forecasting techniques can also warn wind farm operators if extreme events such as thunderstorms are expected. Windy storms are responsible for many unexpected surges in wind power that force system operators to curtail it.

Weather forecasting models are usually applied to large geographical areas to estimate the aggregate wind farm output, but many models can be used to forecast conditions at individual wind farms as well.

Numerical weather prediction is the same technology that meteorologists use to forecast weather for the public. Meteorologists model the atmosphere as a three-dimensional grid of large cells, typically measured in kilometers. Computer input includes pressure, temperature, and humidity data from sensors and satellites. The software incorporates equations that simulate the atmosphere's behavior and predict what will happen.

Traditional numerical models designed for weather prediction, however, are too large-scale to characterize wind accurately. To forecast wind speed, the model must account for the topography or terrain of the area in question. Increasing the resolution of the 3-D grid by breaking it up into smaller cells is one way to improve prediction of near-surface wind speed and other weather conditions. Higher resolution enables the software to analyze smaller topographical features.

Using sophisticated weather forecasts and statistical analysis, today's models can predict wind power generation at 5-minute to 1-hour intervals up to three days ahead. For a regional collection of wind farms, these forecasts are accurate to within 5% for 1–4 hours ahead and on the order of 10% for a day ahead. Even for a single wind farm looking 36 hours ahead, forecast errors now range from 10% to 20% of the energy output.

Developing better forecasting technology depends on acquiring actual data from turbines in commercial operation, allowing researchers to compare field data with computer model results. For example, wind forecasters need data on turbine availability, curtailment, or other reasons for shutdown to improve forecast accuracy. Wind farm operators often don't want to share this sensitive data, but power system operators have begun to require it in some areas.

Regional forecasting via numerical techniques was pioneered by Risø DTU, Denmark's National Laboratory for Sustainable Energy, which developed WAsP (Wind Atlas Analysis and Application Program). This same program was used to map Europe's wind resources, as described in chapter 5.

The WAsP software starts with national weather forecasts up to 36 hours ahead, which indicate wind flow high in the atmosphere. Then the program characterizes near-surface local conditions using a complex terrain flow model, a roughness change model, and a model for sheltering obstacles. This local information enables the software to estimate wind speeds and energy output for multiple wind farms or single sites. Elements of the program can even account for wake effects among wind turbines.

The U.S. National Center for Atmospheric Research is working with utility Xcel Energy on a nested-grid model design that also achieves high resolution. Instead of modeling all the 3-D cells in the same size, the program overlaps grids of three different sizes. Also, the U.S. Department of Energy and the National Oceanic and Atmospheric Administration are working with private partners on a program to improve wind forecasting that was launched in 2011.

Lawrence Livermore National Laboratory is working on a wind forecasting model that focuses on ramping events. Using data from a West Coast wind farm, the model is drilling down in resolution to the turbulence scale, just 10 meters (30 feet), to characterize small areas of swirling, chaotic winds.

References

Airlie, Catherine, and Lundgren, Kari. 2011 (March 2). UK Government Ready to Pay Power Users to Switch off in 'Negawatt' Plan. Bloomberg. http://www.bloomberg.com/news/2011-03-02/u-k-ready-to-pay-power-users-to-switch-off-in-negawatt-plan.html.

American Wind Energy Association. 2010. U.S. Wind Industry Annual Market Report. http://www.awea.org/documents/reports/Annual_Market_Report_Press_Release_Teaser.pdf.

American Wind Energy Association. Wind Power—Clean AND Reliable. http://www.awea.org/documents/factsheets/Reliability_Factsheet.pdf.

American Wind Energy Association. 20% Wind Energy by 2030: Wind, Backup Power, and Emissions. http://www.awea.org/documents/factsheets/Backup_Power.pdf.

California Independent System Operator. 2010 (August 31). Integration of Renewable Resources: Operational Requirements and Generation Fleet Capability at 20% RPS. http://www.caiso.com/2804/2804d036401f0.pdf.

Castano, Ivan. 2011 (April 7). China Installing Wind-power Capacity as Fast as It Can. *Renewable Energy World*. http://www.renewableenergyworld.com/rea/news/article/2011/04/china-installing-wind-power-capacity-as-fast-as-it-can?cmpid=WindNL-Thursday-April21-2011.

Corbus, Dave, et al. 2010 (November). Oahu Wind Integration and Transmission Study. NREL/TP-5500-48632. http://www.nrel.gov/wind/systemsintegration/pdfs/2010/owits_summary_report.pdf.

Economist. 2010 (March 4). Smarting from the Wind. http://www.economist.com/node/15582251.

European Wind Energy Association. 2011 (February). Wind in Power. http://www.ewea.org/fileadmin/ewea_documents/documents/statistics/EWEA_Annual_Statistics_2010.pdf.

European Wind Energy Association. 2011. Fact Sheets: Electricity Markets. http://www.ewea.org/index.php?id=1611.

European Wind Energy Association. 2009 (February). Integrating Wind: Developing Europe's Power Market for the Large-Scale Integration of Wind Power. http://www.trade-wind.eu/fileadmin/documents/publications/Final_Report.pdf.

Fairley, Peter. 2008 (April 17). Scheduling Wind Power: Better Wind Forecasts Could Prevent Blackouts and Reduce Pollution. *MIT Technology Review*. http://www.technologyreview.com/energy/20646.

Global Wind Energy Council. 2011 (March). Global Wind Report: Annual Market Update 2010. http://www.gwec.net/fileadmin/documents/Publications/Global_Wind_2007_report/GWEC%20Global%20Wind%20Report%202010%20low%20res.pdf.

Global Wind Energy Council, Greenpeace International. 2010 (October). Global Wind Energy Outlook 2010. http://www.gwec.net/fileadmin/documents/Publications/GWEO%202010%20final.pdf.

Hagens, Nate. 2010 (September 18). The Fake Fire Brigade—How Can Renewable Sources Support Our Current Energy Delivery Expectations? *Oil Drum*. http://www.theoildrum.com/node/6957.

Heller, Arnie. 2010 (April). Extracting More Power from the Wind. Lawrence Livermore National Laboratory, Science and Technology Review. https://str.llnl.gov/AprMay10/mirocha.html.

International Energy Agency. 2010 (July). IEA Wind Energy 2009 Annual Report. http://www.ieawind.org/AnnualReports_PDF/2009/2009AR_92210.pdf (ISBN-0-9786383-4-4).

Irish Wind Energy Association. 2011 (March 1). Report Launched on Impact of Wind on Pricing within the Single Electricity Market. http://www.iwea.com/index.cfm?page=viewnews&id=73.

Johnson, David. 2010 (May 11). NCAR's Xcel Energy Project Overview. http://ral.ucar.edu/projects/wind_energy_workshop/presentations/Xcel_Energy_Project_Overview_Johnson_5.pdf.

Jones, Lawrence E. 2011 (December 19). Strategies and Decision Support Systems for Integrating Variable Energy Resources in Control Centers for Reliable Grid Operations: Global Best Practices, Examples of Excellence and Lessons Learned. Alstom Grid Inc. and U.S. Department of Energy. http://www1.eere.energy.gov/wind/pdfs/doe_wind_integration_report.pdf.

Kanellos, Michael. 2011 (May 19). Tornadoes, Otter Pops and Other Unlikely Energy Technologies. Greentech Media. http://www.greentechmedia.com/articles/read/tornadoes-pot-pies-and-other-unlikely-energy-technologies/.

Kirby, Brendan, and Milligan, Michael. 2009 (July). Capacity Requirements to Support Inter-Balancing Area Wind Delivery. NREL/TP-550-46274. http://www.nrel.gov/docs/fy09osti/46274.pdf.

Kitasei, Saya. 2011 (February 17). Powering the Low Carbon Economy: The Role of Natural Gas in a Shifting Power Industry. Worldwatch Institute. Saya Kitasei 2011-02-17_Flexible_Power_Symposium_(SSK)[1].pdf.

Li, Junfeng, et al. 2010 (October). China Wind Power Outlook 2010. http://www.greenpeace.org/raw/content/eastasia/press/reports/wind-power-report-english-2010.pdf.

Melnyk, Markian M.W., and Andersen, Robert M. 2009. *Offshore Power: Building Renewable Energy Projects in U.S. Waters*. Tulsa: PennWell. (ISBN 978-1-59370-211-3).

Milligan, Michael, and Kirby, Brendan. 2010 (July). Utilizing Load Response for Wind and Solar Integration and Power System Reliability. NREL/CP-550-48247. http://www.nrel.gov/docs/fy10osti/48247.pdf.

———. 2009 (July). Calculating Wind Integration Costs: Separating Wind Energy Value from Integration Cost Impacts. NREL/TP-550-46275. http://www.nrel.gov/docs/fy09osti/46275.pdf.

Monteiro, C., et al. 2009 (November 6). Wind Power Forecasting: State-of-the-Art 2009. INESC Porto and Argonne National Laboratory. ANL/DIS-10-1. http://www.dis.anl.gov/pubs/65613.pdf.

Muckosy, Pamela. 2010 (June 9). True Cost of Wind Turbine Operations and Maintenance. *Renewable Energy World*. http://www.renewableenergyworld.com/rea/partner/first-conferences/news/article/2010/06/true-cost-of-wind-turbine-operations-maintenance.

National Conference of State Legislatures. 2009. Integrating Wind Power into the Electric Grid. http://www.ncsl.org/documents/energy/Wind_Integration.pdf.

National Renewable Energy Laboratory and GE Energy. 2010 (May). Western Wind and Solar Integration Study. NREL/SR 550-47434. http://www.nrel.gov/wind/systemsintegration/pdfs/2010/wwsis_final_report.pdf.

North American Electric Reliability Corp. 2009 (April). Accommodating High Levels of Variable Generation. http://www.nerc.com/files/IVGTF_Report_041609.pdf.

Northwest Power and Conservation Council, http://www.nwcouncil.org/.

———. 2011 (August 17). The Effects of an Increasing Surplus of Energy Generating Capability in the Pacific Northwest. http://www.nwcouncil.org/library/2011-09.pdf.

Palmer, Adrian. Delivering Power to Meet Britain's 2020 Targets. *Utility Automation Engineering*. http://www.elp.com/index/display/article-display/4283246046/articles/utility-automation-engineering-td/volume-16/issue-2/features/eye-on-europe-feature-delivering-power-to-meet-britains-2020-targets.html.

Parsons, Brian, et al. 2006 (July). Grid Impacts of Wind Power Variability: Recent Assessments from a Variety of Utilities in the United States. NREL/CP-500-39955. http://www.nrel.gov/docs/fy06osti/39955.pdf.

Porter, K., and Rogers, J. 2010 (April). Status of Centralized Wind Power Forecasting in North America. NREL/SR-550-47853. http://www.nrel.gov/wind/pdfs/47853.pdf.

Potomac Economics. 2010. 2009 State of the Market Report for the Midwest ISO. http://www.potomaceconomics.com/uploads/midwest_documents/2009_State_of_the_Market_Report.pdf.

Rowan, Tony. 2011 (August 24). July 2011 Wind Curtailments. Midwest Independent System Operator. https://www.midwestiso.org/Library/Repository/Meeting%20Material/Stakeholder/RSC/2011/20110823-24/20110823-24%20RSC%20Item%2014b%20Wind%20Curtailments.pdf.

St. John, Jeff. 2011 (November 23). Viridity, Enbala Try Negawatts to Balance Pennsylvania's Grid. Greentech Media. http://www.greentechmedia.com/articles/read/viridity-enbala-try-negawatts-to-balance-pennsylvanias-grid/.

Totaro, Philip. 2011 (February 18). Five Trends to Watch in Wind. Greentech Media. http://www.greentechmedia.com/articles/read/guest-post-five-trends-to-watch-in-wind.

Trabish, Herman K. 2011 (March 24). Why Aren't Those Wind Turbines Turning? Greentech Media. http://www.greentechmedia.com/articles/read/why-arent-those-wind-turbines-turning/.

———. 2011 (April 11). At 35 Percent of New Capacity, Wind Moves from Alternative to Mainstream. Greentech Media. http://www.greentechmedia.com/articles/read/is-wind-still-an-alternative-energy/.

———. 2011 (May 18). How Will the California System Operator Cope with 33% Renewables? Greentech Media. http://www.greentechmedia.com/articles/read/how-will-the-california-system-operator-cope-with-33-renewables/.

———. 2011 (May 24). Smackdown: Wind vs. Washington State Grid Operator over Renewable Integration. Greentech Media. http://www.greentechmedia.com/articles/read/smackdown-wind-vs-washington-state-grid-operator-over-renewable-integration/.

———. 2011 (August 18). How Wind Helps Coal Plants. Greentech Media. http://www.greentechmedia.com/articles/read/how-wind-helps-coal-plants/.

———. 2011 (December 13). Grid Integration, Part 2: The Rubik's Cube of Renewable Energy. Greentech Media. http://www.greentechmedia.com/articles/read/grid-integration-part-2-the-rubiks-cube-of-renewable-energy/.

Tuohy, Aidan. 2011. Integrating Wind Generation in the Bulk Electricity System. Electric Power Research Institute. http://mydocs.epri.com/docs/CorporateDocuments/Newsroom/WashingtonUpdate/Tuohy.pdf.

Tweed, Katherine. 2011 (March 17). Demand Response, Rejoice! Negawatts Put on Equal Footing with Generation. Greentech Media. http://www.greentechmedia.com/articles/read/demand-response-rejoice-negawatts-put-on-equal-footing-with-generation.

———. 2011 (March 14). Demand Response and Renewables: Too Good to Be True? Greentech Media. http://www.greentechmedia.com/articles/read/demand-response-and-renewables-too-good-to-be-true.

U.S. Department of Energy, Energy Efficiency and Renewable Energy. 2010. Wind Power Today—2010. http://www1.eere.energy.gov/windandhydro/pdfs/47531.pdf.

———. 2008 (July). 20% Wind Energy by 2030: Increasing Wind Energy's Contribution to U.S. Electricity Supply. DOE/GP-102008-2567. http://www1.eere.energy.gov/wind/pdfs/41869.pdf.

U.S. National Renewable Energy Laboratory, Wind Powering America. 2005 (May). Wind Energy Myths. DOE/GO-102005-2137. http://www.nrel.gov/docs/fy05osti/37657.pdf.

Utility Wind Integration Group. 2010 (August). Frequently Asked Questions About Utility-Scale Wind Power. http://www.uwig.org/UWIG-WINDFAQ.pdf.

Western Electricity Coordinating Council, http://www.wecc.biz/.

Wiser, Ryan, and Bolinger, Mark. 2010 (August). 2009 Wind Technologies Market Report. http://www1.eere.energy.gov/windandhydro/pdfs/2009_wind_technologies_market_report.pdf.

———. 2011 (August). 2010 Wind Technologies Market Report. http://www1.eere.energy.gov/windandhydro/pdfs/51783.pdf.

Yuanyuan, Liu. 2011 (April 26). China's Wind Power Sector Expected To See Tremendous Growth. Renewable Energy World. http://www.renewableenergyworld.com/rea/news/article/2011/04/chinas-wind-power-sector-expected-to-see-tremendous-growth??cmpid=WNL-Friday-April29-2011.

Zervos, Arthouros. 2009 (March 19). Large Scale Integration of Wind Energy into Electricity Grids. Global World Energy Council. http://www.gwec.net/index.php?id=146.

8

Transmission and Energy Storage: Facilitating Wind Power Integration

As we've seen in chapter 7, the challenges involved in integrating wind power into electrical grid systems around the world are real, and the process is not without cost. Global power networks must be extended to reach remote wind farms, and transmission lines must be beefed up to handle more wind power without becoming unstable or congested. However, most industry experts agree that transmission system upgrades are overdue, even without wind energy supplies coming online.

Technologies to store large amounts of electricity until it's needed could facilitate the integration of wind power and other renewable energy sources. However, economical storage is still out of reach at the utility grid scale, except where rivers are dammed or water is pumped to store hydroelectric power.

Transmission

Introduction

A high-voltage electric transmission line is often, literally, the missing link that prevents wind power from being developed. Siting and paying for new transmission lines are key barriers to expanding power systems and accommodating wind energy in developed countries.

The best wind resources are often found far away from where electricity is needed most, in densely populated cities (called "load centers" in industry jargon). When choosing a place to live, people tend to avoid really windy spots, so many wind farms need a new transmission line to connect them to the power system via the nearest utility.

Some excellent wind resource sites have been bypassed in favor of less windy locations because of transmission constraints—either a connecting line can't be built, or existing lines are already operating at full capacity, possibly dedicated to a large coal or nuclear plant. In industrial countries, most power lines that were built decades ago are sorely in need of capacity upgrades (reinforcements) just to improve reliability and handle growing demand, never mind adding wind power.

Wind developers used to look for sites with good wind, but now they have to look for sites with wind, that also have access to good electrical transmission lines.

—Gary Moland, GL Garrad Hassan

But building transmission capacity is enormously expensive these days. In America, new transmission lines on the West Coast are estimated to cost more than $10 million/mile (5 million euros/km). Also, it takes longer to build a transmission line to connect new wind energy than it does to build the wind farm itself. High-voltage transmission lines can take anywhere from 3 to 10 years to acquire rights-of-way and finish construction, versus 1 to 5 years for a wind farm, depending on its site, size, and other factors.

Part of the reason for this lengthy process is that regulatory approval procedures are fragmented and time-consuming, especially in the United States. In other parts of the world, governments and regulators work with utilities and industry toward common goals. The U.S. transmission development process could improve with regulatory changes (see "Sharing the cost" below).

Transmission planners must consider communities' concerns and their potential stance when siting new high-voltage lines, since opposition from people along the proposed right-of-way will delay the project. The familiar not-in-my-backyard argument ("nimby") has expanded to build-absolutely-nothing-anywhere-near-anything ("banana") and even in some cases to nowhere-on-planet-earth ("nope").

One controversy in siting new transmission capacity to serve wind power and other renewable resources is the fact that any power line carries whatever mix of electricity happens to be passing through. Opponents say this means that the transmission line isn't "green" and could transport fossil fuel–derived power.

This argument is based on the misconception that electricity can be sourced and isolated. In reality, there are no green electrons—power flows throughout the system regardless of its origin or destination.

Electrons don't have colors. If you don't have [the transmission line], it can't accommodate renewables. If you have it, it will carry everything.

—Yakout Mansour, California Independent System Operator

Another objection to building new transmission is that other, less costly alternatives might not always be explored thoroughly first. Distributed power, more efficient dispatching, and other improvements should be examined before turning to expensive construction projects that get government incentives and have the advantage of being rate-based (costs are spread throughout the utility's ratepayers). Some consumers call this a "high-voltage gravy train."

Building and upgrading transmission lines could boost employment in some skilled jobs, and utilities are getting better at communicating this message to people who would be affected by a transmission project. However, other jobs might be lost if wind power displaces substantial amounts of fossil-fuel generation in the power mix (see chapter 11).

Access to transmission

Interconnection. In addition to complying with grid codes and other utility requirements, wind farm planners have to apply early for a position in the interconnection queue (projects waiting to connect with the transmission system). In some countries, the waiting list has a backlog due to lack of transmission capacity or slow regulatory approval. Not all projects end up getting built, but the queues indicate developer interest.

About 260 GW of potential new wind energy in the United States had applied for interconnection as of 2011 (fig. 8–1), amounting to more than six times the country's installed capacity. Streamlining interconnection procedures and boosting the capacity of the transmission network can help more wind power come online more quickly. Transmission construction projects under way (see "America's plans" below), if completed on schedule, could reduce the backlog by about 29 GW by 2015.

In the European Union, each nation determines how the grid is accessed. Great Britain took steps to speed transmission access by implementing a new "connect and manage" approach in 2010. The country's goal is to produce 15% of its total energy from renewables by 2020, which could require about 30% of its electric power to come from renewables.

In the new scheme, UK regulators will allow wind farms and other new generating units to connect to the grid as soon as local work is complete, rather than waiting for transmission system upgrades. If power output exceeds transmission capacity, the system operator will pay certain units to ramp down, and the cost of transmission constraints will be spread across all power suppliers on an output basis.

In Germany, a catch-22 needs to be resolved. Although power system operators are required by law to connect wind farms, they often won't guarantee a connection until the wind farm secures financing—but the banks want to see a guaranteed connection before they'll finance the wind farm.

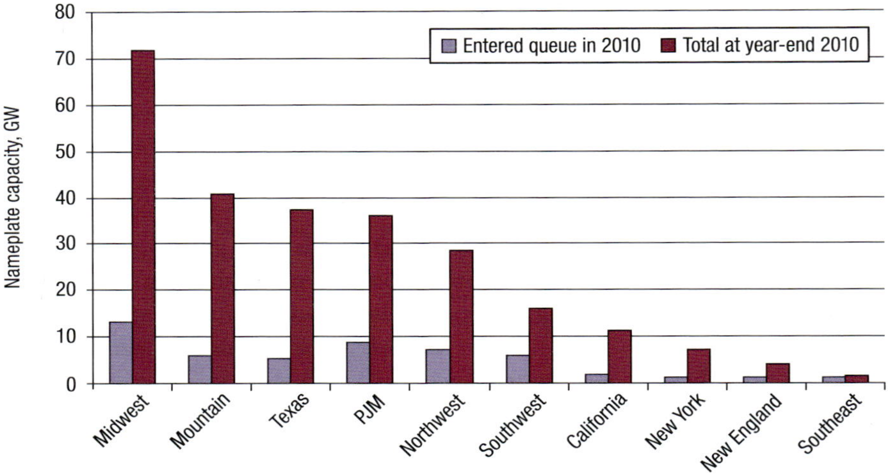

Fig. 8–1. Wind power in selected U.S. interconnection queues. *Data courtesy of Exeter Associates and U.S. Department of Energy.*

The cost of interconnection is another factor affecting the integration of wind power. Wind farm developers have borne most of the cost burden traditionally, but transmission costs have begun to be allocated among stakeholders. So why shouldn't interconnection costs also be allocated among those who benefit from the additional wind energy? It's a regulatory issue that remains to be resolved.

Transmission services. Power buyers and sellers typically have to purchase transmission services as part of their delivery package. Transmission can be provided as a firm, conditional, or interruptible service, either point-to-point (generator to load) or networked (accommodating multiple generators).

Firm service provides reserved capacity on the transmission line, while conditional firm service can be curtailed if system reliability is threatened, but not for economic reasons. Interruptible service is provided when it's available. In large, real-time power markets, transmission customers can purchase the right to bypass congested points without a reservation (see "Congestion").

Power system support

Some of today's wind farms can be actively controlled, like the Horns Rev example mentioned in chapter 7, allowing them to adjust their output up or down when called upon. This gives power system operators time to manage other generating units more efficiently.

Wind energy also supports the power system in other ways. Today's "grid-friendly" turbines are equipped with power electronics and other features that enable them to respond to changing conditions on the grid and help keep it stable.

For example, wind turbines provide low-voltage ride-through, which supports the grid when the utility needs generating capacity to maintain voltage on its system, as noted in chapter 6. Even small, split-second disturbances can alter the system voltage, causing a utility fault. Wind farms are now designed to keep operating throughout these events.

More generally, grid-friendly turbines can supplement the system's ancillary services rather than consuming more of them. New wind power plants are required by regulators in many countries to provide ride-through, power quality control (voltage and frequency), reactive power support, and other features. Older wind farms can be retrofitted with equipment that performs these functions for multiple turbines.

HVDC technology

High-voltage direct-current transmission technology (HVDC) affords higher power flows with lower losses, compared to AC transmission. HVDC also lets the power system operator control active and reactive power independently, which makes the system easier to manage. However, costly equipment is required at each end of the system to convert power from AC to DC and back again.

One way to make the most of HVDC technology is to mesh the advantages of AC and DC—use AC for collecting, switching, and distributing electricity, and use DC for long-haul, bulk transmission.

A key advantage of HVDC technology is that it is more controllable than AC, particularly in dedicated, point-to-point applications, where electricity is not tapped along the way for local delivery. HVDC's flexibility of control makes it easier to accommodate variations in wind energy output. A DC system synchronizes the AC electricity produced by wind farms.

Also, HVDC lines have a smaller footprint. They require fewer sets of conductors, because they are single-phase versus AC's three phases. So the rights-of-way required for HVDC are narrower than for AC lines, and they occupy less land. One DC line can replace triple AC lines in the same available space, or AC and DC lines can work alongside each other within the same existing, permitted corridors.

One HVDC supplier claims that a 380-kV DC line can deliver 300% more power than a 220-kV AC line with the same tower, foundation, and right-of-way. This means that substantially less materials and land are needed by DC to deliver equal amounts of power.

The chief trade-off lies in the distance of the transmission line. Traditional AC lines lose more power the longer they are, so HVDC is preferable for long-haul lines. Also, the longer the distance, the easier it is to justify the cost of power conversion at each end.

Another advantage of HVDC is that the cable can be buried underground or underwater in the seabed. AC transmission lines lose substantially more energy in these conditions than above ground. In fact, when buried, they can lose all of their power at distances as short as 200 kilometers (100 miles).

As an alternative to HVDC, a technology called flexible AC transmission uses power electronics and other static equipment to enhance the controllability of the grid network and boost its power transfer capability.

Many HVDC transmission lines are being installed in Europe offshore, where underwater cables are often the best solution, or the only one, for getting power to shore. However, HVDC projects are becoming more commonplace on land as well. China built an 800-kV "ultra"-HVDC transmission line to carry up to 7.2 GW of hydropower about 2,000 kilometers (1,200 miles) to Shanghai. Transmission losses are projected to be less than 7%.

More than 145 HVDC projects worldwide are operational or pending. However, very few HVDC projects have been attempted in the United States, where the technology historically has accounted for only 2% of the transmission market. Analysts estimate that HVDC could represent 8% of the U.S. cumulative transmission investment through 2020. Globally, investment in HVDC could grow 44% by 2015.

Some advanced HVDC transmission systems incorporate new technology called *high-temperature superconducting* cables. These have to be kept cold to operate, but can dramatically increase capacity and reduce line losses during transmission, even more than HVDC alone.

Transmission constraints

Power transmission systems in most developed countries have some points of congestion, or bottlenecks. During times of peak demand on the system, electricity from less expensive resources farther away from big cities can't be delivered because of congestion, so more expensive local generators have to come online to serve the load.

To manage congestion, power system operators tack on extra charges for transmitting through a congested connection or, conversely, they might pay transmission customers for relieving congestion.

Another way to manage bottlenecks is to hold capacity auctions that increase the value of power from generating units located in transmission-constrained areas or reward new transmission lines that reduce congestion. Market-clearing prices can vary enormously by location (fig. 8–2).

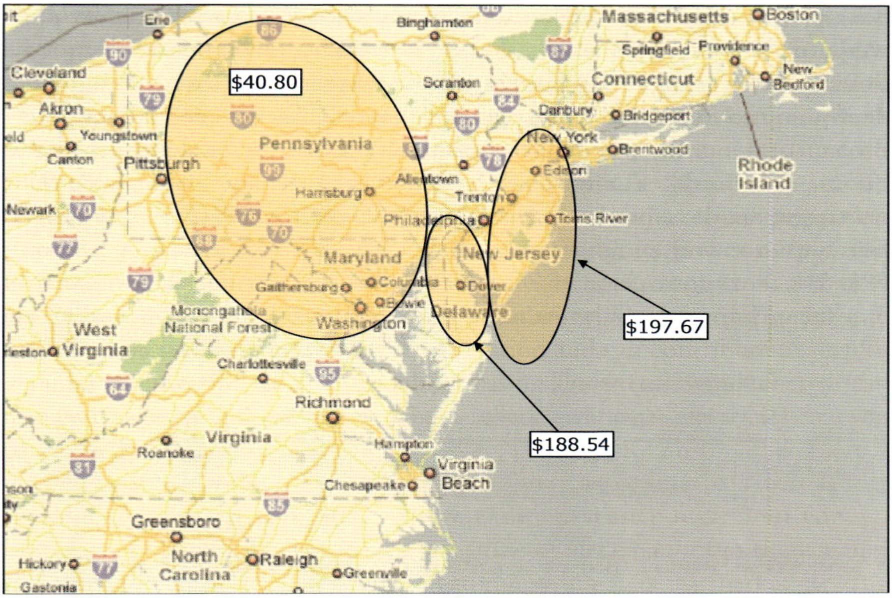

Fig. 8–2. Prices within a U.S. transmission system ($/MW-day). *Image courtesy of Melnyk and Andersen.*

Europe. The European Wind Energy Association, in a project called TradeWind, analyzed cross-border wind power transmission and market design at the pan-European level. As Europe's wind power penetration level increases, so will cross-border energy exchanges, and existing bottlenecks could get worse.

By 2020–2030, Europe's transmission system could become increasingly congested at several points—the borders of France with its neighbors, the boundary between Great Britain and Ireland, and at points along several Swedish, German, and Greek borders.

Transmission constraints hinder access to cheaper generation resources such as wind power, resulting in higher electricity prices across the board. Yet reinforcements in the works by European transmission system operators will not be enough to alleviate existing bottlenecks, much less accommodate greater amounts of offshore wind energy.

The TradeWind study identified 42 onshore interconnection points for upgrades that would enable the European power system to integrate more wind energy. In the 2020–2030 time frame, these upgrades could save some $2.1 billion per year (1.5 billion euros) in total operating costs, justifying an investment on the order of $27 billion (20 billion euros). European consumers would benefit economically from the upgrades, even without substantial new wind power capacity.

Integrating 265 GW of wind power into European grids by 2020, versus no more growth in wind energy capacity, could save $57.1 billion (41.7 billion euros) in electricity costs annually, assuming a fully functioning power market. This translates to a savings of $15 (11 euros) for every megawatt-hour generated, not just from wind turbines.

United States. Four U.S. areas were identified by the federal government as being critically congested or at a level of concern in 2009, one in the east and three in the west:

1. From mid-state New York southward along the Atlantic coastal plain to northern Virginia
2. Southern California's urban areas
3. Northern California's San Francisco peninsula
4. The Seattle-Portland area (Washington state to Oregon)

Size matters

As noted earlier, large, well-connected power systems are more stable and can accommodate higher penetrations of wind energy with less disruption. That's one reason Europe is planning a "supergrid" to harness the vast wind resource off the continent's northern shores. Similar plans are underway off America's Atlantic coast (see chapter 9).

Europe's progress. Much of the European grid is already synchronized, and national system operators have gradually increased coordination with neighboring grids. Transmission systems could be upgraded in the future at the pan-European level through a harmonized planning and permitting process.

However, the lack of cross-border transmission links, or interconnectors, has hampered competition in Europe's power market. Much of the electricity generated in Europe is consumed within the same country that produces it.

Making it easier to import and export electricity among European countries would help integrate the continent's wind energy resources and improve power system efficiency by encompassing a larger geographic spread. Also, if wind farms are pooled together, their overall power level might stay more consistent, minimizing costs.

Europe's goal for renewable sources to supply 20% of energy consumption by 2020 will require transmission work to deliver wind power to urban areas and to develop a more liquid, interconnected energy network. The European

Network of Transmission System Operators for Electricity (Entso-e) is committed to developing onshore and offshore power grids, in particular the North Sea Countries Offshore Grid Initiative.

Entso-e represents 41 system operators from 34 European countries. The group serves 532 million customers over a transmission network totaling 305,000 kilometers (190,000 miles). Members can exchange up to 380 TWh of electricity. Entso-e will start a three-year study in 2012 on how to develop an integrated, near-zero carbon grid by 2050.

Although much of Europe's proposed supergrid would be new offshore links, the continent's overall transmission system requires reinforcement to deliver wind power to Europe's cities. An estimated 45,000 kilometers (28,000 miles) of new or upgraded lines will be needed through 2020, according to the European Commission.

Barriers to integrating larger amounts of wind energy into Europe's electricity grid are related to infrastructure and markets rather than technology. The commission identified four priorities for improving Europe's grid infrastructure—strategic planning, more regional cooperation, faster permitting, and a stable framework for financing, including cost allocation.

European utilities are forging ahead with intercountry links. For example, Norwegian and Danish utilities started contracting for another HVDC transmission line connecting the two countries' power grids. The 500-kV, 240-kilometer (150-mile) link, by adding 700 MW of transmission capacity to three existing lines, would make it faster and cheaper for Denmark to export its excess wind power to Norway, which has loads of flexible hydropower to store the electricity. The transmission line should be commissioned in 2014.

Similarly, Norwegian and UK utilities began studying the possibility of a 2-GW HVDC line between the two countries. UK wind power would be exported to Norway and stored by its hydro resources. A joint French-UK study suggested that a 500-MW twin-cable, 75-km (47-mile) DC interconnector might be installed safely inside the existing tunnel across the British Channel. And Germany is planning an HVDC system to connect North Sea wind farms to the mainland grid.

In 2011, a 1-GW, 250-kilometer (160-mile) HVDC line went live connecting Britain and the Netherlands. The "BritNed" line allows 100% third-party access and will transmit power in either direction, depending on price differences between the two markets. Capacity will be auctioned in a full range of time frames, from intraday to annual.

In Germany—Europe's largest power market—the network of four high-voltage grids might need up to 3,600 kilometers (2,200 miles) of new lines by 2020 to connect offshore wind power in the north to factories in the south, according to the German Energy Agency. The work could cost more than

$13 billion (10 billion euros), or up to three times as much for underground lines. The government's plan to phase out its nuclear plants makes the project more compelling, and the energy ministry might use fast-track powers to jump-start the work.

America's plans. The concept of a "green power superhighway" has been proposed in the United States as part of the government's goal of 20% wind power penetration by 2030. The highway would require an investment of about $20 billion (15 billion euros) to add 20,360 kilometers of transmission (12,650 miles) to the nation's system of more than 320,000 kilometers (200,000 miles), an increase of about 6%. A comparable utility study estimated that 31,000 kilometers (19,000 miles) of new 765-kV lines would be needed and would cost $26 billion (19 billion euros).

During 2010, a total of 14,000 kilometers (8,800 miles) of transmission lines were built, another 5,000 kilometers (3,100 miles) were under construction, and 63,000 kilometers (39,000 miles) were projected through 2020, according to the U.S. Department of Energy. About 130 projects were under development, representing more than $180 billion in potential investment (130 billion euros).

Progress on expanding America's transmission system to accommodate wind power includes several key projects:

- **Southwest.** The One Nevada transmission project won a federal loan guarantee in 2010 for $343 million (250 million euros). The line is part of the Southwest Intertie Project, a 500-kV AC, 378-kilometer (235-mile) system intended to carry up to 2 GW of wind and solar power from Idaho and Nevada (with an additional 3-GW, 500-kV DC link to Wyoming) to markets in Las Vegas and California.

- **West coast.** A 4.5-GW transmission project started up its first three segments, which carry 700 MW from California's Tehachapi wind farms to customers in Los Angeles. The Tehachapi project's total of 278 kilometers (173 miles) of new lines plus upgrades, to be finished in 2015, could cost nearly $2 billion (1.5 billion euros). Also, the Sunrise Powerlink began construction in 2010. The 500-kV AC, 117-mile, $1.9 billion line (118-kilometer, 1.4 billion euros) will carry at least 1 GW of wind power and other renewables to San Diego.

- **Texas.** This state is building out its transmission network using a novel cost-allocation scheme (see "Sharing the cost" below). The expanded network could bring almost 9.9 GW of new transmission capacity online by 2013. In addition, a 344-kilometer (214-mile) transmission line was built to connect a large wind farm in western Texas with the state's urban markets toward the east. The 1-GW merchant line took less than two years to plan and build because it was privately funded by a utility, which paid landowners directly rather than relying on governmental eminent domain. This allowed the utility to bypass state regulation and public processes.

In addition to these projects, ambitious private plans are in the works to connect the three largest U.S. power grids at a single point in New Mexico. The Tres Amigas hub would link the Western, Eastern, and Texas networks, which now are virtually unconnected, with a 5–30 GW HVDC system. This would help integrate Western and Midwestern wind power with markets on both the East and West coasts and eastern Texas.

The project, estimated to cost more than $1 billion (700 million euros), was scheduled to begin construction in 2012 and start operation in 2014. The Tres Amigas station will use a 750-MW/345-kV HVDC converter and could incorporate grid-scale battery storage to accommodate variable energy supplies (see "Energy Storage" below).

Investment needed

The International Energy Agency estimates that by 2030, more than $1.8 trillion (1.3 trillion euros) must be spent on transmission and distribution systems in the 34 countries worldwide that are members of the Organisation for Economic Co-operation and Development (OECD), which does not include China, Russia, India, or Brazil. Expanding and upgrading existing systems would be needed regardless of whether wind energy increases its share of the world's power mix.

China also recognizes the need to build out its transmission system and upgrade existing lines if the country is to harness its vast wind resource. China's windiest regions lie in the "three norths"—the north, northeast, and northwest—mostly far from urban load centers along the heavily populated eastern coast.

During 2006–2010, the state grid invested $3 billion (20 billion yuan, or 2 billion euros) in ultra-high voltage transmission and about $6 billion (40 billion yuan, or 4 billion euros) to facilitate wind power integration, according to government data. During 2011–2015, the state plans to spend $78 billion (500 billion yuan, or 57 billion euros) to upgrade China's grid. Despite only modest incentives, China's grid operators have agreed to connect 80 GW of wind power by 2015 and 150 GW by 2020.

China's total wind capacity reached 44.7 GW in 2011 (see chapter 11). About two-thirds of this (29.6 GW) was connected to the grid. Up to 10% of new capacity was unconnected due to grid constraints. China's electricity regulators reported that about 2,800 GWh of wind power was wasted in the first half of 2010 due to insufficient transmission and grid connection.

Much of China's wind power development has focused on the northwest in two autonomous regions, Inner Mongolia and Xinjiang Uygur, and in Gansu province. Northwestern areas have accounted for up to 75% of China's unconnected wind power.

However, a fundamental misunderstanding could be exaggerating China's grid connection problems. According to the Global Wind Energy Council, China counts a wind farm as "connected" only when the last turbine is connected. In other countries, many turbines are connected and generating power earlier during the construction process, and all grid-connected turbines are counted, regardless of whether the wind farm is complete.

In America, investment in transmission lagged behind generation investment for many years. When more power was needed, natural gas-fueled units were installed close to load centers without building major new transmission capacity.

Utilities began to invest in transmission again around 2000 to increase reliability and to connect intra- and interregional markets for trading purposes. Grid reliability is the primary driver of new transmission construction in North America, accounting for 50% of additions. Integration of wind power and other renewables accounts for 27%.

Recent spending levels of about $8 billion per year (6 billion euros) should continue to increase as the industry plays catch-up for years of under-investment. The U.S. government estimates that an additional $3 billion per year (2 billion euros) would be needed to integrate large amounts of wind power and other renewables.

Another analysis predicts that $40–$70 billion in transmission investment (30–50 billion euros) will be needed during 2010–2025 to meet existing state-level renewable energy standards. At least two-thirds of U.S. states have some kind of renewables standard (see chapter 11). If a federal standard were enacted, the transmission investment required could increase to $80–$130 billion (60–95 billion euros).

Sharing the cost

To overcome financing hurdles, the cost of building a new transmission line can be shared among all the stakeholders in the power system that would see benefits from the additional capacity, rather than a single utility, state, province, or country bearing the entire cost.

The European Commission in 2010 allocated more than $1.2 billion (900 million euros) to nine interconnection projects, including links between Spain and France, an interconnection between Sweden and the Baltic countries, and reinforcement of a Finland-Estonia link.

Federal U.S. rules that were approved in 2010 call for a cost-sharing approach among all parties who benefit from new transmission capacity. The cost of building or upgrading transmission lines that boost regional reliability, offer widespread benefits to the public, and deliver renewable energy would be allocated among participants in what is called a multivalue project (MVP).

Regional MVP upgrades should minimize energy losses during transmission, facilitate long-term planning, and spur competition among various generation resources. They could also help prevent blackouts, diversify the power mix, and reduce wind power curtailment.

The cost-sharing approach was pioneered in California and formalized in Texas, which established competitive renewable energy zones (CREZs). The zones contain wind energy or other renewable resources that need more transmission capacity to reach markets. State regulators identify the zones and award development projects to utilities.

Construction projects underway in these Texas zones include about 3,900 kilometers (2,400 miles) of new 345-kV AC transmission lines that could cost more than $6 billion (4 billion euros). By 2014, the Texas system could be capable of delivering 18.5 GW of wind power to consumers.

In the Midwest region, however, the federal MVP rules have encountered opposition. Some states to be served by new transmission capacity claim they will receive less financial benefit than other states, and they'd rather generate their own renewable energy—even if it costs more.

Federal legislation might overturn the cost-recovery mechanism by defining the term "benefit" more narrowly. The rules have also been criticized by the Coalition for Fair Transmission Policy.

Proponents of expanding transmission systems point out that the investment needed to integrate wind power and other renewables would be paid off in seven years or less by lower electricity costs to consumers and by other quantifiable benefits such as greater grid reliability, lower carbon dioxide emissions, and in the case of HVDC lines, reduced power losses.

Energy Storage

Introduction

Storing very large amounts of energy to deliver electricity on demand is virtually impossible to do economically, except where dams can be built on rivers to store hydroelectric power or where water can be pumped and stored to create artificial hydropower reserves. Engineers are developing and testing many other types of bulk power storage.

Energy storage allows power system operators to balance loads by decoupling the instantaneous supply of electricity from the variable nature of demand. Essentially, storage transforms energy, which is dynamic and transient, into capacity, which is static and can be dispatched. Storage is a tool for making the grid more flexible, but it's not an indispensable one. Wind power and other renewables can be integrated into power systems without energy storage.

Global electric generating capacity is more than 4,500 GW, while total installed energy storage capacity is only around 3% of that. Europe has relatively more storage than America, where interest in grid-scale storage has intensified. According to Pike Research, new global energy storage capacity could increase 100-fold within 10 years, exceeding 12 GW in 2021. Slower growth is forecast by Lux Research, which sees an annual market for grid-based storage of about $1.5 billion in 2015.

Storage is often called the "holy grail" or "silver bullet" of the green energy industry—something that would allow us to exploit unlimited amounts of wind and solar power. But energy storage is most valuable when used as a resource for the benefit of the entire power system, rather than serving just wind farms or solar plants.

Generally, it's the system's net load that needs to be balanced, not an isolated supply. Storage as a system resource, linked to network controls, can respond to operators' needs for grid support or extra capacity.

In addition to hydropower, electricity can be stored in large-scale battery banks and *capacitors*, as well as *compressed-air* systems and devices such as *flywheels*. Power networks worldwide have begun to embrace energy storage by treating it as just another type of reserve that adds to system flexibility (as shown earlier in fig. 7–2).

By its nature, storage requires one form of energy to be converted to another form and then back again (or to another form of energy). Physics tells us that energy losses are inevitable. With more technology advances—particularly in materials science—multi-megawatt energy storage could become less expensive and more practical for utilities and power systems to integrate into their networks. If costs come down, storage could become an element of electric energy delivery systems worldwide.

Storage functions and costs

In today's power systems, storage serves three basic categories of functions:

1. Energy management—to balance supply and demand by charging up during periods of low demand, when power is cheap, and discharging when demand rises and prices increase. This function, also called load-leveling or *time-shifting*, requires continuous discharge for several hours

or more. System operators with long-duration storage can also defer the need to build or upgrade their transmission capacity. This category also includes price arbitrage, where suppliers make money by selling stored electricity at peak prices.

2. Power quality—to maintain grid stability by providing near-instantaneous frequency regulation and other ancillary services. Stored power can also be used to slow down generators' ramping rates, giving the system operator extra time to balance loads.

3. Bridging power—to provide continuous electricity service over a period of seconds or minutes while generation is being switched from one source to another. This category includes spinning reserves, load-following, and backup power for unit commitment errors and forecasting uncertainties.

These and other functions are listed in table 8–1, along with the duration, capacity requirements, and economic benefits for each function. Although high-value applications make tempting targets, much of the market demand and revenue opportunities could be in some of the lower value applications.

Table 8–1. Functions of grid-scale energy storage in power systems. *Data courtesy of Sandia National Laboratories.*

Function	Discharge duration		Power capacity		Value, $/kW	
	Low	High	Low	High	Low	High
Electric energy time-shift	2 hr	8 hr	1 MW	500 MW	400	700
T&D system upgrade deferral	3 hr	6 hr	250 kW	5 MW	481	687
Time-of-use energy cost management	4 hr	6 hr	1 kW	1 MW	1,226	1,226
Demand charge management	5 hr	11 hr	50 kW	10 MW	582	582
Renewable energy time-shift	3 hr	5 hr	1 kW	500 MW	233	389
Electric supply reserve capacity	4 hr	6 hr	1 MW	500 MW	359	710
Electric service power quality	10 seconds	1 min	0.2 kW	10 MW	359	978
Electric service reliability	5 min	1 hr	0.2 kW	10 MW	359	978
Load-following	2 hr	4 hr	1 MW	500 MW	600	1,000
Area regulation	15 min	30 min	1 MW	40 MW	785	2,010
Electric reserve capacity	1 hr	2 hr	1 MW	500 MW	57	225
Voltage support	15 min	1 hr	1 MW	10 MW	400	400
Transmission support	2 seconds	5 seconds	10 MW	100 MW	192	192
Congestion relief	3 hr	6 hr	1 MW	100 MW	31	141
Renewable capacity firming	2 hr	4 hr	1 kW	500 MW	709	915
Wind power integration, short	10 seconds	15 min	0.2 kW	500 MW	500	1,000
Wind power integration, long	1 hr	6 hr	0.2 kW	500 MW	100	782

The option to store energy must always be economically justified, and the value of the various storage options to system operators can be hard to quantify. The Electric Power Research Institute (U.S.) is working on a way to identify and account for the benefits of storage beyond its obvious, immediate value.

Sometimes, the value of a single benefit deriving from energy storage might not exceed the cost, but the value of two or more benefits combined might make the project economical. When examining costs and benefits, power system operators should consider the overall value of benefits afforded by storage.

For example, storage can time-shift wind power that's generated off-peak, holding it for release during on-peak hours. This might not justify installing storage, but if the system can time-shift baseload power as well, the combined benefits might make a storage project feasible financially.

California's system operator qualified certain storage resources such as batteries and flywheels as reserves, acknowledging the ability of fast-response storage to provide regulation service just as conventional generation does. State regulators are also looking at how much storage capacity utilities might have to procure in the future. California plans to allow energy storage devices to bid into the market for regulation services in 2012.

U.S. federal regulators implemented a rule in 2011 that will require system operators to pay more for regulation service from storage reserves such as flywheels that ramp up within seconds. One study indicated that for providing fast-response regulation service, 30–50 MW of storage is just as effective as a 100-MW combustion turbine, if storage-specific controls are used.

However, the high cost of grid-scale batteries and other alternative storage technologies is the main reason they haven't been more widely applied before now. It's still cheaper to build natural gas–fueled generation capacity than to store electricity.

Regulated utilities will probably be among the last to invest heavily in grid-based storage because of their risk aversion and their need to justify capital spending to regulatory agencies that are charged with protecting the ratepayers. . . . Over time, the most successful technologies will build a long enough track record of reliability to take a direct run at utilities and transmission system operators, but it's not reasonable to expect the utility and transmission markets to develop rapidly over [the years 2012–2016].

—John Petersen, Fefer Petersen & Cie

As with wind power systems, the cost of storage is analyzed as either capital or life-cycle cost:

- Capital costs are measured in two ways, whether for power applications or for energy applications. The capital cost of storage for power applications such as frequency regulation that depend on quick response is measured in cost per megawatt, reflecting the amount of electricity that can be absorbed or supplied at any given instant. Storage for energy applications such as load-leveling that depend on the capacity being stored is measured in cost per megawatt-hour, which reflects the length of time that the storage system can provide a given amount of power.

- The life-cycle cost of storage is calculated by adding the capital cost plus maintenance costs, divided by the number of cycles the storage technology can deliver over its useful life.

For utility grid-scale storage to be economical, it would need a capital cost less than $250 per kilowatt-hour (180 euros) for energy applications or $1,250 per kilowatt (913 euros) for power applications (assuming the equipment can deliver rated power for 5 hours). A practical storage technology would also need a service life of more than 15 years, a charge-discharge cycling life of 15 years or 3,900 full cycles, and an 80% full-cycle efficiency, as well as low maintenance costs and high safety standards.

In addition to cost, other factors in choosing a storage technology include the application, efficiency, charge and discharge rates, size and weight, location constraints, and *energy density* (basically, storage capacity per unit of volume). Table 8–2 list some general characteristics of energy storage technologies, which are compared in table 8–3.

Table 8–2. Characteristics of grid-scale energy storage technologies. *Data courtesy of Hawaiian Electric Co.*

Characteristic	Pumped hydro	Compressed air	Batteries	Capacitors	Flywheels
Energy storage capacity	22 GWh	2.4 GWh	50-250 MWh	0.5 kWh	25 kWh-1 MWh
Power storage capacity	up to 4 GW	50-300 MW	50 kW-50 MW	200 kW	100 kW-30 MW
Duration of discharge	20 hours or more	4-24 hours	1-8 hours	1-30 seconds	2-15 minutes
Response time	30 seconds-15 minutes	2-12 minutes	4 milliseconds	4 milliseconds	near-instantaneous
Round-trip efficiency	70-85	85	65-90	95	80-85
Useful life	30 years	30 years	5-15 years	up to 20 years	up to 20 years

Table 8–3. Comparison of storage technologies and their applications

Technology	Relative advantage	Relative disadvantage	Power application	Energy application
Pumped hydro	High energy, low cost	Special site requirements	Not practical or economical	Fully capable and reasonable
Compressed-air	High energy, low cost	Special site requirements	Not practical or economical	Fully capable and reasonable
Flywheels	High power	Low energy density	Fully capable and reasonable	Not feasible or economical
Capacitors	Long cycle life	Very low energy density	Fully capable and reasonable	Not feasible or economical
Batteries				
Lead-acid (conv.)	Low capital cost	Limited cycle life	Fully capable and reasonable	Feasible but not yet practical/economical
Lead-acid (adv.)	Low capital cost	Low energy density	Fully capable and reasonable	Fully capable and reasonable
Sodium-sulfur	High power and energy density	Cost; high-temp. operation	Fully capable and reasonable	Fully capable and reasonable
Lithium-ion	High power and energy density	Cost; temp. monitoring/control	Fully capable and reasonable	Reasonable for this application
Zinc-bromine	Independent power/energy	Medium energy density	Reasonable for this application	Fully capable and reasonable
Vanadium redox	Independent power/energy	Medium energy density	Reasonable for this application	Fully capable and reasonable

As a footnote, natural gas is sometimes considered energy storage because most utilities have gas-fueled generators on their system. Gas is routinely stored underground, with extra gas in pipelines counting as additional supply.

Storage Technologies

Hydropower

Not all utilities and power system operators are lucky enough to have hydroelectric power on tap, because it requires rivers and natural differences in elevation—in other words, the opportunity to build a dam and reservoir or to harness an existing waterfall. Areas that do have a lot of natural hydropower—like Canada and Norway—usually have low electricity prices, since hydro is cheap to operate once it has been built.

Most good sites for large hydropower projects have already been developed in industrial countries. Less-developed nations such as China and Brazil continue to exploit hydropower with massive new projects.

Pumped hydropower capacity can be developed where natural features are lacking. Water is pumped from a lower elevation to a higher one, mimicking the damming of a river. Typically, this is done using cheap electric power during periods of low demand. The water is stored in a reservoir, then released to flow through turbines that generate electricity when needed, as shown in figure 8–3.

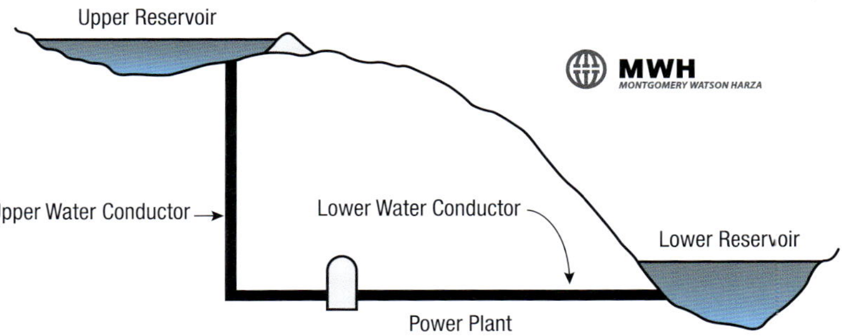

Fig. 8–3. Pumped hydropower. *Image courtesy of Montgomery Watson Harza.*

Pumped storage plants typically reach efficiencies of 70%–85% ("round trip," over the entire charge/discharge cycle) and can release energy for 20 hours or more. They cost little to operate and, like natural hydropower, they are flexible to control and versatile, providing a wide range of grid support from ancillary services to load leveling.

However, pumped storage plants are fairly capital-intensive to build—more than \$1–\$1.5 million per megawatt (0.7–1.1 million euros)—and take a long time to develop, at least seven years. They also require a significant amount of land and water for the upper and lower reservoirs, involving a multitude of regulatory permits. Once the facility is up and running, pumped hydro storage costs are relatively low, \$125 per kilowatt-hour (91.3 euros).

Pumped hydro is the most mature, proven energy storage technology. In terms of capacity, pumped hydro constitutes the vast majority of energy storage in use. Most of it was built decades ago to serve nuclear plants, which can't easily adjust their output. More than 150 facilities are in operation worldwide, with total capacity of over 100 GW.

Europe has 38 GW of pumped hydro capacity, and North America has more than 40 facilities, including 38 in the United States with capacity of about 20 GW. China plans to complement wind power development by boosting its pumped hydro capacity from 14 GW in 2010 to 21 GW by 2015 and 41 GW by 2020.

Despite formidable siting and permitting issues, almost 40 more pumped hydro facilities could be built in the United States, potentially boosting storage capacity by about 31 GW. Among these, four facilities totaling more than 1 GW are being planned (along with new transmission lines) for exporting wind power from Wyoming. And in Hawaii, a 300-MW pumped hydro system will help absorb new wind power on a grid that has only 1.2 GW of peak demand. In a novel twist, the ocean will serve as the lower reservoir.

Batteries

Battery storage technology has been around for more than 200 years, but today's batteries wouldn't be recognized by their early inventors. A great variety of batteries can be used for storing energy on electrical grids, with different types better suited to different applications.

Conceptually, a battery is nothing more than a bottle that stores electricity. The term "energy" describes the total amount of electricity you can put into the bottle. The term "power" describes how quickly you can empty or fill the bottle. The basic problem with energy storage is that batteries are thousands of times more expensive than the electricity they store.

—John Petersen, Fefer Petersen & Cie

Fast-response. Several types of batteries can respond within seconds and discharge for up to an hour. These include nickel-metal hydride, lithium-ion, and nickel-cadmium batteries, as well as the familiar lead-acid type used in cars and trucks. A subtype of lithium-ion batteries uses a lithium-iron-phosphate chemistry.

Some battery storage equipment is being marketed to utilities as a service, rather than sold as a capital expense. The battery system can be housed in a mobile container and moved around. One vendor offers a 1-MWh portable battery system that can produce 1–2 MW of AC power. Also, some companies specialize in developing storage, serving as a one-stop shop for the whole process.

However, fast-response batteries are practical mainly as bridging power rather than for frequency regulation and other power quality applications. That's because the constant cycling required would erode the battery's life prematurely. Most demonstration projects using large banks of fast-response batteries are focused on providing short-term reserves, until a backup generator can come online.

Chile has a 12-MW lithium-ion battery bank being used as spinning reserve and is planning a second one with 20 MW of capacity. An 8–MW lithium-ion battery pack was operating in New York State, with plans to expand it to 20 MW by 2012. Another 32-MW system began operating in 2011 at a 98-MW wind farm in West Virginia. And China build a 36-MWh lithium iron-phosphate battery station to complement 140 MW of wind and solar power.

One U.S. utility is planning a 36-MW battery storage system, scheduled for start-up in 2012, at its 153-MW wind farm in Texas. The Xtreme Power battery material is a dry solid with virtually no electrical resistance, which minimizes energy loss during cycling (see "Energy storage for Hawaii's wind power" below). Another U.S. utility is considering a 400-MW, 1,600-MWh battery storage system in New York as an alternative to additional generation.

A California utility was building a lithium-ion battery facility in the Tehachapi region that could provide 32 MW of power and 8 MWh of energy to smooth out local wind power. However, while it could function well on a minute-by-minute basis, the area's multiple wind farms, with an average daily production swing of more than 200 MW, would need storage on a GWh scale, not MWh. Also, the storage will likely be quite expensive in terms of capital costs. A similar lithium-ion battery project to supply the Russian grid could be even more costly.

High-energy. This type of battery includes high-temperature batteries and *flow batteries*. Sodium-sulfur (NaS) batteries are the most mature high-temperature, high-energy type (fig. 8–4). They have liquid (molten) sodium and sulfur as electrodes, thus the high temperature, and are made of inexpensive materials.

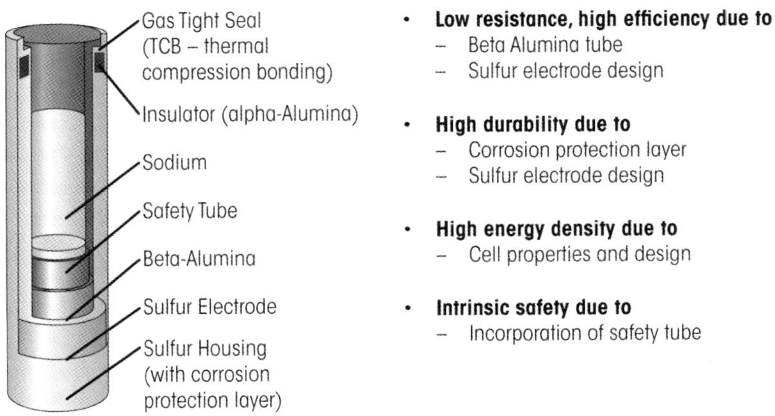

Fig. 8–4. Sodium sulfur battery. *Image courtesy of NGK Insulators, Ltd.*

Sodium-sulfur batteries are ideal for load-leveling and other energy management purposes because they have a high energy density, high charge/discharge efficiency, and long cycle life. Also, they have performed safely even under extreme conditions, although a battery fire in 2011 raised safety questions.

Japan pioneered sodium-sulfur battery development, though Korea is working on similar technology. Japan has a total of almost 300 MW of these batteries in use. The largest single unit, with 34 MW of storage capacity in northern Japan, delivers 245 MWh of energy. The Japanese technology has been tested at a small wind farm in Minnesota.

Many other projects are under way around the globe, including a 1-GW sodium sulfur battery system planned for Mexico that is estimated to cost nearly $4 billion (3 billion euros). This so-called energy warehouse could be built near the Mexico-U.S. border, where it might sell storage capacity to both countries' grids.

Flow batteries use a liquid electrolyte that flows across a membrane. The electrolyte is stored in external containers and circulated through the battery as needed. Flow batteries are sometimes called regenerative fuel cells. Their key advantage is that the power component and energy component can be sized independently—that is, they can offer low power and lots of storage or high power and less storage, or any mix of the two parameters.

Flow batteries include vanadium redox and zinc-bromine types, as well as polysulfide-bromine. A small zinc-bromine system began operating in Australia in 2010, and a 1-MW zinc redox flow battery storage system was planned for a wind farm in Minnesota. Other new chemistries are being developed, such as a technology that uses iron instead of vanadium, an expensive rare-earth metal, and a zinc-air battery.

Advanced battery concepts. Lead-acid batteries in the future might use carbon additives to improve their longer term performance. Some innovators are working on lead-carbon batteries with features similar to capacitors.

One idea for making a large-capacity battery is similar to the sodium-sulfur technology, but might be cheaper. The "liquid metal battery" concept, based on the way an aluminum smelter works, uses molten antimony separated by an electrolyte from molten magnesium. The antimony is forced to behave like a nonmetal. The concept is in early development, with government and energy industry funding.

Flywheels

A flywheel is a spinning wheel or disc with a fixed axle (fig. 8–5). An electric motor/generator accelerates the flywheel to extremely high speeds.

Because of inertia, flywheels resist any change in their rotational speed, so energy is stored in the rotating mass and discharged by slowing the flywheel down and powering the generator.

Flywheels have been around for thousands of years. In modern history, they have been used as an uninterruptible power supply in applications such as data centers, where a split-second of downtime can be catastrophic. Today's flywheels are made of carbon-fiber composites with a very high strength-to-density ratio.

Flywheels feature near-instantaneous response and high efficiency, making them well-suited for frequency regulation and other power quality services. As a rule of thumb, 1 MW of flywheel regulation capacity can substitute for about 2 MW of conventional regulation service. Flywheels can also discharge energy more slowly if desired for backup power. They are not much affected by temperature changes, they take up relatively little space, their maintenance needs are lower than batteries, and they're very durable.

However, friction must be kept to an absolute minimum to prolong the duration of energy storage, so flywheels are often operated in a vacuum chamber. The use of superconducting electromagnetic bearings can virtually eliminate frictional losses.

The New York State power system is testing a flywheel storage facility, designed to provide regulation service, that reached 20 MW of capacity in 2011. The system comprises 200 1-ton flywheels. Technical and financial problems forced the flywheel vendor into bankruptcy, but the plant's new owner plans to keep it running and to build a second 20-MW storage project. Flywheel manufacturers, like battery makers, are seeking to market their technology as a service for providing frequency regulation.

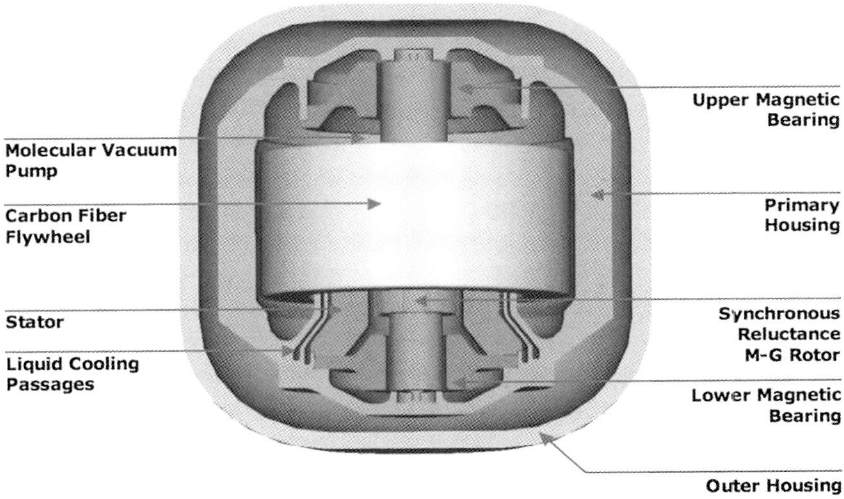

Fig. 8–5. Flywheel energy storage. *Image courtesy of Pentadyne Power Corp.*

Capacitors

Capacitors are devices that store electricity in an electrostatic field (fig. 8–6). Advanced capacitors (ultracapacitors or supercapacitors) have some of the fastest response times of any energy storage device. They're typically used in power quality applications, such as providing transient voltage stability, and for bridging power.

Capacitors can be cycled hundreds of thousands of times, but their energy density is relatively low. This means that their power is available for only a short time, preventing their use in applications that require long discharge durations.

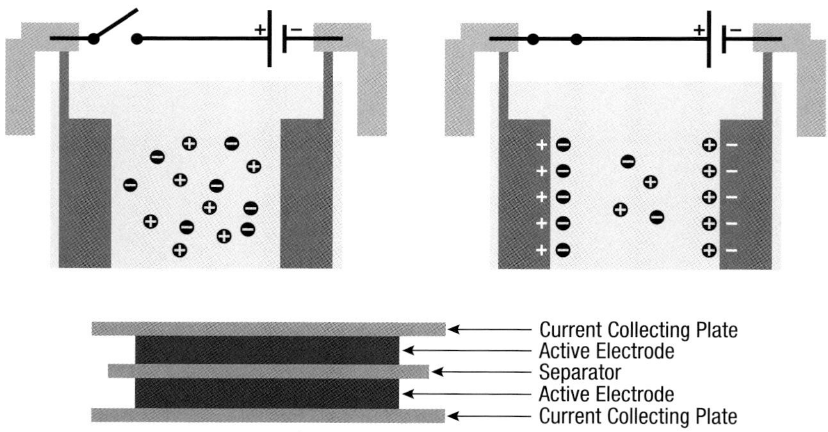

Fig. 8–6. Capacitor energy storage. *Image courtesy of Maxwell Technologies, Inc.*

Compressed air

This hybrid generation/storage technology uses an electric motor/generator to store energy by compressing air in an underground mine or natural cavern. To extract the energy, the process is reversed. The air is withdrawn, heated, and expanded through a high-pressure turbine that captures some of its energy. Then the air is mixed with fuel and burned, with the hot exhaust being expanded through a low-pressure turbine. Both of these turbines are hooked up to generators that feed into the grid, as shown in figure 8–7.

A simpler way to understand compressed-air energy storage is to think of it as a natural gas–fueled power plant that consumes much less gas because its intake air is precompressed. What's more, cheap, off-peak electricity is used to compress the air, increasing the system's overall fuel and cost savings.

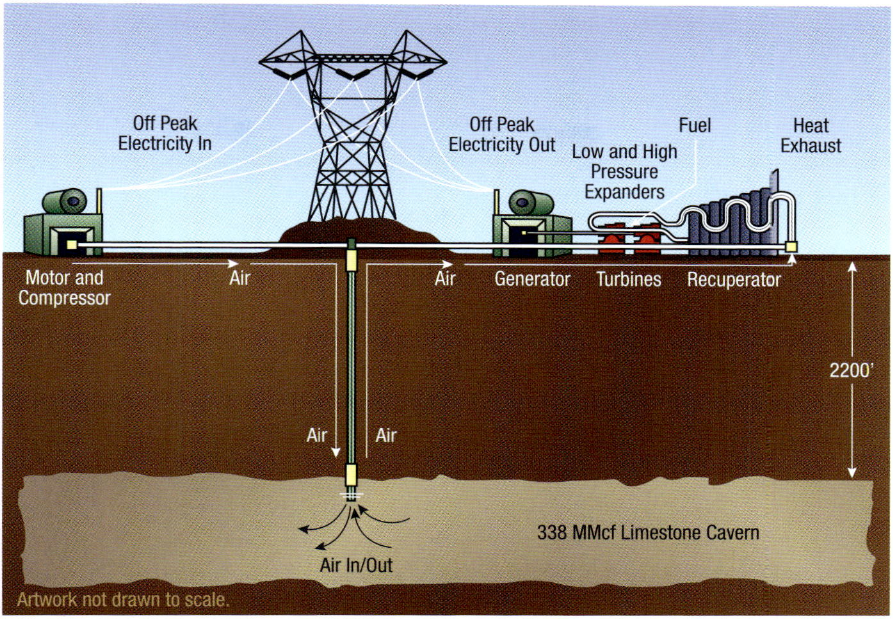

Fig. 8–7. Compressed air energy storage. *Image courtesy of Sandia National Laboratories.*

Compressed-air technology has the advantage of being familiar to electric utilities, who operate gas power plants routinely. Also, the equipment is commercial, reliable, and reasonably priced, making it easier to finance.

Only two large compressed-air storage systems are operating commercially, plus several more in niche applications. A 150-MW demonstration of an advanced system in New York State will enable the utility to integrate wind energy. Another utility is looking at a 300-MW site in California.

Besides its relative complexity, the primary drawbacks of compressed-air energy storage (often called CAES for short) are the need for an underground storage cavern and the system's reliance on natural gas fuel. New designs are being investigated that would help mitigate the latter disadvantage. For example, recovering the heat lost during compression, saving it, and reusing it to preheat combustion air could reduce or eliminate the need for natural gas.

A U.S. company, General Compression, has developed a 2-MW compressor/expander that might store energy without burning fuel. The compressed-air storage process yields a round-trip efficiency of 70%–75% and a response time of less than 6 seconds. Modules can be linked together for larger storage capacities and longer durations, up to 300 hours. A commercial project should start construction by 2012, with government funding and private investment.

Another innovator in the CAES sector is SustainX Energy Storage Solutions, which is developing a way to improve the thermal efficiency of the process. Novel technologies are also being developed by a private U.S. start-up, LightSail, and by the UK's Isentropic Energy.

Other storage technologies

Engineers are researching several technologies that also hold promise for grid-scale energy storage. For example, superconducting magnetic technology stores energy in a magnetic field created by the flow of direct current through a large coil of superconducting material that has been supercooled (fig. 8–8). This technology can be used for power quality, bridging power, and grid voltage support.

Superconducting magnetic technology can respond almost instantaneously, with a very high power output for a brief period of time. It's very efficient (more than 95% round-trip efficiency), but its capacity is limited, and materials are expensive. Superconducting magnetic storage systems cooled by liquid helium are commercially available. Systems cooled by liquid nitrogen to less cold temperatures are being developed.

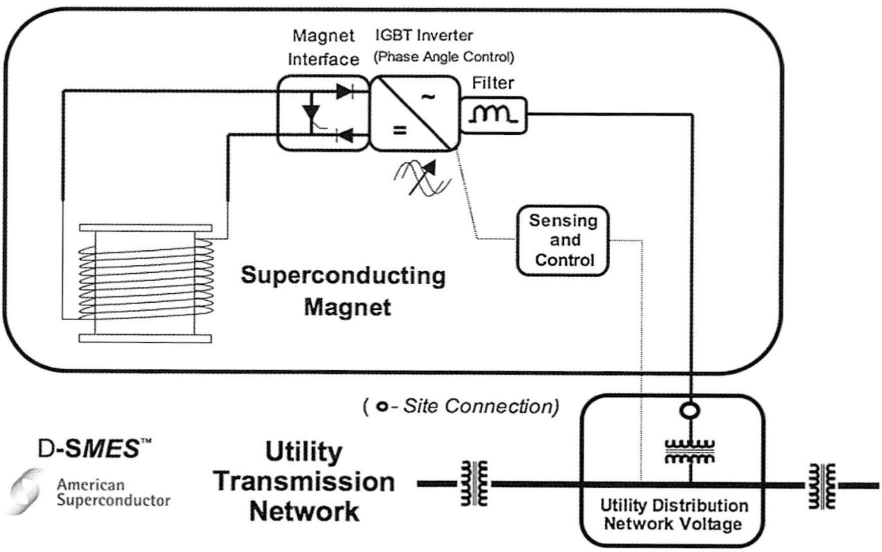

Fig. 8–8. Superconducting magnetic energy storage. *Image courtesy of American Superconductor.*

Another option, fuel cells, are commercial in relatively small, stationary applications. These electrochemical devices use hydrogen to generate power. The hydrogen could be supplied by using off-peak wind power to split water into its oxygen and hydrogen components. However, fuel cells are still quite expensive, and most are not large enough for grid-scale purposes. Some German companies are using wind energy–derived hydrogen as a fuel to supplement biogas. Hydrogen can also be stored in compounds called metal hydrides that are being investigated.

Many proponents of renewable energy hope that fleets of electric vehicles worldwide could offer a nighttime market for off-peak wind power. Essentially, each vehicle stores electricity at night while charging up and consumes it during the day. But millions of vehicles would have to connect with utility grids to make a big difference to power system operation.

Another use for off-peak wind power might be to synthesize liquid transportation fuels. Conversion processes for making "windfuels" are proven, but doing so economically might be challenging.

One ingenious approach to energy storage uses the force of gravity on a million-ton rock. The idea is sort of like pumped hydro, but upside down. The huge stone sits in a tight space where water can be pumped into the bottom, lifting the stone. When a valve is opened, the water is forced at high pressure through a turbine. A similar concept is in the works that uses a huge cylindrical weight encased underground in concrete. Water is the propulsive force for moving the cylinder up or down to store or release energy.

Researchers in Denmark have succeeded in storing small amounts of wind energy in giant synthetic bags that are buried under sand dunes using wind-powered compressors.

Energy storage for Hawaii's wind power

Experts say that energy storage need not serve individual wind power plants, but Hawaii presents an unusual situation. For one thing, the state aims to rely on clean energy for 70% of its electricity and ground transport by 2030, an ambitious goal. Overall, Hawaii hopes to get 40% of its energy from renewables by that time.

Also, Hawaii's eight islands are isolated from other grid networks, so its system can't import electricity when consumption spikes or export it when demand is low. And the state enjoys a bountiful wind resource as well as plenty of solar energy, yet pays dearly for conventional fuels that must be shipped in to generate power.

So Hawaii has dived into energy storage as a way to manage its power system. The technology of choice is a storage system by Xtreme Power that's capable of providing load-balancing and other grid support services.

The company's PowerCell technology, which produces 1 kWh at a 3-hour rate, is packaged with power electronics and smart controls into 1.5-MW storage modules that supply 1–4 MWh of energy. The modules are combined into multi-megawatt systems with dedicated communications to the utility operations center.

Hawaii's grand plan is called Interisland Wind, a project to upgrade the electric grid on the island of Oahu, which is densely populated, and import wind power from the smaller, neighboring islands of Molokai and Lanai via an underwater cable. Wind farms on these islands have yet to be approved.

Meanwhile, three Hawaiian wind farm/storage projects—collaborations between Xtreme Power and developer First Wind—were completed or under way.

A 1.5-MW storage system that delivers 1 MWh of energy was tested in 2009–2010 at the 30-MW Kaheawa wind farm on the island of Maui. The battery storage was used to control the ramp rate of 3 MW of wind power to plus-or-minus 1 MW per minute.

Kaheawa was being expanded in 2011 with another 21 MW of wind power and a 10-MW storage system with a 20-MWh energy capability. The battery system will store excess wind power and deliver it when needed, as well as providing ramp control and frequency and voltage services.

On Oahu island, a wind farm began generating power in 2011, the 30-MW Kahuku project, which has 15 MW of storage that delivers 20 MWh of energy. The battery system controls turbine ramp rates to within 1 MW/min, which is required because the wind farm is located at the end of a 12.47-kV radial line. Additional storage could be installed if curtailment becomes an issue.

References

American Wind Energy Association. 2010. U.S. Wind Industry Annual Market Report. http://www.awea.org/documents/reports/Annual_Market_Report_Press_Release_Teaser.pdf.

———. Building a Green Power Superhighway. http://www.awea.org/documents/Green_Power_Superhighways_Factsheet.pdf.

American Wind Energy Association. Green Power Transmission and Consumer Savings. http://www.awea.org/documents/factsheets/Transmission_and_Consumer_Savings.pdf.

American Wind Energy Association. Wind Power and Energy Storage. http://www.awea.org/documents/factsheets/Energy_Storage_Factsheet.pdf.

Appleyard, David. 2011 (February 6). Where Next for the Supergrid? *Renewable Energy World*. http://www.renewableenergyworld.com/rea/blog/post/2011/02/where-next-for-the-supergrid?cmpid=WNL-Wednesday-February9-2011.

Baxter, Richard. 2005. *Energy Storage: A Nontechnical Guide*. Tulsa: PennWell. (ISBN 978-1-59370-027-0).

Blau, John. 2011 (March 29). Germans Push for More Renewable Energy; Offshore Wind to Benefit. *Renewable Energy World*. http://www.renewableenergyworld.com/rea/news/article/2011/03/germans-push-for-more-renewable-energy-offshore-wind-to-benefit.

Brattle Group. 2011 (January 25). Brattle Estimates That Approximately 1,000 MW of Additional Generator Regulation Service Will Be Needed to Integrate 50,000 MW of Wind Generation in Large Power Systems. http://www.brattle.com/NewsEvents/NewsDetail.asp?RecordID=902.

Castano, Ivan. 2011 (April 7). China Installing Wind-power Capacity As Fast As It Can. Renewable Energy World. http://www.renewableenergyworld.com/rea/news/article/2011/04/china-installing-wind-power-capacity-as-fast-as-it-can?cmpid=WindNL-Thursday-April21-2011.

Coalition for Fair Transmission Policy, http://www.thecftp.org/.

Comfort, Nicholas. 2011 (June 1). Merkel Faces Achilles Heel in Grids to Unplug German Nuclear. Bloomberg. http://www.bloomberg.com/news/2011-05-31/merkel-faces-achilles-heel-in-grids-to-unplug-german-nuclear.html.

CREZ Transmission Program Information Center, http://www.texascrezprojects.com/.

Crowe, Robert. 2011 (August 17). Energy Storage Industry Grows to Integrate Wind, Solar. Renewable Energy World. http://www.renewableenergyworld.com/rea/news/article/2011/08/energy-storage-industry-grows-to-integrate-wind-solar.

Doty, F. David, 2011 (February 2). Kicking Oil Addiction with Windfuels. Greentech Media. http://www.greentechmedia.com/articles/read/guest-post-kicking-oil-addiction-permanently-with-windfuels.

Electrical Power Systems Technology. 2010 (November 25). HVDC Transmission Market to Grow 44% by 2015. http://eps-technology.blogspot.com/2010/11/hvdc-transmission-market-to-grow-44-by.html.

European Network of Transmission System Operators for Electricity, https://www.entsoe.eu.

European Wind Energy Association. 2010 (November). Powering Europe: Wind Energy and the Electricity Grid. http://www.ewea.org/fileadmin/ewea_documents/documents/publications/reports/Grids_Report_2010.pdf.

———. 2011. Fact Sheets: The Power Grid. http://www.ewea.org/index.php?id=1611.

Eurotunnel Group. 2011 (May 28). Groupe Eurotunnel and Star Capital Partners Establish Joint Venture to Develop an Electricity Interconnector between Great Britain and France. http://www.eurotunnelgroup.com/uploadedFiles/assets-uk/Media/Press-Releases/2011-Press-Releases/110526ELECLINK-UK.pdf.

Eyer, Jim, and Corey, Garth. 2010 (February). Energy Storage for the Electricity Grid: Benefits and Market Potential Assessment Guide. Sandia National Laboratories. SAND2010-0815. http://www.goodcompanyassociates.com/files/manager/SAND_Sandia_BenefitsAndMarketGuide_FINAL_Feb.pdf.

Farrell, John. 2011 (April 26). FERC's High Voltage Gravy Train. Energy Self-Reliant States. http://energyselfreliantstates.org/content/fercs-high-voltage-gravy-train.

———. 2011 (May 23). Michigan to Grid Operator: We Prefer to Generate Our Own Renewable Energy. Renewable Energy World. http://www.renewableenergyworld.com/rea/blog/post/2011/05/michigan-to-grid-operator-we-prefer-to-generate-our-own-renewable-energy??cmpid=WNL-Wednesday-May25-2011.

Galbraith, Kate. 2010 (September 9). Fighting the Power Lines to Protect Hill Country Vistas. The Texas Tribune. http://www.texastribune.org/texas-energy/wind-energy/fighting-power-lines-protect-hill-country-vistas.

Global Wind Energy Council. 2011 (March). Global Wind Report: Annual Market Update 2010. http://www.gwec.net/fileadmin/documents/Publications/Global_Wind_2007_report/GWEC%20Global%20Wind%20Report%202010%20low%20res.pdf.

Global Wind Energy Council, Greenpeace International. 2010 (October). Global Wind Energy Outlook 2010. http://www.gwec.net/fileadmin/documents/Publications/GWEO%202010%20final.pdf.

Gross, Martin. 2011 (January). The Transmission Imperative for Renewables. *Utility Automation Engineering*. http://www.elp.com/index/display/article-display/1065111962/articles/utility-automation-engineering-td/volume-16/issue-1/feature/the-transmission-imperative-for-renewables.html.

International Energy Agency. 2010 (July). IEA Wind Energy 2009 Annual Report. http://www.ieawind.org/AnnualReports_PDF/2009/2009AR_92210.pdf (ISBN-0-9786383-4-4).

Jones, Jackie. 2011 (March 1). Europe's Policymakers Sign up for a New Offshore Transmission Grid. *Renewable Energy World*. http://www.renewableenergyworld.com/rea/news/article/2011/03/along-the-right-lines??cmpid=WNL-Wednesday-March2-2011.

Kanellos, Michael. 2011 (January 12). Energy Storage Originators: A New Type of Company? Greentech Media. http://www.greentechmedia.com/articles/read/energy-storage-originators-a-new-type-of-company.

———. 2011 (February 11). New Storage Technology: A 100 Million Ton Stone. http://www.greentechmedia.com/articles/read/new-storage-technology-a-100-million-ton-stone.

———. 2011 (April 21). Alstom Gets Contract for High Voltage DC at Tres Amigas. Greentech Media. http://www.greentechmedia.com/articles/read/alstom-gets-contract-for-high-voltage-dc-at-tres-amigas/.

———. 2011 (June 15). Water or Rocks: What's the Secret Ingredient for Energy Storage? Greentech Media. http://www.greentechmedia.com/articles/read/water-or-rocks-whats-the-secret-ingredient-for-energy-storage1/.

KEMA. 2010 (June 23). KEMA Sees Opportunities for Storage Applications in High-Penetration Renewables. http://www.kema.com/news/pressroom/press-releases/2010/KEMA-sees-opportunities-for-storage-applications-in-high-penetration-renewables.aspx.

Li, Junfeng, et al. 2010 (October). China Wind Power Outlook 2010. http://www.greenpeace.org/raw/content/eastasia/press/reports/wind-power-report-english-2010.pdf.

Link, David. 2010 (September 8). Pumped Hydro: Part of the Solution to Our Energy Storage Needs? Pike Research. http://www.pikeresearch.com/blog/articles/pumped-hydro-part-of-the-solution-to-our-energy-storage-needs.

Melnyk, Markian M.W., and Andersen, Robert M. 2009. *Offshore Power: Building Renewable Energy Projects in U.S. Waters.* Tulsa: PennWell. (ISBN 978–1-59370-211-3).

North American Electric Reliability Corp. 2009 (April). Accommodating High Levels of Variable Generation. http://www.nerc.com/files/IVGTF_Report_041609.pdf.

———. 2010 (October). 2010 Long-Term Reliability Assessment. http://www.nerc.com/files/2010%20LTRA.pdf.

Palmer, Adrian. Delivering Power to Meet Britain's 2020 Targets. *Utility Automation Engineering.* http://www.elp.com/index/display/article-display/4283246046/articles/utility-automation-engineering-td/volume-16/issue-2/features/eye-on-europe-feature-delivering-power-to-meet-britains-2020-targets.html.

Patel, Sonal. 2010 (February 1). The Age of the 800-kV HVDC. Power. http://www.powermag.com/issues/departments/global_monitor/The-Age-of-the-800-kV-HVDC_2450.html.

Petersen, John. 2011 (March 8). Two Stocks for Grid Storage: ZBB Energy and Axion Power. *Seeking Alpha.* http://seekingalpha.com/article/257113-two-stocks-for-grid-storage-zbb-energy-and-axion-power.

———. 2011 (March 23). Grid-Based Energy Storage: Widely Misunderstood Challenges and Opportunities. *Seeking Alpha.* http://seekingalpha.com/article/259651-grid-based-energy-storage-widely-misunderstood-challenges-and-opportunities?source=email_alternative_energy_investing.

———. 2011 (April 10). Wind Power Investors Get Another Reality Check. *Seeking Alpha*. http://seekingalpha.com/article/262713-wind-power-investors-get-another-reality-check?source=email_alternative_energy_investing.

———. 2011 (December 15). Hype Busters From Lux Research Explain Grid Based Energy Storage. Renewable Energy World. http://www.renewableenergyworld.com/rea/news/article/2011/12/hype-busters-from-lux-research-explain-grid-based-energy-storage?cmpid=WindNL-Thursday-December15-2011.

Pfeifenberger, Johannes. 2010 (October 12). Renewable Energy Development and Transmission Expansion—Who Benefits and Who Pays. http://www.brattle.com/_documents/UploadLibrary/Upload887.pdf.

Rouda, Jennifer, and Yanez, Silvia. 2010. Connecting Renewable Energy to the Electric Transmission Grid: Trending Toward Innovative HVDC Technology. *North American Clean Energy*. http://www.nacleanenergy.com/index.php?option=com_content&view=article&id=3676:connecting-renewable-energy-to-the-electric-transmission-grid-&catid=45:solar-feature-articles&Itemid=112.

Saugrain, Jean-Maxime. 2010 (October). Successfully Operating 200 kV Superconducting HVDC Transmission Highlights Reliability, Efficiency. Utility Automation Engineering. http://www.elp.com/index/display/article-display/3784736836/articles/utility-automation-engineering-td/volume-15/issue-10/features/successfully-operating-200-kv-superconducting-hvdc-transmission-highlights-reliability-efficiency.html.

Totaro, Philip. 2011 (February 18). Five Trends to Watch in Wind. Greentech Media. http://www.greentechmedia.com/articles/read/guest-post-five-trends-to-watch-in-wind.

Trabish, Herman K. 2010 (September 10). Wind's Future Ally: Better Transmission. http://www.greentechmedia.com/articles/read/winds-future-ally-better-transmission.

———. 2010 (October 7). Energy Storage for the Grid: Better Than the Holy Grail. Greentech Media. http://www.greentechmedia.com/articles/read/energy-storage-for-the-grid-better-than-the-holy-grail/.

———. 2010 (December 9). The Texas CREZ Plan to Get Wires for Wind. http://www.greentechmedia.com/articles/read/the-texas-crez-plan-to-get-wires-for-wind.

———. 2010 (December 24). Federal Regulators Approve 100 Percent Cost Sharing for New Transmission. https://www.greentechmedia.com/articles/read/federal-regulators-approve-100-percent-cost-sharing-for-new-transmission.

———. 2011 (January 28). The Transmission Quandary of California. http://www.greentechmedia.com/articles/read/the-transmission-quandary-of-california.

———. 2011 (August 8). Can Germany's North Sea Winds Blow Away Nuclear? Greentech Media. http://www.greentechmedia.com/articles/read/can-germanys-north-sea-winds-blow-away-nuclear/.

———. 2011 (December 13). Grid Integration, Part 2: The Rubik's Cube of Renewable Energy. Greentech Media. http://www.greentechmedia.com/articles/read/grid-integration-part-2-the-rubiks-cube-of-renewable-energy/.

Tweed, Katherine. 2010 (May 5). Tehachapi Renewable Transmission Project Completes Phase One. http://www.greentechmedia.com/articles/read/phase-one-completed-in-tehachapi-renewable-transmission-project.

U.S. Department of Energy. 2009 (December). National Electric Transmission Congestion Study. http://congestion09.anl.gov/documents/docs/Congestion_Study_2009.pdf.

U.S. Department of Energy, Energy Efficiency and Renewable Energy. 2008 (July). 20% Wind Energy by 2030: Increasing Wind Energy's Contribution to U.S. Electricity Supply. DOE/GP-102008–2567. http://www1.eere.energy.gov/wind/pdfs/41869.pdf.

———. 2010.
Wind Power Today—2010. http://www1.eere.energy.gov/windandhydro/pdfs/47531.pdf.

U.S. National Renewable Energy Laboratory, Wind Powering America. 2005 (May). Wind Energy Myths. DOE/GO-102005-2137. http://www.nrel.gov/docs/fy05osti/37657.pdf.

Varma, K.J.M. 2011 (February 17). Wind Power Wasted in China Due to Poor Grid Connectivity. http://www.powergenworldwide.com/index/display/wire-news-display/1361827140.html.

Wesoff, Eric. 2010 (November 9). Gravity Power's New Take on Pumped-Hydro Energy Storage. Greentech Media. http://www.greentechmedia.com/articles/read/Gravity-Power.

———. 2011 (March 9). Update: Gigawatt-Scale Energy Storage by Rubenius. Greentech Media. http://www.greentechmedia.com/articles/read/update-gigawatt-scale-energy-storage.

———. 2011 (March 10). MIT's Star Prof. Don Sadoway on Innovations in Energy Storage. Greentech Media. http://www.greentechmedia.com/articles/read/mits-don-sadoway-on-energy-storage-and-his-liquid-battery.

———. 2011 (March 24). Xtreme Blends 15MW of Energy Storage with Wind. Greentech Media. http://www.greentechmedia.com/articles/read/xtreme-power-on-hawaii/.

———. 2011 (April 18). Ionex Has a 1 MW Battery Ready for the Grid. Greentech Media. http://www.greentechmedia.com/articles/read/ionex-has-a-1-mw-battery-ready-for-the-grid/.

———. 2011 (August 17). Beacon Power's Spin on Energy Storage: It's a Service. Greentech Media. http://www.greentechmedia.com/articles/read/beacon-power-a-new-spin-on-energy-storage/.

———. 2011 (September 13). AES Brings Real Utility-Scale Energy Storage to the Grid. Greentech Media. http://www.greentechmedia.com/articles/read/aes-bringing-real-utility-scale-energy-storage-to-the-grid/.

Widman, Miriam. 2012 (February 6). Germany Invests in Hydrogen Technology for Renewables Storage,Vehicles. Renewable Energy World. http://www.renewableenergyworld.com/rea/news/article/2012/02/germany-invests-in-hydrogen-technology-for-renewable-storage-vehicles?cmpid=WNL-Wednesday-February8-2012.

Wiser, Ryan, and Bolinger, Mark. 2010 (August). 2009 Wind Technologies Market Report. http://www1.eere.energy.gov/windandhydro/pdfs/2009_wind_technologies_market_report.pdf.

———. 2011 (August). 2010 Wind Technologies Market Report. http://www1.eere.energy.gov/windandhydro/pdfs/51783.pdf.

Xtreme Power, http://www.xtremepower.com/.

Yuanyuan, Liu. 2011 (April 26). China's Wind Power Sector Expected to See Tremendous Growth. *Renewable Energy World*. http://www.renewableenergyworld.com/rea/news/article/2011/04/chinas-wind-power-sector-expected-to-see-tremendous-growth??cmpid=WNL-Friday-April29-2011.

9

Offshore Wind: Entering the Mainstream

Introduction

Compared with onshore wind power, offshore technology and industry experience are less mature, and cost reductions have yet to materialize as once hoped. The technical challenges offshore are formidable, and so are the risks. That might sound like a gamble, but it's one that many developers, investors, and policy makers are willing to take.

Advantages

Offshore wind makes sense from many perspectives. Number one, winds are stronger and steadier offshore, so turbines not only deliver more wind energy, but they also need not cope with as much turbulence and wind shear. This means that turbines suffer less wear and tear, so they need less maintenance and should last longer.

Second, the seas are vast in area (oceans cover about 70% of the planet), offering plenty of wide-open space for siting wind farms. This is especially attractive to densely populated countries where land is at a premium. Third, massive wind power system components are easier to transport over water by ship or barge, so larger turbines can be installed offshore than on land, where transportation logistics have begun to limit wind turbine size.

Offshore turbines are more likely to supply wind power during times when it's needed most, on hot summer afternoons, thanks to the sea breeze effect. When the air over land heats up and rises, the cooler, heavier air over the ocean moves in, creating strong winds offshore and along coastlines. In contrast, the energy output from onshore wind farms often doesn't match well with consumers' demand for electricity.

When sited offshore, wind farms can deliver power to large cities where it would be impossible to build an onshore wind turbine. This means that offshore wind can help relieve bottlenecks in the power system by supplying electricity directly to load centers nearby, without cross-country transmission and energy losses. This has a rippling effect throughout regional power markets, enabling them to function more efficiently.

Like onshore turbines, offshore machines are compatible with some existing uses such as recreational fishing, which can continue within a wind farm boundary. Also, offshore turbine foundations create artificial reefs that support marine life.

Disadvantages

Compared with land-based wind, offshore wind power development is immature, as noted above. Technologies are improving, but building a wind farm offshore still costs much more than it does on land. Turbine foundations, electrical systems, and other wind power system components can cost two or three times as much as those on land.

Higher capital cost means that all of the financing hurdles for wind farm developers described in chapter 6 are that much more problematic. Also, operation and maintenance costs are greater. As a result, wind energy produced by offshore wind farms can cost twice as much as land-based wind power.

The environment offshore is also more challenging. Strong winds are good, up to a point. While high wind speeds enable more energy to be captured, turbines must be able to withstand more powerful gusts offshore, not to mention hurricanes, cyclones, and the occasional tsunami wave. Overall, wind turbines must be more robust and reliable to thrive in offshore conditions, and many opportunities remain for tailoring existing technology to better exploit offshore wind.

Maintenance is much more difficult offshore, where engineers must get to the turbine by ship or helicopter. Storms frequently delay access to turbines that need work. Also, the salty sea air and water can cause turbine components to corrode, unless they are designed for this harsh environment.

But the size of the world's offshore wind resource is so large that it's hard to comprehend (see chapter 5). If global potential could be tapped efficiently, the offshore resource won't be a limiting factor in wind power development. Europe's offshore winds could supply about seven times its current demand for electricity. For America, this figure is roughly four times the nation's electric generating capacity.

The Offshore Wind Industry

The first commercial offshore wind farm was built in Denmark's waters in 1991, so engineers have had more than 20 years of experience with offshore wind power. But until recently, many offshore wind farms had reliability problems that, along with a shortage of suitable turbines, restrained development. Now the offshore wind sector has begun to move into the mainstream, building utility-scale wind farms and capturing a growing share of the wind power market.

Most offshore wind turbines today are installed along coastlines in relatively shallow water, but the industry is extending its reach. Wind farms are being built in water depths up to 40 meters (over 100 feet), some turbines are operating as far as 100 kilometers from shore (more than 50 nautical miles), and a *floating wind turbine* is being tested in much deeper water (more than 200 meters, or 700 feet). Design standards also are evolving for offshore wind farms, including the 2011 revision by Det Norske Veritas (DNV) covering turbine structures.

Market snapshot

Eleven countries had wind farms operating offshore as of 2011, nine of them in Europe—Belgium, Denmark, Finland, Germany, Ireland, the Netherlands, Norway, Sweden, and the United Kingdom. Japan had three shallow-water, near-shore wind farms, and the 11th country, China, commissioned its first offshore wind farm in 2010. The capacity of offshore wind farms in operation worldwide totaled more than 3 GW.

By 2015, at least 16 GW of offshore wind capacity could be added globally. If significant growth in China comes online as expected, along with a modest contribution from America, global offshore capacity could shoot up to 75 GW by 2020, according to BTM Consult. Another analysis, by Pike Research, projects about 60–70 GW worldwide by 2018, including more than 11 GW in Chinese waters.

The European Union is by far the world's leader in offshore wind power deployment (fig. 9–1 and table 9–1), accounting for 96% of global capacity. Among Europe's countries, the United Kingdom has the most offshore wind power capacity, with a total of 1.34 GW, after installing 458 MW in 2010 alone.

UK wind farms that were under construction or approved (1.6 and 2.1 GW, respectively) would bring this total to around 5 GW. If preapproved projects are counted, the United Kingdom has more than 40 GW of offshore capacity planned. Of this, 12 GW could come online by 2017, and the country's total could reach as much as 37 GW by 2020.

Denmark commands the second-largest fleet of offshore wind farms in terms of capacity. The country installed 207 MW in 2010, the second-biggest addition after the United Kingdom. Belgium was third in new installations, adding 165 MW of capacity as of 2011.

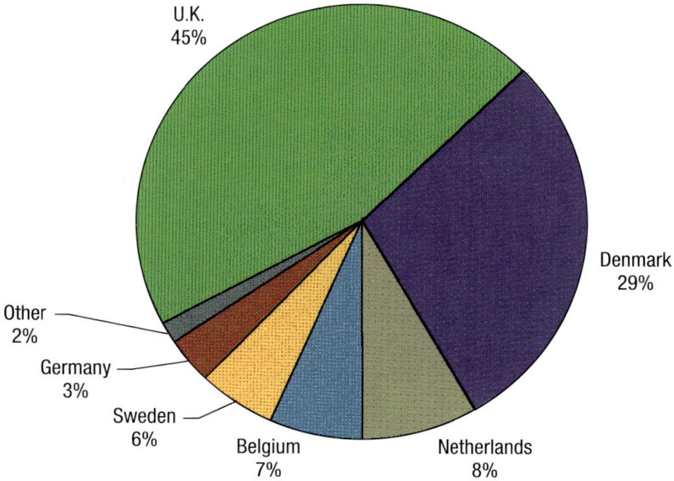

Fig. 9–1. European offshore wind power market shares

Table 9–1. Europe's offshore wind power capacity by country

Country	Capacity, MW	% of Total
UK	1,341.2	45.52
Denmark	853.7	29.00
Netherlands	246.8	8.38
Belgium	195.0	6.62
Sweden	163.7	5.56
Germany	92.0	3.12
Finland	26.3	0.89
Ireland	25.2	0.86
Norway	2.3	0.08
Total	2,946.2	100.00

Germany is emerging as one of the world's biggest markets for new offshore wind power, after a relatively late start. More than 1 GW of capacity was under construction and about 8.5 GW more had been approved. Germany could reach almost 10 GW of offshore wind capacity when these projects are completed.

A 48.3-MW, 21-turbine wind farm started up in 2011 in the Baltic Sea off Germany, operated by utility EnBW. An 80-turbine project was planned to come online by 2014, also in Germany's sector of the Baltic. The country's plans to reduce reliance on nuclear energy could accelerate offshore wind power development. At least 23 wind farms totaling 5.65 GW in capacity were approved for the North Sea, plus three in the Baltic totaling 1.04 GW.

France has lagged its neighbors, with no offshore wind farms in operation. However, in 2011 the government said it would build 6 GW of offshore wind power capacity by 2020. Plans are to invest $14 billion (10 billion euros) in 2–3 GW of offshore wind power, with contracts being tendered in 2011. About 30 French projects were in early stages with a capacity totaling 8 GW, potentially exceeding the government's target.

Table 9–2 lists the world's largest offshore wind farms. Not surprisingly, all of them are in northern Europe. At least five new projects have been sited in the United Kingdom and Germany that will each have capacities greater than 500 MW. One UK site under development, the London array, could become the world's largest offshore wind farm, at 630 MW, when completed in 2012. Appendix G provides details on offshore wind farms in Europe and worldwide.

European offshore installations set a record in 2010, growing by 51% and accounting for 9.5% of Europe's new wind power. Total capacity approached 3 GW as of 2011, consisting of 1,136 grid-connected turbines operating in 45 wind farms. The average wind farm size more than doubled in 2010 to 155.3 MW from 72.1 MW in 2009.

Installation of another 1–1.5 GW of offshore wind capacity in Europe was slated to be finished and grid-connected in 2011. Ten wind farms were being built, totaling about 3 GW of capacity. When all of these are completed, Europe's installed capacity will exceed 6 GW.

Table 9–2. World's largest offshore wind farms

Name	Country	Capacity
Walney	UK	367.2
Thanet	UK	300.0
Horns Rev 2	Denmark	209.3
Nysted 2/Rødsand 2	Denmark	207.0
Robin Rigg	UK	180.0
Gunfleet Sands	UK	172.8
Nysted 1/Rødsand 1	Denmark	165.6
Belwind	Belgium	165.0
Horns Rev 1	Denmark	160.0
Prinses Amalia	Netherlands	120.0
Lillgrund	Sweden	110.4

If you count output from wind farms that are fully approved, the total reaches 19 GW in the near term. Looking farther out, Europe's offshore wind capacity might reach 110 GW by 2020, 150 GW by 2030, and 350 GW by 2050.

Although most offshore wind farms are in relatively shallow water, the trend is moving deeper. The average water depth increased 43% in 2010, to 17.4 meters (57.1 feet). New projects being built lie in waters averaging 25.5 meters deep (83.7 feet), a 47% increase over 2010.

Similarly, new wind farms are being sited farther offshore. The average distance rose 88% in 2010, to 27.1 kilometers (14.6 nautical miles). This will increase to 35.7 kilometers (19.3 nautical miles) for projects under construction. The developers of Bard 1 installed the first cluster of turbines about 100 kilometers off Germany (more than 50 nautical miles).

Key players

Operators. Globally, the leading offshore wind farm operators are Vattenfall (Sweden), E.on (Germany), and Dong Energy (Denmark), which installed 308, 305, and 65 MW, respectively, in 2010. Other major developers include Scottish and Southern Energy (UK) and Bard, EnBW, and RWE (Germany).

Nearly all of the world's offshore wind capacity has been controlled by about 10 companies or consortiums, though that number is growing. The Belwind consortium entered the market in 2010, installing 165 MW off Belgium. The group is made up of Dutch and Belgian investors.

Turbine manufacturers. Twenty-one turbine suppliers introduced 29 new models for offshore duty in 2010. Commercial offshore turbine ratings can exceed 6 MW (table 9–3).

Vestas (Denmark) and Siemens (Germany) are the world's top suppliers of offshore wind turbines, together accounting for more than 90% of installations (with 1.391 and 1.357 GW of capacity, respectively). Vestas installed 555 MW of wind turbines offshore in 2010, followed by Siemens with 278 MW.

However, new players are muscling into the market rapidly, such as REpower, Bard (both German), and Areva (France), as well as China's Sinovel and Finland's WinWinD. Several more Chinese manufacturers plan to enter the offshore wind market.

Although direct-drive technology is penetrating the market for large offshore turbines, geared machines won't be phased out of the offshore market for a while, mainly because they're cheaper. Among commercial machines, Sinovel's 6-MW turbine uses a gearbox, while XEMC's is direct-drive.

The world's turbine suppliers are developing more new products for the offshore market (table 9–4), while super-size turbines are also in the works (see "Bigger turbines").

Table 9–3. Commercial offshore turbines

Manufacturer	Capacity, MW	Rotor diameter, m (ft)	Hub height, m (ft)
REpower	6.15	126 (413)	85–95 (279–312)
	5	126 (413)	90–100 (295–328)
Sinovel	6	128 (420)	n/a
	3	90–113 (295–371)	80–110 (262–361)
Areva	5	116 (381)	90–130 (295–427)
Bard	5	122 (400)	90 (295)
XEMC	5	115 (377)	65–80 (213–262) project-specific
GE	4	113 (371)	n/a
Siemens	3.6	107–120 (351–394)	80–90 (262–295) site-specific
Doosan	3	91.3 (300)	80 m (262)
Sany	3	110 (360)	90–100 (295–328)
Vestas	3	90–112 (295–367)	n/a
WinWinD	3	90–109 (295–358)	80–100 (262–328)
Nordex	2.5	90 (295)	70–80 (230–262)

Table 9–4. Offshore turbines under development

Manufacturer	Capacity, MW	Rotor diameter, m (ft)	Introduction
Alstom	6	n/a	2012–2015, 40-MW demo. at Belwind
Bard	6.5	120 (390)	2-unit demo. at Rysumer Nacken
Doosan	6	n/a	n/a
Gamesa	5	128 (420)	2013, preseries
	6–7	n/a	2014, preseries
Hyundai	5.5	n/a	prototype by 2012
Nordex	6	150 (490)	2014, 70-unit demo. in Baltic Sea
Samsung	7	n/a	n/a
Vestas	7	164 (540)	prototypes by 2013; series in 2015

The 6.5-MW Bard turbine, with a hub height of 90 meters (about 300 feet), squeezes extra power out of the 5-MW model within the same physical size, using many of the same parts—tower, foundation, and rotor—and weighing only 2% more.

The drivetrain is split, with one main gearbox connected via twin driveshafts to two 3.4-MW permanent-magnet synchronous generators. This means that replacement parts are more readily available from the 3-MW generator market.

The 7-MW Vestas machine will also be geared rather than direct-drive. It's designed for a minimum hub height of 105 meters (344 feet) and water depths up to 45 meters (150 feet).

Chinese manufacturers, including Goldwind and Mingyang, are also working on turbines of 5 MW or more. Sinovel established an R&D center in China to develop offshore technology, including turbines bigger than 10 MW. Goldwind intends to shift some of its capital from onshore to offshore wind. However, Chinese machines will likely be deployed and tested thoroughly in Asian markets before being introduced on a global scale.

Supply chain. The supply chain for offshore wind turbines—the companies that manufacture, distribute, and deliver products to customers—is narrower than the supply chain for land-based wind farms. That's because of the bigger size of the components and the extremely high quality required for offshore duty.

However, the offshore supply chain has expanded substantially with the buildup of installed capacity in Europe and the huge number of projects in the global development pipeline. Many suppliers are setting up manufacturing plants, training centers, and test laboratories in the United Kingdom, where offshore wind power development has accelerated, particularly in Scotland.

One example is the Centre of Engineering Excellence for Renewable Energy at Glasgow's University of Strathclyde, where Siemens will collaborate with utility Scottish and Southern Energy to establish commercial agreements (technical specifications and prices) during the turbine design process. This reflects an industry trend for wind farm developers and turbine manufacturers to test new designs together. Scotland established a fund of $56 million (35 million pounds, or 41 million euros) to invest in production of new wind turbines.

Iberdrola, a Spanish renewables leader, sited its global offshore wind headquarters in Glasgow. Korea's Samsung will design and test a new wind turbine in Fife, Scotland. Gamesa will also headquarter its offshore operations in the United Kingdom, with plans to invest up to $200 million (150 million euros) by 2014. The facility will include a technology center and a blade manufacturing plant. Other UK ports will serve as Gamesa's turbine maintenance bases.

Siemens, Mitsubishi, GE, and Clipper also could base offshore turbine and blade manufacturing facilities in the United Kingdom. Although Vestas closed a blade factory in Britain, it was building an R&D center there. Vestas also plans to develop a facility at a UK port for manufacturing the company's 7-MW machine. Doosan will locate an R&D center near Glasgow, to be followed by assembly and manufacturing facilities.

More than 200 major players were involved in supplying balance-of-plant products and services for the offshore wind industry. These include transformer suppliers, seafloor surveying, cable laying, foundation construction, and specialty installation and service vessels.

Vessels. Shipbuilders in Korea, Poland, and Germany dominate the market for specialty vessels. Some offshore wind power developers team up with shipbuilders to ensure that vessels will be available, such as the agreement between Areva and Germany's Beluga Hochtief Offshore, which operates a 60-vessel multi-purpose fleet.

About 28 specialty vessels were in operation in 2011 capable of installing turbines or foundations, or both, with enough capacity to handle 3–4 GW/yr. But 23 more vessels were due to be delivered by 2014, so an oversupply could develop for a few years as the fleet's annual installation capability rises to nearly 2,000 turbines. Day rates could drop, and some vessels could be diverted to other industries or restricted to operation and maintenance. Ports with quays designed for loading/unloading foundations and turbines could become more of a bottleneck.

The cable-laying market is limited to a handful of players, and cable installation could snag offshore wind farm deployment. Demand for cable-laying vessels is expected to increase, with the European market alone requiring up to 24 such vessels equipped with dynamic positioning by 2016. Globally, more than 30 cable layers could be needed by 2020.

Also, the volume of transmission cable needed to export wind power to land could explode by 1,300% during 2011–2019, according to ODS-Petrodata. Pike Research predicts pressure on the cable supply chain, with offshore transmission projects increasing to more than 350 by 2020. HVDC cable installation will grow faster than its AC counterpart, Pike said.

If America's offshore wind industry opens for business, an opportunity could emerge for companies to build ships, barges, and other custom vessels for installing and maintaining offshore turbines. U.S. law precludes using foreign-made or foreign-operated ships. Although existing oil and gas vessels can be retrofitted, they have not performed as well as custom-built vessels for offshore wind turbine installation.

One American developer plans to build a turbine installation vessel for a proposed wind farm off Delaware, and Gamesa established a partnership with America's largest shipbuilder, Northrop Grumman, to launch Gamesa's 5-MW turbine. Another U.S. ship operator partnered with a UK seafloor surveying company to develop a marine surveying research vessel.

Expanding Beyond Europe

China

China's long coastline—about 14,500 km (9,000 miles)—could give the country an enormous offshore wind resource, estimated at a potential 200–700 GW. China is revving up its offshore wind industry at an impressive speed. By 2020, offshore wind might account for 12% of China's total wind capacity. The government's plans call for 32.8 GW of offshore wind power to be installed by 2020 (table 9–5).

The country commissioned its first offshore wind farm in 2010, a 102-MW project at Shanghai's Donghai Bridge, comprised of 34 Sinovel 3-MW turbines. Already, this gives China sixth place in global offshore wind power capacity (see appendix G). The second stage of the project will add 116.6 MW of capacity. China is also planning to build four offshore wind farms with a total capacity of 1 GW along the eastern coast off Jiangsu Province, which should be finished by 2014.

The four new projects were awarded via auction, giving winners a chance to participate in price-setting for offshore wind power projects. Previously, the government established four feed-in tariff rates for land-based wind power (see chapter 11). The offshore bidding results will serve as input for regulators to determine future prices for offshore wind. The winning prices were $97–$116 per megawatt-hour (620–740 yuan, or 71–85 euros).

In other Asian countries, South Korea plans to build a 2.5-GW wind project (500 5-MW turbines) more than 20 km offshore (12 miles), and Japan plans to test six 2-MW floating wind turbines.

Table 9–5. Offshore wind power capacity planned for China's coastal provinces. *Data courtesy of China Wind Power Outlook 2010.*

Province	Planned Capacity, MW 2015	2020
Shanghai	700	1,550
Jiangsu	4,600	9,450
Zhejiang	1,500	3,700
Shandong	3,000	7,000
Fujian	300	1,100
Other (tentative)	5,000	10,000
Total	15,100	32,800

United States

America has a strong incentive to develop offshore wind power. Recent assessments place U.S. offshore wind energy resources at about 4,150 GW (see chapter 5). In general, however, wind speeds are not as high as Europe's, and many U.S. offshore areas are prone to hurricanes and extreme wind speeds, which would increase insurance costs.

In the continental U.S., 28 states have a coastal boundary, including some in the freshwater Great Lakes region. These states consume 78% of America's electricity, and most of them have enough offshore wind potential to meet all of their power demand. What's more, U.S. seaboards have lots of densely populated cities where transmission capacity is constrained and retail electric rates are high.

Offshore wind power could relieve power system congestion and perhaps contribute to lower electric prices in the long run. While offshore wind projects might have a fixed-price contract at a high rate, wind farm operators can still bid into the power system hourly, usually at a much lower rate. So offshore wind power could displace some premium-priced power in short-term markets while the wind farm's contracted power remains in the system operator's dispatch ladder.

Projects. Several offshore wind farms are in various stages of development. Cape Wind off the Massachusetts coast—a 468-MW project—has all of its permits and should start construction in 2013, after nearly a decade of regulatory delays and at least 13 lawsuits. The waves in Nantucket Sound where the turbines are sited are low compared to open ocean conditions. Two power purchase agreements were in place, and the project might start up as early as 2015.

Two other offshore wind farm projects had power purchase agreements—NRG Bluewater Wind in the mid-Atlantic off Delaware (for 200 MW) and Block Island off Rhode Island (30 MW). However, Bluewater Wind was unable to finance its project and terminated the agreement while retaining development rights.

Other American states are vying to become the site of America's first offshore wind farm, betting that they can get turbines installed before Cape Wind's. Texas is planning a 300-MW project off Galveston on the Gulf Coast. Fishermen's Energy hopes to install 24 MW off New Jersey initially, then an additional 350 MW. The state provides renewable energy credits specifically for offshore wind.

Meanwhile, the state of Maine is pioneering U.S. deepwater wind technology (see "Deepwater floating turbines" below). In addition to validating design tools, researchers will test lightweight composite materials and structural components at a new research laboratory at the University of Maine. The state won $20 million in federal funding (15 million euros) to develop deepwater wind power.

States along the freshwater U.S. Great Lakes also are planning to exploit wind power. A 20-MW project in Lake Erie off Ohio has a lease option, and the developer is evaluating the area to site five GE turbines. Ohio's goal is to install up to 1 GW of wind energy capacity in its lake waters, which are shallow and close to load centers.

I've watched the offshore wind industry in the United States inch forward for the past ten years, with individual developers and state initiatives leading the way. It seems like we've finally reached a tipping point where all these disparate activities are starting to come together.

—Fara Courtney, U.S. Offshore Wind Collaborative

Policy. A long-term, stable energy policy at the national level could give offshore wind developers the certainty they need to obtain project financing. But the U.S. government is unlikely to reward wind power or other renewable resources with more incentives any time soon.

However, federal policymakers learned from Cape Wind's experience. One reason the development took so long to get approved was that no single government entity is in charge. Instead, a myriad of national, state, and local agencies have some control over siting and permitting offshore wind farms. National agencies alone are responsible for executing as many as 19 different laws dating back to 1899.

That won't change much, but two federal departments—Energy and Interior—are stepping in to coordinate and accelerate the development process, as well as reduce wind energy costs. Their targets are 10 GW of offshore capacity by 2020 at a cost of 10 U.S. cents per kilowatt-hour (7 euro-cents) and 54 GW by 2030 at 7 U.S. cents (5 euro-cents).

Federal agencies control leasing in most U.S. waters more than 5.6 kilometers from shore (3 nautical miles), which includes the Outer Continental Shelf. Individual states have authority over activities inside that distance.

One Interior Department program, Smart from the Start, has targeted four Mid-Atlantic wind energy areas for regional environmental reviews, instead of individual site reviews. The program is targeting similar areas for review along the northern and southern Atlantic coast. Leasing could begin in 2012, with bidding on block areas of about 5 by 5 kilometers (3 by 3 miles).

Regulators hope that regional environmental reviews will slash permitting time by up to two years. The federal process eliminates conflicts with shipping, military operations, air-traffic radar, fishing, and recreation. Offshore wind farm installation in these areas could begin by 2016.

In addition to streamlining siting and permitting, Smart from the Start fast-tracks areas that are already being assessed and should also accelerate transmission line development. An offshore transmission line has been proposed (see "Offshore Grids" below) and would be considered as part of the energy delivery system for new wind projects.

Smart from the Start focuses on the Atlantic coast because waters are relatively shallow, less than 30 meters deep (100 feet). Nearly 213 GW of wind resource is considered available along the coast, after subtracting areas that might be restricted from development because of environmental or socioeconomic factors (table 9–6).

Early signs suggest that opponents to offshore wind might become more willing to compromise. A group of 122 environmentalists, conservationists, and elected officials publicly urged support for swift development of Atlantic offshore wind, given strong environmental safeguards. Table 9–7 summarizes Atlantic offshore wind projects in the pipeline.

Projects are also being considered on the Pacific coast, where deep waters still pose a technical challenge, and offshore wind farms are proposed in tropical waters prone to hurricanes, such as the U.S. Gulf of Mexico and southern Atlantic coast. New design guidelines have been developed for foundations in such waters, based on existing standards established by the International Electrotechnical Commission and the American Petroleum Institute.

Table 9–6. U.S. Atlantic coast offshore wind resource (< 50 km/30 miles out and <30 m/100 ft deep). *Data courtesy of the U.S. National Renewable Energy Laboratory (NREL).*

Region/states	Wind resource, GW		
	Total	Limited	Available
New England and New York	100.2	60.12	40.10
Mid-Atlantic (New Jersey–North Carolina)	298.1	178.86	119.24
South Atlantic (South Carolina–Florida)	134.1	80.46	53.64
Total	532.4	319.44	212.98

Table 9–7. U.S. Atlantic coast offshore wind projects. *Data courtesy of National Wildlife Federation and American Wind Energy Association.*

State	Advancing projects	Proposed projects
Maine	Demonstration (25 MW) Monhegan Isl. deepwater test site (110 kW)	
New Hampshire		Test turbine (10 kW)
Massachusetts	Cape Wind (468 MW)	Hull (12–15 MW)
Rhode Island	Block Isl. (30 MW)	Rhode Island Sound (1 GW)
New York	Long Isl.-New York City (350–700 MW)	
New Jersey	Garden State (345 MW) Fishermen's Energy-Atlantic City (24 MW) Fishermen's Energy (350 MW) NRG Bluewater (348 MW)* Offshore MW 1 (700 MW)	OffshoreMW 2 (700 MW)
Delaware	NRG Bluewater (300–450 MW)*	
Maryland		Ocean City (450 MW)
Virginia		Apex Hampton Roads (n/a) Seawind Renewable Energy (480 MW) Hampton Roads demonstration (3 test turbines)
North Carolina		Outer Banks Ocean (>1 GW)
South Carolina		Palmetto Wind/Clemson University (monitoring)
Georgia		Southern Winds (monitoring)

Notes: Proposed projects have been publicly announced and are being planned.
Advancing projects have taken additional steps such as leasing, permitting, and/or power purchase agreements.
* Projects on hold as of 2012.

Offshore Grids

Europe

Europe envisions an offshore supergrid that will evolve slowly as national offshore grids and interconnections are expanded. Most offshore wind power is being developed in shallower waters along the northern coasts, so Europe's plans focus mainly on the North, Baltic, and Irish Seas.

At least 11 individual offshore grids were operating in European waters, with another 21 being built or considered by system operators in the Baltic and North Seas as of 2011. European utilities are already expanding their intercountry links in the North Sea, including subsea HVDC transmission lines connecting Norway, Denmark, the United Kingdom, and the Netherlands, as noted in chapter 8.

Responsibility for developing and operating Europe's offshore grids lies with Entso-e, the European Network for Transmission System Operators for Electricity. Entso-e has formed subgroups to deal specifically with offshore grids and transmission. Issues to be resolved include the differing regulations and markets in individual countries and the lack of rules for giving wind power priority feed-in versus bidding in day-ahead trade.

Europe also has regional efforts under way, such as the North Sea Countries' Offshore Grid Initiative. This group, which includes governments, regulators, and transmission system operators, will create a common political and regulatory basis for offshore infrastructure development among the 10 countries surrounding the North and Irish Seas (Belgium, France, the Netherlands, Luxembourg, Germany, the United Kingdom, Ireland, Denmark, Sweden, and Norway).

By 2013, the group will recommend measures on grid design and integration, market and regulatory issues, and procedures for planning and authorizing projects, according to an agreement signed in 2010 by all 10 countries and the European Commission.

Europe has mapped out a blueprint for expanding its grid to accommodate wind power and other renewable energy, with specific goals for each decade. The plan establishes priorities for infrastructure construction and will accelerate the permitting process for key projects with a time limit on the final decision. Europe envisions a 100% renewables-powered grid by 2050, with wind power supplying up to half of Europe's electricity demand.

Some groups oppose the development of an offshore supergrid for Europe, citing a study by Pöyry which found that better connectivity would not greatly smooth out the variability of wind power and other renewables. Several countries across Europe can get hit by the same weather at the same time, which could idle most wind turbines within the grid or, conversely, overwhelm the system with wind power.

Also, although electricity prices would fall in some countries, they would rise in others, creating winners and losers. The analysis suggested that wind's variability can be better addressed by demand response.

United States

Private developers and investors, including high-profile Internet giant Google, plan to build a transmission supergrid along the Atlantic coast, the focus of America's near-term offshore wind power development. The

560-kilometer (350-mile) backbone, called the Atlantic Wind Connection, would be a 6-GW HVDC line running along the U.S. East Coast from New Jersey to Virginia.

The Atlantic grid would link offshore wind power to the strongest nodes on PJM, the largest U.S. system operator and the one that controls the mid-Atlantic coast. Offshore hubs along the line would convert AC power from wind farms to DC for transmission to shore. The project cost is estimated at $5 billion (4 billion euros). However, building dedicated connections from each offshore wind farm could cost far more, perhaps $9–$15 billion (7–11 billion euros).

The Atlantic grid would let developers site projects more efficiently, where the wind energy resource is the best, rather than worrying about transmitting wind power to the nearest city or the highest price market. Wind power transmitted to shore could also reinforce weak spots in the onshore grid and relieve congestion in the region.

Mixing output from several wind farms could make wind energy's supply less variable, minimizing integration costs. When offshore winds are calm and output falls, the grid can carry conventional power. By moving electricity along the coast and leveling out price differences, the offshore grid could save consumers $1.5 billion per year (1.1 billion euros) in lower marginal energy prices.

The process of obtaining regulatory approvals began in 2011 and will likely take several years. Proponents of the Atlantic transmission line include some coastal states seeking to exploit offshore resources and some wind farm developers who are unaffiliated with utilities, along with a prominent international ocean conservation group, Oceana.

Opponents include utilities, other electricity suppliers, and their partners in wind power development, as well as states with low electricity prices and ratepayer advocates concerned with offshore wind energy's high cost.

Offshore Wind Farm Development

The fundamentals of onshore and offshore technology are the same, but construction methods and wind farm operations are very different, with greater attention offshore to reliability and maintenance issues. As the offshore wind industry grows, the process of development will probably become more different from its onshore counterpart.

Challenges offshore include resilience to tropical storms, with extreme winds and high waves, and coexistence with marine life, shipping, recreational boating, and other activities. At greater distances offshore, engineers must figure out how to deal with deeper water, exposure to more extreme weather in the open ocean, long-distance electrical transmission on high-voltage submarine cables, and turbine maintenance at sea.

Experience in the oil and gas industry is valuable in offshore wind technology development. Drilling platforms are similar to wind turbines in some ways—they're massive structures that must remain stable during ocean storms. Also, offshore wells rely on subsea electrical cables and pipelines, much like offshore wind turbines require a cable network on the seabed to transmit power.

Assessment and resources

Wind speeds are typically higher offshore than on land because the ocean's surface, by comparison, is flat and free of obstacles, so developers don't usually need to microsite each turbine. The wind is also less turbulent because of the sea's smooth surface. Strong, steady winds make offshore resources very compelling to prospectors and developers.

Technologies for assessing the wind resource offshore are similar to those used for onshore assessment, with some additional issues to consider. These include technological factors such as the maximum water depths and distance from land that are feasible. They also include different environmental and regulatory constraints such as marine conservation areas, busy shipping lanes, and the distance from shore needed to minimize visual impact.

As with onshore assessment, the wind resource offshore can be modeled using existing measurements from coastal weather stations, ships, ocean buoys, and satellites. However, prevailing winds differ onshore and offshore, and they vary within the coastal zone. Also, atmospheric boundary layers differ between coastal and marine environments.

To get a better picture, developers should record offshore measurements at the site using a meteorological mast. Offshore met tower platforms are commercially available that can be sited in water depths up to 45 meters (150 feet).

Lidar technologies are being tested offshore, as noted in chapter 5. Figure 9–2 shows Natural Power's ZephIR lidar unit operating on a wind energy research station in the German sector of the North Sea. Another ZephIR was installed to collect wind speed data at heights up to 200 meters (660 feet) at the UK's Dogger Bank site about 125 km from shore (67.5 nautical miles).

An advanced lidar system, the SeaZephIR, will be used to assess the wind resource in the northern Atlantic off New Jersey. This system, mounted on a floating buoy and powered by solar and wind energy, measures ocean movements and wave heights as well as wind characteristics.

Fig. 9–2. ZephIR deployed offshore on the FINO 1 wind energy research platform. *Image courtesy of Natural Power.*

Another lidar technology, Catch the Wind's Vindicator, will be integrated with floating buoys made by Axys Technologies. The assessment buoy will validate its data using GL Garrad Hassan's protocol. One of the buoys, to be tested near an onshore met tower, will be deployed off New Jersey and will measure bird activity as well as weather.

Siting and permitting

Despite a few notable exceptions, it can be easier to site a wind farm offshore than on land. As mentioned above, much more open space is available. And even when wind farms are in shallow water, they can appear to be smaller and less intrusive than on land. The turbines' distance from shore also mitigates any noise they create while operating.

Opponents to offshore wind farms are generally concerned less with appearance and noise than with the possible impact on marine life, commercial fishing, and recreational boating, as well as the possibility of higher electric rates. Many environmental and socioeconomic factors must be considered:

- **Marine animal populations.** European studies suggest minimal impact, and offshore wind farms so far have not substantially affected bird or bat populations or habitats.
- **Visual effects.** Coastal residents react in different ways to the prospect of visible wind turbines.
- **Property values.** Experience on land shows little or no impact on real estate prices, but studies are less extensive in coastal communities.
- **Noise.** European experience underscores the need to reduce construction noise, specifically pile driving.
- **Tourism.** Some places depend on beach vacationers for revenues and taxes, but the evidence of offshore wind's impact is ambiguous, and actual effects seem minimal.
- **Marine safety.** If a ship crashes into a turbine, fuel could leak, and people could be endangered if the turbine collapses. No such incidents have yet occurred.

Prudent siting strategies that get all the stakeholders involved can go a long way toward relieving such concerns. Chapter 12 discusses offshore wildlife and environmental issues in more detail.

In general, European countries have been more amenable to offshore wind development, largely to meet government goals for reducing carbon emissions by exploiting renewable energy resources. Denmark has one of the simplest development processes for offshore wind farms, with the Danish Energy Agency acting as a one-stop shop for permitting and licensing.

The UK's rounds process for tendering bids has greatly accelerated offshore project approvals, which can take as little as 9–12 months, the fastest in Europe. The United Kingdom also offers generous financial incentives for offshore wind.

Financing and economics

Offshore wind farms typically cost about 30%–50% more to build than onshore, and some can cost up to twice as much. Norway's energy minister has said that building offshore wind farms is too expensive; instead, the country should increase its hydropower.

Nonetheless, many European offshore wind projects have been supported by governmental or quasi-governmental agencies such as the European Investment Bank and export credit agencies in Denmark and Germany, according to Navigant.

Offshore wind's costs could be partially offset by higher energy yields—30% or more, resulting from stronger, steadier winds. Also, offshore wind farm costs should fall as developers gain experience and as the technology matures. Designing and building offshore wind power cost-effectively is still at the learning stage.

Besides higher costs, investors are less enthusiastic about funding offshore development because a single general contract for a project is the exception rather than the rule. Instead of having one contractor in charge of engineering, procurement, and construction, multiple suppliers are involved. Each supply contract holds the manufacturer liable for its own product, so there's no single deep pocket of liability. Banks prefer a general contract. Also, the parameters of offshore projects are more diverse, making cost estimates more difficult.

However, the investment community has begun to show more confidence in the offshore wind market and its maturing technology. According to the European Wind Energy Association, the number of banks willing to take offshore wind risk is steadily growing. Germany's largest offshore wind farm obtained a loan for $754.1 million (550.5 million euros), and the UK's first and second rounds of tenders for offshore work seem to be attracting even some risk-averse investors.

Capital costs. Offshore wind projects are analyzed in terms of their initial capital cost (including construction) as well as their *life-cycle energy cost* over 20–25 years, also called the levelized cost of energy, as discussed in chapter 6. It would seem logical that offshore wind farm capital costs would rise with water depth and distance from land, and would fall with the size of the project, thanks to economies of scale. However, data from existing projects does not yet confirm this logic.

Capital costs started trending upward in 2007, reaching an estimated average of $4.26 million per megawatt (3.11 million euros) for projects proposed for 2010–2015. However, these estimated costs range broadly from $2 to $6.39 million per megawatt (1.5 to 4.66 million euros). The wind turbine contributes less to the cost of a wind power system offshore than on land—about 44% of the total, versus about 70%.

As an example of the capital required for offshore wind, one project under development in 2011 will cost about $1.8 billion (1.3 billion euros). This project involves the installation of 295 MW of wind power off Belgium and refinancing an existing 30-MW wind farm at the same site.

Life-cycle energy costs. In contrast to capital costs, the life-cycle cost of energy includes not only wind farm installation but also financing, operation and maintenance (O&M), and decommissioning, as well as projections of wind power system performance in terms of energy output and long-term reliability. Life-cycle energy costs are less certain than capital costs because they're estimated over a longer time period, with more variables. The wind turbine itself contributes less than 30% to total life-cycle costs (fig. 9–3).

O&M costs are typically higher offshore, mainly because it's harder to get to the turbines and to perform the work. Rough estimates say that O&M costs are two to three times greater, accounting for 20%–30% of the life-cycle cost of energy. In general, energy produced from offshore wind farms costs twice as much as land-based wind energy, due mainly to higher O&M costs and also rising capital costs.

Europe's wind farm developers look for synergies among projects—for example, where one expensive vessel and crew can service more than one wind farm. This helps boost the volume and scale of a single operator's projects.

Offshore wind farm economics are sensitive to the cost of financing. This suggests that even if capital costs continue to climb, energy production costs could be reduced if developers can get favorable financial terms for their projects.

Much of the financing cost is based on the perception of financial risk and project uncertainties. More research might help alleviate these perceived risks, but investor confidence built upon real-world deployment experience will have a greater impact, not only on financing but also on other factors affecting the cost of energy from offshore wind farms.

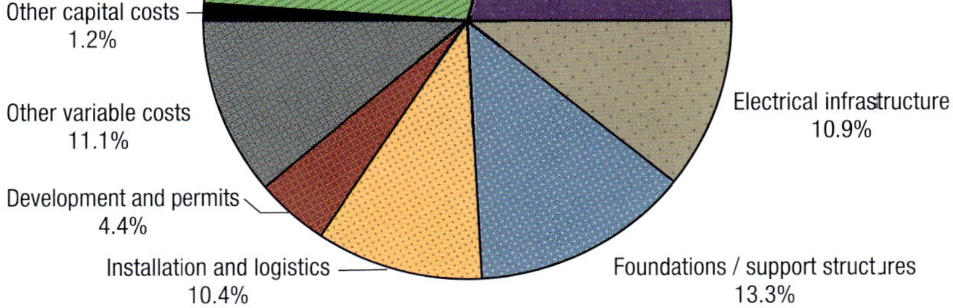

Fig. 9–3. Life-cycle cost of energy for typical offshore wind project. *Data courtesy of NREL.*

If offshore wind power is to compete in price with conventional generation, its cost of electricity to the consumer must be reduced by at least 50%. Proponents say this will happen as companies climb the learning curve, as contractors become more familiar with offshore installations, as port infrastructure is developed, and as the supply chain expands.

However, building the infrastructure implies one-time investments in custom vessels, port and harbor upgrades, new manufacturing facilities, and workforce training. Europe has already invested in much of this infrastructure, while new entrants to the offshore wind industry will have to establish similar capabilities. Making such an investment requires a long-term financial commitment.

Risk assessment

Although the basic process of designing and developing a wind farm offshore is similar to that on land, certain factors demand more attention, such as risk assessment. Compared with offshore wind farms, the world has many more wind turbines on land that have been operating for many more years. This means that the risk of something unpredictable going wrong is higher offshore. The early stages of a project should include a detailed front-end engineering and design study to mitigate risk and reassure investors.

Developers of offshore wind farms should also conduct a comprehensive risk assessment and revisit it regularly during the development process. Risk factors include government policies and regulations. Is there a chance that halfway through development, a financial incentive will disappear for some political reason? Strong political support for offshore wind can help alleviate these risks.

Another risk factor is the anticipated revenue from an offshore wind farm. Countries with feed-in tariffs that guarantee a price for wind power are the most hospitable, reducing developers' risk. Regulations that allow wind farm owners to sell renewable energy credits are also favorable.

Construction risk is substantially greater for offshore wind farms, since less experience has been gained so far. Laying electrical cables on the ocean floor is more complicated than stringing them overhead on land or even burying them underground, and the best turbine installation methods and vessels to use are still a subject for debate. Also, projects are more complex than onshore, requiring more careful management, especially concerning communication among all parties involved.

Similarly, O&M expenses are less predictable, due largely to the shorter history of offshore wind power's deployment on a large scale. Offshore weather and other environmental factors not only increase stresses related to turbine damage, but also raise issues of accessibility. For example, when vessels can't access turbines, standby charges can escalate into cost overruns that are not accounted for adequately in financial projections.

Turbine selection

In appearance, offshore wind turbines differ little from their onshore counterparts. Offshore towers are a little taller than those onshore, averaging around 90 meters (300 feet), and rotor diameters are greater to capture more energy from stronger winds. Because noise is less of an issue, offshore turbines can rotate faster, at lower torque. This can translate to lighter, less costly drive-train components.

However, because of O&M risks and costs, one key to a successful offshore wind farm project is reducing the need for turbine maintenance. Reliability has become the paramount concern for offshore turbines. They need to work from the start and throughout their lifetimes. The possibility that a turbine might fail within a few years is simply not built into the business case any more.

In the future, new offshore wind farms could increasingly specify direct-drive turbines if they prove to be more reliable and easier to maintain. Direct-drive machines have no gearbox, which is the top cause of wind turbine failures. Replacing a gearbox offshore can cost more than $1.4 million (1 million euros).

One new direct-drive turbine will be tested at the world's northernmost offshore wind farm, Havøygavlen, off Norway. Operator Statoil and manufacturer Siemens will run the machine for 17,000 hours, equivalent to two years of operation, to evaluate its potential for reducing maintenance. Also, a 6-MW model being developed by Alstom will use a direct-drive, permanent-magnet generator, said to be the world's largest for a wind turbine. The wind power system will be demonstrated at the 40-MW Belwind II wind farm off Belgium.

The average turbine in European offshore wind farms produces 3.2 MW, and many machines that generate 5 MW have been installed. Two wind farms that were financed in 2010 will use turbines of 5 MW or greater, signaling that investors have confidence in these upsized technologies.

European manufacturers are developing prototypes that deliver 6–7 MW, including dedicated offshore turbines (see table 9–4 above). At least four manufacturers are developing 10-MW machines (see "Bigger turbines" below).

Figuring out the best layout for an offshore wind farm involves many trade-offs, as it does on land. For example, one layout might produce the most wind energy, but would cost more to install electrical cables, and the cost of offshore electrical systems is already greater than on land (see "Electrical systems").

Energy prediction

Estimating how much wind power an offshore farm will produce is essentially the same as for land-based wind farms. But because the ocean has no topographical features to disrupt wind flow, measurements from a met mast can characterize conditions over a larger area.

However, wake-effect losses are likely to be higher offshore, partly because of the steadier ocean winds. Offshore, the air behind the turbine is less mixed by ambient turbulence, so it takes longer to reenergize the slow-moving wake air, and the wake effect lasts longer (fig. 9–4). Classic wake modeling techniques fail to account for these differences, but are being improved for offshore applications.

Also, wind turbines offshore experience more downtime, directly affecting the amount of energy produced. New computer models are being used to analyze offshore turbine availability more accurately, since it's a critical factor in appraising the economics of a wind farm project.

Offshore, downtime results mainly from lack of access for maintenance. Production lost from inaccessible turbines is often the greatest cost penalty of an offshore wind turbine fault. Nevertheless, offshore wind farms operate at higher capacity factors than on land, averaging around 40%.

Fig. 9–4. Wake effect at Denmark's Horns Rev wind farm. *Image courtesy of GL Garrad Hassan.*

Designers can increase the reliability of a wind power system to minimize maintenance, but reliability is expensive, and it can sometimes reduce accessibility. For example, a turbine nacelle can be tightly sealed to protect equipment from corrosive moisture. This makes the system more reliable, but then it's harder to get to the components inside. Conversely, the nacelle can be designed to cost less and be easier to access. But without an effective seal, equipment inside can corrode and fail sooner.

As a rule of thumb to minimize the cost of energy, the general consensus is that it pays off to avoid maintenance by investing in the reliability of the wind power system rather than investing in equipment or design to facilitate access.

Foundations

The foundation of an offshore wind turbine can cost two or three times as much as foundations on land. Overall, the support structure accounts for about 20% of the capital cost of an offshore wind farm and 13% of life-cycle energy costs.

What drives the price of offshore foundations is the height of the ocean's waves. Wave heights are predicted over time frames of decades. Europe follows rules established by the International Electrotechnical Commission, which predicts an extreme wave height over a 50-year time frame.

In contrast, America follows regulations that predict 100-year extreme wave heights, which are higher because a longer time period is considered. This could make offshore turbine foundations even more expensive in U.S. waters.

Japan's 14-MW Kamisu wind farm, located less than 50 meters from land (165 feet), survived a tsunami wave in 2011 with no ill effects. The tsunami reached 5 meters in height (over 16 feet). The seven 2-MW downwind turbines, with 60-meter hub heights (200 feet), have *monopile* foundations (see below) that are sunk 25 meters deep (82 feet) into the seabed. The transformers are enclosed in plastic on a jetty, so they stayed dry.

The type of foundation needed to keep offshore turbines stable depends on how deep the water is—shallow, up to about 30 meters (100 feet), or medium, 30–60 meters (100–200 feet). *Fixed-bottom* foundations, which attach to the seabed, are used in shallow and medium waters, while turbines in deeper waters use floating devices for stabilization (see "Deepwater floating turbines" below).

Monopile. The simplest type of fixed-bottom foundation is the monopile, a single steel tube with a diameter of about 4–6 meters (13–20 feet). Monopiles are a proven technology—about 65% of the world's offshore wind turbines have monopile foundations. Monopiles are used mostly in shallow water, but some have been installed in water as deep as 37 meters (120 feet).

However, the transition piece that links the foundation to the turbine tower must be carefully trued in and grouted (which is also required for other types of foundations that use transition pieces). Grout failures have occurred on many existing offshore turbines with monopile foundations. Grouting technology has worked on drilling rigs but apparently needs more attention when applied to securing turbine towers to their foundations. DNV's recent design standards update addresses grouted connections.

A variation of the monopile is the tripile, using three monopiles, which can be installed in medium-depth waters. This concept, developed by Bard, was tested on a 5-MW turbine in shallow water off Germany and is being used on four 5-MW turbines at another Bard wind farm in water depths of 40 meters (130 feet), 100 kilometers from shore (54 nautical miles).

Gravity base. A concrete *gravity-base* foundation is suitable for shallow water. This type of foundation, which can require careful seabed preparation, relies on the structure's weight to keep the turbine from overturning. Gravity-base foundations are used in about 25% of existing offshore wind turbines at water depths of less than a meter up to 27 meters (89 feet).

The next generation of gravity-base foundations under development should be easier to install and could be mass-produced and assembled onshore at quayside, then towed to the site. The units might be installed in one or two days, making the job easier to schedule within a fair-weather window. The foundations could become available by 2014.

Jackets and tripods. At medium water depths, offshore wind turbines require a fixed-bottom base stabilized by a *jacket* or a *tripod*. The jacket foundation—very similar to an offshore oil rig—looks like a conventional, four-legged steel lattice or truss tower, but with most of the tower submerged. Both jacket and tripod foundations are made of steel tubes, typically 0.5–1.5 meters in diameter (1.5–4.9 feet), welded together and anchored by piles up to 2.5 meters in diameter (8.2 feet). A transition piece connects the turbine to the foundation.

Only about 8% of offshore wind turbines have jacket foundations, but they could become more competitive in deeper waters. Britain's Beatrice wind farm in the North Sea consists of two 5-MW turbines that have a tube tower on top of a submerged, lattice-type jacket foundation. These are located in 40 meters of water (130 feet), 25 kilometers from shore (14 nautical miles).

Germany's Alpha Ventus wind farm has six 5-MW turbines with jacket foundations, as well as six 5-MW turbines on tripods, the earliest commercial application for this type of foundation. The wind farm lies 43 kilometers from land (23 nautical miles), in water depths of 30 meters (100 feet). The performance of the two types of turbines is also being studied, with one design being direct-drive and the other having a traditional gearbox.

"With wind turbine sizes now scaling up to 5 MW and beyond, the top-head mass is virtually double that of the current mainstream sizes. A corresponding increase in the major static and dynamic loadings imposed on the underlying support structure necessitates a new type of thinking on turbine foundation design."

—*Alan Tricklebank,* Offshore Wind Construction and Installation Report

---

Electrical systems

In addition to the wind turbine generator, electrical system components can include a power-collection network of interturbine array cables, an offshore substation, turbine transformers, a main transmission export cable to shore, and an onshore substation/grid connection, which is included in the offshore wind farm cost. (On land, costs are counted only up to the grid connection.) Offshore electrical systems can cost three times more than onshore.

Offshore substations and transformers reduce electrical losses by stepping up the wind power's voltage before it's exported to shore, from the 30–36 kV medium level in the interturbine array to the 100–220 kV or higher voltage level for transmission to the grid connection point. This boost in voltage reduces the number of cables needed to transmit the wind power. Each export cable typically is rated at 150–200 MW.

Offshore substations aren't needed if the project is small (less than 100 MW) or close to shore (less than 15 kilometers, or 8 nautical miles), or if the onshore grid voltage is the same as the wind farm supply voltage.

As wind farm size and distance from shore both increase, a higher voltage transmission system becomes more practical. Many large offshore projects already use 132-kV or 220-kV AC cables. But power from very large wind farms, or those that are very far away from a grid connection, can't be transmitted via AC, which becomes too costly or simply impossible due to line losses.

That's why many new, large offshore wind farms are designed with HVDC technology, as described in chapter 8. If HVDC is selected, converter stations are also needed as part of the electrical system.

The offshore substation can function as a central service and control facility for the wind farm. It might have a helicopter landing pad, control room, SCADA system, crane, rescue boat, and communications station, along with firefighting equipment, emergency diesel generators, and temporary living quarters.

For example, the high-voltage substation to be installed at Germany's 288-MW Baltic 2 wind farm will contain transformers, high- and medium-voltage switchgear, auxiliary power supply, substation automation, and turbine protection and control systems, as well as boat landings and a heli-hoist. The substation will be used as a logistics service platform during turbine installation and operation.

Subsea cables for offshore wind farms are the same type used in other applications—a single cable containing all three electrical phases and optical fiber for communications, plus an insulating gas for deep-sea cables. Interturbine array cables are typically rated at 30–36 kV and installed in single lengths from one turbine to its neighbor, forming a collection circuit that feeds the substation. Collection cables typically are rated up to 30 MW.

The long-term reliability of the interturbine and main transmission cables is a major concern in offshore wind farm design and construction (see "Installation" below). To specify a reliable cable system, the developer must identify potential hazards ahead of time, characterize the geology of the seabed, select the best route, and protect the cable from the scouring effects of seawater.

Cables must be buried deeply enough that they can't be damaged by trawlers and anchors or exposed to hydrodynamic loads. Another common problem is interturbine cable terminations, which can rub against the bottom of the tower, causing them to short out.

The depth under the seabed at which the cables must be buried varies depending on the wind farm. Sometimes, developers plan on a certain burial depth, but their insurance company might believe that deeper is better, or they might encounter unexpected challenges in the seabed. Another problem is that measurements taken at each turbine foundation site might not be accurate enough for designing cable layouts that can be several kilometers long.

Installation

The installation process is a major driver of offshore wind farm costs, especially the construction of the turbine foundations. Installing the substation is also a major cost, but it's a one-time event as opposed to installing multiple turbines and foundations. Running the electrical transmission cables costs relatively little in terms of overall investment, but cables can cause major problems when the wind farm is operating. Specialty vessels are needed for all of these installation procedures.

Building offshore wind farms could become easier and less expensive if harbor facilities are developed that are strategically located. For example, assembling turbine components onshore or near shore in sheltered waters and towing the machine to the site can substantially reduce installation costs. The industry also needs new equipment such as piling hammers and cable plows that are custom-designed for installing offshore wind turbines, foundations, and cables.

Cables. Installing the subsea cables is one of the most important jobs in constructing an offshore wind farm. Cabling accounts for up to 10% of project costs, and as noted above, developers must identify hazards in advance to avoid problems at the installation site.

But more importantly, cable laying must simply be done right the first time. Cables have historically been a prime cause of offshore turbine downtime and the biggest liability at offshore wind farms, accounting for up to 80% of insurance claims (fig. 9–5).

As noted earlier, the cable-laying industry is small, in terms of what might be required for offshore wind, and it could choke development.

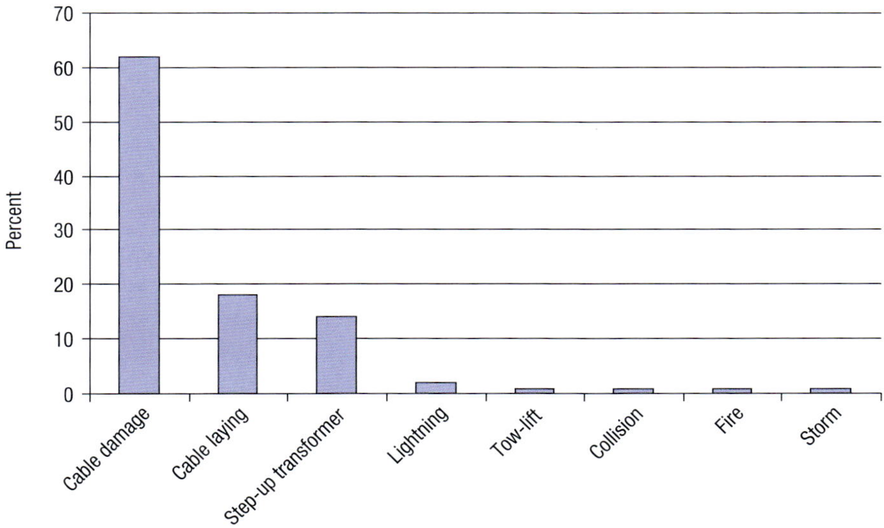

Fig. 9–5. Offshore wind farm insurance claims (based on 27 claims, 12 projects totaling 1 GW, payments of $43 million/31 million euros). *Data courtesy of GL Garrad Hassan.*

Vessels. Specialty ships and barges are needed to install turbines and serve the offshore wind farm market. Most of these vessels must use dynamic positioning to stay in place while work is under way or to move around slightly as required.

Jack-up ships and barges, traditionally used in the oil and gas industry, motor to the site and then extend their legs to the sea floor while raising their platform well above the water line. An onboard crane piledrives the foundation, erects the turbine tower sections, and installs the nacelle and rotor. Leasing a jack-up vessel and crane for a day can cost anywhere from $50,000–$200,000 (35,000–150,000 euros).

Offshore installation vessels must have enough room to carry massive turbine foundations and enough power for pile driving. They must also transport long, heavy turbine tower segments and huge cranes capable of lifting nacelles. One Danish ship owner has permanently mounted a telescopic crane on one of its jack-up barges. Telescopic cranes have an extendable, relatively compact boom.

Several installation vessels have been custom-built for the European offshore wind industry (see "Key players" above), and more jack-up vessels are about to enter service, capable of carrying more than one foundation and/or multiple turbines. One vessel will incorporate a pedestal-mounted crane with a 1,200-ton lifting capacity, destined for duty on the German North Sea's 288-MW Dan Tysk wind farm. Another new vessel with two cranes (one 1,000-ton and one 50-ton) is being used to build the UK's 630-MW London Array wind farm.

But Europe's expanding fleet of specialty vessels isn't available worldwide. If new ships are needed to install offshore wind farms in other countries, they can take two to three years to be delivered. Each vessel costs about $150–$220 million (110–160 million euros).

Maintenance

Obviously, it's much more difficult to get to a wind turbine offshore and work on it for either routine servicing or emergencies. Engineers must get to the turbine, mainly by boat—in shallow waters, typically a purpose-built aluminum catamaran—or by helicopter for less accessible wind farms.

Offshore maintenance is often delayed while crews wait for a break in the weather. Wave heights above 1.5 meters (5 feet) and wind speeds above 12 m/s (27 mph) are considered unsafe for transferring workers from a boat to an offshore turbine. Also, mobilizing a crane vessel for major repairs can take weeks or even months, while the wind farm operator loses revenue every day from the idle turbine. In cold climates, sea ice can also delay maintenance.

The existing fleet of service vessels is not well matched to working on the wind farms of the future, which will be sited farther offshore with higher waves. Safely getting technicians from the vessel onto the turbine will be harder and more hazardous. Vessels won't be able to maintain position simply by thrusting against the turbine foundation, as they have in the past. Special equipment such as gangways will be needed to transfer people back and forth.

At some wind farms farther out, maintenance crews live in temporary offshore facilities ("flotels") near the turbines. So wind farm operators need large vessels to carry crews to their accommodation, but still need a fleet of smaller service vessels for transporting workers from turbine to turbine.

As with land-based wind farms, reducing unplanned maintenance is key to profitability. But with offshore projects, this is even more crucial. Many of today's offshore wind turbines are equipped with features for load control and condition monitoring (described in chapter 6). These features can improve the economics of each wind power system by reducing unscheduled downtime.

More opportunities exist to reduce O&M costs by minimizing work done at sea, but this is usually a trade-off against higher capital costs. Better quality assurance for components, more qualification testing, and more reliable designs equal greater initial costs. Offshore wind farm developers have to weigh higher up-front investment against lower maintenance costs down the road.

Developers must also decide which drivetrain components should be maintained individually by removing them from the nacelle and repairing them onshore. The alternative is to replace the whole nacelle and drivetrain.

Corrosion and contamination

One challenge unique to the offshore environment is corrosive saltwater and moisture-laden air. Turbines designed for offshore duty are equipped with coatings, cathodic protection, and other measures to prevent corrosion. Problems can still occur on individual machines, but today's offshore wind turbines are designed to withstand the corrosive effects of the ocean environment.

Turbine manufacturer Siemens recognized this problem when the company installed the first offshore wind farm in 1991. Siemens discovered that pitch bearings in its first 25 turbines were being corroded due to a supplier defect. The problem was detected early and fixed before the machines malfunctioned. Since then, Siemens has taken steps to control corrosion by condition monitoring of key components.

Still, the effects of salty air and water can be especially devastating to offshore transformers and gearboxes. Salt can contaminate the transformer's coolant and the gearbox's lubricating oil, making them as thick and useless as chocolate mousse. In an offshore wind farm, transformers can be replaced individually, but depending on the turbine design, replacing a gearbox might require swapping out the entire nacelle.

Advanced Offshore Technologies

Deepwater floating turbines

The concept of siting a wind power system in deep water—more than 60 meters (200 feet)—depends on developing turbines that need no fixed foundation on the seabed. Instead, the turbine is installed on a floating structure that is moored, or tethered, to the ocean floor.

Floating, deepwater wind farms could greatly increase the usable area of the world's offshore wind resource. They would also be ideal for countries surrounded by deep water such as Japan, Italy, and Norway, as well as parts of America's western coast.

Deepwater turbines rely on their floating structure to keep the turbine from overturning into the sea. The structure not only supports the turbine's weight, but also stabilizes the turbine—that is, restrains its pitch, roll, and heave motions, keeping them within acceptable limits.

The floating structure can remain stable by using ballast weight, mooring lines, or buoyancy, or by a combination of these mechanisms. Figures 9–6 and 9–7 show the three basic types of floating structures and the ways that they stabilize the turbine.

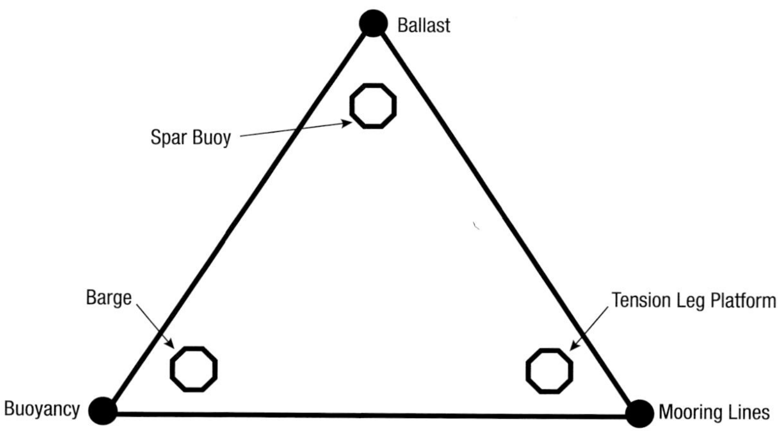

Fig. 9–6. Classification of deepwater wind turbine stabilizing mechanisms. *Image adapted from NREL.*

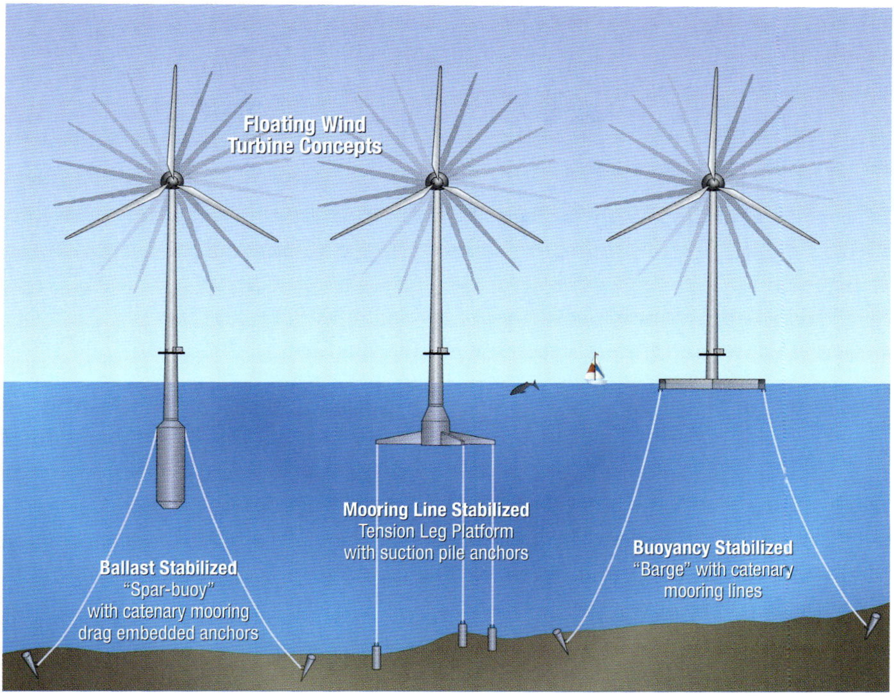

Fig. 9–7. Illustration of deepwater floating structures. *Image courtesy of GL Garrad Hassan.*

Most of the stabilizing technologies being used for floating wind turbines were pioneered in the deepwater oil and gas drilling industry, which has proven that floating structures can survive over many decades. Deploying thousands of offshore oil rigs has been financially rewarding, but the economics of floating wind turbines has yet to be determined.

It should be noted that in drilling lingo, "deep water" usually means more than 2,000 meters (6,000 feet), which is much deeper than envisioned for floating wind farms.

Spar buoy. The world's first full-scale floating wind turbine began operating off Norway in 2009. The 2.3-MW Hywind turbine was being tested for two years at a water depth of about 220 meters (720 feet), 10 kilometers offshore (just over half a mile). The Hywind design could be practical at ocean depths of 120–700 meters (400–2,300 feet).

Developed by Norway's Statoil and partners, the Hywind turbine was assembled in sheltered waters and towed to the open ocean. The Siemens turbine's hub height is 65 meters (213 feet), with a rotor diameter of 82.4 meters (270 feet).

The Hywind turbine sits on top of a submerged *spar*, a buoy-type float that stabilizes the turbine with its ballast weight. The spar, a slender steel cylinder, has a draft of 100 meters below the waterline (330 feet). The spar's diameter is 6 meters at the waterline (20 feet) and 8.3 meters in the submerged section (27 feet).

The spar is moored to the sea floor by three catenary lines (curved outward below the structure) with anchors that are embedded by dragging them into place. France's Technip built the support structure and installed the floating wind turbine.

Hywind's developers plan to commercialize the concept, possibly at a site in Scotland or the United States, for installation around 2015–2017.

Tension leg platforms. Deepwater turbines might also be stabilized on a *tension-leg platform*, a tripod with three vertical, anchored mooring lines that stabilize the turbine. An 80-kW floating wind turbine of this type was operated for several months by Blue H Technologies (Netherlands).

In the Blue H design, the turbine is stabilized by a modified tension-leg platform that is semi-submerged and tethered to a seafloor counterweight. The prototype structure was located 21.3 kilometers off Italy (11.5 nautical miles) in a water depth of 113 meters (371 feet). Plans are to test a 2-MW turbine in 2012, followed by a larger, preproduction model in 2014.

Another type of tension-leg platform was planned for deployment off the U.S. northeast coast. The PelaStar technology was developed by Glosten Associates, with support from the Carbon Trust (see "Other R&D" below). The platform is designed to support wind turbines of 5–6 MW but could be scaled to accommodate sizes of 10 MW or more.

Barges and platforms. A third type of deepwater turbine being developed would be mounted on a barge or platform, whose buoyancy at the water line stabilizes the turbine. Semisubmersible platforms can be held in place by catenary mooring lines.

A 2-MW Vestas turbine integrated with a full-scale semisubmersible WindFloat platform was deployed off Portugal in 2011 for a 2-year trial, which could be followed by tests with larger turbines. The triangular, three-column structure dampens wave- and turbine-induced motion by using water-entrapment heave plates at the base of each column. The platform is moored by four to six drag-embedded anchor lines.

The WindFloat platform, being developed by Principle Power (U.S.), has a hull weight of 1,080–1,600 tons, total displacement (including ballast) of 5,500 tons, and a hull draft of about 20 meters (70 feet). It can be deployed in water deeper than 60 meters (200 feet) but is suitable for medium depths as well.

Other concepts. Norway's Sway company is developing a floating tower that could support a 5-MW turbine in water depths of 80–400 meters (260–1,300 feet). In Sway's concept, the lower portion of the tower is submerged to a depth of about 100 meters (300 feet), with about 2,000 tons of ballast for stabilization. The tower is tethered to the seabed with a single pipe and suction anchor.

In contrast to nearly all horizontal-axis turbines, Sway's would operate as a downwind machine, rather than upwind. The turbine is designed to tilt 6–8 degrees during operation and to swivel as the wind direction changes. A wire bar or tension rod, projecting outward from the tower above the waterline, adds enough strength to avoid tower fatigue.

Sway plans to test a 5-MW Areva turbine using this floating concept 7 kilometers off Norway (4 nautical miles). Sway is also designing and building a 10-MW test turbine for either fixed foundations or floating structures.

The UK's Energy Technologies Institute, in its Nova project (novel offshore vertical axis), looked at the feasibility of a 5–10 MW vertical-axis wind turbine design for offshore applications. The Aerogenerator X is lighter in weight and not as tall as comparable horizontal-axis machines, and its weight is concentrated in its base. Wind Power Ltd. plans to develop the concept, possibly scaling up to a 20-MW size.

Another project also aims to develop a floating vertical-axis wind turbine. The Vertiwind concept will be designed mainly by Technip, which participated in engineering Statoil's Hywind project. Plans are under way to test a 2-MW prototype.

A publicly-funded research team led by Germany's Fraunhofer Institute aims to develop a 10–20 MW floating wind turbine, starting with a 1-MW test platform in Spain's Bay of Biscay, to be ready before 2015. The concept is called HiPRWind.

The DeepC Wind consortium, affiliated with the University of Maine, plans to test floating turbines (10-kW and 100-kW) and demonstrate a 25-MW deepwater wind farm. Maine law requires that the floating wind turbines be sited at least 19 kilometers from shore (10 nautical miles) in water at least 90 meters deep (300 feet). Ultimately, DeepC hopes to build up to 4 GW of deepwater wind capacity.

The project focuses on developing more durable, lightweight composite materials and integrating them into a platform design capable of supporting a wind turbine up to 250 kW in capacity. Researchers also aim to reduce installation costs by assembling as much of the turbine and platform near shore as possible and towing it to the demonstration site.

Poseidon's floating power plant is a hybrid wind/wave technology that integrates three standard wind turbines with a platform that converts wave energy into water pressure for powering a turbine generator. The structure is stabilized by an anchor buoy. A prototype was being tested at sea in 2010, prior to commercial design.

Deepwater wind technology depends on overcoming several challenges. For example, direct-drive generators feature high reliability and low maintenance, but they're quite heavy. Lighter-weight generators would make the technology easier to deploy in deep water. In fact, all the materials that go into a deepwater wind turbine need to be lighter.

In deep water, instruments for monitoring turbine performance are restricted to the structure itself. Also, the cost of moorings and anchors needs to be reduced, the industry must develop standards for deepwater machines, and engineers must figure out ways to assemble as much of the structure as possible before it is towed to the site.

Bigger turbines

Turbine size growth. As logistics restrain the growth of turbine size in land-based wind farms, future size growth will be paced by offshore development. Several manufacturers offer commercial turbines for offshore duty in the 5–6 MW scale, and 7-MW machines will soon be out (see tables 9–3 and 9–4).

Several wind turbines with capacities of 10 MW or more are being developed (table 9–8), and some could soon be tested and deployed. Research suggests that 20-MW machines are feasible.

Industry experts believe that turbines with double-digit megawatt ratings will become common around 2020. Physical barriers that limit size to certain thresholds have been overcome in the past by technology development.

The critical issue in developing ultra-large wind turbines is that the physical laws of scaling do not allow some components to be made larger without changing the fundamental technology. Advanced, size-enabling technologies include stiffer, lighter weight materials; new ways to manufacture composite materials; lighter, lower maintenance drivetrains; lighter direct-drive generators; and gearbox and bearing systems that can tolerate slower rotational speeds.

Table 9–8. Super-size offshore wind turbines

Manufacturer/developer	Capacity, MW	Rotor diam., m (ft)	Name/description
American Superconductor	10	190 (623)	SeaTitan
Areva/Sway	10	145 (476)	submerged tower
Clipper Windpower	10	175 (574)	Britannia
Gamesa consortium	15	n/a	n/a
GE/U.S. DOE	10–15	n/a	n/a

Europe's UpWind consortium of 40 partners is working on design tools to optimize blades and other components of large wind turbines (8–10 MW). The project addresses the aerodynamic, aero-elastic, structural, and material design of rotors, as well as critical analysis of drivetrain components. Research is aimed at reducing turbine weight and loads by identifying materials with extreme strength-to-mass ratios.

The UpWind project explored the design limits of scaling wind turbines to 20 MW and concluded that this size could be feasible. Rotor diameters would reach about 200 meters (more than 600 feet).

However, innovative design would have to overcome scaling problems, such as reducing fatigue load on blades to make them longer and lighter. One way to do this would be dividing the blade into two sections with separate controls. This would also make it easier to transport blades.

Other R&D. The UK government set up the Carbon Trust, which operates the Offshore Wind Accelerator program with funding of $48 million (30 million pounds, or 35 million euros). The trust analyzed six offshore wind turbine foundation concepts for feasibility studies and selected four of them for demonstration.

The trust also selected several new concepts for transferring engineers to offshore wind turbines in wave heights up to 3 meters (10 feet) as far as 300 km offshore (200 miles). In addition to foundations and O&M logistics, the trust will focus on wake effects and electrical systems.

The U.S. Department of Energy (DOE) is funding 41 projects to spur offshore wind energy. Funding of $43 million (31 million euros) during 2011–2016 will help advance turbine technologies, reduce environmental risks, and develop the manufacturing and supply chain infrastructure.

Projects will also address transmission planning and interconnection, as well as logistics readiness such as ports and vessels. Previously, DOE funded six projects to develop next-generation drivetrains, largely for offshore applications.

University researchers are proposing a "wetlab" off Massachusetts where new wind turbines and other offshore technologies could be tested without permitting. Called the National Renewable Energy Innovation Zone, the wetlab concept has been funded by federal grants. Another test site is being considered off Delaware.

Chapter 10 discusses other wind turbine research, development, and testing.

References

Alstom, http://www.alstom.com/power/renewables/wind/.

American Wind Energy Association. Offshore Wind Energy. http://www.awea.org/documents/factsheets/Offshore_Factsheet.pdf.

Appleyard, David. 2011 (February 16). Record Growth for EU Offshore Wind in 2010. *Renewable Energy World*. http://www.renewableenergyworld.com/rea/news/article/2011/02/wind-power-market-statistics??cmpid=WNL-Wednesday-February16-2011.

Areva, http://www.areva.com/.

Atlantic Wind Connection, http://atlanticwindconnection.com/.

Bard, http://www.bard-offshore.de/.

Blau, John. 2011 (May 11). Baltic Sea Could Be Home to 1 GW of Wind Power Capacity. *Renewable Energy World*. http://www.renewableenergyworld.com/rea/news/article/2011/05/baltic-sea-could-be-home-to-1-gw-of-wind-power-capacity?cmpid=WNL-Wednesday-May11-2011.

———. 2011 (May 18). Europe Leads Offshore Wind Market. *Renewable Energy World*. http://www.renewableenergyworld.com/rea/news/article/2011/05/wind?cmpid=WNL-Friday-May20-2011.

Bonafoux, John. 2011 (March 10). Getting into Ship Shape for Far Offshore. *Renewable Energy World*. http://www.renewableenergyworld.com/rea/news/article/2011/03/getting-into-ship-shape-for-far-offshore??cmpid=WNL-Friday-March11-2011.

BTM Consult. 2010 (November 22). Offshore Wind Power 2010. http://www.btm.dk/news/offshore+wind+power+2010/?s=9&p=&n=39.

Butterfield, S., et al. 2007 (September). Engineering Challenges for Floating Offshore Wind Turbines. NREL/CP-500-38776. http://gisceu.net/PDF/U444.pdf.

Carbon Trust, http://www.carbontrust.co.uk/.

Dawei, Yu, and Yue, Yang. 2010 (June 18). Cool Start for Offshore Wind Energy in China. http://english.caing.com/2010-06-18/100153345.html.

Det Norske Veritas, http://www.dnv.com/.

EnBW, http://www.enbw.com/content/en/index.jsp.

Enercon, enercon.de/en/_home.htm.

Ernst & Young. 2011 (February). Renewable Energy Country Attractiveness Indices. http://www.ey.com/Publication/vwLUAssets/Renewable_energy_country_attractiveness_indices_-_Issue_28/$FILE/EY_RECAI_issue_28.pdf.

European Wind Energy Association 2009 (February 13). Wind Energy—The Facts, Executive Summary. http://www.ewea.org/fileadmin/ewea_documents/documents/publications/WETF/1565_ExSum_ENG.pdf.

———. 2010 (November). Powering Europe: Wind Energy and the Electricity Grid. http://www.ewea.org/fileadmin/ewea_documents/documents/publications/reports/Grids_Report_2010.pdf.

———. 2011 (January). The European Offshore Wind Industry Key Trends and Statistics 2010. http://www.ewea.org/fileadmin/ewea_documents/

documents/00_POLICY_document/Offshore_Statistics/110120_Offshore_stats_Exec_Sum.pdf.

———. 2011 (March). New Research Shows 20 MW Wind Turbines Are Feasible. http://www.ewea.org/index.php?id=60&no_cache=1&tx_ttnews%5Btt_news%5D=1901&tx_ttnews%5BbackPid%5D=1&cHash=b4462d717ccc0ee61e55922ce236f70c.

———. 2011. Operational Offshore Wind Farms in Europe, end 2010. http://www.ewea.org/fileadmin/ewea_documents/documents/statistics/110214__public_offshore_wind_farms_in_Europe_2010.pdf.

———. 2011. Fact Sheets: Offshore. http://www.ewea.org/index.php?id=1611.

———. 2011. Fact Sheets: The Power Grid. http://www.ewea.org/index.php?id=1611.

———. Wind Energy—the Facts. http://www.wind-energy-the-facts.org.

Farrand, Tim. 2011 (September 27). Scotland to Invest 35 Million Pounds in Wind Turbine Production. Bloomberg. http://www.bloomberg.com/news/2011-09–27/scotland-to-invest-35-million-pounds-in-wind-turbine-production.html.

Fisher, Curtis, et al. 2010. Offshore Wind in the Atlantic: Growing Momentum for Jobs, Energy Independence, Clean Air, and Wildlife Protection. National Wildlife Federation. http://www.nwf.org/News-and-Magazines/Media-Center/News-by-Topic/Global-Warming/2010/~/media/PDFs/Global%20Warming/Reports/NWF-Offshore-Wind-in-the-Atlantic.ashx.

Gamesa, http://www.gamesa.es/en/.

Gassert, Jochen, and Heckmann, Wilhelm. 2010 (January). Too Many Question Marks? GL Garrad Hassan. Energize Renewables. http://www.gl-group.com/pdf/energize-renewables_KOMPLETT.pdf.

GE, http://www.gepower.com/businesses/ge_wind_energy/en/index.htm.

Global Wind Energy Council. 2011 (March). Global Wind Report: Annual Market Update 2010. http://www.gwec.net/fileadmin/documents/Publications/Global_Wind_2007_report/GWEC%20Global%20Wind%20Report%202010%20low%20res.pdf.

Global Wind Energy Council, Greenpeace International. 2010 (October). Global Wind Energy Outlook 2010. http://www.gwec.net/fileadmin/documents/Publications/GWEO%202010%20final.pdf.

Glosten Associates, http://www.glosten.com/.

Goldwind, http://www.goldwindglobal.com/web/index.do.

Hastings, Rob. 2010 (September 21). Wind Energy Development in the UK. Presentation at the Offshore Wind Best Practice Summit, Boston.

International Energy Agency. 2010 (July). IEA Wind Energy 2009 Annual Report. http://www.ieawind.org/AnnualReports_PDF/2009/2009AR_92210.pdf (ISBN-0-9786383-4-4).

Knight, Matthew. 2010 (September 21). Offshore Wind Electrical Infrastructure. Presentation at the Offshore Wind Best Practice Summit, Boston.

Kremer, Josiane. 2011 (March 17). Norway Says Offshore Wind Too Expensive to Develop, Aftenbladet Reports. Bloomberg. http://www.bloomberg.com/news/2011-03-17/norway-says-offshore-wind-too-expensive-to-develop-aftenbladet-reports.html.

Lawson, James. 2011 (August 22). What Are the Challenges and Innovations in Offshore Wind Design and Development? *Renewable Energy World*. http://www.renewableenergyworld.com/rea/news/article/2011/08/what-are-the-challenges-and-innovations-in-offshore-wind-design-and-development?cmpid=WNL-Wednesday-August24-2011.

Leirgulen, Svein Inge. 2011 (September 27). DNV Releases Updated Standard for Offshore Wind Turbine Structures. Det Norske Veritas. http://www.dnv.com/press_area/press_releases/2011/DNV_releases_updated_standard_for_offshore_wind_turbine_structures.asp.

Li, Junfeng, et al. 2010 (October). China Wind Power Outlook 2010. http://www.greenpeace.org/raw/content/eastasia/press/reports/wind-power-report-english-2010.pdf.

London Array, http://www.londonarray.com/.

Melnyk, Markian M.W. 2011 (Spring). Offshore Wind Could Get Its Backbone. *Offshore Wind: A Supplement to North American Wind Power*. 8(2): 14–15.

Melnyk, Markian M.W., and Andersen, Robert M. 2009. *Offshore Power: Building Renewable Energy Projects in U.S. Waters*. Tulsa: PennWell. (ISBN 978-1-59370-211-3).

Musial, Walt. 2010 (January 13). Trends in U.S. Wind Industry Technology. Presented at the Maine Wind Seminar, January 12-14, 2010, Portland, Maine.

Musial, Walt, and Ram, Bonnie. Large-Scale Offshore Wind Power in the United States. NREL/TP-500-40745. http://www.nrel.gov/wind/pdfs/40745.pdf.

Nordex, http://www.nordex-online.com/en.

Pike Research. 2011 (April). Offshore Wind Power: Market Opportunities and Challenges, Technology Issues, Key Industry Players, and Global Capacity and Production Revenue Forecasts. http://www.pikeresearch.com/research/offshore-wind-power.

Pike Research. 2011 (July 21). Submarine Electricity Cable Demand to Place a Significant Strain on Supply During the Next 10 Years. http://www.pikeresearch.com/newsroom/submarine-electricity-cable-demand-to-place-a-significant-strain-on-supply-during-the-next-10-years.

Poseidon. http://www.floatingpowerplant.com.

Pöyry. 2011 (March 30). Groundbreaking Study into the Impact of Wind and Solar Generation on Electricity Markets in North-West Europe. http://www.poyry.com/media/media_2.html?Id=1301471113.html&C=0&Y=2011.

Principle Power, http://www.principlepowerinc.com/.

Renewable Energy World. 2010 (December 1). Spanish Companies Plan a 15-MW Wind Turbine. http://www.renewableenergyworld.com/rea/news/article/2010/12/spanish-companies-plan-a-15-mw-wind-turbine.

REpower, http://www.repower.de/index.php?id=347&L=1.

Rock, Mary, and Parsons, Laura. 2010 (October). Offshore Wind Energy. Environmental and Energy Study Institute. http://www.eesi.org/files/offshore_wind_101310.pdf.

Sany Group, http://www.sanygroup.com/products/en-us/turbine/Offshore.htm.

SBI Energy. 2011 (February 3). Offshore Wind Farm Manufacturing Worldwide. SBWire. http://www.sbwire.com/press-releases/sbwire-76729.htm.

Siemens, http://www.energy.siemens.com/hq/en/power-generation/renewables/wind-power.

Sway, http://www.sway.no/.

Sykes, Benj. 2010. Supporting Innovation. *Carbon Trust*. http://www.taplondon.co.uk/bwea31/files/benjaminsykes.pdf.

Trabish, Herman K. 2010 (June 10). Deepwater Drilling for U.S.? Try Deepwater Wind. Greentech Media. http://www.greentechmedia.com/articles/read/deep-water-drilling-no-deep-water-wind.

———. 2010 (November 24). Offshore Wind Moves to Full Speed Ahead. Greentech Media. http://www.greentechmedia.com/articles/read/offshore-wind-moves-to-full-speed-ahead.

———. 2010 (December 30). The Emerging Opportunity in Offshore Wind Vessels. Greentech Media, http://www.greentechmedia.com/articles/read/the-emerging-opportunity-in-offshore-wind-vessels.

———. 2011 (February 24). Will Google's Offshore Wind Transmission Be Slowed by Regulatory Red Tape? Greentech Media. http://www.greentechmedia.com/articles/read/will-googles-offshore-wind-transmission-be-slowed-by-regulatory-red-tape.

———. 2011 (February 3). The Truth About the Cape Wind Turbines. Greentech Media. http://www.greentechmedia.com/articles/read/the-truth-about-the-cape-wind-turbines.

———. 2011 (April 19). Goldwind: Will Wind Be China's Next Big Export to the US? Greentech Media. http://www.greentechmedia.com/articles/read/goldwind-will-wind-be-chinas-next-big-export-to-u.s/.

———. 2011 (June 10). From Oil and Gas to Offshore Renewables. Greentech Media. http://www.greentechmedia.com/articles/read/from-oil-and-gas-to-offshore-renewables/.

———. 2011 (June 15). Will German Offshore Wind RAVE On? Greentech Media. http://www.greentechmedia.com/articles/read/will-german-offshore-wind-rave-on/.

———. 2012 (January 6). Floating the Future of Offshore Wind. Greentech Media. http://www.greentechmedia.com/articles/read/floating-the-future-of-offshore-wind/.

UpWind. http://www.upwind.eu/default.aspx.

Vestas, http://www.vestas.com/.

Wickless, Andy. 2011 (July 27). Financing Offshore Wind Farms in the U.S. Navigant Consulting. http://www.renewableenergyworld.com/rea/news/article/2011/07/financing-offshore-wind-farms-in-the-u-s?cmpid=WNL-Wednesday-July27-2011.

Williams, Andrew. 2011 (April 15). Direct Drive Turbines: Developers Square Up to Large-Scale Offshore Turbine Challenge. Wind Energy Update. http://social.windenergyupdate.com/industry-insight/direct-drive-turbines-developers-square-large-scale-offshore-turbine-challenge?utm_source=newsletter&utm_medium=email&utm_campaign=weunewsletter1904.

———. 2011 (May 3). A Buoyant Future for Floating Wind Turbines? *Renewable Energy World*. http://www.renewableenergyworld.com/rea/news/article/2011/05/a-buoyant-future-for-floating-wind-turbines?cmpid=WNL-Wednesday-May4-2011.

Wind Energy Update. 2010 (September). The European Offshore Wind Energy Standards, Permitting And Markets Report: A Country-by-Country Comparison and Practical Guide. http://www.windenergyupdate.com/standards/report-summary.shtml.

———. 2011 (May 2). Report: Offshore Wind Investment to Surge. http://social.windenergyupdate.com/weekly-brief/weekly-intelligence-brief-april-26-may-02.

WinWinD, http://www.winwind.com/.

10

Advanced Technologies: R&D, Testing, and New Twists on Old Ideas

Introduction

Testing of wind power systems and components helps manufacturers evaluate the anticipated lifetime performance of the equipment and identify weaknesses in the design before moving to commercial production. Tests are also required to meet wind turbine design standards and to reduce the technical and financial risk of deploying new mass-produced wind turbine models.

Wind turbine research focuses on improving the energy output and reliability of large, horizontal-axis machines and on making them cheaper to build and operate. Some large, well-established wind turbine manufacturers can afford to conduct their own research and build their own test laboratories.

But government research facilities are needed for smaller manufacturers to compete and for the wind industry to thrive. Many national laboratories share research results with industry and provide equipment for manufacturers to test emerging technologies and new components.

Beyond the research community and the manufacturing industry, many brave engineers are working on designs that might eventually revolutionize wind energy technology—if they can be made to work cost-effectively on a large enough scale. Many of these ideas are based on concepts that have been around for decades, but only now have a chance of becoming practical with new materials and computer design tools.

European Research and Testing

Although Europe leads the world in wind power capacity, until recently European governments devoted little of their R&D funding to wind energy. The European Union (EU) spent less than 1% of its energy research budget on wind energy before 2002. As recently as 2007, the European Commission provided only 3% of Europe's total R&D investment in wind power, while 76% came from the private sector.

However, the EU's current framework program (2007–2013) devotes 65% of its energy research budget ($188 million, or 137 million euros) to renewable resources, with wind energy being one of the priorities. The UpWind consortium discussed in chapter 9 is just one of the programs receiving EU funding.

The EU also launched a wind energy research roadmap in 2010 that sets priorities for spending more than $8 billion (6 billion euros) on wind sector R&D through 2020. The European Wind Initiative will fund resource mapping, turbine test facilities, and prototypes of next-generation turbines and new offshore structures, as well as new manufacturing processes and grid integration demonstrations.

Denmark's national lab

Test facilities. Risø DTU, Denmark's National Laboratory for Sustainable Energy, was planning another wind turbine test center in Østerild, Denmark. The center will provide a site for testing seven wind turbines up to 250 meters tall (820 feet), each with a capacity up to 15 or 20 MW. Wind speeds and behavior are relatively unknown at such heights. Risø DTU's existing facility at Høvsøre can test wind turbines up to 160 meters tall (525 feet).

Turbine manufacturers can buy or lease a test stand at Østerild, erect their wind turbines, and take measurements, with assistance from private contractors if needed. Risø DTU's researchers will also carry out a wide range of complex measurements in collaboration with the manufacturers. The first turbine was scheduled to be installed in 2011–2012.

The Østerild site has forested areas as well as open land, so at two of the test stands, researchers can analyze turbine performance in conditions that are prevalent in woodlands. In contrast, the Høvsøre facility is all flat, open land. The new test facility will have seven measuring masts 150 meters tall (490 feet), one for each test stand, plus two 250-meter (820-foot) masts.

Forest trees strongly affect wind flow conditions. At Østerild, Risø DTU will be able to analyze the effects of vertical wind shear and turbulence. Researchers can use this knowledge to improve aerodynamic models, reducing uncertainties in future wind turbine designs.

Also, the new facility's grid connection will be able to handle current from up to seven wind turbines, each with a 16-MW capacity, for a total of 112 MW. Researchers will be able to subject the wind turbines to changes in grid frequency, observe how they react, and identify methods for preventing turbines from disconnecting when grid frequency varies.

Blade testing. A three-year project under way at Risø DTU is improving the way that wind turbine blades are tested by mimicking the loads actually experienced during turbine operation. The new test method, as opposed to conventional vertical and horizontal pulling, turns the blade at the same time, applying a combined pull to the edge and the flap. This stresses the glued joints between the shells and other areas of the blade structure that are known to be critical.

In the new method, steel anchor plates are mounted directly on the blades, allowing the blades to deform and twist when they're pulled. An optical system observes even the slightest blade deformation, and acoustic emission measurements also monitor deformation during the test to prevent fatal cracks. Ten partners are collaborating with Risø DTU on the blade testing project, including Vestas, LM Wind Power, Dong Energy, and Det Norske Veritas (DNV).

Research and development. Engineers at Risø DTU pioneered the use of turbine-mounted lidar units to scan and characterize wind before it hits the rotor (described in chapter 6). Taking this concept another step forward, they are testing lidar technology mounted on wind turbine blades to mitigate loads and improve control over the turbine's yaw and rotational speed.

Optical fibers are mounted along the blade for communication with small lidars attached along the blade's leading edge. This could provide data on the wind speed and turbulence that each blade experiences as it rotates.

To control flapwise bending of wind turbine blades—which can result in a catastrophic tower strike (see chapter 3)—Risø DTU researchers have performed tests in a wind tunnel on a prototype trailing-edge flap made of molded rubber. The flap can be controlled by compressed air or hydraulics.

Tests confirmed that the flap can supplement pitch control by adjusting the blade in response to wake-effect turbulence and other local wind conditions. As a bonus, the flap gives the blade a very sharp trailing edge, which boosts energy output and reduces noise.

LM Wind Power's facility

Denmark's LM Wind Power started its own testing program in 1999 to avoid long waiting times for other facilities. Now the company is accredited by Danak (the Danish Accreditation and Metrology Fund) to test wind turbine blades and to serve as the basis for approvals by DNV, GL Garrad Hassan, and other certifying agencies.

The LM Wind Power facility has five test beds for dynamic tests and one for static tests. The largest test bed has room for blades up to 80 meters long (260 feet). Its base is an anchor block, one of the strongest cast concrete structures in the world, to accommodate the blade root's bending moment, which more than doubles in relation to the blade's length.

Static testing subjects blades to extreme tensile loads in all directions, and data are collected on force, strain, and deflection. Dynamic testing applies fatigue loads to the blade that simulate 20 years of normal wear and tear. The blade is oscillated at its natural frequency 5 million times edgewise and another 5 million flapwise.

The larger the blade, the lower the frequency, and the longer the test takes. A 61.5-meter blade (202-foot) is left oscillating for a whole year. Strain gauges measure stresses, collecting about 2 gigabytes of data per blade, while an infrared camera looks for small breaks or cracks in the laminate. Static tests are repeated after dynamic testing.

If the blade uses new materials or incorporates other significant changes, it might be subjected to the ultimate strength test, where extreme loads are applied statically until the blade breaks. The point of fracture is analyzed thoroughly to find the fault.

LM Wind Power also performs nondestructive infrared scans of adhesive joints, ultrasound scans of blade laminates, and laser scans of blade molds, as well as four-wire resistance measurement to identify faults in lightning conductor systems.

The company also operates its own wind tunnel at test conditions that can match the operating conditions of multi-megawatt turbine blades. The wind tunnel allows precise aerodynamic testing of airfoils to evaluate different ways to actively control loads, such as by using flaps.

The tunnel can mimic conditions ranging from stall through various angles of attack. Almost 300 sensors characterize the airfoil flow, including differential pressure sensors and a load cell system.

Lelystad test site

This site in the Netherlands was scheduled to start testing its first eight turbines in 2011, with four more to be added in 2012. The 1,200-hectare (3,000-acre) site, being operated by Ecofys, is on the campus of the Wageningen University and Research Centre, where wind speeds average 6.6 m/s (15 mph) at a height of 80 meters (260 feet).

Six of the turbine positions at Lelystad will be able to test turbine prototypes with blade tip heights up to 200 meters (650 feet); the other six will be capable of 150-meter (500-foot) heights. Lelystad will also have several met masts. The site is designed for a maximum installed capacity of 30 MW.

Manufacturers will be able to get their turbines certified to comply with the International Electrotechnical Commission (IEC) and the Measuring Network of Wind Energy Institutes (Measnet). Measurement services at Lelystad will be provided by Germany's Windtest Grevenbroich. Some manufacturers will focus on certification, while others might seek to improve blade performance or analyze the effect of new designs such as generator modifications.

Wieringermeer test site

The Energy Research Centre of the Netherlands (ECN) manages this facility, which has five 2.5-MW wind turbines with 80-meter (260-foot) rotor diameters. The test site also has four prototype locations that can be leased for testing turbines up to 6 MW in capacity with hub heights up to 100 meters (330 feet) and rotor diameters up to 120 meters (390 feet). Three met masts measure conditions at 100 and 108 meters (330 and 350 feet). To analyze wake effects, the Wieringermeer site has a miniature wind farm of ten 10-kW machines.

ECN also participates in Europe's UpWind project (see chapter 9), and ECN's Sirocco project (silent rotors by acoustic optimization) aims to reduce turbine noise by 3–6 decibels without sacrificing power.

Together with Delft University of Technology, ECN established the Knowledge Centre WMC, which focuses on wind turbine materials, components, and structures. The centre performs fundamental research as well as full-scale blade testing and component verification tests for industry.

Other European facilities

A test center at Germany's Fraunhofer Institute for Wind Energy and Energy System Technology (Fraunhofer IWES) can handle blades up to 70 meters long (230 feet). Other components and materials are also tested, such as adhesives and laminates.

Fraunhofer opened a second test center in Bremerhaven in 2011 for blades up to 90 meters (300 feet). This facility is equipped to simulate cyclic loads on wind turbine blades using a hydraulically powered, tiltable mounting block. Fraunhofer is a founding member of Germany's new National Wind Energy Research Center.

Fraunhofer's research program also includes a countrywide wind measurement network of 30 30-meter tall met masts (100-foot), four 50-meter masts (160-foot), and a 200-meter mast (660-foot), as well as three mobile lidar systems.

To develop concepts such as flexible, smart dynamic rotors and controls, researchers have several facilities for testing blades at Duwind (Delft University of Technology's wind program). These include a closed-loop wind tunnel made of 10 interchangeable octagonal test sections and an open-jet wind tunnel. The closed-loop facility can test rotors at a range of wind turbulence intensities, while the open-jet facility offers wind speeds up to 30 m/s (67 mph).

In addition to smart dynamic controls, Duwind's research program, which is part of its aerospace engineering department, focuses on unsteady aerodynamic loads, offshore design, wind power system design methodology, and electrical conversion technology.

Several more facilities are also operating in European countries:

- **United Kingdom.** The National Renewable Energy Center in Blyth can perform static and dynamic fatigue and destructive tests on blades up to 70 meters (230 feet). Also in Blyth, the Energy Technologies Institute was building a rig for testing complete drivetrains and nacelles up to 15 MW. And a gearbox R&D and testing center was being developed in Yorkshire by manufacturer David Brown. The center will feature a 3.6-MW test rig for developing new gearing technology.
- **Spain.** The National Renewable Energy Center (CENER, Centro Nacional de Energías Renovables) can perform static and dynamic fatigue and destructive tests on blades up to 75 meters (250 feet). The facility also has a drivetrain laboratory with four test benches.
- **Germany.** REpower established a testing center for large offshore turbines that will focus on new materials, new designs, and improvement of existing models.
- **Denmark.** Siemens plans to set up two Danish research facilities to focus on direct-drive technology and lightweight blades.
- **Norway.** Two research centers in Norway focus on offshore wind energy, while another analyzes issues such as integration of wind power into utility grids.
- **Finland.** Finland operates an icing wind tunnel to research and test materials and instruments for cold climates.
- **Sweden.** The Swedish Wind Power Technology Center at Chalmers Institute of Technology focuses on the complete design of optimal wind power systems, accounting for interactions among all of the components.
- **Switzerland.** Ecole Polytechnique Fédérale de Lausanne constructed a 30-meter (100-foot) wind tunnel for analyzing wake effects and other inefficiencies. Plans are to build a 100-meter (330-foot) tunnel in the United Arab Emirates.

Offshore testing and demonstration

Lindø Offshore Renewables Center. This test facility, being designed with Risø DTU and Europe's leading offshore wind industry companies, will be built at the former Lindø shipyard in Denmark. The new facility will transform the shipyard's existing infrastructure and equipment—harbor, cranes, steel and pipe production, etc.—into a nexus for offshore wind power development in the North and Baltic Seas.

Lindø's test facility will be able to verify the performance of nacelles up to 10 MW before turbines are installed by simulating wind loads and other extreme conditions over a 50-year period. The facility will also simulate grid code tests such as low-voltage ride-through.

Major industry players including Vestas, Siemens, Dong, and Vattenfall are on board, and the project won public funding of $16 million in 2011 (12 million euros).

Spain's Zèfir test station. This site, to be operated by Spain's Catalonia Institute for Energy Research (IREC, Institut de Recerca en Energia de Catalunya) will offer space for testing 12 wind turbines offshore. The test station will be built in two phases (table 10–1)—the first for turbines with fixed foundations, and the second for floating structures. Gamesa asked the Institute to reserve four turbine spaces, two of each type.

The Zèfir test station is intended to bolster the credibility of offshore wind power technologies in deeper waters and help reduce installation costs. The facility will also collaborate with academic researchers and will offer training in all phases of offshore wind farm development.

Table 10–1. Spain's Zèfir offshore test station. *Data courtesy of IREC.*

	Phase 1	Phase 2
Water depth	40–45 m (130–150 ft)	> 100 m (330 ft)
Distance to shore	3 km (2 miles)	20 km (12 miles)
Number of turbines	4	8
Power installed (max.)	20 MW	50 MW
Type of foundation	fixed-bottom	floating
Scheduled completion	2013	2015–2018

Gamesa/Northrop. Gamesa teamed with America's largest shipbuilder, Northrop Grumman, to build an Offshore Wind Technology Center in Virginia, which opened in 2011. The center focuses on developing Gamesa's next-generation 5-MW prototype turbine, as mentioned in chapter 9.

Alpha Ventus. During installation of Germany's Alpha Ventus offshore wind farm, scientists tested a layered curtain of bubbles to reduce underwater noise from pile driving. Although difficult to operate, the curtain reduced sound by about 12 decibels in the direction of the sea current.

Alpha Ventus is also equipped for testing and monitoring a variety of other offshore wind turbine operational parameters. For example, the performance of direct-drive versus gearbox technologies is being compared.

Egmond aan Zee. This offshore wind farm was extensively instrumented when installed in 2006–2007, yielding data on availability, energy production, corrosion, lightning, and scouring, as well as information on bird and marine life. The turbines are linked to a comprehensive monitoring and evaluation program that will run until 2012.

Heden. A group of engineers, the association Heden (Haveneiland duurzame energie North Sea, or Offshore Port Sustainable Energy North Sea), came up with a concept for an artificial port on a man-made island to support installation and maintenance of future offshore wind farms that lie far from shore.

At an offshore island port, wind turbine components can be shipped in, stored, assembled, tested, and transported a short distance to the installation site. This would not only speed up construction, but would also make repairs faster and easier.

The island could also support the offshore electrical infrastructure. The concept envisions an island with a diameter of 1 kilometer (about 1,000 yards). Heden researchers will investigate and develop the concept.

Fraunhofer offshore. The institute operates four offshore sites for testing the effect of weather conditions on wind power system materials.

U.S. R&D

Government programs

NREL. The National Wind Technology Center, operated by the U.S. National Renewable Energy Laboratory (NREL) at its headquarters in Colorado, conducts tests to evaluate turbine loads, performance, power quality, and noise emissions. The tests help determine compliance with IEC and American standards. Research and testing at three universities is also funded by NREL's wind program, including the deepwater project at the University of Maine (see chapter 9).

NREL operates America's only facility capable of full-scale testing on wind turbine blades up to 50 meters long (160 feet). To expand its testing capabilities, NREL built a new blade testing facility in Massachusetts that opened in 2011. The new facility allows researchers to analyze the effect of stresses on prototype blades up to 90 meters (295 feet).

Blades are mounted horizontally on a concrete test stand, then subjected to stresses by steel cables that pull the blade in different directions. Static measurements evaluate how well the blade is weighted and balanced. Fatigue tests predict whether the blade will last for 20 years. The facility can test two blades at a time.

Also, a 5-MW drivetrain test facility was to be completed in 2012 at NREL's center. A wind industry partner will conduct up to four months of testing for the lab, followed by up to four months of customer-specified testing. The center already has a 2.5-MW dynamometer. Previously, NREL conducted extensive dynamometer and field testing of gearboxes, a prime source of wind turbine failures and downtime.

To accommodate larger wind turbine drivetrains, up to 15 MW, NREL is involved in a university project to build a test facility at the Charleston Naval Complex on South Carolina's coast. The facility will feature power control equipment capable of performing highly accelerated endurance testing (see "Private research").

NREL also tests complete turbines. A 2.3-MW Siemens turbine was being heavily instrumented at NREL's center to produce a constant stream of data on aerodynamics, power characteristics, vibration, system fatigue, noise, and other key measurements.

Siemens will use the results to optimize turbine structures, mitigate loads, increase power output, test safety systems, develop and validate controls, improve system and component reliability, and better understand wind turbine aerodynamics.

Weather researchers plan to equip the 2.3-MW machine with lidar technology that scans a broad wedge of air around the turbine. This will enable them to characterize turbulence and other wake effects, study the impact of turbine ramping during sudden gusts, and analyze data on what happens downstream when winds shift direction rapidly.

A 1.5-MW GE turbine also was being instrumented to allow researchers to assess the aerodynamics of its design and the effects of turbulence on its loads and performance.

A 3-MW Alstom turbine was installed at NREL in 2011. A team of engineers will evaluate the turbine's drivetrain design and technology. The turbine will also be used to develop advanced controls and other technologies related to offshore wind power.

NREL operates two 600-kW turbines that are highly instrumented for testing innovative control systems. The turbines' array of wind speed sensors can be modified to meet the needs of test engineers and partners such as Risø DTU and GL Garrad Hassan. The work aims to develop advanced feedback and feed-forward controls that can work on commercial turbines to mitigate loads and improve performance.

Government researchers have also developed wind turbine simulation codes that enable designers to build virtual models of blades and full systems. Using these publicly available codes, engineers can predict a wind power system's design performance in different environments before prototypes are built and tested. The codes include FAST, AeroDyn, and ADAMS.

To reduce the cost of wind power system foundations and electrical infrastructure, NREL custom-designed and instrumented foundations for two multi-megawatt test turbines to provide load data. Electrical cables were installed in various types of soils to measure their heat dissipation and optimize their thermal performance.

Sandia. To reduce loads on wind turbine blades, researchers at Sandia National Laboratories have investigated an elegant concept: designing a blade that twists as it bends, thus passively adapting itself to changing loads.

Basically, *twist coupling* lets the blade adjust its angle as aerodynamic loads fluctuate—that is, the blade feathers itself to relieve some of the load. The technology has been tested on small machines that have no active pitch control system. GE's 4-MW turbine uses a bent-twist blade, and some small wind turbine blades also have curved tips.

Sandia is also improving the tools used by turbine designers to develop the next generation of rotors at the lowest possible cost. One project aims to design a blade that is 100 meters long (330 feet), using aerospace-derived software called HyperSizer to analyze complex composite structures. The key is to minimize weight while maximizing structural integrity.

Sandia researchers are evaluating a wide range of new materials, control techniques, airfoil designs, and blade manufacturing processes. Some examples include the following:

- Automated manufacturing to ensure consistent product quality and boost labor productivity
- Optimal carbon fiberglass hybrid designs that reduce blade weight while increasing strength and flexibility

- Blade sensors that enhance reliability and load control, extending blade life
- Thick airfoil designs that integrate structure and aerodynamics, along with slenderized blade geometries
- Adaptive structures such as active-aero devices

Sandia is also working on a 9-meter (30-foot) sensor blade that incorporates condition-monitoring instruments such as accelerometers and strain gauges. These could improve blade performance and alert operators to excessive loads being applied to the blade. Sandia will field test the sensor blade, then NREL will subject it to full-scale structural tests.

Other labs and partners. Argonne National Laboratory is exploring chemical and lubrication technologies for wind turbine drivetrain components that could reduce friction, increase load-bearing capacity, and increase resistance to wear, scuffing, and fatigue-related failures of gears and bearings.

Los Alamos National Laboratory is developing an instrument that can make detailed inflow, blade, and outflow velocity field measurements in full-scale wind turbines. The goal is to improve understanding of blade loading and load control in a variety of wind conditions. Researchers are also designing advanced systems for structural condition monitoring.

To improve wind turbine design tools, Los Alamos is developing Windblade software that simulates the interactions between rotating turbine blades and atmospheric conditions such as wind shear.

The U.S. Department of Energy (DOE) is funding three projects to develop mid-size wind turbines, including a 200-kW vertical axis machine and a 500-kW downwind turbine that can be tilted up and guy-wired in place without cranes.

Also with DOE funding, Oak Ridge National Laboratory was building a facility devoted to carbon fiber technology. The laboratory's Carbon Fiber Composites Consortium aims to accelerate the use of new carbon fiber and composite materials in wind energy application and other key markets. The new facility, to start up before 2013, will include a pilot plant that can produce up to 25 tons/yr of carbon fiber materials from several different precursors.

Private research

The world's largest wind turbine drivetrain testing facility was being built on South Carolina's coast in a former Navy warehouse next to rail and shipping infrastructure. Scheduled to begin operating in 2012, the facility will be able to test next-generation drivetrains and direct-drive generators with a 30% current overload capacity for up to five minutes.

The facility will have two test rigs, one for drivetrains up to 7.5 MW and the other up to 15 MW. The larger rig will be able to accommodate complete nacelles. Both rigs will use hydraulic actuators to apply radial and axial loads and rotational moments. The allows the rigs to simulate the static and dynamic loads that impact a wind turbine drivetrain under normal, cyclic conditions and under extreme conditions caused by violent winds.

Most of the drivetrains to be tested will be for offshore wind power applications, but no turbine manufacturers contributed to the project. Tests will be performed impartially, although results can be kept proprietary. Clemson University's Restoration Institute and private partners funded the $98 million facility (72 million euros) with support from state and federal governments. The university might expand the facility into an energy systems testing park, including a bearings test lab and a grid simulator for testing interconnection.

A wind energy R&D center was being built in Ohio to develop ultra-large bearings and seals. The $11.8 million center (8.6 million euros) was funded largely by Timken, a bearings manufacturer, along with support from local agencies and Stark State College, which will use the facility for research and certification.

Boulder Wind Power is designing a direct-drive wind turbine generator aimed at reducing the cost of wind energy by up to 30%. The drivetrain incorporates a new type of air-core, axial-gap, permanent-magnet generator and a related power converter. The drivetrain could deliver a 50% improvement in torque density over existing technologies, says the company. Also, the generator uses a type of rare earth metal that is less expensive and more common than those used in other designs.

Unlike other direct-drive designs, the generator has no iron in its core. This enables a lighter weight, less expensive wind power system. Also, the stator is segmented into small individual units that can be handled by a standard two-person crew. This would let wind farm operators repair and replace stator segments up-tower, without expensive cranes and extended downtime. Boulder Wind Power anticipates having operational prototypes in 2013.

Danotek also is developing permanent-magnet, direct-drive generator technology that will be scalable from about 5 to 8 MW. Danotek's design features a unique rotor-stator interface. Both Danotek and Boulder Wind Power have captured millions of dollars in venture capital.

Other Facilities

Canada

The Wind Energy TechnoCentre in Quebec's Gaspé region helps the industry develop, test, and commercialize wind energy products. The facility acquired two 2-MW REpower turbines to allow the manufacturer to test the cold climate version. The TechnoCentre won $1.9 million in funding (C$1.9 million, or 1.4 million euros) from Canada Economic Development in 2011.

Another site, operated by WEICan (Wind Energy Institute of Canada), can accommodate testing of several wind turbines on its 15 hectares (38 acres). The location on Prince Edward Island's North Cape in the Gulf of St. Lawrence offers 300-degree exposure to a strong wind resource with relatively low turbulence, a harsh marine environment with high corrosion rates, icing events in the winter, and large winter/summer temperature differences. The WEICan facility can be used for certification, R&D, training, and public outreach.

Mexico

A prototype rugged turbine was developed by a team of researchers from Mexico and Japan, where the durability of conventional turbines has been a problem due to high wind speeds. The 300-kW prototype is designed to withstand extreme winds (turbulence intensity up to 20%) and lightning strikes, as well as seismic hazards. The turbine was being tested in the extreme environment of La Ventosa by Mexico's Regional Wind Technology Center.

Asia

China. Vestas established an R&D Center in Beijing in 2010 with an investment of $50 million (37 million euros) over five years. The R&D center will cover areas such as high-voltage engineering, aerodynamics, and development of materials and software.

Chinese turbine manufacturers generally lack the research capabilities of multinational suppliers, but they could catch up soon, given strong government support and the companies' commitment to internal investment in R&D.

Sinovel, China's largest turbine manufacturer, operates the National Offshore Wind Power Technology and Equipment R&D Center, which has developed technology to support turbines of 3–6 MW (for both land and offshore applications). The center is researching capacities above 10 MW.

Goldwind has operated a 1.5-MW turbine for R&D purposes in Bohai Bay since 2007. The machine uses permanent-magnet direct-drive technology, which has been applied to the company's 2.5-MW onshore/offshore turbine. Results are being used to design a 5-MW offshore turbine product line.

Goldwind also operates R&D centers in Germany, Beijing, and Xinjiang. The German center in Neunkirchen serves as the company's global platform for technology innovation. The Beijing center focuses on designing power converters, turbines, and SCADA systems, while the Xinjiang facility is supported by the Chinese government as a national center.

Other Asian countries.

India. A test center for wind turbine components opened in 2011 in Chennai. Vestas operates the center, which is adjacent to its technology R&D office.

Korea. The Korean government began developing an offshore test bed near Jeju Island that will accommodate a total capacity of 100 MW by 2012. The first turbine tested was a 3-MW machine made by Doosan.

Japan. Seven 2-MW downwind turbines were operating off Japan for testing purposes at the near-shore Kamisu wind farm.

International

The International Energy Agency supports cooperative R&D among its 28 member companies, including several tasks with activities that are related to wind power:

- Cost of wind energy (Task 26)
- Offshore wind technology and deployment (Task 23)
- Comparison of dynamic codes and models for offshore wind energy (Task 30)
- Operation of power systems with large amounts of wind power (Task 25)
- Integration of wind and hydropower (Task 24)
- Analysis of wind tunnel measurements and improvement of aerodynamic models (Task 29)
- Labeling small wind turbines for consumers (Task 27)
- Wind energy in cold climates (Task 19)
- Base technology information exchange (Task 11)
- Social acceptance of wind energy projects (Task 28)

Alternative Turbine Designs

As noted above, most wind energy research focuses on traditional horizontal-axis turbines. But concepts that have been tried in the past and failed are getting a new lease on life, as persistent engineers work to make them practical and profitable.

These ideas range from shrouded turbines to jet-propelled blades and from airborne machines to rotors suspended from bridges. Some of these futuristic designs might seem bizarre, but many are winning venture capital from investors and even some government funding.

Wind-accelerating turbines

What if you could speed up the wind yourself, before it enters the turbine, whenever Mother Nature's wind isn't blowing fast enough? That's essentially what a certain type of wind turbine does.

Known as *compact wind-acceleration turbines* (CWATs), these machines use structures surrounding the turbine to boost wind speed before the wind reaches the rotor. If commercialized, these turbines could make wind energy viable in areas with subpar wind speeds, yet can keep operating at higher cut-off speeds as well.

The surrounding structure, commonly called a shroud, encloses an area around the turbine, funneling wind into it. Think of the end of a trumpet—a large-diameter opening that gets narrower. As wind enters the shroud, the design increases the wind speed by forcing the wind through a smaller space, where it encounters the turbine rotor.

This is the same principle used in aircraft jet engines, which are surrounded by a metal shroud with a wide front opening. However, some wind turbine shrouds are less extensive, surrounding just the circumference of the rotor, around the blade tips. In some shrouded turbines, the blades are connected to the shroud, sort of like a wagon wheel.

The wind acceleration concept has been around for decades, known as diffuser-augmented. A *diffuser* makes an air stream more uniform by spreading it out evenly. Historically, shrouded wind turbines, which have also been called ducted turbines, have failed to perform economically.

In addition to the extra cost of the shroud, the yawing mechanism in some designs must handle heavier duty, because the whole shrouded structure has to pivot to face the wind. Today, however, several companies are exploring the idea again, with funding from private and government investors. Alternative wind turbine concepts using the shrouded approach are summarized in table 10–2. The WindTronics product is shown in figure 10–1.

Table 10–2. Shrouded turbine concepts being developed

Developer	Capacity	Rotor diameter m (ft)	Hub height m (ft)	Notes
Enflo	0.5–500 kW	n/a	n/a	Turbines look like fans with a wide band around the rotor edge; made by Windtec
FloDesign	n/a	n/a	n/a	Has a working prototype
Kyushu University	n/a	112 (352)	n/a	"Wind lens" diffuser shrouds the rotor circumference
Mass Megawatts	n/a	n/a	24 (80)	Prototypes have been tested
Master Flo	1-1.5 kW	1.8 (6)	n/a	Looks like a basket-type coffee filter with vertical slits in the shroud
OptiWind	150–300 kW	n/a	61 (100)	A metal cylinder has shrouded turbine rotors attached in pairs to opposite outside edges
V2 Wind	100 kW–7 MW	n/a	n/a	Accelerate wind by passing it through a tube shaped inside like an hourglass; modular design
WinFlex	200 kW	17-21 (56-69)	26–48 (85-157)	Downwind, six-blade design; flexible composite material inflated to form a wheel around the blades; prototypes include 10 and 200 kW
	1 MW	36–52 (118–171)	40–70 (131-230)	
WindTamer	1.3 kW	1.4 (4.5)	n/a	Looks like a megaphone with the rotor enclosed in the smaller end; diameters with shroud are 4–5 m (12–15 ft)
	4.8 kW	2.4 (8.0)	n/a	
WindTronics	n/a	1.8 (6)	varies	Shroud encircles the rotor with electromagnetic components that function as a generator

Blade aerodynamics

Wind turbine blades could be fundamentally redesigned by applying *blown wing* technology. A blower in front of the turbine compresses air and blows it into the turbine, then the air is released over the trailing edge of the blade.

This dynamically redirects the airflow, allowing designers to change the airfoil shape to increase aerodynamic lift. It's sort of like having mechanical flaps on an airplane wing. By capturing more lift, the blade could make wind power economical in low wind-speed areas.

Building a smarter wind turbine blade is the goal of researchers at Rensselaer Polytechnic Institute in New York State, where winds are gusty and turbulent. Small jet actuators are incorporated into the wind turbine blades. The jets release puffs of air that change the air flow around the blade. Also built into the blade are sensors that monitor air flow and vibration strains, then automatically activate the jets.

Fig. 10–1. Small shrouded wind turbine. *Image courtesy of WindTronics.*

Other unusual designs

The counter-rotating turbine design (also called contra-rotating) uses two sets of two-blade horizontal-axis rotors that spin in opposite directions. The second rotor takes advantage of the air flow disturbed by the rotor in front of it.

Other concepts being developed seek to apply suspension bridge technology to wind turbines. For example, a tension cable could be strung across a valley to help support multi-megawatt turbine towers with hubs at 200 meters (660 feet).

Alternatively, turbine rotors could be suspended on cables, allowing them to be pushed by the wind in a loop back and forth. Another concept is to perch turbines on thin towers, connected by tension cables, which are part of a curved-span bridge.

Several researchers have proposed vertical-axis wind turbines that use magnetically levitated bearings. "Maglev" turbines have been designed and exhibited, but are not commercial except for a small wind/solar hybrid technology developed by China's ZK Energy.

Other wind power system R&D

At the California Institute of Technology, researchers observed that when fish are swimming in a large school, some vortexes behind them rotate clockwise, others counter-clockwise, and the vortexes form a staircase pattern.

This has led to a field demonstration project where six vertical-axis wind turbines on mobile platforms will be tested in various layouts to take advantage of the positive aerodynamic interference among them.

Also to optimize wind turbine siting, researchers at Australia's University of Adelaide are using biology-inspired algorithms that mimic the process of evolution, step by step, developing solutions that become incrementally better each time. The method accounts for wake effects, land requirements, wind characteristics, and turbine aerodynamics.

Airborne Wind

Introduction

Why worry about foundations and hub heights, when your turbine could simply take to the skies? As mainstream researchers work to improve large horizontal-axis turbines based on land or sea, some engineers are trying to develop futuristic airborne wind technologies that might liberate designers from some constraints while imposing others. These concepts include gliders and wings, kites and sails, blimps, and even unmanned aircraft, like drones.

Atmospheric winds are stronger and steadier than wind closer to the ground. At 600 meters up (2,000 feet), wind speeds are about two to three times greater than ground-level speeds. In much of the Northern Hemisphere at that altitude, wind speeds exceed 8 m/s (18 mph) more than 40% of the time, and power densities rival the best wind resources on land. To avoid conflicts with low-flying aircraft, airborne wind systems might be deployed far offshore.

At commercial flight altitudes of about 9 kilometers (5–6 miles), wind-capturing technology could tap into the jet stream, where wind velocities reach 67 m/s (150 mph). At that level, 40 to 80 times more wind energy is available (densities of 20,000–40,000 W/m^2, versus around 500 W/m^2 on the ground).

Technology and design

Aircraft cost roughly 100 times as much per unit weight as wind turbines do, but they are designed to be more reliable. Engineers hope that by marrying the two technologies, aerospace and wind power, they can get the best of both worlds—the low cost of wind turbines and the high reliability of aircraft.

Materials science engineers play a role, too, in this synergistic quest. Aircraft and turbine designers are always looking for stronger, lighter weight materials. The fabrics used in kites and sails have to be extremely resilient and durable. Cables that tether flying structures to the ground have to be incredibly strong to handle very high stresses.

Airborne wind power concepts can be classified into two general categories: flying structures with turbines on board, which send electricity to the ground via the tethering cable, and flying systems that capture wind energy without turbines and generate electricity on the ground.

Onboard turbines add to the aircraft size and cost, so flying systems without turbines—such as kites, sails, and gliders—are less expensive and lighter weight. They convert the wind's kinetic energy to mechanical power by pulling their tether cable farther and farther out.

Reeling the cable back in consumes some energy but not much, compared to the amount generated on the ground. It works sort of like a yo-yo, which bounces back up when thrown down.

One nagging problem with any airborne machine is the prospect of a crash to earth. Conventional wind turbines weigh almost 100 tons/MW, while airborne turbines could weigh just a fifth of that, 20 tons/MW.

But that would still be a heavy-duty crash, if it happened, and even flying structures without turbines could cause damage. Engineers are working on ways to avert the hazard by incorporating guidance and recovery systems.

Many entrepreneurs believe that harvesting high-altitude wind is not only possible, but could eventually be profitable. Several companies are developing and testing prototypes, while others are at the conceptual stage. Table 10–3 summarizes various types of airborne wind turbine concepts. Figure 10–2 shows the Ampyx glider, and figure 10–3 shows the SkySails kite.

Table 10-3. Airborne turbine concepts being developed

Developer	Capacity	Dimensions	Altitude, m (ft)	Notes
Gliders and wings		**Wing span, m (ft)**		
Ampyx Power	10 kW	5.5 (18)	300 (1,000)	Tethered glider produces electricity on the ground; prototype PowerPlane has been tested
	100 kW	15 (50)	350 (1,150)	Planned for test by 2013
	1 MW	34 (110)	400 (1,300)	Conceptual
Joby Energy	30 kW	n/a	n/a	Looks like a biplane with no body; wings carry generator arrays that transmit electricity to the ground; prototypes have been tested
	100–300 kW	150–600 (500–2,000)		Plans to test and build commercial systems by 2013
Makani Power	10 kW	n/a	n/a	Glider-like wing carries generators that transmit electricity to the ground; prototype has been tested
	1 MW	35 (115)	200–400 (660–1,300)	Conceptual
Kites and sails		**Kite/sail area, m^2 (ft^2)**		
KitEnergy	n/a	n/a	500–1,000 (1,600–3,300)	Kite flies in figure-8 pattern
KiteGen	5–30 kW	n/a	800 (2,600)	C-shaped kite prototype has been tested for almost 1,000 hr
	3 MW	n/a	n/a	Conceptual
SkySails	n/a	320 (2,400)	100–420 (300–1,400)	C-shaped kite flies in figure-8 pattern; prototype has been tested; Cargill will test SkySails on a 30,000-ton cargo ship by 2013
Windlift	12 kW	40 (430)	n/a	Sail-like airfoil flies in figure-8 pattern; tether extends up to 150 m (500 ft)
Helium-lifted		**Diameter, m (ft)**		
Altaeros Energies	n/a	10 (30)	600 (2,000)	Shrouded "aerostat" looks like a puffy cylinder with rotor in the middle; prototype to be tested
Magenn Power	n/a	n/a	120–300 (400–1,000)	Looks like a blimp that rotates on a horizontal axis; generators at each end transmit electricity to the ground; conceptual
Unmanned aircraft				
Honeywell	<2 MW	n/a	n/a	Conceptual; onboard generators; tether would be >1 km (0.6 mile)

Fig. 10–2. Ampyx winch and PowerPlane. *Image courtesy of Ampyx Power.*

Fig. 10–3. SkySails kite powering 10,000-ton Beluga cargo ship. *Image courtesy of SkySails.*

References

Alpha Ventus, http://www.alpha-ventus.de/index.php?id=80.

Altaeros Energies, http://www.altaerosenergies.com/.

Ampyx Power, http://www.ampyxpower.com.

Angelou, Nikolas. 2010 (October 14) .Predicting the Future Using Lidar. *Renewable Energy World*. http://www.renewableenergyworld.com/rea/news/article/2010/10/lidar-forecasts-boost-wind-harvests.

Boulder Wind Power, http://www.boulderwindpower.com/.

Caitin, http://www.caitin.com.

Catalonia Institute for Energy Research (IREC), http://www.irec.cat/.

Cohen, Yoni. 2011 (March 10). An Air Compressor That Can Improve Wind Turbines? http://www.greentechmedia.com/articles/read/an-air-conditioner-that-can-improve-wind-turbines.

Energy Research Centre of the Netherlands, http://www.ecn.nl/units/wind/.

Enflo Windtec, http://www.enflo-windtec.ch/english/turbine.html.

FloDesign, http://www.flodesign.org.

Fraunhofer IWES, http://www.iwes.fraunhofer.de/en/.

Heden, http://www.haveneilandopzee.nl/.

Home Energy International, http://www.home-energy.com/int/windpower.htm.

Infinite Wind Energy, http://www.infinitewindenergyllc.com/index.html.

IREC Zèfir Test Station, http://www.irec.cat/index.php/en/zefir-test-station.

Joby Energy, http://www.jobyenergy.com.

Kanellos, Michael. 2010 (September 8). Power From the Atmosphere: Is It Real? Greentech Media. http://www.greentechmedia.com/articles/read/power-from-the-atmosphere-is-it-real.

———. 2010 (September 28). Will the Netherlands Rule High-Altitude Wind? Greentech Media. http://www.greentechmedia.com/articles/read/will-the-netherlands-rule-high-altitude-wind.

———. 2010 (September 29). Honeywell's Take on Wind Turbines and the Flying Donut from MIT. Greentech Media. http://www.greentechmedia.com/articles/read/honeywells-take-on-wind-turbines-and-the-flying-donut-from-mit.

Kirsner, Scott. 2011 (April 17). A Generator That's Lighter Than Air—and Relatively Light on the Wallet. *Boston Globe*. http://www.boston.com/business/technology/articles/2011/04/17/a_generator_thats_lighter_than_air__and_relatively_light_on_the_wallet/?page=full.

KiteGen, http://www.kitegen.com/en/.

KITEnergy, http://www.kitenergy.net/.

Knowledge Centre WMC, http://www.wmc.eu/.

LaMonica, Martin. 2010 (November 17). Start-Up Squeezes Flowing Wind to Boost Turbine Power. Green Tech. http://news.cnet.com/8301-11128_3-20023117-54.html#ixzz15ePYQwG2.

Lindø Offshore Renewables Center, http://www.lorc.dk/Home.

LM Wind Power, http://www.lmwindpower.com/.

MacCleery, Brian. 2011 (January). The Advent of Airborne Wind Power. Wind Systems. http://windsystemsmag.com/article/detail/187/the-advent-of-airborne-wind-power.

Magenn Power, http://www.magenn.com.

Makani Power, http://www.makanipower.com.

Mass Megawatts Wind Power, http://www.massmegawatts.com/index.php.

Master Flo Technology, http://www.masterflowind.com/Home_Page.html.

NASA. 2011 (February 1). NASA Engineer Finds Answer to Green Energy in the Air. NASA Tech Briefs. http://www.techbriefs.com/component/content/article/9168.

National Wind Technology Center, http://www.nrel.gov/wind/nwtc.html.

Oak Ridge Carbon Fiber Composites Consortium, http://www.cfcomposites.org/index.shtml.

O'Brien, George. 2008 (April 29). FloDesign Has Innovation Down to a Science. Economic Development Council of Western Massachusetts. http://www.westernmassedc.com/news/flodesignhasinnovationdowntoascience_469.

OptiWind, http://www.optiwind.com.

Orlowski, Aaron. 2011 (February 7). Largest Wind Turbine Testing Facility Nears Opening in Charlestown. *Boston Globe*. http://www.boston.com/yourtown/news/charlestown/2011/02/largest_wind_turbine_testing_f.html.

Patel, Sonal. 2011 (April 1). Changing Winds: The Evolving Wind Turbine. Power. http://www.powermag.com/renewables/wind/Changing-Winds-The-Evolving-Wind-Turbine_3557_p4.html.

Relich, Natalie. 2010 (December 29). A Profile of China's Involvement in the U.S Wind Energy Industry. In News. http://leadenergy.org/2010/12/a-profile-of-chinese-involvement-in-the-u-s-wind-energy-industry/.

Rensselaer Polytechnic Institute. 2011 (January 28). Rensselaer Researcher Working to Make Smarter Wind Turbine Blades. http://news.rpi.edu/update.do?artcenterkey=2815.

Risø DTU. 2010 (August 17). A New Wind Turbine Test Centre at Østerild in Northwestern Jutland to Ensure Denmark's Leading Position Within Wind Energy. http://www.risoe.dtu.dk/News_archives/News/2010/0817_Oesterild.aspx?sc_lang=en.

Sandia National Laboratories Wind Power Technologies, http://windpower.sandia.gov/index.htm.

Sinovel National Offshore Wind Power Technology and Equipment R&D Center, http://www.sinovel.com/en/RandDcenter.aspx.

SkySails, http://www.skysails.info/index.php?id=472&L=2.

Titchen, John. 2010 (May). Direct Drive Permanent Magnet Wind Turbine Experience. Goldwind. http://www.cleanenergyweek.com.au/mediaObject/

events/2010–conference/presentations/1500-Titchen/original/1500%20Titchen.pdf.

Trabish, Herman K. 2010 (November 18). Will Goldwind Rule the Wind World? Greentech Media. http://www.greentechmedia.com/articles/read/will-goldwind-rule-the-wind-world/.

———. 2011 (September 15). Can Wind Generate Electricity at $0.04 per Kilowatt-Hour? Greentech Media. http://www.greentechmedia.com/articles/read/how-can-wind-generate-electricity-at-four-cents-per-kilowatt-hour/.

———. 2011 (June 15). Will German Offshore Wind RAVE On? Greentech Media. http://www.greentechmedia.com/articles/read/will-german-offshore-wind-rave-on/.

U.S. Department of Energy, Energy Efficiency and Renewable Energy. 2010. *Wind Power Today—2010.* http://www1.eere.energy.gov/windandhydro/pdfs/47531.pdf.

———. Wind and Water Program. http://www1.eere.energy.gov/windandhydro.

V Squared Wind, http://www.vsquaredwind.com.

WEICan, http://www.weican.ca/.

Weiner, Jon. 2010 (May 17). Schooling Fish Offer New Ideas for Wind Farming. Caltech University. http://media.caltech.edu/press_releases/13347.

Williams, Andrew, 2011 (March 17). Seeking Wind Power Innovation: Europe's Largest Wind Turbine Test Facility Now Open. *Renewable Energy World.* http://www.renewableenergyworld.com/rea/news/article/2011/03/seeking-wind-power-innovation-europes-largest-wind-turbine-test-facility-now-open.

Windlift, http://www.windlift.com/.

WindTamer, http://www.windtamerturbines.com/.

WindTronics, http://www.windtronics.com/home.

WinFlex, http://winflex.co.il/.

ZK Energy, http://www.zkenergy.com/en/index.html.

11

The Global Wind Industry: Markets, Business, and Government Policy

Wind power is one of the world's fastest growing sources of energy in general and electricity in particular. Wind energy supplied an estimated 1.92% of the world's electricity in 2011. By 2020, wind power could meet 9% of global electric demand.

Today's wind industry is ever more dominated by growth in China, where wind power capacity is rising rapidly and domestic turbine manufacturers are thriving. China's government policies have contributed to the staggering expansion of the country's wind energy market.

Nonetheless, traditional strongholds of wind power are also growing, and the world's leading turbine manufacturers are adapting to a changing marketplace.

Market Snapshot

Introduction

Overall, the global wind power market is increasing at a healthy rate, roughly doubling every fourth year since the mid-1990s (fig. 11–1). However, this pace was maintained in 2010 only because faster expansion in China compensated for slower growth in some traditional wind powerhouses, such as America and parts of Western Europe.

For the first time in the industry's history, more new wind capacity was added in developing economies than in established markets. In other words, wind power's center of gravity is shifting from West to East.

Global wind power capacity totaled 197 GW as of 2011, according to the Global Wind Energy Council (GWEC). During 2010, 38.3 GW was installed, translating to a growth rate of 24% (annual capacity increase divided by previous year's total). About 41 GW of capacity was installed in 2011 (a 21% growth rate), bringing the worldwide total in 2012 to 238 GW, said GWEC.

The World Wind Energy Association's tally for 2010 was marginally lower, but reflects the same trends. Industry analyst BTM Consult reported slightly higher figures—about 200 GW total, 39.4 GW installed, and a 25% increase.

In any case, the wind power market's growth rate was slower than 2009's, and 2010 was the second-slowest growth of the past decade (fig. 11–2). Also, new installations fell for the first time in 20 years (fig. 11–3), with 7% less capacity added.

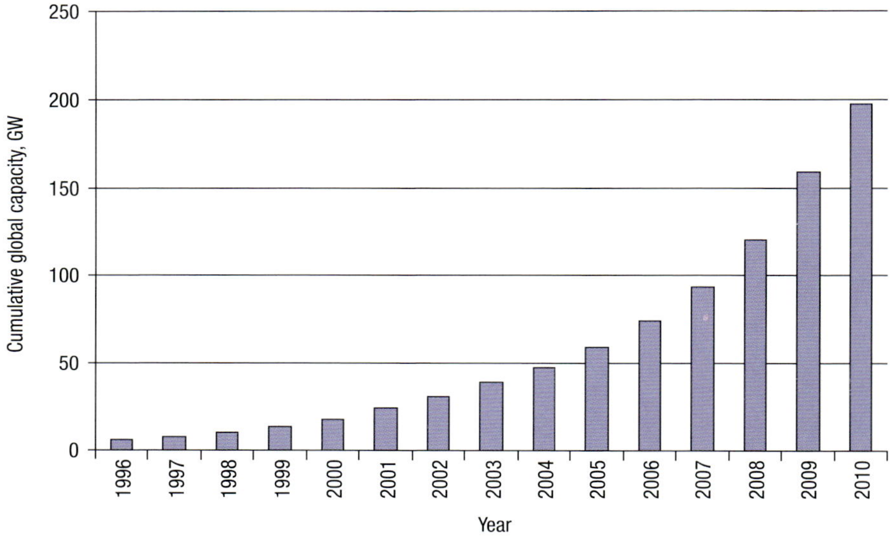

Fig. 11–1. Global installed wind power capacity. *Data courtesy of Global Wind Energy Council.*

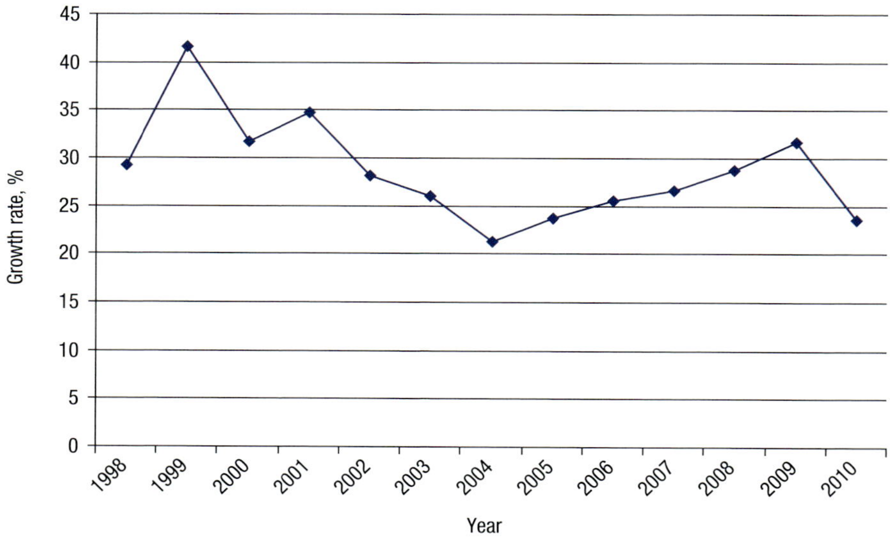

Fig. 11–2. Wind power market growth slows down. *Data courtesy of World Wind Energy Association.*

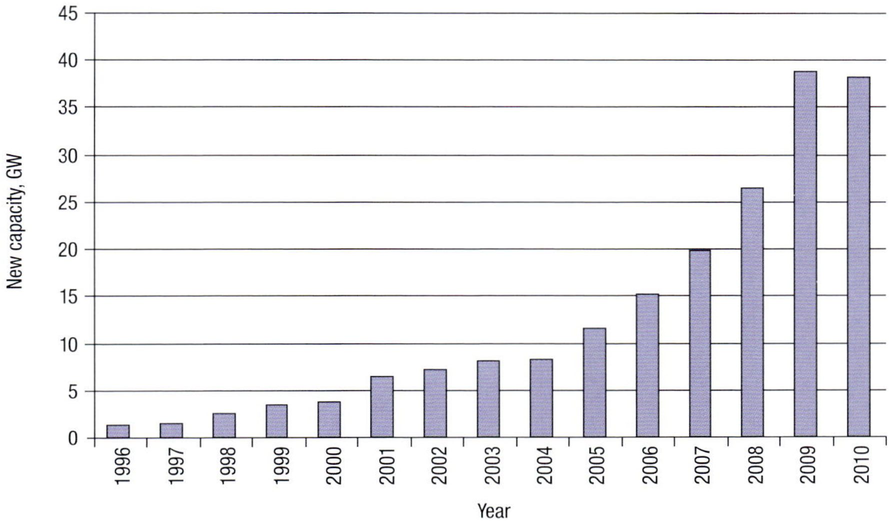

Fig. 11–3. Annual wind power installations. *Data courtesy of Global Wind Energy Council.*

Leading countries

In contrast to offshore wind energy, which is dominated by Western Europe, the global market encompassing land-based wind power is much more diverse. More than 70 countries have at least 1 MW of wind power, and almost 40 of these have wind farm capacities totaling 100 MW or more.

Table 11–1 shows the 20 countries with the largest total capacities as of 2011, according to GWEC. The top five of these account for nearly 75% of worldwide wind power (fig. 11–4). Appendix H lists capacities and other details for all individual countries and from additional sources.

Total capacity isn't the only way to measure a country's investment in wind power. Statistics are available that compare the amount of capacity per person, per unit of land area, and per unit of economic output (GDP, or gross domestic product).

By all of these measures, Denmark is the top country in the world, with Germany, Portugal, and Spain near the top in varying orders. Interestingly, large countries with lots of wind power capacity rank lower. By megawatts per person, America is number 8, China 24, and India 35; by megawatts per unit of GDP, China is 7, India 13, and America 16.

Table 11–1. Global wind power capacity by country (top 20). *Data courtesy of GWEC.*

Country	Total capacity 2010, GW	Capacity added 2009, MW	Growth rate 2009–2010, %
China	44.733	18,928	73.4
United States	40.180	5,115	14.6
Germany	27.214	1,493	5.8
Spain	20.676	1,516	7.9
India	13.065	2,139	19.6
Italy	5.797	948	19.6
France	5.660	1,086	23.7
United Kingdom	5.204	962	22.7
Canada	4.009	690	20.8
Portugal	3.898	345	10.3
Denmark	3.752	327	9.4
Japan	2.304	221	10.6
Netherlands	2.237	32	1.4
Sweden	2.163	604	38.7
Australia	1.880	167	9.8
Ireland	1.428	118	9.0
Turkey	1.329	528	65.9
Greece	1.208	123	11.3
Poland	1.107	382	52.7
Austria	1.011	16	1.6

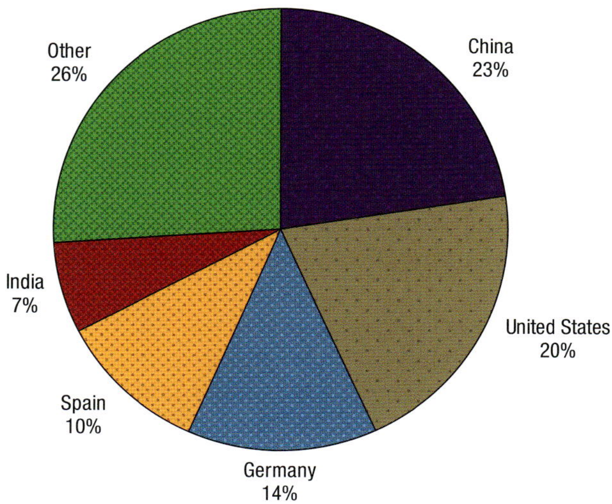

Fig. 11–4. Top five countries account for almost 75% of global wind power capacity. *Data courtesy of GWEC.*

China. China overtook the United States in wind power capacity in 2010, moving from second to first place, and its growth rate of over 73% dwarfs America's, where new capacity grew by less than 15%.

China's installations during 2010 accounted for almost half of the world market for wind power (fig. 11–5), driven by a record level of investment that exceeded $20 billion in 2009 (15 billion euros). Without China, the global wind power market would have shrunk substantially.

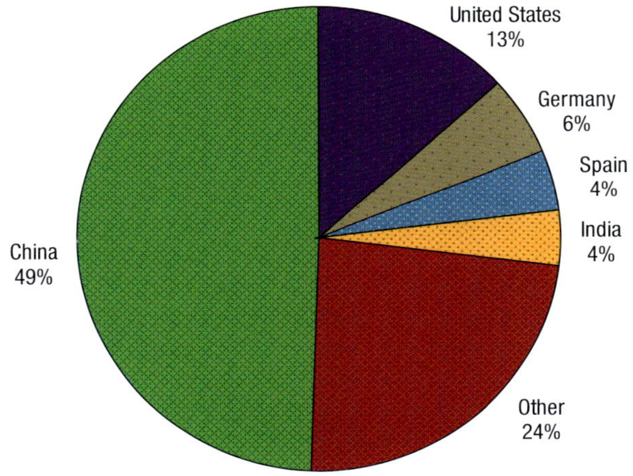

Fig. 11–5. Nearly half of new wind power went to China in 2010. *Data courtesy of GWEC.*

China added almost 19 GW of new capacity in 2010—the biggest volume ever by one country in a single year. As a portion of net annual power generation capacity additions, wind power accounted for nearly 20%, almost as much as hydropower. Another 18 GW of wind capacity was installed in 2012, giving China a total approaching 63 GW.

Europe. This region still rules the world in total wind power capacity, with more than 86 GW—greater than China and the United States combined in 2010. Since 1995, the market growth of wind power in Europe has averaged 17.6% (fig. 11–6).

Germany kept its top position in 2010, followed by Spain (see table 11–1). In terms of electricity generated by wind, however, Spain's exceeded Germany's, reaching almost 43,000 GWh (43 TWh).

As recently as year 2000, only three countries accounted for 85% of Europe's wind capacity—Denmark, Germany, and Spain. Since then, wind power has become much more widespread in Europe, with these top three countries accounting for only 36% of the total in 2010.

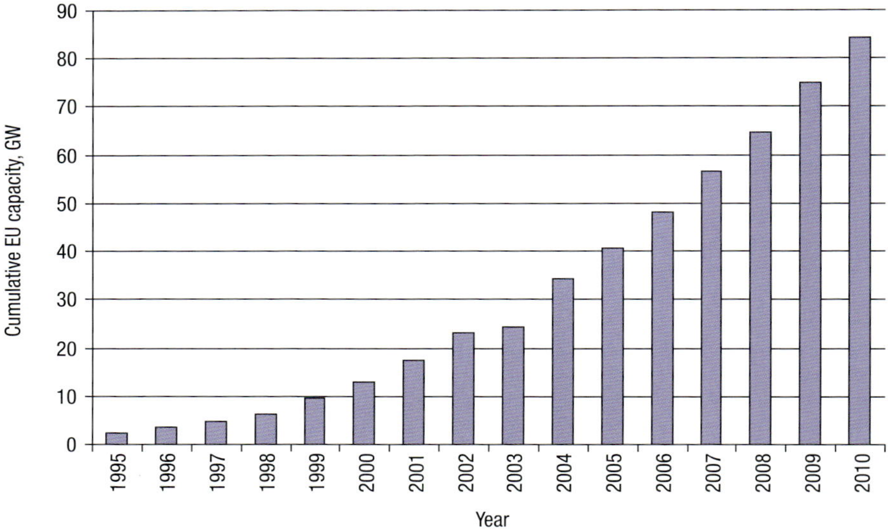

Fig. 11–6. Cumulative wind power growth in Europe. *Data courtesy of European Wind Energy Association (EWEA).*

Wind power markets expanded by more than 20% in France and the United Kingdom, while Belgium's increased by 62.2%. Spain installed the most new capacity, followed by Germany, France, the United Kingdom, Italy, and Sweden. Much of the UK's market is focused in Scotland, which accounted for almost 46% of the country's total capacity and a third of its additions in 2010.

Much of Europe's market growth is driven by offshore wind power. While the onshore market fell by more than 13% compared to 2009, the offshore market thrived in some countries, notably Belgium, the United Kingdom, Germany, and Denmark. Overall, Europe's offshore installations set a record in 2010, growing by 51% and accounting for 9.5% of new capacity (see chapter 9).

Similarly, while some Western European wind markets have begun to stagnate, Central and Eastern Europe's are taking off. Among these countries, Turkey and Poland had the biggest amounts of wind power, as well as outstanding growth rates in 2010, while Romania, Croatia, and Bulgaria rapidly expanded their wind capacities.

Despite this growth, however, wind power was eclipsed by natural gas as a source of new power generation capacity in Europe, for the first time since 2007. Wind power accounted for 17% of new capacity, versus 51% for gas.

North America. The U.S. market saw new wind power installations drop by almost 50% in 2010, resulting in overall market growth of less than 15%, as noted earlier. Of 50 states, 38 have utility-scale wind farms, and 14 of those have more than 1 GW installed.

Texas leads the nation, with 10.1 GW of capacity—about half-way between that of India and Italy. Iowa was a distant second, with 3.7 GW, followed by California, Minnesota, and Washington State. Appendix H contains details on individual state wind power capacities.

In terms of installations during 2010, Texas also was on top with 680 MW of new capacity. The next four states leading in installations were Illinois, California, South Dakota, and Minnesota.

The U.S. wind farm construction pipeline remained in good shape in 2011, with 8.3 GW being built. Power purchase agreements with utilities were in place for about two-thirds of megawatts under construction. An estimated 6.8 GW of new capacity was completed in 2011, according to GWEC's estimates.

Despite 2010's slump, wind power accounted for more than one-fourth of America's new electricity supplies that year and has averaged 35% of new generation capacity during 2005–2010. Wind power supplied 2.3% of America's electricity in 2010, up from 1.8% in 2009.

Canada joined the world's top 10 countries in 2010, ranking ninth in terms of total wind energy capacity at about 4 GW, even though its market was down compared to 2009. Also, by installing 690 MW, Canada attained the ninth position for new capacity as well.

Ontario Province leads the country's wind development, with more than 1.6 GW installed, followed by Alberta and Quebec (806.5 and 759.4 MW, respectively, according to the Canadian Wind Energy Association). This country intends to triple its wind power production during 2011–2016. Canada's total capacity approached 5.3 GW as of 2012, GWEC reported.

Expansion elsewhere

In Asia, wind power markets in 2010 grew in India (at a 19.6% rate), Taiwan (19.0%), Japan (10.6%), and South Korea (8.9%). India set a record for new installations in 2010, and the country's 13 GW of capacity supplies 7.6% of India's renewable energy and 2.9% of its electricity.

Japan might focus more in the future on wind energy and other renewables after a tsunami wave in 2011 damaged some of its nuclear power plants. Wind power installations slowed somewhat in Australia and New Zealand in 2010 (growth rates of 9.8% and 1.8%).

Latin America is a prime target for wind power development, with its rapidly growing economies and rising demand for electricity. As a region,

Latin America had more than 2 GW of wind capacity as of 2011, led by Brazil and Mexico, with 931 and 519 MW, respectively.

Brazil expects to build another 4–5 GW of wind power before 2014, much of which was already contracted. At auctions held in 2011, wind energy captured the largest share of new power capacity—49% at the first auction and 81% at the second. Another auction was planned for 2012. Several international wind equipment manufacturers have decided to build production facilities in Brazil, many of them already under construction.

Brazil is highly dependent on hydropower, which has been compromised by droughts, resulting in electricity shortages. This gives the country impetus to develop wind power as a complementary alternative. Ambitious government plans call for 11.5 GW by 2020. However, the public financing process is slow, and many wind projects face other delays as well.

Energy analysts project that Brazil could capture 69% of Latin America's wind power market by 2025. According to IHS Emerging Energy Research, Brazil's capacity could grow in that time frame to 31.6 GW, followed by Mexico at 6.6 GW.

Africa possesses rich wind resources on its coasts and in the eastern highlands, but so far, wind power development is significant only on the northern Mediterranean coast, with 97% of the continent's capacity. Egypt, with 550 MW, and Morocco, with 286 MW, plan to boost their wind power capacity substantially.

Egypt was gearing up to tender contracts for four 250-MW wind farms in the Gulf of Suez, according to the transition government that took over in 2011. The country also has set aside 7,600 sq km of desert land (2,900 sq mi) for more wind farms and intends to spend $220 million (160 million euros) to build grid-connection infrastructure for wind power, using World Bank financing.

Morocco plans to build a 300-MW plant, which has a power purchase agreement. The country hopes to reach 2 GW of capacity as part of its goal for renewables to supply 20% of electricity consumption by 2020.

Although the Middle East in general has lots of oil and gas, some countries lack domestic resources and have to import fuels. But wind power in the region is in its infancy, and only Iran has significant capacity (92 MW), which could increase to 400 MW if its plans succeed.

Two countries have renewable energy targets, Jordan and Syria, but neither specifies the amount that wind power would contribute. Syria was planning two wind farms totaling 130 MW in capacity.

Global trends

Given China's role in today's wind power market, industry growth could continue to swing toward the developing world. Although China would drive this shift, other emerging markets could contribute to wind power expansion.

In developed countries, transmission constraints remain a threat to future wind industry growth. A robust, resilient grid infrastructure is crucial to support development of tomorrow's wind farms. As discussed in chapter 8, countries that have decided to support wind power are taking steps to upgrade their existing transmission networks, build out their power systems, and interconnect grids to form large regional markets.

Wind power's growth will also hinge on whether governments worldwide step up with favorable support schemes and financial incentives, as well as economic and legal climates that foster financing and grid infrastructures that are adequate to deliver wind power to consumers. Some governments such as Germany's are being swayed away from additional nuclear capacity after Japan's experience.

Looking ahead, global wind power capacity could reach 449 GW by 2016, GWEC estimates, with the market expanding at an annual average rate of 18.2% in terms of total capacity and 11.1% for installations. The council concedes that this forecast is modest compared to the wind industry's past growth, but cites continuing uncertainty surrounding the North American market as a reason to be conservative.

Pike Research forecasts that wind power capacity will rise to 562.9 GW by 2017. In a more bullish forecast by BTM, installations could increase by 15.5%/yr through 2015 and by 11.5%/yr in 2016–2020. This translates to global capacity totaling more than 500 GW in 2015 and close to the 1,000-GW milestone by 2021. BTM cites Asia's economic growth as a reason for optimism, along with encouraging signs in U.S. and European markets.

Yet more confident is the World Wind Energy Association, which sees a global capacity of 600 GW by 2015 and more than 1,500 GW by 2020.

Industry Snapshot

Introduction

The lingering effect of the 2008 financial crisis strangled the wind business in many developed countries in 2010, as the decline in orders worked its way through the system. New installations represented capital of about $65 billion (47.3 billion euros), said GWEC.

Industry analysts expected 2011 to improve as rebounding investment in wind power equipment—much of it destined for China—would translate to more projects coming online in 2011–2012. However, persistently low natural gas prices in America and reduced subsidies for wind energy in some countries could depress investment, slowing business again later in the wind power development cycle.

Sales of about $55 billion in 2010 (40 billion euros) were recorded by the World Wind Energy Association, compared with $68 billion in 2009 (50 billion euros). Lower turbine prices, especially in China, contributed to falling investment (see "Turbine prices and energy costs" below). However, the global wind market value could rise from $91.5 billion in 2011 to $153 billion in 2015 (from 66.8 to 111.7 billion euros), according to BTM.

Financial closings for U.S. wind farms totaled $11.1 billion in 2010 (8.1 billion euros), the American Wind Energy Association reported. Of these, 20 were tax equity projects representing $2.7 billion (2 billion euros), and 29 were debt capital deals worth $8.4 billion (6.1 billion euros). Of the tax equity deals, 85% (17 of 20) took advantage of the federal government's cash grant option (see "Government Policy" below).

Turbine manufacturers

The global turbine market is worth an estimated $45 billion (33 billion euros). Denmark's Vestas kept its position as the world's leading wind turbine supplier in 2010, increasing its market share from 12.5% to 14.8%, while Sinovel overtook GE for second place (table 11–2).

Among the top 10 manufacturers, four are Chinese. A second tier includes China's Mingyang and XEMC. When all components of wind power systems are counted (not just turbines), China is the world's biggest producer of wind energy equipment.

Table 11–2. Top 10 wind turbine manufacturers' market share. *Data courtesy of BTM Consult.*

Manufacturer (Country)	Market share, 2010
Vestas (Denmark)	14.8
Sinovel (China)	11.1
GE (U.S.)	9.6
Goldwind (China)	9.5
Enercon (Germany)	7.2
Suzlon (India)	6.9
Dongfang (China)	6.7
Gamesa (Spain)	6.6
Siemens (Germany)	5.9
United Power (China)	4.2
Total	82.5

China. China's turbine manufacturing capacity is on the multi-gigawatt scale, supplying most of Chinese demand. New turbines, made by at least 86 manufacturers, have an average of 95% domestic content, versus about 90% in Europe's market and more than 50% in America's.

Several leading international manufacturers have subsidiaries in China, including Vestas, Gamesa, Suzlon, Nordex, and GE. Other companies have targeted China's market through joint ventures. Participation in China's wind power boom requires foreign manufacturers not only to assemble turbines there, but also to train Chinese workers and buy components from local suppliers.

As China's home-grown enterprises have strengthened, they've increasingly displaced foreign turbines in new domestic wind farms (fig. 11–7). Sinovel, Goldwind, and Dongfang dominated China's market in terms of both new installations and cumulative capacity in 2009, and in 2010, 90% of new capacity was made in China. Chinese manufacturers have access to low-interest loans and get preferential treatment from government-owned wind power developers.

As recently as 2005, Gamesa controlled more than a third of the Chinese turbine market. Although its market share has dwindled, China's demand is so huge that Gamesa still sells more turbines to China than ever, said the *New York Times*. The company has deals with Chinese power companies for 900 MW's worth of turbines.

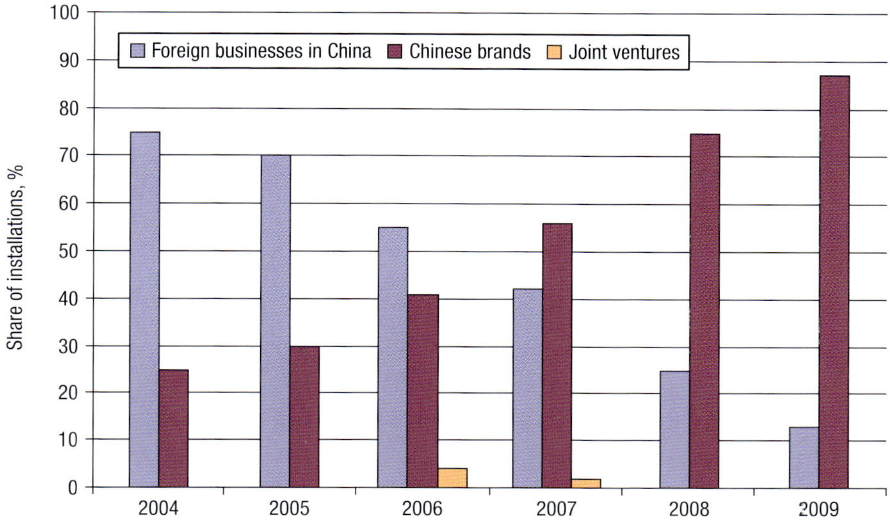

Fig. 11–7. Domestic vs. foreign market share of new installations in China. *Data courtesy of China Wind Power Outlook 2010.*

Goldwind has nine manufacturing plants in China that can produce 4,000 1.5-MW machines per year. However, China's supply chain is challenged at the upper end by bottlenecks in small-scale equipment production, and the country lacks a fully developed network of related services such as certifying agencies and dedicated R&D.

Chinese companies continue to use some outdated technologies, and manufacturers have not developed their own core competencies. The Chinese government wants to correct this situation. For example, many of China's manufacturers could be forced out of the market unless they learn to build turbines with capacities of 2.5 MW or greater, according to technical standards introduced in 2011. New manufacturers must be capable of producing 1 GW of turbine capacity per year and will have to take a 30% stake in new plants.

To overcome their limitations, China's turbine manufacturers are forging business relationships with companies such as American Superconductor, which licenses and codevelops turbines in China and supplies power electronics and control systems. Most of American Superconductor's revenues come directly from its Chinese customers, and its new product for connecting wind farms to power grids was introduced in China. However, Sinovel's long-standing partnership with American Semiconductor deteriorated in 2011 over an industrial espionage dispute.

Demonstrating the economic muscle of China's turbine makers is their foray into the financial mainstream. Goldwind raised $917 million (669 million euros) via an initial public offering (IPO) on the Hong Kong exchange in 2010, and Sinovel raised $1.5 billion (9.5 billion yuan, or 1 billion euros) on the Shanghai exchange in 2011. Mingyang is listed on the New York exchange.

Chinese manufacturers have begun to export their machines and compete intensely in the global marketplace. Some Chinese products are capturing market leadership, such as Goldwind's direct-drive generator, with 6,000 units in operation worldwide. Goldwind and Sinovel have established U.S. offices. Sinovel also has divisions in Canada, Spain, Australia, and Brazil, while Goldwind also operates in Australia and Africa.

As recently as 2009, exports of complete wind turbines from China totaled only 20 units with a combined capacity of 28.75 MW. That has begun to change. Goldwind was installing 600 MW at a Texas wind farm (240 2.5-MW units) and 106.5 MW in Illinois (71 1.5-MW). Both projects rely on China Development Bank funding, but Goldwind also provided equity. Several other small American wind farms use Chinese machines or have plans to buy them.

China's manufacturers are expanding beyond domestic markets because they must, in order to grow. Some analysts perceive that Chinese demand for wind power equipment could begin to weaken. Thus the manufacturers' move into global markets including the United States, even though their subsidiaries must deal with issues such as domestic content and public resistance to channeling government incentives to Chinese companies.

Product quality, too, remains a concern, since Chinese turbines have a shorter track record. This makes it hard to get project financing, a serious obstacle to their expansion unless manufacturers continue to fund their own projects. If existing Chinese equipment worldwide performs successfully for another couple of years, they could win more recognition for reliability among investors and wind farm developers.

Another problem with Chinese turbines is their lack of low-voltage ride-through capability, which enables wind power to remain stable and connected during transient faults on the grid. China's electric regulatory authority reported three incidents in 2011 when hundreds of wind turbines disconnected in Hebei and Gansu Provinces.

The regulatory authority urged that existing turbines without low-voltage ride-through be upgraded soon. Several of China's turbine manufacturers were already working on upgrades. Wind farms can be retrofitted with equipment that provides this feature and other grid-support functions for multiple turbines.

Also, in the wake of Japan's 2011 earthquake and tsunami, China's turbine manufacturers were reevaluating their technology for withstanding natural disasters. Chinese machines have a less robust anti-quake design than Japan's and should be improved, especially for offshore duty.

China's government regulations protect its wind industry by requiring domestic content for turbines manufactured there. The extent of China's protection prompted America's steel workers' union to file a complaint of unfair trade advantage with the World Trade Organization. China agreed in 2011 to end subsidies to turbine component manufacturers. U.S. wind tower manufacturers have also complained of unfair subsidies and pricing.

Another controversy is China's corner on the market for rare earth metals (more than 90% of world production), which are used by the ton in wind turbine permanent-magnet generators, as well as other products such as electric vehicles. China has restricted exports of these metals, raising international concerns as prices rose.

Other countries. In America, GE retained the largest share of the wind turbine market in 2010 (49.7% of new capacity), followed by Siemens (16.2%) and Gamesa (11%), but at least 15 other manufacturers supplied turbines to U.S. wind farms. This increasingly competitive marketplace will require further development of the U.S. supply chain to reduce transportation costs, mitigate currency risk, and improve just-in-time turbine availability, product quality, and service (see "Manufacturing" below).

About 2,900 turbines were installed in America in 2010, bringing the cumulative number to 35,600. About half of new U.S. turbines are 1.5–1.6 MW models, but the market grew for turbines exceeding 2 MW. The average size increased slightly in 2010, to 1.79 MW from 1.74 in 2009, said the U.S. Department of Energy (DOE). Since 1998, nameplate capacity has risen 151%. Hub heights and rotor diameters also have scaled with time, increasing by 43% and 76%, respectively, during 1998–2010.

India's growing wind power market has spawned domestic turbine manufacturing, with global leader Suzlon commanding almost 50% of the country's market. Several other turbine manufacturers are pursuing a share of India's wind business, including four of the top ten global leaders (Vestas, Gamesa, Siemens, and GE).

Gamesa plans to invest $83 million (61 million euros) through 2012 to open an R&D center and turbine production plants in India, including a blade factory in Gujarat State. India's market accounted for about 8% of Gamesa's global sales in 2010. The company expects its Indian manufacturing capacity to exceed 800 MW and its sales there to more than double by 2013.

However, India's business climate might leave something to be desired, if Enercon's experience is anything to go by. A dispute between the turbine manufacturer and its local partner over a joint venture and patents led Enercon to write off its subsidiary there and exit from the Indian market in 2011.

Elsewhere in Asia, prominent heavy-machinery manufacturers in South Korea, including Samsung, Hyundai, and Daewoo, are getting into the wind turbine market. Korea's Doosan markets a 3-MW turbine and is developing 6-MW technology (see chapter 9).

In Brazil, Enercon has dominated the turbine market, but Impsa, Alstom, and GE are part of the country's manufacturing base. Other international suppliers are seeking entry and establishing manufacturing plants in Brazil, as noted above. The country promises to become a major production hub for Latin American wind industry operations. In addition to turbine assembly, an even larger opportunity is opening to develop the country's wind power component supply chain.

Turbine industry trends

Repowering. As wind farms age, the market is growing for repowering old machines—replacing them with bigger, more efficient equipment, as noted in chapter 6. This market accounted for 183 MW in 2010. Repowering a wind farm costs substantially less than siting and permitting a new project.

In India alone, as much as 1.3 GW (10% of the installed base) consists of undersized, rusted machines. Gamesa and Vestas are seeking to gain a foothold in this potential repowering market. In America, about 1.5 GW of capacity dates back to the 1980s and could be ripe for replacement.

However, when turbines are repowered, wind power price agreements and subsidies usually have to be renegotiated, and disposal of old equipment can be problematic. Some wind farm operators choose to extend the useful life of their existing turbines rather than replace them prematurely.

Low wind speed designs. Another notable trend in the turbine business is development of machines designed for low to moderate wind speeds (averaging 7.5–8.5 m/s, or 17–19 mph). Manufacturers are targeting markets where the best wind resources are either already developed or impractical, leaving areas with less wind energy, but enough to capture economically.

In general, low/moderate wind speed turbines are simply conventional models with longer blades, but other design factors have to be considered. Several manufacturers have introduced such turbines, including Alstom (2.7-MW), Gamesa (4.5 and 2-MW), REpower (3.2 and 1.8-MW), Vestas (3-MW), GE (2.5 and 1.6-MW), Nordex (2.4-MW), Siemens (2.3-MW), and Goldwind (1.5-MW), and more are planned.

Direct-drive generators. Another shift in the wind turbine business is toward direct-drive, gearless technology, as discussed earlier. Direct-drive turbines accounted for 17.6% of wind turbine supply in 2010. Although many of these have been deployed offshore, gearless machines are being selected for onshore wind farms as well.

New markets. Taking advantage of a lull in sales, some turbine manufacturers are focusing not only on reducing costs, but also expanding into nontraditional markets. Wind farm developers will doubtless remain the biggest customers, but Vestas, Suzlon, and others are looking at new applications for turbines, such as power supply along railroad lines and at computer server farms. Carbon-conscious companies are also potential customers.

Turbine prices and energy costs

Prices for onshore wind turbines have continued to experience fierce downward pressure worldwide, due to greater manufacturing scale, better efficiency, and some overcapacity in the supply chain. Prices fell below \$1.4 million per megawatt (1 million euros) in 2010, which was 7% lower than 2009 and 19% down from their peak in 2007–2008, according to Bloomberg New Energy Finance.

This trend was more pronounced in countries without feed-in tariffs or other guaranteed price support, where wind power competes against conventional electricity derived from natural gas and coal. Turbine prices in the United States averaged $1.27 million per megawatt (0.93 million euros), said Bloomberg, while DOE cited a range of $0.9–1.4 million/MW (0.7–1 million euros). In China, wind turbine prices were even lower, ranging around $0.6–$0.9 million per megawatt (4–6 million yuan, or about 0.5–0.7 million euros).

Falling turbine costs trickled down to result in lower prices for wind power. Wind energy costs worldwide reached record lows in many areas, especially in countries with direct price competition from conventional electricity.

Several wind power projects had a life-cycle, or levelized, cost of energy (including capital and O&M, but excluding the impact of subsidies) below $68 per megawatt-hour (50 euros), Bloomberg reported. Levelized costs were driven down not only by lower turbine prices, but also by higher energy output per megawatt of capacity.

The rule of thumb for prices in U.S. wind power contracts has been around 7 U.S. cents per kilowatt-hour (5 euro-cents), close to competing with fossil fuels. In some areas, price agreements were signed for 5–6 U.S. cents/kWh (4 euro-cents). This could make it cheaper to add wind turbines than to build natural gas or coal plants equipped with the latest technology. According to DOE, average wind power prices won't fully reflect falling turbine prices until 2012 and beyond.

For the past few years, wind turbine costs went up due to rising demand around the world and the increasing price of steel. Behind the scenes, wind manufacturers were reducing their costs, and now we are seeing just how cheap wind energy can be when overcapacity in the supply chain works its way through to developers.

—Michael Liebreich, Bloomberg New Energy Finance

Other key players

Wind power projects enjoy good access to low-cost financing in China, and the country's power generation market is dominated by a handful of large state-owned enterprises, which together account for more than 80% of the wind power market.

The three largest wind farm operators are Guodian (Longyuan Electric Group), Datang, and Huaneng. Each added 1–2 GW of capacity in 2010. Huaneng expects to increase its capacity by 45% in 2011 to reach a total of 5.1

GW by 2012. Huaneng's and Datang's wind units raised capital in 2010–2011 with IPOs on the Hong Kong exchange.

China Resources New Energy Group (the renewables arm of China Resources Power, or CRP) is another large operator. It holds stakes in 42 wind farms totaling 646 MW, as well as 631 MW of wind power capacity being built and 970 MW of projects approved by the government. Other operators include Huadian, Guohua, and China Guangdong Nuclear Power.

Europe's national utilities operate many of the bigger wind farms or parks, though some of Europe's onshore capacity is community-owned. Spain's Iberdrola Renovables is one of the world's top ten operators. The company planned to install 1.5 GW in 2011, accumulating more than 16 GW of installed capacity by 2012. Portugal's EDP Renováveis (with U.S. subsidiary Horizon) and Italy's Enel also are major players.

Another global wind energy leader is Germany's E.on, with more than 70 onshore wind farms in Europe and America, totaling almost 3 GW in capacity. E.on is also one of Europe's leading offshore wind farm developers, along with Vattenfall and Dong Energy, as noted in chapter 9. E.on intends to operate about 10 GW of offshore capacity by 2015.

NextEra Energy Resources is America's biggest owner of wind power capacity, ending 2010 with more than 8 GW under management. The company plans to add 2 GW of wind power by 2013. Iberdrola will spend $3.9 billion (2.8 million euros) on new U.S. wind farms in 2011–2012. MidAmerican has invested or committed $5.4 billion to wind (3.9 billion euros) and had 2.9 GW of wind power operating as of 2012. Table 11–3 shows the top five U.S. wind farm owners, which control nearly half of installed capacity.

Unlike MidAmerican, some other regulated U.S. utilities take a more conservative approach to wind power. Typically, they're concerned about rate increases, uncertain environmental policies, and load-balancing services, as well as transmission constraints. Many utilities prefer to buy wind power or credits rather than own and operate wind farms themselves. The decision depends on the utility's size, expertise, business strategy, and other variables.

Table 11–3. Top five U.S. wind farm owners control 49% of installed capacity. *Data courtesy of AWEA.*

Owner	Capacity, GW (2010)
NextEra Energy Resources (U.S.)	8.077
Iberdrola Renewables (Spain)	4.301
Horizon-EDPR (Portugal)	3.141
MidAmerican & PacifiCorp (U.S.)	2.316
E.on Climate and Renewables (Germany)	1.920
Total	19.755

Manufacturing

The explosive growth of wind power worldwide has placed enormous pressure on manufacturers of system components such as bearings, generators, gearboxes, blades, and towers. Suppliers are building factories around the world to meet demand.

To compete, the leading turbine manufacturers are expanding their operations into growing markets such as China, India, and Brazil as well as the U.S. Midwest where it makes sense to be near the heartland's wind farms for strategic and logistical reasons. Similarly, companies with an eye on the offshore turbine market are also expanding, particularly into Scotland and other parts of the United Kingdom and along Europe's northern coast, as discussed in chapter 9.

In general, establishing factories near markets makes sense in the wind business for three reasons—transportation costs, inventory costs, and currency fluctuations. Local wind power equipment plants can supply heavy and/ or large components more efficiently to customers nearby. Just-in-time manufacturing minimizes inventory, and sealing deals is more predictable when currency values don't have to be accounted for.

Asia. LM Wind Power was building a manufacturing facility in Jiangyin (near Shanghai) that will produce wind turbine blades up to 100 meters long (330 feet). The facility will include a technology center for fatigue testing the blades and a three-mold manufacturing workshop. The location along the Yangtze River will allow shipping the blades to China's offshore Jiangsu region and other Asian sites.

Siemens invested $80 million (60 million euros) in a rotor blade and nacelle plant, which was also sited in Shanghai as a strategic coastal location for shipping. Gamesa invested $120 million (90 million euros) to build five manufacturing plants in China over the past decade, with a sixth scheduled to open by 2012. Vestas's largest integrated manufacturing plant is in Tianjin, China, and the company established an R&D center in Beijing.

India is also emerging as a major manufacturing hub for wind turbines in Asia. Annual production capacity is 7.5 GW, with 17 manufacturers operating in the country. Capacity could exceed 17 GW by 2013, according to India's World Institute for Sustainable Energy.

North America. Historically, North American wind power manufacturing capabilities have been weak, and most equipment has been imported. However, component suppliers continue to enter the U.S. manufacturing sector, and domestic content has risen above 50%. Production capacity looks ample to supply a resurgence in demand, if America's economy and wind power market recover.

Most new American factories are greeted with enthusiasm by local communities and journalists, especially in the decaying rust belt where green jobs are still a novelty. U.S. states compete with each other to attract wind power manufacturers with tax deals, loan guarantees, streamlined permitting, and other incentives, as well as a skilled workforce. Occasionally a plant closes, underscoring the precarious nature of the wind business in a country with unpredictable support for renewable energy.

In 2010, 14 plants opened, giving the country a total of more than 400 wind equipment manufacturing facilities sprinkled among 44 states, though most are concentrated in the Midwest's Great Lakes region. These include plants making or assembling towers (22), turbines (12), nacelles (12), and blades (11). Traditional industries also capture a piece of the action, supplying steel, concrete, fiberglass, adhesives, copper, and aluminum to new plants.

Most towers and blades in new U.S. wind farms are now domestically produced, and most nacelles are assembled locally. However, the supply of U.S.-made nacelle components lags demand, and certain pieces have no domestic supplier. Some factories are planning a shift in production from geared turbine nacelles to direct-drive machines with fewer parts. The impact on labor requirements is unclear.

However, the most common type of U.S. wind power plant by far makes turbine gearboxes, which incorporate a lot of different components that can be made in America. This has helped beef up the U.S. supply chain, which includes manufacturers of nuts and bolts, gears, lubricants, fasteners, and a spectrum of other small parts and materials.

As a result, half of all wind turbine components were sourced domestically in 2010, up from one-fourth in 2005. Factoring in labor and transport raises domestic content. Some manufacturers and developers hope to reach 80%–90%. However, not enough foundries exist to cast utility-scale turbine parts that can weigh more than 30 tons.

In an unusual twist, some mainstream manufacturers are adding wind turbine production lines to their existing factories and cross-training their employees or hiring more people. Examples include a printing press operation in New Hampshire, which trained its workers to produce mid-range turbines (225 and 750 kW), and a prefabricated housing manufacturer in South Carolina, which will hire up to 100 workers to make small turbines (less than 2-meter rotor diameter, or 6 feet).

Offshore wind power, which requires the largest turbines, could drive local sourcing due to high transportation costs. The U.S. supply chain would need to expand to accommodate large-scale development offshore, offering opportunities to manufacturers in coastal states.

In Canada, about 200 companies are involved in developing and manufacturing wind turbine components and operating wind farms. Siemens and GE are among their customers. Manufacturers of blades, nacelles, towers, and other large components typically outsource many smaller fabricated parts such as bolts and fasteners.

Canada's wind power manufacturers are concentrated in two of the provinces with the most wind power and populations, Ontario and Quebec. Siemens and Korea's Samsung together planned to open a turbine factory in Ontario by 2012. Activity was also spreading to other provinces, such as a tower manufacturing plant being built in Nova Scotia by Daewoo.

Jobs

Worldwide, the wind industry employed more than 630,000 people in 2010, versus just 235,000 in 2005. Positions encompass a broad range of skills, from factory assembly, technicians, and engineers through managers and experts in financial, legal, and environmental issues. The world's biggest employers that are focused solely on the wind industry are Vestas and Gamesa, with about 20,000 and 6,700 workers, respectively.

China has more wind sector employees than any other country, at about 150,000, according to the Renewable Energy Policy Network (Ren21). Among emerging economies, Brazil employs 14,000 people in the wind business and India, 10,000.

Europe. The European wind energy sector employs more than 200,000 people, most of them in turbine and component manufacturing. Employment could grow to 446,000 by 2020 and 479,000 by 2030, due mainly to expanding offshore wind development, said EWEA. The offshore sector accounted for 19% of wind power jobs in 2010 but could reach 35% in 2020 and 61% in 2030. Offshore wind could create 28,000 jobs by 2020 in Scotland alone.

Germany has the most wind power-related jobs in Europe, at roughly 100,000, followed by Spain at 40,000. The Italian wind sector employs more than 28,000 people, about 10,000 of them directly. The French wind business has about 11,000 employees spread through more than 180 countries.

Denmark has more than 20,000 workers in the wind business. Vestas cut 3,000 jobs in 2010, closing four production units in Denmark and another in Sweden, after closing a UK blade factory in 2009. More job cuts were announced in 2012, amounting to about 10% of Vesta's workforce.

United States. Wind industry employment has fallen along with construction of new capacity. Jobs totaled 75,000 in 2010, down from a peak of 85,000. If the wind business recovers along with the economy, wind farm construction jobs could boot back up quickly. As a rule of thumb, each 10 MW of new wind power capacity needs one technician for service, NREL estimates.

In U.S. wind industry manufacturing, employment increased more than sevenfold in five years, from 2,500 jobs in 2004 to 20,000 in 2010, mostly in assembly. Although manufacturing employment dipped in 2009 due to the financial crisis and excess inventory, an additional 14,000 jobs were still in the pipeline as of 2011. State renewable energy standards and other policies play a key role in attracting new factories and jobs.

European studies suggest that offshore wind development could create 20.7 direct jobs annually for each megawatt in U.S. waters, according to NREL. If America's offshore wind resource is developed to a capacity of 54 GW (the target for 2030), more than 1.1 million jobs/yr would be required to manufacture and install the turbines.

Offshore wind power development could also establish more than 43,000 permanent jobs in manufacturing, construction, engineering, and O&M—jobs that can't be outsourced overseas. Construction and operation of offshore wind farms could generate about $200 billion in new economic activity (150 billion euros).

In general, however, wind power and other renewable technologies are less labor-intensive than conventional generation, according to one utility study. Coal and nuclear power plants require about six times as many jobs as wind, while natural gas plants need about twice as many workers if you count drilling and pipeline transport.

Although many jobs in the wind industry require a four-year college degree or better, people with a high-school education or associate degree can get well-paid positions (table 11–4). In *Businessweek*'s list of best-paid green jobs, wind turbine technician ranks number 12, with median pay of $53,600 (39,100 euros), a bit higher than shown in the table.

Table 11–4. Wind industry jobs and U.S. salaries. *Data courtesy of PayScale and Clean Edge.*

Job title	Median annual pay, $ (euros)	Typical job level	Typical education required
Wind business development manager	92,300 (67,400)	mid-level	bachelor's
Wind farm/plant operations manager	83,400 (60,900)	mid-level	bachelor's
Construction superintendent	76,700 (56,000)	senior	high school/associate
Electrical transmission engineer	70,000 (51,000)	entry	bachelor's engineering
Wind turbine mechanical engineer	63,300 (46,200)	entry	bachelor's engineering
Wind turbine technician	48,300 (35,300)	entry	high school/associate

Training

Many jobs in the wind industry are not specific to wind power, such as steel workers, crane operators, heavy machinery mechanics, assembly line workers, and ditch diggers. Similarly, professionals in the financial, legal, administrative, and environmental fields can work in the wind business with little training.

But for some wind power jobs, the skill set is specialized, and people need to become qualified to work on a wind energy project. Academic institutions around the world have stepped in to provide training for wind industry workers.

Europe's wind energy sector could face a shortage of candidates to fill existing and future job vacancies, according to EWEA. This is particularly true of positions that require years of study and experience, such as engineers, O&M technicians, and site managers.

Siemens regards turbine technician training as one of its biggest challenges in the service business. The company has 1,500 people in its global wind service organization, but might need to triple that number by 2015. Siemens' UK training facility was being expanded to help accommodate this growth.

Numerous universities and colleges around the world offer programs designed for the wind industry, which can't be listed exhaustively here. In general, Europe has the most graduate programs in wind technology design and manufacturing, as well as more experienced educators in the offshore wind sector.

Renewable Energy Jobs maintains a Web page with links to training and education providers in eight countries—Austria, Canada, France, Germany, Greece, Spain, the United Kingdom, and the United States.

More than 150 educational programs in America offer certificates, degrees, or coursework related to wind energy, including about 40 focused on technician training. Educational programs are split mainly between four-year universities (39%) and community colleges, technical schools, and training centers (46%).

For turbine technician training, the American Wind Energy Association established a seal of approval program. Standards cover general skills, safety, and electrical and mechanical components. Three colleges (two community and one four-year) had earned the certificate as of 2011, along with one technical institute, and ten more programs had applications pending.

Government Policy

Introduction

Politicians around the world say a lot of great things about wind power, but their follow-through doesn't always match up. Even well-meant promises often run aground in the real world, when push comes to shove in terms of political willpower and funding.

The market for wind power, like other nascent energy sources, is driven largely by government mandates and financial incentives. Government policy, in turn, is driven by concerns such as climate change and environmental sustainability, as well as the desire to diversify energy sources, improve security of supply, and reduce exposure to volatile fuel prices.

In terms of reducing emissions, global wind energy achieved more than one-fourth of the reduction in carbon dioxide pledged by industrial nations under the Kyoto Protocol. Wind energy had avoided 255 tons of carbon dioxide as of 2011, equivalent to 26% of industrial countries' commitment. By 2020, wind power could avoid 46%–69% of the emissions pledged in the Cancun agreement.

Financial incentives. Financial incentives to drive wind power market growth include tax credits for production and investment, loan guarantees, tradable renewable energy credits, and guaranteed prices or rates, as discussed in chapter 6. Production tax credits are based on the amount of wind power generated, while investment tax credits are taken upfront against the capital expense. In addition to financial incentives, governments can support wind industry R&D by establishing turbine guidelines and standards and by easing financing.

Feed-in tariffs. This is not a tariff in the sense of a tax levied on imports or exports. Instead, the feed-in tariff is a rate or payment for electricity that a wind farm or other renewable resource "feeds" into the grid, based on the cost of generation.

These tariffs (sometimes called FITs) drive market growth by giving wind farm developers and owners a long-term contract for their electricity (typically 15–20 years), a guaranteed connection to the grid, and a price high enough to recover a reasonable return on their investment. A tariff reduces financial risk by providing investors with rates that are clear and generally stable, making wind farms easier to develop and finance (see chapter 6).

Feed-in tariffs are one of the most widely used policies in the world for accelerating development of wind and solar power. The tariffs have motivated the installation of more renewable generation capacity than either tax incentives or government mandates. About 45% of global wind power projects resulted from feed-in tariffs, including 55 GW in the European Union during 2000–2009. More than 40 countries worldwide have instituted feed-in tariffs, most of which apply to wind power (table 11–5).

Feed-in tariff rates can be set at a fixed price independent of market rates or can be pegged to prevailing electricity prices. Typically, costs are spread throughout the utility rate base. Many feed-in tariffs pay higher rates in the earlier years of a project, then decrease on a sliding scale. In some areas with explosive growth in renewable energy markets, governments are scaling back their initial tariff rates.

Table 11–5. Selected countries with feed-in tariffs for wind power

Country	Country
Bulgaria	Luxembourg
Canada	Portugal
China	Slovakia
Cyprus	Slovenia
Czech Republic	South Africa
Denmark	Spain**
Ecuador	Switzerland
France	Taiwan
Germany	Thailand
Greece	Turkey
India	Uganda
Iran	Ukraine
Ireland	United Kingdom
Israel	Vietnam
Japan*	

* Effective in 2012.

** Temporarily suspended in 2012 for new applications.

Some analysts believe that feed-in tariffs discourage competition, but experience in Europe has shown that instead, companies scramble for market share, said NREL. In Germany, Spain, and France, feed-in tariffs have driven R&D and innovation in the private sector, leading to lower costs and better technology.

Renewable standards. Another important policy tool is clean energy targets and renewable portfolio standards—government regulations that require a certain percentage of electricity supply or energy consumption to come from renewable sources by a certain date. At least 119 countries worldwide have established some form of clean energy targets or supportive policies, according to Ren21. Many have set standards specifically for electricity, requiring utilities to generate certain amounts of their power from wind and other renewable sources.

Private investment

Renewable energy standards, feed-in tariffs, and other government incentives encourage private investment in clean energy technologies. This investment comes from various sources, including the following:

- Asset financing (balance-sheet, debt, equity, and loans), for installing capacity
- Publicly listed equity, for raising capital during start-up
- Venture capital, for supporting innovation

China attracted $54.4 billion (39.7 billion euros) of private investment in clean energy in 2010, setting a world record, according to the Pew Environment Group. Investment in Germany's renewables sector totaled $41.1 billion (30.0 billion euros) and in America's, $34.0 billion (24.8 billion euros). When a country's GDP is considered, Germany and Italy have the highest clean energy investment from private sources.

About 72% of China's private investment was devoted to wind power, compared to 43% of America's, 39% of Italy's, and only 12% of Germany's.

One measure of a country's attractiveness to wind power investment is the wind index calculated by Ernst & Young, based on analysis of market potential, wind energy pricing, resource quality, project size, taxes, grants/soft loans, and installed base, as well as the country's infrastructure (electricity markets, grid connection, and financing). The wind index scores attractiveness to wind power on a scale of 0 to 100 (table 11–6).

Table 11–6. Countries' attractiveness to wind power. *Data courtesy of Ernst & Young.*

Overall		Onshore		Offshore	
Country	**Score**	**Country**	**Score**	**Country**	**Score**
China	76	China	78	UK	79
UK	68	India	71	Germany	74
U.S.	67	U.S.	70	China	69
Germany	66	Canada	65	Belgium	58
India	64	Italy	65	France	57
Italy	62	UK	64	Ireland	57
Canada	60	Germany	63	Denmark	56
France	60	France	62	U.S.	56
Ireland	57	Spain	60	Netherlands	55
Spain	56	Portugal	59	Italy	54

Note: Where scores are tied, countries appear alphabetically.

China

China's government is clearly committed to embracing wind power and other renewables as a way to fuel its rapidly growing economy, diversify its energy resources, and spur development in poor rural areas. The government plans to invest $736 billion in renewable energy (537 billion euros) through 2020.

China's political system might be susceptible to disputes over the best way to develop and integrate massive amounts of wind energy. And as China steps further into the global spotlight, its protective regulations and questionable trade practices could be subject to greater scrutiny and criticism. The country's business environment and financial markets might need to become more transparent if China is to realize its potential as a renewable energy powerhouse.

Nevertheless, China seems determined to seize the economic opportunity of wind power by enacting policies that drive its wind market and thrust the country into the role as a world leader in wind industry manufacturing.

To develop its wind power resource, the government established the so-called "wind base" program, which is well under way. Its goal is to build almost 70 GW of wind capacity through 2015 and 138 GW through 2020 (table 11–7).

The Chinese Renewable Energy Industry Association and other analysts expect that installed capacity will surpass state goals to total 80–130 GW by 2015 and 150–230 GW by 2020, even if the country's wind industry growth rate slows down. The government's official goals were boosted in 2012. The new roadmap calls for China's wind power capacity to reach 100 GW by 2020, 400 GW by 2030, and 1,000 GW by 2050, when wind could supply 17% of the country's electricity demand.

Table 11–7. Capacity planned for China's wind power bases. *Data courtesy of GWEC.*

Wind Power Base	2010 installed GW	2015 planned GW	2020 planned GW
Heibei	4.2	9.0	14.1
Inner Mongolia East	4.2	13.2	30.8
Inner Mongolia West	3.5	18.0	38.3
Jilin	3.9	10.1	21.3
Jiangsu	1.8	5.8	10.0
Gansu Jiuquan	5.2	8.0	12.7
Xinjiang Hami	0.0	5.0	10.8
Total	22.7	69.1	138.1

China's astounding wind power market expansion has its roots in years of government policies, including a mandate for nonhydro renewables to supply 3% of the country's electric power mix by 2020. According to China's five-year plan for 2012–2016, 11.4% of the country's primary energy will come from non-fossil fuels by 2016 and 15% by 2020. China's Renewable Energy Law was amended in 2009 to require grid operators to buy fixed amounts of renewable energy.

Also in 2009, China introduced a four-tier feed-in tariff for wind power, with prices based on a region's wind resource. Price categories range from about $80 to $95 per megawatt-hour (510–610 yuan, or 58–70 euros). China also made offshore wind energy development a top priority (see chapter 9).

To upgrade China's electrical grid, the government plans to invest $80 billion (500 billion yuan, or 60 billion euros) in 2011–2015, as noted in chapter 8. Grid operators are required to accommodate growing amounts of wind power and have agreed to connect 80 GW by 2015 and up to 150 GW by 2020.

In the past, however, subsidies for purchasing wind power have been too low for grid operators to jump on board, and rules don't include punishment for not accepting renewable power. Also, the government has had little control over grid connections with small wind projects (less than 50 MW), which are approved at the local level.

At China's largest wind farm, the Jiuquan project in Gansu Province, more than 5 GW of turbine capacity was installed, but only 1.15 GW was connected, reportedly due to transmission constraints. The Jiuquan project is scheduled to total more than 12.7 GW by 2015.

Chinese wind power projects approved by the United Nations as clean development mechanisms (CDMs) to satisfy Kyoto Protocol requirements have contributed to developers' income. A wind farm in Inner Mongolia was the 3,000th CDM registered, a milestone for the program. However, glitches in the definition of additional projects have to be worked out, and the scheme could evaporate in 2013 when the international emissions agreement expires.

India

The central government plans for 10.5 GW of new wind energy to help meet India's power generation requirements of 78.7 GW through 2012. India offers an either/or incentive—a developer can opt for accelerated depreciation (up to 80% of project cost within the first year of operation) or a feed-in tariff called the generation-based incentive of $11 per megawatt-hour (500 rupees, or 8 euros).

Depreciation has been the incentive of choice and could be responsible for much of India's wind power market growth in 2010. Both options expire in 2012. In addition to these central government policies, 18 of India's 25 states offer feed-in tariffs for wind power, and 17 states mandate renewable energy purchases. India also plans to create a national green bank to fund development of wind energy and other clean sources.

However, India's weak grid infrastructure could limit wind power's growth in the future, as operators already struggle to accommodate existing amounts.

Other Asia

After the earthquake and tsunami in 2011, Japan reconsidered its plans to build more nuclear plants and decided to expand the country's renewable energy industry. Japan has a good wind resource, but mostly in the north and south, which lack transmission lines to load centers in central Japan.

The country's weather and geography are challenging, too. Frequent typhoons can damage turbines, and Japan is studded with mountains, which are not the best terrain for conventional wind power, except for ridgelines. Also, much of Japan is surrounded by deep water, requiring the most advanced technologies for offshore wind farms.

Still, the Japanese Wind Energy Association thinks that 11.1 GW of wind capacity by 2020 is achievable, with an ultimate goal of 50 GW by 2050 (half on land and half offshore). Japan's feed-in tariff for wind was to be reinstated in 2012, which could help restart the domestic wind industry.

Taiwan, which lacks domestic energy resources and imports nearly all of its fuel, introduced a feed-in tariff in 2009 and aims for renewables to comprise 16% of power generation capacity by 2025. In 2011, the country boosted its offshore wind tariff by 33% to about 19 U.S. cents per kilowatt-hour (5.56 Taiwan dollars, or 14 euro-cents). Onshore wind tariffs were raised almost 10%, to 9 U.S. cents/kWh (2.61 Taiwan dollars, or 6 euro-cents).

Australia unveiled a renewable energy investment plan that devotes about $29 billion (A$28 billion, or 21 billion euros) to wind farms and other utility-scale power projects. Australia's goal is to get 20% of its electricity from renewable energy by 2020.

South Korea instituted a renewable portfolio standard and has targeted development of 2.5 GW of offshore wind power by 2020.

Europe

About half of Europe's electricity comes from fossil fuels, generating carbon dioxide and other greenhouse gases. The European Union plans to cut these emissions by 2050 to 80%–95% of 1990's level, according to a roadmap released in 2010. Wind power could play a significant role in working toward this target, which might require virtually eliminating carbon emissions from power generation.

Renewable energy goals. Thanks to the European Union's strong support for renewable energy, its wind power market has enjoyed spectacular growth for decades. However, Europe missed its 2010 target of 21% renewable sources in its electric power mix, getting just 18%. The next goal is for renewables to supply 20% of overall energy consumption by 2020.

Each European Union country has a specific target that is legally binding. Countries must write plans to achieve their targets and report on progress every two years. Germany, which plans to phase out its nuclear plants, raised its goal for renewables to provide at least 35% of electricity supply by 2020, up from 30% previously.

Given Europe's overall goals, wind power and other renewable sources could account for as much as one-third of Europe's electricity consumption in 2020. According to EWEA, wind power alone could supply 14.2%–16.7% of Europe's electricity demand by 2020.

Feed-in tariffs. Analysts attribute much of Europe's thriving wind power and solar markets to feed-in tariffs, which have been used for about 20 years in the European market. The tariffs spurred development of at least 55 GW of wind power in the European Union, as noted above.

Germany's feed-in tariff for offshore wind increases 15% in 2012, approaching 21 U.S. cents per kilowatt-hour (15 euro-cents), while the "starter" bonus rises almost 25% and onshore tariffs are maintained, reported Renewable Energy World.

In countries like Germany, in addition to a guaranteed price and grid connection for wind power, the feed-in tariff streamlines administrative procedures, which can help shorten lead times, reduce bureaucratic overhead, and minimize project costs. Many European policies extend eligibility to anyone who can invest, including homeowners, businesses, government agencies, utilities, and private groups.

Denmark, in addition to its feed-in tariff, reinstated special incentives for community ownership of wind turbines. Historically, much of wind power development in Europe, especially Denmark and Germany, resulted from government policies that encouraged the public to join wind cooperatives.

Funding. The European Union established a wind initiative to advance research, development, and demonstration by providing more than $9 billion (6 billion euros) during 2010–2020. Its objectives are the following:

- Maintain Europe's technology leadership in onshore and offshore wind power

- Make onshore wind the most competitive energy source by 2020, with offshore following by 2030
- Enable wind energy to supply 20% of Europe's electricity by 2020, 33% in 2030, and 50% in 2050

Also, the United Kingdom's renewable energy plan emphasizes wind power, with funding to be allocated by the Green Investment Bank, to open in 2013. Although investment capital could initially total almost $5 billion (3 billion pounds, or 3.5 billion euros), loans would be the likely vehicle for offshore wind power development after 2015.

North America

United States. Political and regulatory uncertainty persists in America, where renewable energy is mandated only by individual states. Wind power sometimes gets federal financial incentives, but a nationwide renewables standard is unlikely for the foreseeable future. DOE has estimated that wind power could supply 20% of America's electricity needs by 2030, but probably not without new policies and government support.

The federal production tax credit, which has been turned on and off for years, expires again at year-end 2012. The federal investment tax credit (also expiring then), can be converted to a cash grant, making it more popular than the production tax credit. To qualify, however, wind farms must begin construction by 2012 and enter service by 2013. Another, more complex tax credit is tied to projects that invest in low-income communities.

One bright spot emerged in 2011 when federal tax rules allowed 100% depreciation of wind turbines and some transmission assets (full tax write-off in the first year of operation). Equipment placed in service by 2012 is eligible for the bonus depreciation, which can shave up to 4.45% off the capital cost of wind power. Equipment in service by 2013 qualifies for half the bonus.

Given the vacuum in federal policy, individual states are left to decide how to encourage development of wind power and other renewable energy resources. Most states rely on some kind of standard that requires investor-owned utilities to get a percentage of their electricity supply from renewable sources. These regulations vary quite a bit, and some are voluntary (table 11–8). In 2010, 58% of new wind power capacity was built in states with renewable standards.

In addition to renewable standards, several states, cities, and utilities have experimented with feed-in tariffs, starting with California in 1978, or have considered them since then. Several feed-in tariffs and/or similar incentives are in place in California, Hawaii, Oregon, Vermont, and Florida.

Table 11–8. U.S. states with renewable energy requirements

State	Amount	Year
Mandatory		
Arizona	15%	2025
California	33%	2020
Colorado	20%	2020
Connecticut	23%	2020
Delaware	20%	2019
Hawaii	20%	2020
Illinois	25%	2025
Kansas	20%	2020
Maine	40%	2017
Maryland	20%	2022
Massachusetts	15%	2020
Michigan	10%	2015
Minnesota	25%	2025
Missouri	15%	2021
Montana	15%	2015
Nevada	20%	2015
New Hampshire	23.8%	2025
New Jersey	22.5%	2021
New Mexico	20%	2020
New York	24%	2013
North Carolina	12.5%	2021
Ohio	25%	2025
Oregon	25%	2025
Pennsylvania	18%	2020
Rhode Island	16%	2019
Washington	15%	2020
Washington, DC	20%	2020
Wisconsin	10%	2015
Voluntary		
Indiana	10%	2025
North Dakota	10%	2015
Oklahoma	15%	2015
South Dakota	10%	2015
Utah	20%	2025
Vermont	10%	2013
Virginia	12%	2022
West Virginia	25%	2025

Notes: Iowa and Texas capacity (MW) requirements have been met.
Some standards include alternative energy as well as renewables.

Almost half of U.S. states are considering feed-in tariff legislation, according to the Alliance for Renewable Energy, and America's federal regulations recently became more favorable to the tariffs, said Renewable Energy World.

States offer a variety of other incentives and regulations that support wind power, including grid interconnection standards, net metering, access laws, contractor licensing, equipment certification, corporate and sales tax incentives, and mandatory or voluntary green power purchasing programs.

However, these policies are less effective in encouraging wind power development than a long-standing renewable standard. An online tool developed by the U.S. DOE could help state and local policy makers evaluate the various options for supporting community wind power and smaller systems.

Canada. The national government's goal is to produce at least 20% of the country's electricity from wind by 2025. Ontario's feed-in tariff for onshore wind is 13.6 U.S. cents per kilowatt-hour (C$0.135, or 9.96 euro-cents). The province also provides incentives for community wind farms. Nova Scotia Province introduced a wind feed-in tariff of 13.2 U.S. cents/kWh for systems over 50 kW (C$0.131, or 9.66 euro-cents).

Latin America and Africa

Policymakers in Latin America are less concerned with the environment and more interested in obtaining a secure energy supply from domestic resources. Latin American governments are looking for proven technologies that compete economically and spawn local industries. Wind fits the bill.

In addition to wind power's success in Brazil's power auctions as noted earlier, the Brazilian Development Bank offers loans and financing for up to 80% of a wind project's value. And the country has plans to expand transmission capacity. Working against wind power in Brazil, however, are low contract prices for wind energy and a lack of long-term incentives, which could inhibit investment in the supply chain.

In northern Africa, Egypt plans to produce 2.4% of its electricity from wind energy by 2020 (12% of a 20% renewables target). This will mean building and connecting about 7.2 GW of wind power. A World Bank loan will finance grid infrastructure. Egypt intends to introduce a feed-in tariff in 2012, according to the Egyptian Wind Energy Association.

Consumer demand for wind?

Regardless of government policies, many people want to buy environmentally sustainable products, which could create a visible presence for wind power in retail markets. More than 90% of consumers believe that renewable energy is a good way to address climate change, and many would buy products that are made using wind energy, even at a higher price, according to GWEC.

Armed with this data, the council launched its WindMade initiative, which identifies such products with a global consumer label. Companies will be able to differentiate their brands in the marketplace and demonstrate their commitment to wind energy. EWEA endorses the WindMade initiative as a way to give consumers a choice and make people more aware of wind energy.

Governments are dragging their feet, but consumers want to see change now. The private sector needs to step up to provide the solutions we need to respond to the global energy and climate crises.

—Steve Sawyer, Global Wind Energy Council

References

Alliance for Renewable Energy. 2011. Status of FIT Legislation in Each State. http://www.allianceforrenewableenergy.org/us-reps.html.

American Wind Energy Association. 2011 (April 7). U.S. Wind Industry Annual Market Report. http://www.awea.org/learnabout/publications/reports/AWEA-US-Wind-Industry-Market-Reports.cfm.

———. 2011 (May). 2010 U.S. Wind Industry Annual Market Report: Rankings. http://www.awea.org/learnabout/publications/factsheets/upload/2010-Annual-Market-Report-Rankings-Fact-Sheet-May-2011.pdf.

Appleyard, David. 2010 (September 30). Latin America Warms to Wind. *Renewable Energy World.* http://www.renewableenergyworld.com/rea/news/article/2010/09/latin-america-warms-to-wind.

———. 2011 (June 8). Spring Growth for Wind Turbines. *Renewable Energy World.* http://www.renewableenergyworld.com/rea/news/article/2011/06/spring-growth-for-wind-turbines.

AWEA Seal of Approval Program, http://www.awea.org/learnabout/education/awea_soa.cfm.

Beattie, David. 2010 (December 22). Key Players in the Wind Energy: Pausing for Thought. *Renewable Energy World.* http://www.renewableenergyworld.com/rea/news/article/2010/12/key-players.

Blau, John. 2011 (March 17). Spanish Wind Power Manufacturer Expanding into India. *Renewable Energy World.* http://www.renewableenergyworld.com/rea/news/article/2011/03/spanish-wind-power-manufacturer-expanding-into-india??cmpid=WNL-Wednesday-March23-2011.

Bloomberg News. 2011 (May 12). Huaneng Renewables Said to Revive $1 Billion Hong Kong Initial Share Sale. http://www.bloomberg.com/news/2011-05-12/huaneng-renewables-said-to-revive-1-billion-hong-kong-initial-share-sale.html.

BlueGreen Alliance. 2010 (June). Winds of Change: A Manufacturing Blueprint for the Wind Industry. http://www.bluegreenalliance.org/admin/private_publications/files/BGA-Report-062510_FINAL.pdf.

Bolinger, Mark, and Wiser, Ryan. Understanding Trends in Wind Turbine Prices Over the Past Decade. Lawrence Berkeley National Laboratory. LBLN-5119E. http://eetd.lbl.gov/ea/ems/reports/lbnl-5119e.pdf.

Bradsher, Keith. 2010 (December 14). To Conquer Wind Power, China Writes the Rules. *New York Times.* http://www.nytimes.com/2010/12/15/business/global/15chinawind.html.

BTM Consult, http://www.btm.dk/.

———. 2011 (April). BTM Consult Releases New Wind Report: World Market Update 2010. Wind Today. http://www.windtoday.net/info/articles.html?ID=106940.

———. 2011 (April). Strong Year Predicted as BTM Reveals its Top Ten Leading Turbine Markets in 2010 by Market Share. http://www.btm.dk/news/strong+year+predicted+by+btm-+a+part+of+navigant+consulting/?s=9&p=1&n=44&p_id=1.

Buffett, Warren. 2011. Annual Letter to Shareholders. Berkshire Hathaway. http://www.berkshirehathaway.com/letters/2010ltr.pdf.

Castano, Ivan. 2011 (April 7). China Installing Wind-Power Capacity As Fast As It Can. *Renewable Energy World*. http://www.renewableenergyworld.com/rea/news/article/2011/04/china-installing-wind-power-capacity-as-fast-as-it-can?cmpid=WindNL-Thursday-April21-2011.

Couture, Toby D., et al. 2010 (July). A Policymaker's Guide to Feed-in Tariff Policy Design. NREL/TP-6A2-44849. http://www.nrel.gov/docs/fy10osti/44849.pdf.

Datamonitor. 2011 (January 19). Brazil: Policy Changes Aim to Drive the Domestic Wind Market Forward. http://about.datamonitor.com/cop15/archives/552.

Distributed Wind Policy Comparison Tool, http://www.eformativeoptions.com/dwpolicytool/.

Doris, Elizabeth, and Gelman, Rachel. 2011 (January). State of the States 2010: The Role of Policy in Clean Energy Market Transformation. NREL/TP-6A20-49193. http://www.nrel.gov/applying_technologies/state_local_activities/pdfs/49193.pdf.

Economist. 2011 (March 12). Europe's Climate Policy: Being Ambitious. http://www.economist.com/node/18333149.

———. 2011 (June 2). Japan's Winds of Change. http://www.economist.com/node/18750586?story_id=18750586.

Ernst & Young. 2011 (February). Renewable Energy Country Attractiveness indices. http://www.ey.com/Publication/vwLUAssets/Renewable_energy_country_attractiveness_indices_-_Issue_28/$FILE/EY_RECAI_issue_28.pdf

European Master in Renewable Energy, http://www.master.eurec.be/en/.

European Union. 2010 (April). Roadmap 2050: A Practical Guide to a Prosperous, Low-Carbon Europe. Volume 1, Technical Analysis. http://www.roadmap2050.eu/attachments/files/Volume1_fullreport_PressPack.pdf.

———. 2010 (April). Roadmap 2050: A Practical Guide to a Prosperous, Low-Carbon Europe. Volume 2, Policy Recommendations. http://www.roadmap2050.eu/attachments/files/Volume2_Policy.pdf.

European Wind Energy Association. 2009 (January). Wind at Work: Wind Energy and Job Creation in the EU. http://www.ewea.org/fileadmin/ewea_documents/documents/publications/Wind_at_work_FINAL.pdf.

———. 2010 (June). The European Wind Initiative: Wind Power Research and Development for the Next Ten Years. http://www.ewea.org/fileadmin/ewea_documents/documents/publications/EWI/EWI_2010_final.pdf.

———. 2011 (February). Wind in Power. http://www.ewea.org/fileadmin/ewea_documents/documents/statistics/EWEA_Annual_Statistics_2010.pdf.

———. 2011 (April). 2011 Wind Power Meeting Kyoto Commitments While Climate Negotiations Stall. http://www.ewea.org/index.php?id=60&no_cache=1&tx_ttnews[tt_news]=1904&tx_ttnews[backPid]=1&cHash=d3188cd02e6ec4ad617f42b1000fc17e.

———. 2011. Fact Sheets: Employment. http://www.ewea.org/index.php?id=1611.

Ewing, Jack, and Bajaj, Vikas. 2011 (March 23). German Energy Company Hits Headwinds in India. *New York Times*. http://www.nytimes.com/2011/03/24/business/global/24enercon.html?ref=todayspaper.

Gipe, Paul. 2010 (August 10). NREL Releases Feed-in Tariff Guide. *Renewable Energy World*. http://www.renewableenergyworld.com/rea/news/article/2010/08/nrel-releases-feed-in-tariff-guide.

———. 2011 (February 3). Taiwan Increases Offshore Wind Tariff 30 Percent. *Renewable Energy World*. http://www.renewableenergyworld.com/rea/news/article/2011/02/taiwan-increases-offshore-wind-tariff-30-percent.

———. 2011 (July 25). Germany Passes More Aggressive Renewable Energy Law. *Renewable Energy World*. http://www.renewableenergyworld.com/rea/news/article/2011/07/germany-passes-more-aggressive-renewable-energy-law?cmpid=WNL-Wednesday-July27-2011.

———. 2011 (August 11). How to Design Feed-in Tariffs in the U.S. without Fear of Federal Preemption. *Renewable Energy World*. http://www.renewableenergyworld.com/rea/news/article/2011/08/how-to-design-feed-in-tariffs-in-the-u-s-without-fear-of-federal-preemption?cmpid=WNL-Friday-August12-2011.

———. 2011 (September 13). Nova Scotia Community Feed-in Tariffs to Launch September 19th. *Renewable Energy World*. http://www.renewableenergyworld.com/rea/news/article/2011/09/nova-scotia-community-feed-in-tariffs-to-launch-september-19th?cmpid=WNL-Wednesday-September14-2011.

Global Wind Energy Council. 2011 (March). Global Wind Report: Annual Market Update 2010. http://www.gwec.net/fileadmin/documents/Publications/Global_Wind_2007_report/GWEC%20Global%20Wind%20Report%202010%20low%20res.pdf.

———. 2011 (August 24). 1,928.8MW of Wind Power Contracted at Brazilian Auctions. http://www.gwec.net/index.php?id=30&no_cache=1&tx_ttnews[tt_news]=298&tx_ttnews[backPid]=97&cHash=a56c316695.

———. 2012 (February). Global Wind Statistics 2011. http://www.gwec.net/uploads/media/GWEC-PRstats-2011_20120206_06__1_.pdf.

Gopal, Prashant. 2009 (July 23). Now Hiring: Green-Collar Workers. *Businessweek*. http://www.businessweek.com/managing/content/jul2009/ca20090721_381476.htm.

Green, John. 2011 (April 6.) Change In the Wind. *The Hutchinson News*. http://www.hutchnews.com/Todaystop/dillon-lecture-speaker--1.

IHS Emerging Energy Research. 2010 (May 4). Brazil Leads Latin America Wind Energy Markets to 46 GW by 2025. http://www.emerging-energy.com/content/press-details/Brazil-Leads-Latin-America-Wind-Energy-Markets-to-46-GW-by-2025/29.aspx.

International Energy Agency. 2011. Clean Energy Progress Report. http://www.iea.org/papers/2011/CEM_Progress_Report.pdf

Li, Junfeng, et al. 2010 (October). China Wind Power Outlook 2010. http://www.greenpeace.org/raw/content/eastasia/press/reports/wind-power-report-english-2010.pdf.

Martin, Christopher. 2011 (April 6). Google, Warren Buffett May Be Next Wind-Turbine Buyers, Suzlon, Vestas Say. Bloomberg. http://www.bloomberg.com/news/2011–04-06/google-buffett-may-be-wind-turbine-buyers-say-suzlon-vestas.html.

Morales, Alex. 2011 (February 7). Wind Turbine Prices Fall Below 1 Million Euros Per Megawatt. Bloomberg. http://www.bloomberg.com/news/2011–02-07/wind-turbine-prices-fall-below-1-million-euros-per-megawatt-bnef-says.html.

———. 2011 (April 15). Clean Energy Investment Fell 34% as Incentives Cut in Europe, BNEF Says. Bloomberg New Energy Finance. http://www.bloomberg.com/news/2011–04-15/clean-energy-investment-fell-34-as-incentives-cut-in-europe-bnef-says.html.

Musial, Walt, and Ram, Bonnie. Large-Scale Offshore Wind Power in the United States. NREL/TP-500-40745. http://www.nrel.gov/wind/pdfs/40745.pdf.

Pearson, Natalie Obiko. 2011 (April 28). India's Rusting Wind Farms Help Gamesa, Vestas Expand Foothold. Bloomberg. http://www.bloomberg.com/news/2011–04-28/india-s-rusting-wind-farms-help-gamesa-vestas-expand-foothold.html.

Pernick, Ron. 2011 (May 3). Reassessing the Chinese Clean-Tech Juggernaut (or Why the U.S. Shouldn't Be Written off Quite Yet). Clean Edge. http://www.cleanedge.com/views/.

Pernick, Ron, Wilder, Clint, and Winnie, Trevor. 2010 (October). Clean Tech Job Trends 2010. http://www.cleanedge.com/reports/pdf/JobTrends2010.pdf.

Pew Charitable Trusts. 2011. Who's Winning the Clean Energy Race? http://www.pewenvironment.org/uploadedFiles/PEG/Publications/Report/G-20Report-LOWRes-FINAL.pdf.

Pike Research. 2011 (July 12). Global Wind Power Capacity to Triple by 2017. http://www.pikeresearch.com/newsroom/global-wind-power-capacity-to-triple-by-2017.

REN21 Renewable Energy Policy Network for the 21st Century. 2011 (July 12). *Renewables 2011 Global Status Report*. http://www.ren21.net/Portals/97/documents/GSR/REN21_GSR2011.pdf.

Renewable Energy Jobs, http://www.renewableenergyjobs.com/content/global_guide_to_education_and_training_within_renewable_energy.

Salerno, Liz. 2011 (April). Opportunities in Wind Manufacturing. http://www.americanbearings.org/resource/resmgr/2011_annual/salerno.pdf.

Shankleman, Jessica. 2011 (May 11). Japan Ditches Nuclear Plans Raising Hopes for Wind Industry. *Business Green*. http://www.businessgreen.com/bg/interview/2069773/japan-ditches-nuclear-plans-raising-hopes-wind-industry.

Sharp, Duane. 2010 (December 20). In Depth: Profiling Engineering and Manufacturing in Canada's Wind Energy Sector. World.edu. http://world.edu/worldedu_posts/depth-profiling-engineering-manufacturing-canadas-wind-energy-sector/.

Song, Wayne. 2011 (April). Wind Turbine Market Outlook. North American Clean Energy. http://www.morganlewis.com/pubs/WindMarketOutlook.pdf.

SustainableBusiness.com News. 2011 (February 8). Wind Turbine Prices Fall to Lowest Point in Recent Years. http://www.sustainablebusiness.com/index.cfm/go/news.display/id/21839.

Trabish, Herman K. 2010 (August 13). The Birth of a U.S. Wind Power Manufacturing Industry. Greentech Media. http://www.greentechmedia.com/articles/print/the-birth-of-a-u.s.-wind-power-manufacturing-industry/.

———. 2011 (January 11). The Money in the Wind. Greentech Media. http://www.greentechmedia.com/articles/read/the-money-in-the-wind/.

———. 2011 (January 17). Higher Education Gets Wind's Approval. Greentech Media. http://www.greentechmedia.com/articles/read/higher-education-gets-winds-approval/.

———. 2011 (February 15). U.S. Wind Tech Moving into China? Greentech Media. http://www.greentechmedia.com/articles/read/a-u.s.-wind-company-moving-into-china/.

———. 2011 (March 8). Will China Take over the U.S. Wind Market? Greentech Media. http://www.greentechmedia.com/articles/read/why-is-china-coming-for-u.s.-wind/.

———. 2011 (April 11). At 35 Percent of New Capacity, Wind Moves from Alternative to Mainstream. Greentech Media. http://www.greentechmedia.com/articles/read/is-wind-still-an-alternative-energy/.

———. 2011 (April 19). Goldwind: Will Wind Be China's Next Big Export to the US? Greentech Media. http://www.greentechmedia.com/articles/read/goldwind-will-wind-be-chinas-next-big-export-to-u.s/.

———. 2011 (September 16). A Tale of Intrigue and IP Theft in the Wind Market. Greentech Media. http://www.greentechmedia.com/articles/read/american-superconductor-files-suit-against-china-wind-giant-sinovel/.

———. 2011 (September 26). Wind Power: AMSC Employee Pleads Guilty to Passing Code to Sinovel. Greentech Media. http://www.greentechmedia.com/articles/read/amsc-employee-pleads-guilty-to-passing-code-to-sinovel/.

U.S. Department of Energy, Energy Efficiency and Renewable Energy. 2011. U.S. Installed Wind Capacity and Wind Project Locations. http://www.windpoweringamerica.gov/wind_installed_capacity.asp.

U.S. Department of Energy, Energy Efficiency and Renewable Energy, Wind Powering America. 2011 (April). Wind Energy Benefits. DOE/GO/102011-3226. http://www.windpoweringamerica.gov/pdfs/wpa/2011/wind_energy_benefits.pdf.

WindMade, http://www.windmade.org/.

Wiser, Ryan, and Bolinger, Mark. 2011 (August). 2010 Wind Technologies Market Report. http://www1.eere.energy.gov/windandhydro/pdfs/51783.pdf.

World Institute for Sustainable Energy, http://www.wisein.org/.

World Wind Energy Association, http://www.wwindea.org/home/index.php.

———. 2011 (April). World Wind Energy Report 2010. http://www.wwindea.org/home/images/stories/pdfs/worldwindenergyreport2010_s.pdf.

Yuanyuan, Liu. 2011 (April 26). China's Wind Power Sector Expected to See Tremendous Growth. *Renewable Energy World*. http://www.renewableenergyworld.com/rea/news/article/2011/04/chinas-wind-power-sector-expected-to-see-tremendous-growth??cmpid=WNL-Friday-April29-2011.

———. 2011 (May 19). China Seeks Big Wind in a Push to Standardize Project Size. *Renewable Energy World*. http://www.renewableenergyworld.com/rea/news/article/2011/05/china-to-standardize-wind-power-projects.

———. 2011 (September 22). China Releases Technical Standards for Wind Power Industry. *Renewable Energy World*. http://www.renewableenergyworld.com/rea/news/article/2011/09/china-releases-technical-standards-for-wind-power-industry?cmpid=WNL-Friday-September23-2011.

———. 2012 (January 12). China Increases Target for Wind Power Capacity to 1,000 GW by 2050. Renewable Energy World. http://www.renewableenergyworld.com/rea/news/article/2012/01/china-increases-target-for-wind-power-capacity-to-1000-gw-by-2050?cmpid=WNL-Friday-January6-2012.

Zipp, Kathleen. 2010 (October 1). Wind Power—More Efficient Than You Thought. *Windpower Engineering*. http://www.windpowerengineering.com/policy/wind-power-more-efficient-than-you-thought/.

12

Environmental and Economic Issues: Wind Power Grows Up

Introduction

Most people like the idea of wind power and other renewable energy, at least in theory. But when it comes to their community, it can be a whole different story. What are the reasons behind one town's enthusiasm for wind energy and another one's resistance?

In Europe, 40% of wind power projects end up in court, and another 30% are delayed by local opposition. In the United States, 10%–25% of projects are delayed or abandoned due to environmental concerns, as of 2008. These numbers are no doubt rising as wind power development moves closer to populated areas.

People's reactions to a proposed wind farm can be visceral—an automatic defensive response when faced with a threat. Or they can be analytical, a studied approach to confront the same perceived threat. Either way, a variety of factors play into the mix of emotions and decisions that surface in a community when a wind power development is being considered.

On the larger stage politicians, think tanks, and journalists influence what people believe about wind power and other renewable energy sources. Proponents highlight success stories; opponents point to failures. The trouble is, these confusing signals make it hard for most people to tell how wind energy might affect their lives, or whether it matters at all.

In the global media, opposition to wind power has gathered steam, fueled in part by the deep pockets of the fossil fuel industry. Perhaps this is a response to wind energy's nascent presence in power markets worldwide. Or maybe people in countries that were hit hard by the financial crisis are pushing back against anything perceived as unnecessary right now.

But most international oil and gas companies have financial stakes in renewable energy. Even some electric utilities that have relied traditionally on coal are buying into wind power as a hedge against future carbon emission regulations.

Still, news articles and opinion pieces claim that wind power is unpredictable, it's expensive, it kills birds and bats, it doesn't reduce air pollution, it will crash electrical grids, and worse. Some of these issues, such as grid integration, have been addressed earlier. This chapter will attempt to wade through the global debate on environmental and economic issues.

The wind business is not innocent of wrongdoing by any means. When the industry was young in America, some overconfident developers acted with little or no concern for the environment, though admittedly few rules existed at the time. Soil was eroded by wind farm roads cut into steep hills, and thousands of birds were killed in California.

Wind turbines manufactured in the U.S. were cheaper than European machines, but they were not built to last. Lots of investors got burned by irresponsible developers who were in the wind business mainly to suck up government funding—which didn't require the equipment to actually work, just sticking it in the ground was enough to qualify. And some developers, instead of engaging with a community, offered payments with gag orders to prevent complaints.

But the wind industry is growing up. Technology has matured dramatically. Bankers and other professional investors are convinced that the wind business can generate profits, even offshore. Permitting and other regulations have controlled most abuses. And wind energy is gaining political clout around the world, in both industrial and developing countries.

As a clean source of electricity, wind power reduces air pollution and also addresses the issue of climate change, or global warming (see "Environmental Issues" below). Experts debate the extent to which wind energy displaces fossil fuels and cuts carbon dioxide emissions, but wind turbines don't pollute the air and water, and modern wind farms are sited and developed with the least possible impact on the environment.

When it comes to siting and operating wind power, surely some common ground can be found between sincere opponents who understand the issues and conscientious wind farm developers who try hard to design and run projects that are friendly to wildlife and people alike.

Wildlife Issues

Introduction

The populations of many bird and bat species are declining gradually, due to a variety of factors—habitat loss and fragmentation, invasive species, and other impacts from encroaching civilization. This makes it more vital for wind energy development worldwide to avoid stressing bird and bat populations.

Wind energy can affect wildlife in two ways—first during construction of a wind farm, when land is cleared and roads are built for installing the turbines, and second when the wind farm is up and running. Indirect impacts include the loss or degradation of wildlife habitats, which can be disrupted or broken up into less accessible pieces.

Wind farm construction temporarily affects a land area of 0.2–1.0 hectare per turbine (0.5–2.5 acres), while permanent habitat displacement is about 0.3–0.4 hectare per turbine (0.7–1.0 acre), according to the U.S. Department of Energy (DOE).

The potential effects on wildlife are analyzed extensively while the wind farm is being sited and planned, as part of the environmental review process. Studies can include assessment of existing data, physical inspections, and observation of wildlife behavior. Permits must be obtained from a variety of regulatory authorities, depending on the country, province, or state.

These studies can suggest whether the wind farm might affect birds, bats, or other wildlife at levels of concern to experts. Preliminary work can also help planners avoid sites hazardous to wildlife, such as migrating bird flyways, and design a better project. However, the literature does not show a strong link between preconstruction assessment and postconstruction bird and bat fatalities. The evidence so far is too limited to be conclusive.

The wind industry monitors wildlife issues with professional advice from biologists and other environmental experts. To accurately count bird deaths, some wind farm operators use dogs and/or humans to search for bodies regularly after a wind farm has been built.

U.S. federal regulators are considering tougher guidelines for wind farm development and eagle protection that could require years of wildlife studies before and after construction, possibly including retroactive changes to existing projects. The plan is opposed by some wind industry groups and developers. Analysts estimate that the more stringent guidelines for golden eagles alone could jeopardize 34 GW of potential wind projects.

The Wind-Wildlife Impacts Literature Database (WILD), maintained by the U.S. National Wind Technology Center, collects domestic and international documents from journals, conferences, government publications, books, and utility reports. The database, which is searchable by a number of parameters, is continually updated with new information.

To provide global information, the U.S. Fish & Wildlife Service maintains online links to international wind energy/wildlife guidance documents. One example is the European Commission's guidance on wind power development in accordance with the European Union's nature legislation.

As in North America, Europe's government recognizes that various species of birds, bats, and marine animals can be vulnerable, even though wind farms that are appropriately sited and well-designed generally don't threaten biodiversity. Scientific studies and monitoring of new and existing wind farms should be continued and shared throughout the wind industry and the scientific community.

The Nature Conservancy, a leading international conservation organization, has developed a science-based planning approach called development by design, which is being tested at several projects. Threats to wildlife and habitats are evaluated at regional and site levels, offering developers solutions to minimize harm.

In general, the conservancy recommends siting wind farms in areas that have already been disturbed by agriculture, oil and gas development, or other human activities. In undeveloped areas, sites should target the least sensitive land. Conflicts will arise, as in the U.S. mid-Atlantic region, where Appalachian ridgetops are some of the windiest sites—but also where turbines could endanger migrating bats (see "Bats" below).

To improve wind farm planning and siting, the Nature Conservancy teamed with the American Wind Wildlife Institute to develop a screening tool that combines wind resource data with wildlife and environmental information. The Institute was founded by several conservation and environmental organizations, including the National Audubon Society and the Sierra Club, and major wind companies.

The Landscape Assessment Tool, which is available to the general public, uses wind data from AWS Truepower. It's intended to provide early, general guidance on prospective sites and conservation strategies by helping developers understand local environmental issues and avoid risky areas.

The Audubon Society, a conservation group focusing on birds, wildlife, and habitats, strongly supports wind power as a clean alternative energy source that reduces the threat of global warming. To minimize negative impacts on birds and wildlife, Audubon recommends governmental guidelines including the following:

- Prepermitting studies and guidance on methods and acceptable data sources
- Clear siting criteria that designate areas where wind power should not be allowed
- Clear monitoring and mitigation requirements in permits, with adaptive management if impacts significantly exceed the levels allowed
- Guidance on the assessment and mitigation of cumulative population impacts

Certain prized locations challenge the wind industry to address unusual wildlife issues. One example is Hawaii, a paradise for birds and a world-renowned wildlife sanctuary. Wind power development there is still at a modest scale, but has engendered controversy.

Future wind farms in Hawaii and other unique ecosystems will need to negotiate trade-offs between project economics and protecting the planet's remaining populations of exceptionally diverse wildlife.

Birds

Birds die when they fly into spinning wind turbine blades, or sometimes crash into the tower. At most sites, the rate of bird deaths is fairly low, but in certain areas with a high ratio of raptors to prey—for example, where lots of hawks are competing for relatively few ground squirrels—bird deaths can be much greater. Also, old wind turbines that use outdated technology are thought to kill more birds than modern machines (see "Research and mitigation" below).

Altamont Pass. The danger to raptors—hawks, eagles, falcons, owls, buzzards, and others—was recognized early in the wind industry's history at California's Altamont Pass, one of America's oldest wind farm clusters. The Altamont is home to more than 5,000 turbines crowded within 150 sq km (60 sq mi), and the area is highly attractive to raptor birds, with an abundance of prey on the ground.

Historically, the Altamont has experienced high raptor fatalities, estimated at 0.05–0.1 per turbine per year. Part of the problem was the truss towers that supported early machines (fig. 12–1). Raptors would perch on the trusses behind the blades, spot their prey, then take off into the wind, directly into the blades. Also, hawks and eagles were actually nesting in the trusses.

When local regulators stepped in, wind farm operators began to replace old turbines with larger capacity, lower speed machines, spaced farther apart. New, more powerful turbines can each replace several old ones, and truss towers are being phased out in favor of tubular steel columns. These repowering projects should continue to reduce avian fatalities, not only for raptors but for all other birds as well.

America's experience in the Altamont Pass resulted in a remarkably persistent black eye for wind power. People have called wind farms "cuisinarts" or "raptor-matics" (referring to cheap food-choppers sold on late-night TV). But the Altamont did serve to wake up the wind industry to the issue of bird deaths in general and raptors in particular, spurring operators to mitigate the mortality rates for all birds.

Fig. 12–1. Old-fashioned wind turbines in the Altamont. *Image courtesy of Xah Lee.*

Today, the Altamont Pass is recognized as unusual or even unique because of its abundance of raptors, which naturally increases their death rates (see below). But at most newer wind farms, raptor mortality has been very low or absent altogether.

Bird mortality rates. Some impact on birds has been demonstrated at most U.S. wind farms, but overall bird deaths are minimal, and the level of impact varies among facilities and regions, according to a comprehensive summary of research results by the National Wind Coordinating Committee.

Mortality rates at most sites are well under four birds per megawatt per year, or less than 200–800 birds annually for wind farms of 50–200 MW. Still, wind projects are occasionally canceled due to their potential risk to bird populations, such as a 150-MW project in North Dakota. Table 12–1 lists fatalities at some sites.

Europe's wind power facilities also have experienced low mortality rates for birds, except for some coastal sites that are frequented by migrating flocks. However, Europe and several of its countries, as well as Canada and its provinces, have guidance and/or regulations in place to protect birds and other wildlife. Many of these apply specifically to offshore wind farms (see below). Chinese regulations also require an assessment of bird migration routes.

Table 12–1. Bird deaths at selected U.S. wind farms (estimated fatalities per megawatt per year). *Data courtesy of DOE.*

Wind project and location	Total fatalities/MW-yr
Klondike (OR)	0.95
Vansycle (OR)	0.95
Top of Iowa (IA)	1.44
Wisconsin (WI)	1.97
Foote Creek Rim-2 (WY)	1.99
Foote Creek Rim-1 (WY)	2.50
Combine Hills (OR)	2.56
Mountaineer (WV)	2.69
Nine Canyon (WA)	2.76
Stateline (OR-WA)	2.92
Buffalo Ridge-2 (MN)	3.03
Buffalo Ridge-1 (MN)	3.27
Buffalo Ridge-3 (MN)	5.93
Buffalo Mountain (TN)	11.67

The UK's Royal Society for the Protection of Birds determined that wind farms pose no threat to birds and other animals if they are put up in the right area. The society had previously campaigned against wind energy and succeeded in stopping at least one big wind farm. After further study, the society recommended building more UK wind farms as a way to mitigate global warming, which poses a more critical danger to birds. Left unchecked, climate change threatens many species with extinction, said the society.

Global watchdog organizations such as BirdLife International monitor the wind industry and its impact on all bird species. This group called attention to the deaths of white-tailed eagles at a wind farm in Norway in 2005 and opposed development of a wind farm in Ontario due to habitat concerns.

About three-fourths of birds killed by wind turbines are songbirds (perching birds). This type of bird dominates the world's species, and most of them migrate. Wind turbine strobe lights that warn off aircraft do not appear to affect songbird deaths. Some studies suggest that songbirds are fairly resilient to wind farms, becoming habituated and even increasing in number.

Migrating birds fly above turbine heights if weather allows, but taking off and landing at stopovers near turbines appears to be risky, and casualty rates peak in the spring and fall. Collisions also increase when bad weather forces birds to fly at lower altitudes, especially during migration. However, turbine-related deaths are unlikely to affect most songbird population trends.

Raptor deaths seem to increase along with the number of raptor birds, an obvious conclusion but one that had to be proved. However, different birds die at different rates. Red-tailed hawks and golden eagles in particular forage for prey near turbines, which seems to increase fatalities. Common ravens and other species somehow manage to avoid collisions.

To take advantage of air currents for hunting and migrating, raptors congregate along ridgetops, canyons, and the upwind sides of slopes. Siting wind farms to avoid these areas might reduce raptor collision rates. A high prey density—lots of mice, squirrels, and rabbits—is presumed to attract raptors and increase collisions, but scientifically, this connection is not very well established.

Eagles, both bald and golden, are protected species in the United States. Federal regulators are considering tougher guidelines for protecting eagles from new and existing wind farms. Some wind projects in California have been urged to limit the number of turbines planned, due to concern for golden eagles.

Research and mitigation. Scientists are working on high-resolution maps that will show the migration patterns and flight behavior of eastern golden eagles in areas with a high wind energy resource. The maps can be used to guide safer development and to protect a suite of species similar to golden eagles.

Lattice-steel towers supporting older turbines have been blamed for tempting raptors with lots of places to perch near turbine blades. Tubular steel towers probably mitigate this impact, but have not been proven to do so because their introduction coincided with other improvements in technology and siting practices. Similarly, the transition to larger turbines that revolve more slowly at the blade root might also reduce raptor collisions, but this remains unproven conclusively.

Researchers are working to make wind turbines easier for birds to see and fly around. One way is to paint turbine blades with a pattern of colors that makes them more noticeable. Another method is to coat the whole turbine with ultraviolet paint. Neither has yet been proven to help.

One factor working in birds' favor is that as wind turbines get bigger, they deliver more power more efficiently, so a new wind farm with a given energy output has fewer turbines, spaced farther apart. This should help reduce the risk of collision.

U.S. federal research also is investigating whether artificial intelligence can be applied to data from the weather service's network of 158 next-generation radar stations (Nexrad). Algorithms might be able to distinguish bird echoes in the data, which could help identify migratory flyways.

Commercially available radar systems can detect approaching birds, evaluate their risk of death, and activate mitigation responses, such as idling the turbines if necessary. The radar equipment can be moved seasonally to spots where migrating flocks are likely to arrive. The system also works for bat detection.

Other birds and habitats. Various species of grouse and prairie chickens require extensive grassland and open horizons. They are also loyal to their stomping grounds and reluctant to relocate, but their habitat lies in some of North America's most promising regions for wind energy. The population of these grassland birds (including sage grouse, sharp-tailed grouse, and lesser and greater prairie chickens) has been declining for 30 years, adding to biologists' concerns.

Some research indicates that prairie grouse require contiguous, unfragmented habitat to keep local populations connected. The grouse also tend to avoid any human incursions such as roads, poles, drilling platforms, and housing, so it's likely they'd be displaced by wind turbines as well. Biologists predict that sage grouse and prairie chickens will refuse to nest within a mile of a turbine.

Although most grouse and prairie chickens are not protected or endangered officially, environmental activists in America's heartland are fighting wind power development in the birds' traditional habitats. Some states are setting aside tall-grass prairies from further wind development.

Federal agencies are researching the effect of wind turbines in areas frequented by grouse, prairie chickens, and other grassland species. Wind industry representatives are collaborating in the program.

By mapping the most important sites in the region for greater sage-grouse, we've been able to steer energy development elsewhere—often to other locations that already bear a heavy human footprint. That includes hundreds of sites rich in wind energy potential. Americans are not willing to choose between wind energy and wildlife. Fortunately, with good policy and long-term thinking, they won't have to.

—David Yarnold, National Audubon Society

The big picture. Wind energy advocates argue that humanity's way of life kills birds in many other ways, and in far greater numbers. In fact, the cumulative impact of collisions with wind turbines is several orders of magnitude lower than deaths estimated from other anthropogenic (human-related) causes of songbird mortality. However, as wind farms proliferate, their share of bird deaths will probably rise, relative to other causes.

The top causes of bird deaths are glass windows and buildings, power transmission lines, and house cats. Wind turbines are responsible for fewer than 3 out of every 10,000 bird deaths (table 12–2). Other hazards include hunting, which kills an estimated 100 million birds per year, commercial fishing, oil pits, and land development, as well as the long-term effects of climate change, which could result in the extinction of up to one-fourth of bird species.

Of course, activists can argue that we should eliminate these other causes of bird fatalities, too, along with wind turbines. But at some point you have to ask, how much will anyone change their lifestyle and behavior to protect birds? Very few people would give up windows, electricity, or cars to prevent bird deaths.

Table 12–2. U.S. bird death estimates. *Data courtesy of Western EcoSystems Technology, Inc.*

Cause of death	Deaths per year, millions (est.)	Share of total, %
Windows and buildings	550	58.2
Power lines	130	13.7
Cats	100	10.6
Vehicles	80	8.5
Pesticides	67	7.1
Communication towers	4.5	0.5
Wind turbines	0.0285	<0.01
Airplanes	0.0250	<0.01

Note: Other sources accounting for less than 1.5% of deaths were not calculated.

Bats

Although bats have been saddled with a creepy "vampire" reputation, they are actually beneficial in a lot of ways, for instance by eating mosquitoes and other pests. More importantly, bats are a keystone species, meaning that if their populations are disrupted, the effects cascade throughout the entire ecosystem.

Six bat species across the eastern United States are being devastated by a mysterious fungus (white-nose syndrome) that biologists have not yet found a way to conquer. The disease is causing the most precipitous wildlife decline in the past century in North America.

Bat fatalities and causes. Bats use their own sonic "radar," called echolocation, to detect objects in their environment, enabling them to navigate and forage for food. Echolocation might keep bats from flying into turbine blades and towers, or the bats might not detect the turbine until it's too late. Research is under way to understand these interactions.

Collisions do occur, but less frequently than with birds and mostly when bats are migrating. Scientists suspect that migrating bats sometimes turn off their echolocation, leading to more collisions. At any rate, bat fatalities at wind farms peak during migration in the late summer and early fall, with a smaller spike at the spring migration.

As with birds, the impact of wind turbines on bats has been demonstrated in the United States, with deaths reported at every wind farm studied since 2005, but the level varies geographically. Surprisingly high mortality rates have been recorded at wind farms along ridgetops in the Appalachian Mountains. Counts are high at certain sites in Tennessee, West Virginia, Pennsylvania, and New York.

Europe's wind industry was alarmed by the numbers of bat fatalities in America, especially the death of more than 1,700 bats at a West Virginia wind farm. Bat fatalities have been recorded in 14 European countries, mostly in Germany but also in France, Portugal, and the United Kingdom. As in America, most bat deaths happen in late summer and fall among migratory species.

At most North American wind farms, the numbers of bat collisions is small, and researchers are uncertain about the geographic distribution and causes of bat fatalities. Scientists have documented several species of bats that are known to collide with turbines. Most are migratory tree-dwelling bats (75%) with widespread habitats, including the hoary, eastern red, and silver-haired bat species.

Scientists think that bats might be attracted to turbines for some reason. One theory is that the bats mistake the towers for trees, thus the prevalence of tree-dwelling bat deaths. However, few threatened or endangered bat species have been involved, and it seems unlikely that wind farms could affect bat populations overall.

However, wind turbines can also kill bats when they fly too close to the spinning rotor, which creates pockets of low-pressure and high-pressure air. The sudden change in air pressure can severely damage bats' lungs (this is called *barotrauma*), sort of like what happens when a scuba diver comes up from the water too quickly.

As wind turbines get bigger and taller, and as blades get longer and wider, the turbines create more turbulence in their wake (greater air-pressure differences), possibly making them more dangerous to bats. Research is planned to document the occurrence of barotrauma and to find out how many bats succumb immediately and how many are able to fly away with injuries.

Research and mitigation. One way to lessen bat fatalities is to raise the turbines' cut-in speed—the lowest wind speed at which the blades begin to turn and generate power (typically about 3–4 m/s, or 8–9 mph). This curtails the amount of time that the turbines are running, so the wind farm operator loses some energy output, but fewer bats are killed.

One research project documented reductions in bat fatalities of 53%–87% when turbine cut-in speeds were raised to 5.0–6.5 m/s (11–14.5 mph). Energy losses were marginal, less than 1% of the wind farm's total annual output.

In a subsequent two-year study at a Pennsylvania wind farm owned by Iberdrola, higher cut-in speeds reduced bat mortality by 44%–93%, again with energy losses of 1% or less. Fully operational turbines were found to kill 3.6–5.4 times more bats than curtailed turbines.

One reason this strategy works is that bats prefer to fly when it's not too windy, so they're naturally more active around the times that turbines begin to start up. Turbine curtailment won't be an ideal solution for most locations, but conservation groups hope that wind farm operators in areas with high bat mortality rates, like some northeastern U.S. sites, will raise cut-in speeds at night during bat migration in late summer and early fall.

Bat Conservation International is collaborating with the wind industry and U.S. federal researchers on studies to minimize the impact of wind power on bat species. This effort, called the Bats and Wind Energy Cooperative, is working on preconstruction assessment, postconstruction fatality surveys, and ways to mitigate danger during wind farm operation, such as high-amplitude "jamming" sounds like moths use in nature to deter bats.

Bats could be affected by long-term climate change if warmer temperatures and droughts disrupt their migration and mating patterns. Also, further research is needed to determine the influence of storm fronts and other weather patterns on bat fatalities.

Marine life

Europe's expanding offshore wind industry, and the prospects for offshore wind power growth worldwide, raise questions about the potential impact on marine life and habitats. Europe's historical dominance of the offshore arena has resulted in a wealth of information. The two major problems in Europe's experience are noise during turbine installation, which scares off many marine species, and environmental changes that affect migrating seabirds.

The number and type of turbines affect the level of a wind farm's environmental impact, but not enough research has been done on specific relationships between offshore turbines and the world's various marine species, habitats, and ecosystems. One wind farm in Germany's North Sea was canceled due to concerns about its impact on seabirds and the environment.

Seagoing birds. Offshore wind farms present several hazards to seabirds, including collision, short- and long-term habitat loss, barriers to migration, and fragmentation of ecological units. Data on bird fatalities at offshore wind farms are scarce because it's hard to detect collisions and recover bodies. Also, little is known about the migration routes and flight behavior of many seabird species.

The marine environment was monitored for seven years (1999–2006) at the Horns Rev and Nysted wind farms off Denmark. Most bird species avoided the wind farms, and the effect on overall bird populations was negligible. Similar results were found at Horns Rev in a two-year study of bird life by Denmark's National Environmental Research Institute.

Most flocks of birds (71%–86%) avoided flying between the wind farms' turbine rows, diverting by 1.5 kilometers during the day (0.9 mile) and 0.5 kilometer at night (0.3 mile). Avoidance appears to reduce the risk of collision. The estimated collision rate for common eiders at Nysted was 0.02% (about 45 birds out of 235,000 migrating through the area). Monitoring also confirmed that eiders and other waterfowl flew lower, underneath the rotor blades, at Nysted.

Birds' avoidance of turbines represents an effective habitat loss, but a relatively small one compared to their total feeding area. However, new wind farms should be analyzed for the potential to create greater habitat loss for certain species, due to the effects of avoidance on the flying, resting, and foraging behavior of the birds.

The populations of waterfowl, gulls, and passerines on northwest Europe's coast have not declined significantly. (Passerines are terrestrial perching birds or songbirds that constitute about half of bird species worldwide.) However, researchers have measured collision rates for wintering birds ranging from 0.01 to 1.2 birds per turbine annually. Direct observations at the UK's coastal Blyth Harbour wind farm found that bird collisions with turbine blades were rare events.

BirdLife International claims that offshore wind farms have detrimental impacts on seabirds. The most important is collisions with turbine blades, especially when offshore wind farms are sited near breeding colonies or when migrating birds such as passerines fly by night at low altitudes.

BirdLife also noted that wind farms act as barriers to traveling seabirds, causing the birds to become exhausted. Barriers could also reduce the amount of food that adults are able to bring to nestlings. BirdLife characterized the seaduck's avoidance of Nysted as profound, with a dramatic decline in birds entering the area after construction. Another impact is the loss of places to rest and forage, due to changes in those areas that deter birds from coming back.

When visibility is poor, terrestrial birds such as passerines could be attracted to offshore turbine beacon lights. This risk might be mitigated by reserving high-intensity lights for the perimeter and using lower-intensity lighting elsewhere. Also, continuous lights should be replaced with flashing strobes, which are already being used in new wind farms offshore.

Scientists agree that turbine foundations serve as artificial reefs, where a wealth of mussels and other shellfish can attach themselves and grow. While this new food supply is welcome in the ecosystem, it attracts more fish, which in turn can invite more seabirds to prey on the fish, putting the birds at greater risk of collisions.

Noise. Noise is an issue primarily during offshore wind farm installation. Pile driving is a particular concern, along with noise from boat and helicopter traffic. In contrast, noise from offshore wind turbines in operation (sounds transmitted from the gearbox and generator) barely contributes to underwater noise levels.

According to measurements taken during installation of the UK's North Hoyle wind farm, the underwater noise from hammering foundation piles into the seabed peaked at 260–262 *decibels* at water depths of 5–10 meters (16–32 feet). During pile driving, marine mammals and fish avoided the area (table 12–3).

Table 12–3. Marine life's avoidance of offshore pile driving (calculated distance ranges). *Data courtesy of WindFacts.*

Species	Distance, km (miles)
Cod	5.5 (3.4)
Bottlenose dolphin	4.6 (2.9)
Harbor seal	2 (1.2)
Harbor porpoise	1.4 (0.9)
Salmon	1.4 (0.9)

Mammals. Whales and some other marine mammals are especially sensitive to noise, since they use echolocation (like bats) to communicate, orient themselves, and find food. A 90-decibel sound at certain frequencies might disrupt whales' behavior, and 130 decibels could cause injury. Sharp, sudden sounds, such as foundation pile driving, are more harmful than steady noise.

The ocean environment is already fairly noisy, as established by acoustic monitoring buoys in U.S. northeastern waters off Maine. This research program seeks to find out where whales are distributed before offshore wind farms are deployed. The devices picked up sounds not only from whale communications, but also from boat traffic, fish, and even seaweed "exhaling" oxygen.

Still, offshore wind farms could change the behavior of marine mammals, displace them from noisy areas, mask other noises that they use to survive, and confuse their acoustic interpretation of their environment. This could compromise the viability of individuals, make them more vulnerable to disease, and increase the long-term effects of noise-induced stress.

The seven-year monitoring program at Horns Rev and Nysted revealed different reactions between seals and porpoises. Nysted is sited 4 kilometers (2.5 miles) from the Rosland seal sanctuary. Seals reacted to the construction noise, but were unaffected by wind farm operation.

The population of harbor porpoises at Horns Rev fell slightly during installation but recovered quickly. At Nysted, however, porpoise densities declined significantly during construction, and the population did not recover until the wind farm had been operating for two years.

At the Alpha Ventus wind farm off Germany, construction noise seemed to be keeping harbor porpoises away from their natural habitat. Harm to other marine life was also suspected. Researchers tested a curtain of bubbles to insulate seagoing creatures from the noise, and further studies will find out whether the porpoises might accept the wind farm as a habitat over time.

Fish. Monitoring of fish populations at Horns Rev and Nysted showed that the overall effects of noise and vibration were negligible. Fish can be temporarily disturbed during turbine installation, but scientists expect that fish abundance and species diversity will actually increase around turbine foundations as the artificial-reef habitat becomes better integrated with the marine environment.

Electromagnetic fields. Research on electromagnetic fields emitted by offshore wind farm cables revealed no significant impact on the marine environment. Studies conducted at Denmark's Nysted wind farm on two types of cable showed that the magnetic fields emitted were small or zero. However, scientists recommend more research to determine whether certain species that use the earth's magnetic field for navigation might be affected.

Human Issues

Introduction

To watch TV or read the news, you might think that almost everyone's against wind farm development, except for the businesses that make money from it. Despite the media's portrayal of unhappy wind turbine neighbors, some studies suggest that wind power critics are a small, vocal minority that has become skilled at getting attention for their cause, while the silent majority favor wind power development and want it to succeed.

To cite just one example, a poll of rural and urban residents in the U.S. northwest (Idaho, Oregon, and Washington states) found that people in rural areas overwhelmingly support wind farms within sight of their homes.

Of the 80% supporting development, 50% registered that they "strongly" support it, not just "somewhat." In urban areas, the numbers were even higher, not surprisingly. A UK wind power project received similar results, with just 12% of people opposed and 66% in favor of a new wind farm less than 10 kilometers away (within 6 miles).

However, resistance to wind power developments on land is growing in Europe, even in traditional strongholds like Denmark. As noted in chapter 11, though, Denmark already has more wind capacity per person than any country on earth, as well as the highest density per unit of land area. Germany, Portugal, and Spain also rank near the top, suggesting that parts of Europe's land and coasts are fairly well saturated with wind power.

Here we'll take a look at the hot-button human issues that seem to engender most of the opposition to wind power—noise, health, and safety, as well as the way wind turbines look and how their presence might affect property values.

Community outreach programs can help explain these issues to concerned citizens before wind farms are developed. Some studies indicate that community ownership of small-scale wind farms greatly improves the level of public acceptance. Other evidence also suggests that early opposition to a wind power project wanes with time. Surveys conducted in the United Kingdom and Spain have found that local support increases once the turbines are up and running.

Wind power's public image

Despite ardent opposition to some wind farms, most Americans have a favorable impression of wind energy, according to a consumer survey by Pike Research. Wind ranked second just behind solar energy for the highest level of favorable responses (75%) and got very few poor marks (a total of 5% unfavorable), as shown in figure 12–2. The survey concluded that people consider wind energy to be an important piece of tomorrow's power generation mix.

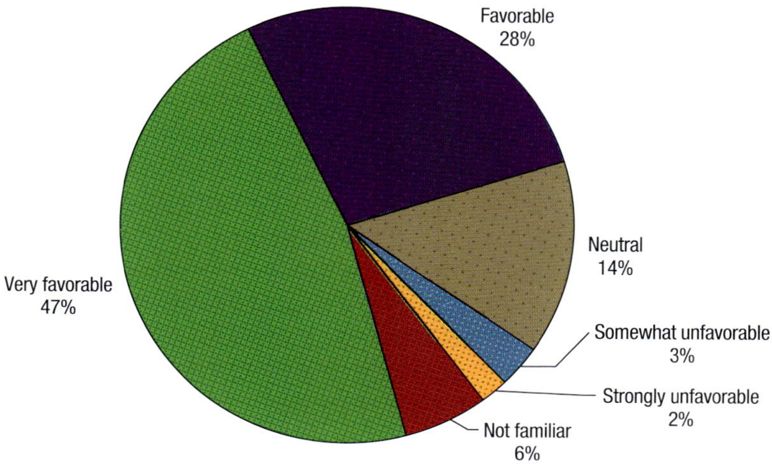

Fig. 12–2. Overall impressions of wind energy. *Data courtesy of Pike Research.*

Demographic analysis of the survey results showed that wind energy is most popular with people 45–64 years old (82%), followed by the 30–44 age group (76%). Little distinction was apparent among other variables such as gender, income, and education, with a large majority viewing wind favorably. The consumer's type of home, electric bill, and other factors showed no significant deviations from the generally popular image of wind.

By way of comparison, hybrid and electric vehicles ranked fairly high in favorable responses (57% and 64%, respectively), ahead of biofuels and clean coal (each at 47%), nuclear power (42%), and smart grid/smart meters (37%).

Appearance

Let's face it—you can't hide a wind turbine. The towers supporting the turbines can be taller than 100 meters (300 feet), or about the height of a 25-story building. The wind industry likes to talk about the size of its equipment, until it comes time to admit how big it looks. When confronted with the possibility of a wind farm nearby, most people want to know right away how the turbines will affect their view and their property values.

Besides being tall to capture the best wind energy, turbines are often sited along ridgelines, on top of hills and mountains, or out in the open plains—all places where the turbines will stand out even more. One obvious solution is to site wind farms in locations where few people live, reducing the number of homes affected. But sometimes these less populated areas are prized for their tranquility, open space, and expansive vistas.

In the modern history of wind power, the visual impact of turbines has been a common denominator defining the opposition. There's no denying that turbines are conspicuous, and they change people's view of the landscape. But whether people see turbines as elegant additions to the picture or an insidious sign of industrial development—that's up to the individual. Some people say they look majestic, others say ugly.

Opinion surveys in the United Kingdom, where wind power capacity has almost tripled since 2005, show that more than 70% of people either like the visual appearance of wind farms or do not mind them.

The visibility of wind turbines might influence national policy in some cases. In Germany, most wind turbines are located in flat northern areas or offshore in shallow waters. The scenic hills of southern Germany remain relatively untouched.

But if the country wants to reduce its dependence on nuclear power, turbines might have to be installed along southern hilltops, new transmission lines would be needed to link the south to offshore wind power in the north, and people would have to accept their skylines being changed.

Wind power developers use computer models to show people what the turbines will look like from various spots around the wind farm. Many concerns about the visibility of turbines can be alleviated through careful design.

Wind farm planners can accommodate objections by adjusting the project layout and relocating some turbines. Sometimes the tower height can be reduced, or trees can be planted to minimize the visual impact. All of these measures involve trade-offs that affect the economic practicality of the wind project.

Another element of visibility is the nighttime lighting required to warn aircraft away from wind turbines. U.S. regulations have eased somewhat, reducing the number of lights needed. Wider spacing is allowed, and lighting is required only around the perimeter of a wind farm. Also, red lights are recommended, which are less annoying than white lights to people living nearby.

Mitigation. Federal researchers in America are developing a screening tool to identify specific land areas where wind power development could be highly visible, increasing the risk of opposition. The tool could help land managers and wind farm planners site projects where they are less likely to face resistance based on aesthetic issues. The research program will also develop ways to minimize the visual impact of potential wind farm sites based on their topography, vegetation, and other site-specific conditions.

Offshore. In terms of visibility, offshore wind farms have a minus and a plus, versus onshore—the turbines are bigger, but they're usually farther away from people. Coastal and near-shore turbines confront the worst of both worlds, as they're large enough to tap into strong ocean winds, but still close to civilization.

In Europe's experience, the visibility of offshore wind farms in the "seascape" (water, coastline, and land) correlates most strongly with distance from shore, but weather and turbine lighting can affect people's perceptions of visibility to a considerable degree.

For highly sensitive seascapes—coasts with an expensive ocean view, for example—a major visual impact can be expected for wind power developments less than 13 kilometers offshore (8 miles), but visual effects should be minor at a distance of 24 kilometers or more (15 miles).

Property values

For many families worldwide, their home is their primary financial asset, so property values are a serious concern. Homeowners in most industrial nations have been hit hard by falling property values. Most people are already anxious about the price they'd be able to get if they had to sell their homes.

So communities have reason to worry about a possible further decline in property values if a wind farm is proposed nearby. Home values are a composite of many factors, so studies of the effects of wind power development must be careful to distinguish those effects from other factors. Some research suggests that if enough people expect that a wind farm will reduce property values, the fear can become self-fulfilling.

Three notable studies of wind energy and home values have been conducted in the United States, the first by the Renewable Energy Policy Project in 2003. This research, which covered 24,000 home sales surrounding 11 wind farms, compared the average selling price of homes near the wind turbines versus those that were at a distance of 8 kilometers or more (5 miles). No clear evidence of falling property values was found. In some communities, housing prices near the wind farm rose faster than those for properties farther away.

A second, smaller study focused on a 20-turbine wind project in New York State. Researchers measured the distance from each home to the nearest turbine and recorded the degree to which residents could see the wind farm. The study found no evidence that the presence of the turbines measurably affected home values, even when concentrating on home sales near the wind farm and/or with a prominent view of the turbines.

A more recent, comprehensive study by Lawrence Berkeley National Laboratory found that the prices of homes surrounding wind farms are not affected measurably, consistently, or significantly by either the view of turbines or the distance of the home from the wind farm. In contrast, property values do fall around landfills, high-voltage transmission lines, and other types of power plants.

However, it's possible that individual homes or small numbers of homes might have dropped in value. If so, either the amount was too small or the occurrence too infrequent to show up in the statistics.

Researchers recommend that future studies concentrate on homes closest to wind farms and analyze a greater number of sales transactions. Also, people who've bought or sold homes close to wind farms should be interviewed to find out how they made their decisions. Research to date might not have adequately reflected the potential for a negative impact on property values for homes near wind farms.

In a follow-up to this study, residents in one community claimed that property values near the wind farm had declined by 25%. But less than a year later, the property tax assessor found that the wind farm did not appear to affect home values. Rather, declining prices were attributed to the economic downturn, not the wind turbines nearby. The assessor pointed out that it's hard to pinpoint a single factor in determining home prices.

For wind farms sited on agricultural land, the potential effect on farmers' crops is less well-known. Some researchers think that wind turbines could keep corn and soybeans cooler and drier, which might help them fend off funguses. One study shows that large wind farms affect near-surface air temperatures. However, farmers worry more that turbines would prevent them from spraying their fields with chemicals.

Noise

Wind turbines made in the 1980s were fairly noisy, but today's wind turbines generally operate more quietly. In windy rural areas, the noise level can be masked by the sound of the wind blowing. The most audible noise is the sound of the wind hitting the rotor blade. The sound made by a spinning turbine is typically described as whooshing or swishing.

However, turbine noise can sound louder in homes that are shielded from prevailing winds, with no masking effect. Similarly, wind turbines can sound noisier when sited in hilly terrain. If homes are located downwind from the turbines in sheltered dips or hollows, the turbine noise can carry farther and be more audible.

Also, although wind turbines revolve more slowly, their blades have become much longer, so the rotating speed at the tip of the blade has stayed about the same. Maximum tip speeds range around 220–320 kph (140–200 mph). Blade tips create high-frequency noise, a sort of whining sound, while the lower frequency thumping that people notice comes from the broader mid-sections of the blades, which are spinning more slowly (see "Types of sound").

Noise from a turbine's mechanical components, such as the gearbox and generator, has been reduced by design improvements. But individual turbines can make banging or knocking sounds if something's wrong inside the drivetrain, like a bad bearing or an impending gearbox failure, or even if a technician just left a safety harness hanging on the tower.

If mechanical problems are ignored, the noise can persist. Sometimes, wind farm operators might choose to keep running the turbine until the component fails, because fixing it before failure would likely cost just as much as replacing it later.

Wind power developers conduct studies to predict sound levels in nearby homes and other buildings. Wind farm designers can avoid excessive noise by providing generous setbacks from places that are especially susceptible (see "Mitigation").

Noise studies are part of the permitting process in many countries, provinces, and states. While it's tempting for developers to simply buy off the opposition—pay people to sign a waiver and/or gag order in advance—that doesn't resolve the noise issue or improve the industry's image.

The wind industry is finding out that noise levels up to 45 or 50 decibels (dB), which have been acceptable in most places, won't work everywhere. Even 40 dB of noise can spark strong opposition. If wind power developers want to avoid more public relations failures, they need to collaborate with homeowners and local governments in finding ways to accommodate people's threshold of annoyance when it comes to turbine noise.

Noise level comparisons. The background noise from a modern wind farm is often no louder than a refrigerator, depending on the turbines' distance. In terms of decibels, wind farms register above the sound level of a bedroom and below that of a busy office (table 12–4).

The best test is to experience the noise yourself, if you have the opportunity. Experts insist that you'll be able to stand under a wind turbine and have a normal conversation, without raising your voice. However, most people aren't in a position to travel conveniently to a wind farm, so they have to rely on second-hand information.

Table 12–4. Noise levels of wind turbine compared to other noise

Noise source	Noise level, dBA
Threshold of hearing	0
Rural background at night	25
Background noise in populated areas	30–45
Quiet bedroom	35
Wind turbine @ 350 m (1,150 ft, or 380 yards)	35–50
Refrigerator	40
Air conditioner	50
Average home or office	55
Car going 65 kph (40 mph) @ 100 m (330 ft)	55
Busy office	60
Truck going 50 kph (30 mph) @ 100 m (330 ft)	65
City traffic	90
Pneumatic drill @ 7 m (23 ft)	95
Jet aircraft @ 250 m (820 ft)	105
Threshold of pain	140

Note: dBA is an A-weighted decibel, which approximates the human ear's response to sound.

Beyond decibels. Simply measuring sound levels doesn't account for subjective factors. Some people are more sensitive to noise in general and turbine noise in particular. People living close to wind farms commonly describe the repetitive whooshing sound as "penetrating." Sleep disturbance is the biggest complaint.

The potential for backlash against a proposed project becomes far greater when the wind power developer fails to accurately describe the quality of sound to be expected, not just how loud it will be—but this is hard to do (see "Types of sound" below).

Also, the way that noise is perceived by people can vary with their attitudes toward wind power. People who are already annoyed with the presence of a wind farm are more likely to hear its noise and think that it's getting louder, whether true or not. This is called the *nocebo* effect (as opposed to its opposite, placebo). If you're already against a wind project, you're more likely to be affected by its noise.

Other studies suggest that relatively low decibel levels can provoke annoyance. In a Scandinavian study, about 25% of people were annoyed at noise levels of 40 dB and 40% at 45 dB. Annoyance was higher in flat, rural locations than in suburban or mixed rural-suburban settings. A Massachusetts study found that people's annoyance with turbine noise was indistinguishable from their annoyance with the sight of a turbine, based on limited evidence.

Nor does distance always fix the problem. According to another project, within 580 meters of a typical wind farm (1,900 feet), 44% of people would be highly annoyed. Only when noise falls below 40 dB, at a distance of about 800 meters (half a mile), does the annoyance level begin to drop substantially.

In one U.S. study, about 50% of residents within 800 meters (half a mile) said noise is a problem, and more than a third of people at distances of 400–800 meters (0.25–0.5 mile) had been awakened from sleep by turbine noise.

Specialists in acoustic science point out that wind farm noise is not that bothersome to most people nearby, and the vast majority of wind farms don't experience serious, ongoing noise issues. However, because of the nature of sounds made by wind turbines, regulations citing simple noise limits are inadequate, and people who are affected have to be accommodated. This is especially true in communities where people have less land, but they value the prevailing quiet, according to the Acoustic Ecology Institute.

Working farmers and ranchers are far less bothered by low and moderate noise levels from wind turbines than are those who live in the country for peace and refuge from the urban and suburban life. This isn't surprising, but it's central to the challenges at the heart of community siting decisions.

—Jim Cummings, Acoustic Ecology Institute

Types of sound. Sound is composed of frequency (hertz) and pressure (decibels). Wind turbines generate four types of sound: broadband, infrasonic, impulsive, and tonal. Virtually all commercial-scale turbines these days are upwind machines, which are less likely to create most types of noise than downwind designs, especially low-frequency and impulsive sounds.

Broadband is a general, steady noise made up of a combination of sound waves. In turbines, broadband noise includes humming and rhythmic whooshing sounds. Most of these turbine sound waves have frequencies higher than 100 hertz, but lower frequencies (20–100 hertz) can be emitted in certain circumstances.

Infrasonic sound (very low-frequency, less than 20 hertz) is always present in the environment but is usually inaudible to the human ear, which begins to detect infrasonic waves at about 90 dB. Examples of infrasonic sound include ocean waves, distant aircraft, and ventilation fans. Infrasonic noise levels range from 71 dB in open fields to 89 dB on a breezy coast.

However, infrasonic waves can rattle windows and cause other structures to vibrate. Infrasonic sound is often blamed for affecting the health of people near wind farms (see "Health and safety" below). Some people feel that the vibration is unbearable, even when the turbine is more than 600 meters away (2,000 feet).

Impulsive sounds are abrupt and brief, like a door slamming. In the case of wind turbines, impulsive sound manifests itself as a regular thump or pulse. This can occur when steady winds synchronize the turbines, causing several to swish at the same time, or during turbulent winds.

Tonal sounds are steady and can have a distinct pitch like a musical note. In wind turbines, the natural frequency (resonance) of rotating mechanical equipment can cause tonal sounds. Turbine manufacturers pay particular attention to preventing resonance in the drivetrain, for many reasons other than noise (see chapter 3).

Mitigation. Turbine designers have already minimized noise to a great degree, and wind power developers provide setbacks to allow sound to dissipate. Although setbacks are a rough measure that cannot capture all the variables of a particular home's location, larger setbacks could lower the maximum noise heard by nearby homes to levels more in line with nighttime quiet.

Researchers are looking at whether the whooshing sound of turbine blades affects people psychologically in a way that noise from highways or airports does not. Also, scientists at Germany's Fraunhofer Institute are investigating vibration absorbers to further control noise from turbine gearboxes.

Flicker and glint

Shadow flicker is a phenomenon that occurs when a turbine's spinning blades pass between the sun and a person who's trying to read, watch TV, or concentrate on some visual task. This moving or flickering shadow, sometimes called the strobing effect, can be irritating.

Usually it happens under certain conditions, frequently at sunset, when the sun, the turbine blades, and the person are all aligned. Shadow flicker is more of a problem in far northern or southern latitudes where the sun's angle is low in the sky more often.

Wind power developers can prevent shadow flicker by calculating precisely where it might occur and how often, then allowing for a setback in those areas or planting trees that block the light.

When sunlight reflects off turbine blades directly into a person's eye, it's called blade glint. All major turbine manufacturers today use a low-reflectivity coating that prevents glint.

There is a clear business opportunity emerging for those companies that take the lead by crafting truly responsive community relations programs. These companies will commit to working with the standards set by local tolerance for new noise sources, rather than pushing local or state authorities to adopt siting standards used elsewhere.

These leading-edge companies may also put their money where their mouth is on property values by establishing programs that compensate landowners for moderate changes in property value . . . and helping create programs that buy and sell homes, so residents who wish to move can do so quickly at fair market value.

These companies will develop reputations as companies that are ready to be good local citizens, and will find that the increase in some costs, and a willingness to forsake some locations altogether, leads to dramatic benefits in terms of long-term stability and acceptance in the communities where they work, and especially, in communities where they propose new projects.

—Jim Cummings, Acoustic Ecology Institute

Health and safety

Health issues. The effect of wind turbines on human health has been debated extensively. Is the so-called "wind turbine syndrome" a real problem? Symptoms include sleep disruption, headaches, ringing ears, blurry vision, memory problems, vertigo, and anxiety attacks. Living with vibration from wind turbines can drive you literally mad, some people claim.

Scientists answer that no evidence supports these fears, but belief in the syndrome persists despite a growing body of research that has failed to document any health risks. Australia's National Health and Medical Research Council found no data linking turbines to adverse health effects, whether from infrasonic sound, flickering shadows, glinting blades, or other concerns.

Similarly, the Massachusetts study found no evidence that exposure to wind turbines causes health effects characteristic of the syndrome. Some researchers have suggested that the stress of worrying about health issues can actually cause symptoms or make them worse.

Claims made in a book about wind turbine syndrome have not been published in a peer-reviewed journal, and acoustic specialists have criticized its assertions. However, peer-reviewed research that was released in 2011 found that turbine noise causes health problems, usually some type of stress disorder.

Scientists recommend further study of health issues associated with wind turbine noise. Ongoing research is seeking to characterize the impact of noise on people's quality of life as well as their health. Although the evidence so far suggests that wind noise is unlikely to broadly affect human health, it's premature to say that the case is closed, according to the Acoustic Ecology Institute.

As wind power advocates point out, ultimately wind energy and other renewable resources can improve human health by reducing fossil fuel emissions, leading to better air quality (see "Air pollution and emissions").

Blades and ice. In the old days, it was not uncommon for wind turbine blades to fly off the machine, but these so-called blade throws occur less often today because of better design and engineering. Modern turbines can operate safely near schools and cities, though these are not necessarily the best sites.

Turbine blades can throw off chunks of ice under certain conditions, but normal setbacks are usually enough to protect people. Moreover, ice buildup impairs the turbine's efficiency, and the control system is designed to stop the turbine if blades become coated with ice.

Tourism

Many wind farms worldwide have become tourist destinations rather than places to avoid. Wind power developers often work with local governments to post signs with information about the wind farm and clearings along the roads for people to pull over and view the turbines. Some towns organize bus tours to local wind farms.

One success story is the coastal Jersey-Atlantic wind farm near a U.S. gambling haven, where five 1.5-MW turbines have been running since 2006 to help power a wastewater treatment plant. Atlantic City's casinos report that some customers ask for rooms with a turbine view, and tours of the wind farm have become popular, attracting about 15,000 people each year, said the *Philadelphia Inquirer.*

In a survey of tourists in wind farm–rich western Scotland, 43% said the turbines made them feel better about the area, while only 8% expressed negative reactions. However, some local European governments have drawn a line, such as the Greek island of Serifos. An 87-turbine project was killed to avoid disturbing an estimated 100,000 tourists who visit each year. Notably, local politicians and ecologists agreed that some wind turbines would be acceptable, but the scale of the project was just too big.

In emerging nations or other places where wind power is unusual, turbines can attract tourists, at least local ones. Novelty can be a magnet.

From June to September . . . a number of tourists visit the Huitengxile wind farms in Inner Mongolia—riding horses, looking at the flowers, eating barbecued lamb, or having a bonfire party. The income to the local herdsmen from tourism could represent as much as half of their total income. The wind farms in Dabancheng of Xinjiang, Guan Ting Reservoir of Beijing, and Bashang of Hebei have also all become popular locations for wedding photographs.

—China Wind Power Outlook 2010

Communications and radar

Wind turbines don't interfere with homeowners' TV or radio signals unless the electronic equipment is in the turbine's line of sight. It's common practice to avoid interference by improving the electronic receiver's antenna or installing relays to transmit the turbines' communications signal around the wind farm.

To warn off air traffic, structures more than 60 meters high (200 feet)—including virtually all utility-scale wind turbines and most meteorological towers, as well as many cellphone towers—are required in America to have aviation warning lights.

In addition, federal regulators must approve proposed turbines and other tall structures within 6.1 kilometers of certain airports (3.8 miles). The project might get a green light, or it might be restricted in height or location unless a solution is proposed.

Wind turbines that are too close to an airport, military post, or weather station can interact with radar operations, including air traffic control. If turbines are in the radar's line of sight, they can reflect radar signals back to the operator, compromising the accuracy of readings.

Radar interference delayed or prohibited about 9 GW of proposed wind capacity in America in 2010, and 6.5 GW was on hold in the United Kingdom. U.S. rules were being relaxed somewhat, but UK regulations do not permit wind farms within 74 kilometers (50 miles) of air defense radar unless developers can demonstrate that the project will not interfere with operations.

The problem is that turbines' radar signatures can look like those of a moving airplane or weather system. The reflected signals can shadow or mask aircraft, clutter operators' displays, or scatter radar waves, which has been reported at Denmark's Copenhagen airport as a result of the Middelgrunden offshore wind farm. In some spots, so many false signals are received from turbines that those zones are ignored, making them effectively blind to radar operators.

Technologically, this problem could be fixed, but would be expensive due to the age and condition of some radar stations. Also, software that distinguishes wind turbines from airplanes would have to be 100% accurate, since a single mistake could be catastrophic if an airplane's presence was ignored.

Federal regulators are funding research to develop a modeling tool that will accurately assess the effects of proposed turbines on existing U.S. radar operations. A retrofit system was tested in 2010 in the United Kingdom and the Netherlands, showing substantial improvement in the probability of distinguishing wind farms from other objects.

Offshore safety

Like onshore turbines near airports, offshore turbines can interfere with coastal and marine radar and weather stations. Because the atmosphere over the ocean differs from that over land, certain standard assumptions don't apply.

To mitigate the effect on radar offshore, vessel routes and distances from wind farms need to be well defined, and new wind farms must allow for certain distances from navigation routes or port approaches. Wind power developers can set money aside at the outset of a project to upgrade nearby radar systems later if needed.

Another safety issue is lighting. If the turbines' lighting is insufficient, they can become hazardous to ships navigating nearby, especially in areas that get heavy fog. A ship colliding with an offshore wind turbine foundation could damage or even destroy the ship, possibly spilling oil or other chemicals.

European researchers have evaluated several collision scenarios with different types of ships and turbine foundations in the North and Baltic Seas off Germany. They concluded that monopile and jacket foundations are safer than tripod structures. Also, while the risk of collision can be reduced, it can't be eliminated.

Safety can be improved by redundant navigation and control systems, better ship traffic management, prohibition of certain unsafe ships in wind farm areas, having tug boats ready for emergencies, and training ship crews thoroughly.

Environmental Issues

Air pollution and emissions

As a clean source of electricity, wind power reduces air pollution, including smog, soot, acid rain, and emissions of greenhouse gases such as carbon dioxide (CO_2) and methane, the main constituent of natural gas.

Air pollution causes serious health problems for people with asthma, emphysema, bronchitis, and other respiratory disorders. It can also affect children, elderly people and, when severe, healthy adults. Greenhouse gases are implicated in global climate change, which could make parts of the world uninhabitable.

Carbon emissions from power generation worldwide totaled 30.6 gigatons in 2010, a record. About 44% came from power plants burning coal, 36% from oil, and 20% from gas. Electricity produced by wind turbines operating on an interconnected grid displaces power from generators that burn fossil fuels, nearly on a per-megawatt-hour basis, avoiding 90%–95% of their emissions.

A fairly straightforward metric for understanding the carbon emission benefits of wind energy is that a single 1.5-MW turbine avoids 2,700 metric tons of CO_2 per year, given the U.S. average power generation fuel mix (about 45% coal, 23% natural gas, 20% nuclear, and 12% hydro and other sources).

The amount of CO_2 displaced would be smaller in countries that have more nuclear and hydropower—generation sources that produce virtually no carbon. Correspondingly, wind power's avoided CO_2 emissions would be larger in countries that rely more on coal and oil to generate electricity, because these fuels release more carbon. The economic price of emissions is discussed below (see "Hidden costs of energy").

Wind energy has one of shortest payback times, in terms of life-cycle emissions, of any energy technology. Within three to nine months of operation, the amount of energy produced by a wind turbine recoups the total amount of energy consumed—and emissions generated—to manufacture its components, transport them and other materials to the site, operate the turbine for 20–25 years, and decommission the equipment.

Some analysts claim that wind power won't reduce CO_2 emissions or, if it does, the amount would be trivial. One study cited to support this view, by Bentek Energy, explains that when wind power comes online in a small power system, the utility must accept it to comply with renewable energy mandates.

This forces the utility to balance wind energy's variable supply by ramping coal-fired plants up and down (also called cycling). Coal plants are not designed for ramping, so their emissions briefly increase, detracting from wind power's overall emissions benefit.

However, this study focused on a single Colorado utility that didn't have enough flexible natural gas–fired capacity to offset wind power's variations, relying instead on coal. As discussed in chapter 7, gas-fired generators are much easier to ramp up and down than coal plants. This study also analyzed Texas's power system, which has more gas-fired plants, and found increases in some emissions, again due to coal plant ramping.

If all of Colorado's electric utilities are considered, emissions were reduced when wind power increased its share of the state's power mix from 2.5% to 6.1% in 2007–2008. This corresponded to the following reductions:

- 4.4% in CO_2 emissions
- 6% in emissions of nitrogen and sulfur oxides
- 3% in coal use
- 14% in natural gas use

A more extensive follow-up study by Bentek concluded that wind power saves very little CO_2 in the U.S. Northwest, California, and Texas power systems, though it is more effective in the Midwest. The study analyzed actual emissions data from 2008–2010. Similarly, a paper in the peer-reviewed journal *Renewable and Sustainable Energy Reviews* analyzed 11 studies and reached similar conclusions, that wind's CO_2 emissions reductions will be less than anticipated.

Nevertheless, emissions reductions similar to Colorado's show up in other states as they install more wind power, and other utilities with high proportions of wind energy on their power systems have documented reductions in emissions.

Four of the seven major U.S. grid operators calculated emissions reductions on their systems in kilograms (pounds) of CO_2 per megawatt-hour of wind energy:

- Texas, 562.9 (1,241)
- Midwest, 579.2 (1,277)
- Mid-Atlantic, 738.4 (1,628)
- New England, 469.9 (1,036)

These figures are comparable to the 600 kg/MWh (1,300 pounds) estimated by the Global Wind Energy Council as an average value for CO_2 reductions obtained from wind energy generation. The council estimates that wind power avoided CO_2 emissions totaling around 1 billion tons worldwide during 2003–2009. This could rise to a cumulative total of 5–10 billion tons by 2020.

In calculating these projections, the council assumed that industrialized nations will keep shifting from coal to natural gas for electricity generation, while developing countries will continue to use coal-fired power plants to a greater extent than elsewhere.

Government statistics and projections are also cited as a criticism of wind energy's emissions benefit. Not surprisingly, the amount of wind power and transmission capacity developed under different scenarios affects the amount of CO_2 emissions.

For example, in the eastern United States, wind power could reduce CO_2 emissions by anywhere from 5% to 33%, depending on how much capacity is added. Other studies cited against wind energy argue mainly that it's too expensive to be used for air pollution control (see "Economic Issues" below).

It's true that wind power mostly displaces electricity supplied from natural gas–fueled plants, which produce less CO_2 than coal plants. Of course, what fuel is displaced depends on the power system's generation mix. Hydropower produces even lower CO_2 emissions than natural gas. Eventually, power system operators will rely more on natural gas and less on coal for balancing supply and demand, resulting in lower emissions overall.

Europe has documented the CO_2 emission reductions associated with its wind power capacity. In 2009, these amounted to 106 million tons. By 2020, wind energy could avoid 333 million tons of CO_2 emissions, almost 30% of Europe's greenhouse gas reduction target.

CO_2 emissions permits, or allowances, are issued by the European Union under a cap-and-trade system, and certified emission reduction credits, or offsets, are issued by the United Nations. These permits and credits are traded on global financial exchanges, setting a market price for CO_2 emissions. Beginning in 2013, permits will be auctioned to large power generators in Europe, shifting more of the cost of greenhouse gas emissions to major sources.

Climate change

Wind power and other renewable energy supplies could help slow down climate change, but is there a chance that climate change could diminish the world's wind resource? The idea is that global warming in the Northern Hemisphere would weaken the wind's strength by reducing the air temperature differences that create and energize wind. As polar regions warm up, the temperature contrast between mid-northern and high northern latitudes decreases.

Notably, the world's three biggest wind power markets—China, Europe, and North America—all lie in northern latitudes. One study estimates that Chinese wind turbine output beyond 2070 could fall by more than 14% if wind speeds decrease by 4%–12%.

Another report documents that wind speeds slowed by 5%–15% in most continents in the northern mid-latitudes during 1979–2008. This study relied on data from 822 surface weather stations that were screened for accuracy from a total of 5,412 locations.

As with most controversial subjects, rebuttal arguments abound. Most of them point to the inaccuracy of historical wind speed data (for a variety of reasons) and the difficulty of using climate models to predict shifts in the location, intensity, and duration of wind resources. Also, time scales of only a few decades are relatively short for predicting weather trends, and surface data are taken mostly at heights of 10 meters (33 feet), whereas wind turbines operate higher up.

Other studies suggest that wind speeds will rise rather than fall as global temperatures warm, at least offshore, or they could remain about the same, at least in areas that matter for wind power development. The bottom line appears to be that no one really knows.

Land use

Wind power development requires large areas of land, and as noted above, wind turbines displace some wildlife habitat permanently. However, the footprint of land that is actually disturbed (occupied by turbines and infrastructure) is less than 5% of wind farms' total land area, according to DOE.

Wind energy can coexist with some land uses such as farming, ranching, and forestry. Housing developments are less compatible with wind power, and planners must take steps to avoid conflict with air traffic, as discussed earlier.

Like any construction project, wind farms must conform to government regulations designed to prevent excessive soil erosion and any damage to local water quality.

Economic Issues

Cost of energy

The cost of wind energy varies widely because it's very site-specific. In general, however, global wind energy costs in 2010 reached record lows in many areas. This was especially true in countries where the price of wind power competes head-to-head with the price of conventional electricity.

Several projects around the world had a levelized (life-cycle) cost of energy below $68 per megawatt-hour (50 euros), compared to estimated average costs of $67/MWh for coal-fired power (49 euros) and $56/MWh for gas (41 euros), reported Bloomberg. This turnaround was largely thanks to lower prices for onshore wind turbines, which fell substantially in 2009–2010. This directly reduces the capital cost of building a wind farm and helps lower its cost of energy.

In the United States in 2009–2011, utilities and other buyers in some areas were getting prices as low as 5–6 U.S. cents per kilowatt-hour (4 euro-cents) for wind power in long-term contracts. In the best wind markets, power purchase agreements were priced in the range of 3 U.S. cents/kWh (2 euro-cents), reported DOE.

This is cheaper than electricity from new coal-fired generators and roughly cost-competitive with new natural gas–fired plants equipped with the latest technology. The range for wind power prices in 2009–2010 was 3.4–11.7 U.S. cents/kWh (2.5–8.5 euro-cents).

In Brazil's 2011 auctions, wind power slightly undercut the price of both natural gas and hydropower, winning about 50%–80% new capacity, as noted in chapter 11. Prices were about $60 per megawatt-hour for wind at the first auction, versus $61 for hydro and $62 for gas (99.6, 102.4, and 103.3 reais, or 43.6, 44.9, and 45.2 euros). Wind power prices went down to about $55/MWh at the second auction (91 reais or 40 euros).

Offshore wind power costs. In general, offshore wind energy costs up to twice as much as onshore, due mainly to higher O&M costs and also rising capital costs, as noted in chapter 9. Power purchase agreements for America's first offshore wind farms have steep prices by U.S. standards. Cape Wind's rate starts at 18.7 U.S. cents/kWh (13.7 euro-cents) and escalates by 3.5%/yr. A project off Rhode Island that calls for a price of 24.4 U.S. cents/kWh (17.8 euro-cents) was being challenged in court.

Some European countries could depend more on offshore wind power in the future, because they're densely populated and have little room for more wind farms on land. However, the UK government argued in 2011 that nuclear power is the country's cheapest low-carbon option for the near term, an opinion supported by its climate advisory panel.

A project by the UK Energy Research Centre estimated that offshore wind energy cost about 23.2 U.S. cents/kWh in 2009–2010 (14.5 pence, or 16.8 euro-cents), compared to only 13 U.S. cents/kWh (8 pence, or 9 euro-cents) for combined-cycle natural gas–fired power.

The center projected as a "best guess" that UK offshore wind power costs could fall about 20% by 2025, to 18.3 U.S. cents/kWh (11.5 pence, or 13.3 euro-cents). A similar analysis by BVG Associates predicted a 15% reduction in UK offshore wind costs by 2022, to 16–19 U.S. cents/kWh (10–12 pence, or 12–14 euro-cents).

Another study pegged UK offshore wind power prices in 2010 at 24–33 U.S. cents/kWh (15–21 pence, or 17–24 euro-cents), versus onshore rates of 13–17 U.S. cents/kWh (8–11 pence, or 9–13 euro-cents). By comparison, natural gas–fired power cost 10–17 U.S. cents/kWh (6–11 pence, or 7–13 euro-cents) and nuclear, 10–13 U.S. cents/kWh (6–8 pence, or 7–9 euro-cents).

Capital vs. fuel costs. Wind farms are more expensive to build than gas-fired plants, which can be installed cheaply and quickly. American utilities installed lots of gas-fired generation capacity near load centers to avoid building transmission lines, which take much longer to approve. That's one reason a utility executive called natural gas the "crack cocaine" of the power industry—because it's cheap and easy to install, but hard to withdraw from when gas prices rise.

In economists' jargon, wind farms are capital-intensive. As much as 75% of the cost of energy from a wind power system is related to its up-front costs. For natural gas–fired power, only 30%–60% of energy costs result from capital expense. However, wind power has an ace up its sleeve, in that wind turbines use no fuel. Wind's marginal cost in power markets is low, and its variable operating cost—fuel—is zero.

Wind power's freedom from fuel shields consumers and governments alike from the threat of escalating fossil-fuel prices in the future. Most industrial nations import fuel from unstable parts of the world, which makes prices more volatile and unpredictable.

Even in countries with healthy domestic supplies, no one can guarantee where market prices will go over the long term. That's why energy futures markets don't offer safe, simple contracts for fossil-fuel delivery 20 years out.

Standard financial methods of calculating energy costs favor plants that are cheaper to build but more expensive to operate, because economists apply the same *discount rate* regardless of fuel-price risk. Discount rates are a measure of the cost of money over time, like an interest rate but more complex.

Using a single discount rate to evaluate power plant economics, a fossil-fuel project is preferred because it might be more profitable if fuel stays cheap. A wind farm is judged to be less profitable, even though its profits are more predictable because they don't depend on fuel prices. Gas-fired power becomes the default choice because it's perceived as having a lower financial risk and is less affected by capital cost.

To account for wind power's value in avoiding fuel-price risk, the European Wind Energy Association developed a methodology for estimating energy costs using different discount rates. When fuel-price risks are considered, gas-fired power costs increase by 85% and the cost of coal power more than doubles.

Benefits to local economies

Wind farm operators pay landowners to lease acreage. Payment amounts can be based on a percentage of project revenues, as well as an annual flat fee per turbine or per megawatt. Amounts can total about $3,000–$6,000 per year (2,000–4,000 euros).

During construction, workers pump money into the local economy for services such as food and lodging. Some permanent jobs are also created, and wind farms add tax revenue to local communities.

As an example, 300 MW of wind capacity was developed in Illinois (102 MW that started operating in 2009 and 198 MW in 2010). The wind farms have a total of 135 landowner leases and 130 neighbor agreements. The operator pays about $2.8 million/yr in property taxes (2.0 million euros). The wind farms employ 23–30 full-time people for operation and maintenance.

Competition with other energy sources

Natural gas. Although wind power and natural gas might battle for investment dollars on the global stage, together they could give the world its best chance for a low-carbon future. Gas's main competitor as wind power's partner is energy storage, but most storage technologies are still more expensive.

As wind power increases its share of the power mix, system operators will need to add flexible generation capacity, not only to balance wind's variability but also to manage growth, diversify power supplies, and improve the grid's overall reliability. A natural gas–fired combustion turbine is one of the cheapest ways to do this.

Combined-cycle gas plants are more expensive, but also more efficient and less polluting. Combined-cycle efficiencies typically range above 50%, and some can exceed 60%. Gas turbine manufacturers are promoting these systems as a solution to enable the integration of wind energy in power systems.

Gradually over decades, wind energy and natural gas could begin to erode coal's position as the leading power generation fuel worldwide. Natural gas–fired power plants could account for 60% of new generation capacity installed in the United States during 2010–2035. At the same time, wind power is projected to almost double its share of total electricity generation.

Some environmental activists object to natural gas on the grounds that even though it's cleaner than coal, it still adds to air pollution and greenhouse gases. Gas-fired power generators produce about half as much CO_2 as coal plants. They can also release a little methane if the fuel isn't burned completely, and gas combustion can produce nitrogen oxides unless the system's design or special equipment controls the emissions.

Another potential environmental problem is hydraulic fracturing, commonly called "fracking"—a technology that drillers use to produce natural gas from shale and other rocks that don't easily release their gas. This drilling technique has been used most widely in the United States, but other countries could discover and exploit gas reserves that have been considered impractical until now.

America's shale gas producers are confronting local resistance, as wind power has, from people who fear that the chemicals used in fracking will affect water supplies. Although impermeable rock usually isolates drinking water from gas wells, the drilling industry has yet to prove that fracking is safe. Also, the fracking process consumes lots of water, which could be a bigger liability in the long run.

The International Energy Agency predicts a global natural gas glut that could last until 2020, due in part to U.S. shale gas development. The surplus could amount to about 5% of world gas demand. Although shale gas has affected the global supply picture, natural gas prices remain high in Asia. Russia and Iran still control about 40% of the world's known reserves, and Middle Eastern countries possess much of the rest.

Looking at the big picture, natural gas and oil might be plentiful for a while, and coal might remain abundant for much longer. But all of our hydrocarbon supplies are limited, and using fossil fuels imposes varying burdens of damage to human health and the environment (see below).

Nuclear power is clean as far as CO_2 emissions go, but it remains expensive despite decades of technology development and government support. Nuclear accidents can cause long-lasting damage, and waste disposal is still a problem.

Hidden costs of energy. Many researchers have attempted to quantify the hidden costs associated with producing and using various forms of energy, both fossil fuels and renewable resources. Economists call these *external costs*, which are not reflected in the market price of the energy.

The prime example of a hidden cost is the damage that air pollution imposes on human health by compromising people's ability to breathe. Power generation can produce sulfur dioxide, nitrogen oxides, ozone, and particulate matter (tiny particles of soot).

Soot is known to cause cancer, while sulfur and nitrogen oxides cause acid rain, which affects timber harvests, other vegetation, buildings, and water resources as well as people and animals.

Industrial nations have controlled acid rain to some extent, especially from newer power plants. Sulfur dioxide emissions are regulated by a cap-and-trade system in the United States. Paradoxically, emissions of sulfur compounds and particulates can shield the earth from global warming by reflecting sunlight.

European and U.S. scientists have studied these hidden costs for many years. More recently, calculations have attempted to account for the cost of climate change, or global warming, associated with emissions of CO_2, methane, and other compounds.

America's National Research Council calculated that air pollutants from coal-fired power plants (not counting CO_2 emissions) caused damages amounting to $62 billion in 2005 (45 billion euros).

This works out to an average of about 3.2 U.S. cents per kilowatt-hour of coal-fired energy (2.3 euro-cents). The council also estimated that coal power's economic impact on climate change in terms of CO_2 emissions ranged from 0.1 to 10 U.S. cents/kWh (0.07 to 7 euro-cents).

Air pollution caused by natural gas power generation is lower, because gas is a cleaner burning fuel. Its hidden costs totaled $740 million in 2005 (540 million euros), for an average of 0.16 U.S. cents/kWh (0.12 euro-cents). Gas's climate impact was about half of coal's, at a range of 0.05–5 U.S. cents/kWh (0.04–4 euro-cents).

Several damaging effects were not quantified in the council's analysis, including mercury emissions associated with coal-fired power, nuclear waste impact, and acid rain's effect on ecosystems.

A similar study by Harvard Medical School estimated that the cost of coal power would double or triple if its life-cycle impacts are considered. The research examined mining accidents and black lung disease, as well as mountain-top removal, which pollutes water. It also accounted for the environmental impact of sludge, slurry, and fly ash ponds. Total costs annually could amount to $300–$500 billion (200–400 billion euros).

Another project identified 39 U.S. coal ash disposal sites where local surface water and groundwater have been contaminated with toxic metals and other pollutants, in addition to the 67 sites already identified by federal regulators.

To quantify the social cost of CO_2 emissions, the E3 Network analyzed the risks associated with 16 climate change scenarios. In contrast to the U.S. government's official cost of $21 per ton of carbon (15 euros), the analysis estimated a wide range of values in 2010, from $28 to $481/ton on average (20–351 euros), depending on discount rates.

The European Commission initiated a project called ExternE (external costs of energy) in 1991, involving more than 50 research teams in over 20 countries. The project attempted to quantify the total economic impact of external costs such as global warming, public health, occupational health, and material damage.

Researchers stressed that the results are specific to location and technology, so no simple generalization is possible. Data published in 2003 are summarized in table 12–5.

Table 12–5. European external power generation costs. *Data courtesy of ExternE.*

Country	Coal	Oil	Gas	Nuclear	Wind
	euro-cents/kWh (U.S. cents/kWh)				
Austria			1-3 (1-4)		
Belgium	4-15 (5-21)		1-2 (1-3)	0.5 (0.7)	
Denmark	4-7 (5-10)		2-3 (3-4)		0.1 (0.1)
Germany	3-6 (4-8)	5-8 (7-11)	1-2 (1-3)	0.2 (0.3)	0.05 (0.07)
Finland	2-4 (3-5)				
France	7-10 (10-14)	8-11 (11-15)	2-4 (3-5)	0.3 (0.4)	
Greece	5-8 (7-11)	3-5 (4-7)	1 (1)		0.25 (0.34)
Ireland	6-8 (8-11)				
Italy		3-6 (4-8)	2-3 (3-4)		
Netherlands	3-4 (4-5)		1-2 (1-3)	0.7 (1.0)	
Norway			1-2 (1-3)		0-0.25 (0-0.34)
Portugal	4-7 (5-10)		1-2 (1-3)		
Spain	5-8 (7-11)		1-2 (1-3)		0.2 (0.3)
Sweden	2-4 (3-5)				
United Kingdom	4-7 (5-10)	3-5 (4-7)	1-2 (1-3)	0.25 (0.34)	0.15 (0.21)

Now that we know how much we could end up paying to endure the impacts of climate change, investing in reducing our emissions is clearly the prudent option. It's the difference between servicing your car, or waiting for it to break down on the highway.

—Frank Ackerman, E3 Network

Saving water. In many parts of the world, fresh water is already a limited resource. About 40% of the world's people live in water-stressed areas, and scientists expect that as global populations grow, water could become more scarce.

Also, rising temperatures associated with climate change could create more localized droughts such as those that cripple hydropower periodically in Brazil and China. Limited water resources will also affect how much food people are able to grow.

After agriculture, the energy sector is the world's biggest water consumer. Nuclear power plants use huge quantities of water, and many of them discharge heated water into nearby oceans, which can threaten aquatic life. Most conventional generators, regardless of fuel, require a lot of water to cool and condense the steam that drives the turbines (fig. 12–3).

In contrast, wind power uses virtually no water and could contribute substantially to conserving water resources worldwide. Every megawatt-hour of wind power that displaces traditional electric generation saves about 2,000 liters of water (528 gallons), according to Vestas's research. DOE estimates that wind power could conserve billions of liters of water in arid western states by displacing conventional electricity.

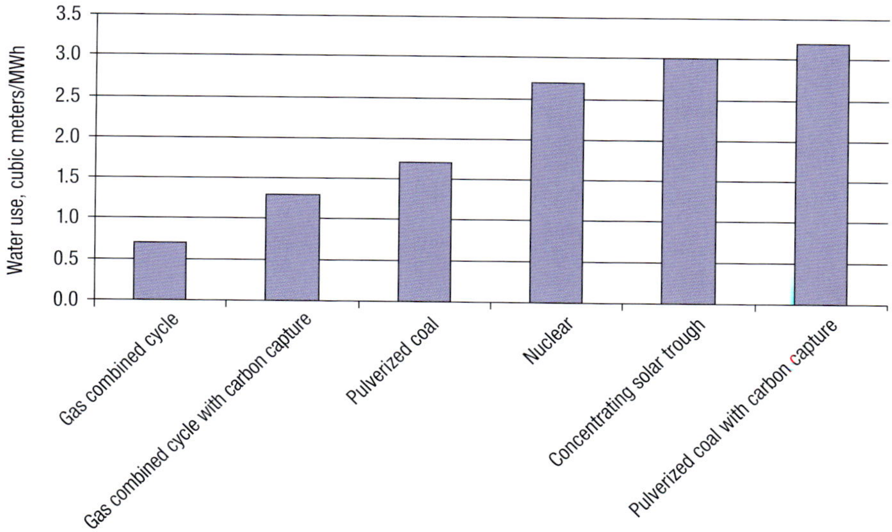

Fig. 12–3. Water consumption of electricity generation. *Data courtesy of Vestas.*

A word on subsidies. Some of the world's sharpest economists have analyzed government subsidies and other types of support, trying to prove whether one form of energy—wind power, for example—gets more than another, such as oil and gas. Not surprisingly, statistics abound to support both sides of the issue, and the numbers are all over the map.

Some forms of government support for energy development have been built into the economy for decades, while others are fresh and glaring. In most industrial countries, hundreds or even thousands of subsidy policies have been implemented incrementally over time. This makes it nearly impossible to compare apples to apples when calculating who gets how much.

The International Energy Agency estimates that fossil fuels received $312 billion (228 billion euros) in subsidies from governments worldwide in 2009, while renewable energy (including biofuels) got $57 billion in government support (42 billion euros). That translates to about 5.5 euros going to fossil fuels for every 1 to renewables. In 2010, global fossil-fuel subsidies rose to $409 billion (nearly 300 billion euros).

Regardless of the amount, every energy technology is subsidized at some point, to some degree. For example, the hydraulic fracturing technology that's now unleashing huge volumes of shale gas was supported by U.S. federal government research funding in the 1980s. It took more than 20 years to prove worthwhile. An analysis by DBL Investors calculated U.S. energy subsidies by type of fuel or energy (fig. 12–4).

Government spending on clean energy research can leverage funding from other sources. Subsidies create markets for new technologies until they can survive on their own, and anyone can argue whether that's good or bad for taxpayers. Would there be a global wind industry today without government support? Probably not. If subsidies were withdrawn, would the wind industry survive? Probably, but we don't know for sure.

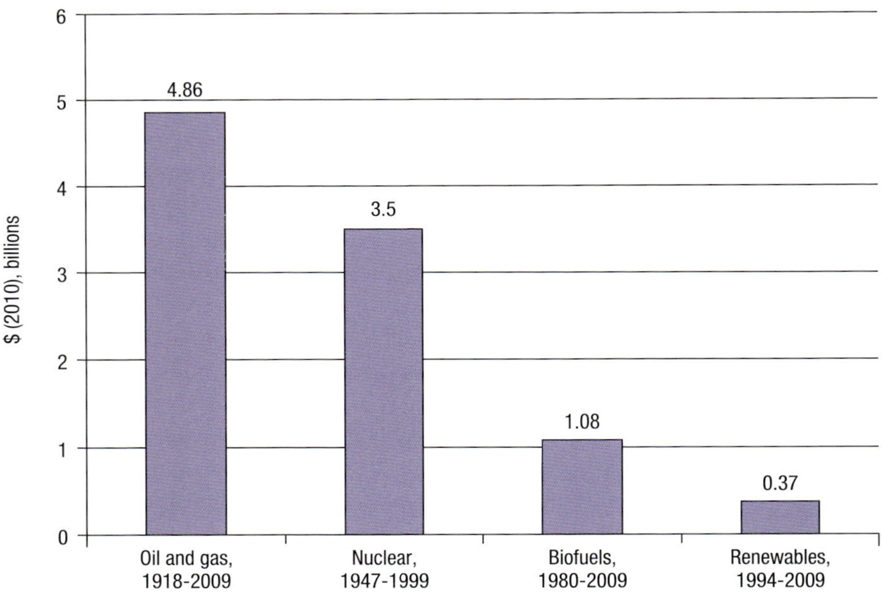

Fig. 12–4. Historical average of annual energy subsidies. *Data courtesy of DBL Investors.*

References

Ackerman, Frank, and Stanton, Elizabeth A. 2011 (July 13). Climate Risks and Carbon Prices: Revising the Social Cost of Carbon. E3 Network (Economics for Equity & Environment. http://www.e3network.org/papers/Climate_Risks_and_Carbon_Prices_executive_summary+full_report+comments.pdf.

Acoustic Ecology Institute. 2011. Wind Farm Noise 2011: Science and Policy Overview. http://acousticecology.org/wind/winddocs/AEI_WindFarmNoise2011.pdf

Acoustic Ecology Institute. 2010. Wind Farm Noise: 2009 in Review. http://acousticecology.org/docs/AEI_WindFarmNoise_2009inReview.pdf.

Alpha Ventus, http://www.alpha-ventus.de/index.php?id=80.

American Wind Wildlife Institute, http://www.awwi.org/.

Arnett, Edward B., et al. 2009 (April). Effectiveness of Changing Wind Turbine Cut-In Speed to Reduce Bat Fatalities at Wind Facilities. Bats and Wind Energy Cooperative. http://www.iberdrolarenewables.us/pdf/curtailment-final-report-05-15-10-v2.pdf.

———. 2010 (November). Altering Turbine Speed Reduces Bat Mortality at Wind-Energy Facilities. Bats and Wind Energy Cooperative. http://www.batsandwind.org/pdf/Arnett%20et%20al.%202011.pdf.

Bajdek, Christopher. 2007 (October 22–24). Communicating the Noise Effects of Wind Farms to Stakeholders. Harris Miller Miller & Hanson Inc. http://www.hmmh.com/cmsdocuments/Bajdek_NC07.pdf.

Bat Conservation International, http://www.batcon.org/.

Bats and Wind Energy Cooperative, http://www.batsandwind.org/.

Bentek Energy LLC. 2010 (April 16). How Less Became More: Wind, Power and Unintended Consequences in the Colorado Energy Market. http://docs.wind-watch.org/BENTEK-How-Less-Became-More.pdf.

———. 2011 (July 19). The Wind Power Paradox. http://www.bentekenergy.com/InTheNews.aspx#Article4701.

Black, Richard. 2011 (May 8). Nuclear "Cheapest Low-Carbon Option" for UK Energy. *BBC News*. http://www.bbc.co.uk/news/mobile/science-environment-13300595.

Boyle, Rebecca. 2010 (October 30). Increasing Wind Turbine Turn-On Speeds Could Help Reduce Bat Deaths, New Study Says. *Popular Science*. http://www.popsci.com/science/article/2010-10/increasing-wind-turbine-turn-speeds-could-help-reduce-bat-deaths-new-study-says.

Bryce, Robert. 2010 (August 27). Wind Energy Gets Huge Subsidies. So Where Are the CO_2 Reductions? Energy Tribune. http://www.energytribune.com/articles.cfm/5103/Wind-Energy-Gets-Huge-Subsidies-So-Where-Are-The-CO2-Reductions?

———. 2010 (September 14). Wind Energy's Real Problems: (Hint: It Has Nothing to Do With The Wall Street Journal). Energy Tribune. http://www.energytribune.com/articles.cfm/5261/Wind-Energys-Real-Problems--Hint-It-Has-Nothing-to-Do-With-The-Wall-Street-Journal.

———. 2011 (July 28). T. Boone's Windy Misadventure and the Global Backlash Against Wind Energy. Energy Tribune. http://www.energytribune.com/articles.cfm/8120/T-Boones-Windy-Misadventure-And-the-Global-Backlash-Against-Wind-Energy.

———. 2011 (August 15). The Party's Over for Big Wind. Energy Tribune. http://www.energytribune.com/articles.cfm/8255/The-Partys-Over-for-Big-Wind.

Cameron, Alasdair. 2011 (June 30). Offshore Wind Costs to Fall 15 Percent by 2022, *Renewable Energy World*. http://www.renewableenergyworld.com/rea/news/article/2011/06/offshore-wind-costs-to-fall-15-percent-by-2022.

Canadian Wind Energy Association. 2006 (August). The Sights and Sounds of Wind. http://www.canwea.ca/images/uploads/File/NRCan_-_Fact_Sheets/7_visual_sound.pdf.

Coates, Christine. 2011 (April–June). Using Radar to Protect Birds and Bats. Wind Today. http://www.bluetoad.com/publication/?i=68530&p=26.

Cummings, Jim. 2011 (February 22). Looking for Wind Industry Leadership in Reducing Noise Impacts. Acoustic Ecology Institute. *Renewable Energy World*. http://www.renewableenergyworld.com/rea/news/article/2011/02/looking-for-wind-industry-leadership-in-reducing-noise-impacts?cmpid=WNL-Wednesday-February23-2011.

Curry, Dick, and Kerlinger, Paul. What Kills Birds. Curry & Kerlinger, LLC. http://www.currykerlinger.com/birds.htm.

Daley, Beth. 2011 (February 16). New Harvard Study Examines Full Cost of Coal. Boston.com. http://www.boston.com/lifestyle/green/greenblog/2011/02/new_harvard_study_looks_at_ful.html.

Davis, Brian, and Wheelock, Clint. 2011 (March). Energy & Environment Consumer Survey: Consumer Attitudes and Awareness About 12 Energy & Environmental Concepts. Pike Research. https://www.pikeresearch.com/wordpress/wp-content/uploads/2011/02/EECS-11-Pike-Research.pdf.Deign, Jason. 2011 (June 13). Will wind energy cut CO2? Wind Energy Update. http://social.windenergyupdate.com/uncategorised/will-wind-energy-cut-co2.

Deese, Heather, and Schmitt, Catherine. 2010 (January 29). Fathoming: How Do Offshore Wind Turbines Affect Marine Life (USA). The Working Waterfront. http://www.workingwaterfront.com/articles/Fathoming-What-are-the-marine-impacts-of-offshore-wind-turbines/13667/printer-friendly/.

Denholm, Paul, et al. 2009 (August). Land-Use Requirements of Modern Wind Power Plants in the United States. NREL/TP-6A2-45834. http://www.nrel.gov/docs/fy09osti/45834.pdf.

DeTect, Inc., http://www.detect-inc.com/index.html.

Drewitt, Allan L., and Langston, Rowena H.W. 2006. Assessing the Impact of Wind Farms on Birds. Ibis. http://www.gflrpc.org/programareas/wind/TechnicalDocuments/AvianImpactsSocietyforProtectionofBirds.pdf.

Earth Track. 2010. Subsidy FAQs. http://www.earthtrack.net/subsidy-faqs.

Economist. 2010 (August 19). Wind Energy and Politics: Not on My Beach Please. http://www.economist.com/node/16846774?story_id=16846774.

EnerNex Corp. 2011 (February). Eastern Wind Integration and Transmission Study. NREL/SR-5500-47078, revised edition. http://www.nrel.gov/wind/systemsintegration/pdfs/2010/ewits_final_report.pdf.

Erickson, Wallace P., et al. 2002 (December). Synthesis and Comparison of Baseline Avian and Bat Use, Raptor Nesting and Mortality Information from Proposed and Existing Wind Developments. Western EcoSystems Technology, Inc. http://www.bpa.gov/Power/pgc/wind/Avian_and_Bat_Study_12–2002.pdf.

Erickson, Wallace P., Johnson, Gregory D., and Young, David P., Jr. 2005. A Summary and Comparison of Bird Mortality from Anthropogenic Causes with an Emphasis on Collisions. USDA Forest Service Gen. Tech. Rep. PSW-GTR-191. http://www.fs.fed.us/psw/publications/documents/psw_gtr191/Asilomar/pdfs/1029-1042.pdf.

European Commission. 2003. External Costs: Research Results on Socio-Environmental Damages Due to Electricity and Transport. ExternE. EUR 20198. http://www.externe.info/externpr.pdf.

———. 2010 (October). Wind Energy Developments and Natura 2000. http://ec.europa.eu/environment/nature/natura2000/management/docs/Wind_farms.pdf.

European Wind Energy Association. 2009 (March). The Economics of Wind Energy. http://www.ewea.org/fileadmin/ewea_documents/documents/publications/reports/Economics_of_Wind_Main_Report_FINAL-lr.pdf.

———. 2011. Fact Sheets: Climate. http://www.ewea.org/index.php?id=1611.

Fairley, Peter. 2011 (January). A Less Mighty Wind. IEEE Spectrum. http://spectrum.ieee.org/green-tech/wind/a-less-mighty-wind/1.

Fraunhofer Institute IWU and Physik Instrumente. 2010 (July). Piezoelectric Vibration Absorbers to Reduce Wind Turbine Noise Emissions. Solar & Wind Power. http://ntbpdf.techbriefs.net/2010/GDM_SW_0710.pdf

Gipe, Paul. 2011. Debunking the Myths About Wind Energy. Wind-Works.org. http://www.wind-works.org/articles/large_turbines.html#Debunking the Myths about Wind Energy.

Global Wind Energy Council, Greenpeace International. 2010 (October). Global Wind Energy Outlook 2010. http://www.gwec.net/fileadmin/documents/Publications/GWEO%202010%20final.pdf.

Goggin, Michael. 2010 (September 1). The Facts about Wind Energy and Emissions. *Renewable Energy World*. http://www.renewableenergyworld.com/rea/news/article/2010/09/the-facts-about-wind-energy-and-emissions.

Hazlett, Maril. 2011 (January 4). Climate Change Could Have Major Impacts on Wind Resources. http://www.nawindpower.com/e107_plugins/content/content.php?content.7130.

Hoen, Ben, et al. 2009 (December). The Impact of Wind Power Projects on Residential Property Values in the United States: A Multi-Site Hedonic Analysis. Lawrence Berkeley National Laboratory. LBNL-2829E. http://eetd.lbl.gov/ea/ems/reports/lbnl-2829e.pdf.

Hull, Dana. 2011 (August 29). Altamont Pass Wind Farm Gets Major Upgrade. *San Jose Mercury News*. http://www.energycentral.com/generationstorage/wind/news/en/21338730/Altamont-Pass-wind-farm-gets-major-upgrade.

International Energy Agency. 2010 (November 9). Recent Policy Moves a Start, but Much Stronger Action Is Needed to Accelerate the Transformation of the Global Energy System, Says the IEA's Latest *World Energy Outlook*. http://www.iea.org/press/pressdetail.asp?PRESS_REL_ID=402.

———. 2011 (October 4). World Energy Outlook 2011: IEA Analysis of Fossil-Fuel Subsidies. http://www.iea.org/weo/Files/ff_subsidies_slides.pdf.

Jones, Gareth, et al. 2009 (May). Scoping and Method Development Report: Determining the Potential Ecological Impact of Wind Turbines on Bat Populations in Britain. University of Bristol. http://www.bats.org.uk/data/files/determining_the_impact_of_wind_turbines_on_british___bats_final_report_29.5.09_website.pdf.

Keane, Brian. 2011 (February 9). Polls: Anti-Wind NIMBYs Are Small Minority. Huffington Post. http://www.huffingtonpost.com/brian-keane/polls-antiwind-nimbys-are_b_819905.html.

Kitasei, Saya. 2011 (February 17). Powering the Low Carbon Economy: The Role of Natural Gas in a Shifting Power Industry. Worldwatch Institute. Saya Kitasei 2011-02-17_Flexible_Power_Symposium_(SSK)[1].pdf.

Kjaer, Christian. 2010 (November 9). More Than $5 Fossil Fuel Subsidies for Every $1 of Support for Renewables. http://www.renewableenergyworld.com/rea/blog/post/2010/11/more-than-5-fossil-fuel-subsidies-for-every-1-of-support-for-renewables.

Komanoff, Charles. 2009 (April 21). Wind Power's Displacement of Fossil Fuels. Komanoff Energy Associates. http://www.komanoff.net/wind_power/Wind_Power's_Displacement_of_Fossil_Fuels.pdf.

Lacey, Stephen. 2010 (August 26). When Is Wind Energy Noise Pollution? *Renewable Energy World*. http://www.renewableenergyworld.com/rea/news/print/article/2011/03/when-is-wind-energy-noise-pollution.

———. 2011 (December 22). Wind Makes Up 80% of Contracts in Brazil's Latest Power Auction. Climate Progress. http://thinkprogress.org/romm/2011/12/22/394455/wind-contracts-in-brazil-power-auction/?utm_source=feedburner&utm_medium=email&utm_campaign=Feed%3A+climateprogress%2FlCrX+%28Climate+Progress%29.

Landscape Assessment Tool, http://wind.tnc.org/awwi/.

Levesque, Carl. 2011 (October 4). Wildlife Groups Team with AWS on Multi-Layered Wind Resource and Wildlife Map. *Renewable Energy World*. http://www.renewableenergyworld.com/rea/news/article/2011/10/wildlife-groups-team-with-aws-on-multi-layered-wind-resource-and-wildlife-map?cmpid=WindNL-Thursday-October6-2011.

Li, Junfeng, et al. 2010 (October). China Wind Power Outlook 2010. http://www.greenpeace.org/raw/content/eastasia/press/reports/wind-power-report-english-2010.pdf.

Massachusetts Department of Environmental Protection and Massachusetts Department of Public Health. 2012 (January). Wind Turbine Health Impact Study: Report of Independent Expert Panel. http://www.mass.gov/dep/energy/wind/turbine_impact_study.pdf.

Milton, Joseph. 2010 (October 17). Why Winds Are Slowing. *Nature Geoscience*. http://www.nature.com/news/2010/101017/full/news.2010.543.html.

Morales, Alex. 2011 (May 9). UK Should Build More Nuclear, Less Offshore Wind, Panel Says. *Renewable Energy World*. http://www.bloomberg.com/news/2011-05-08/britain-should-rely-on-more-nuclear-power-less-offshore-wind-panel-says.html.

National Audubon Society. Audubon's Position on Wind Power. http://policy.audubon.org/wind-power-overview-0.

National Health and Medical Research Council. 2010 (July). Wind Turbines and Health. http://www.nhmrc.gov.au/_files_nhmrc/file/publications/synopses/public_statement_wind_turbines_and_health.pdf.

National Research Council. 2010. Hidden Costs of Energy: Unpriced Consequences of Energy Production and Use. http://www.nap.edu/openbook.php?record_id=12794&page=1.

National Wind Coordinating Committee. 2010 (Spring). Wind Turbine Interactions with Birds, Bats, and Their Habitats: A Summary of Research Results and Priority Questions. https://www.nationalwind.org/assets/publications/Birds_and_Bats_Fact_Sheet_.pdf.

National Wind Coordinating Collaborative. 2002 (August). Permitting of Wind Energy Facilities. http://www.nationalwind.org/assets/publications/permitting2002.pdf.

Pfund, Nancy, and Healey, Ben. 2011 (September). What Would Jefferson Do? DBL Investors. http://www.dblinvestors.com/documents/DBL_energy_subsidies_paper.pdf.

Renewable Energy World. 2010 (November 22). Measuring Wind Turbine Noise. http://www.renewableenergyworld.com/rea/news/article/2010/11/measuring-wind-turbine-noise.

———. 2011 (August 18). Price of Wind Lower Than Gas, Hydro in Brazil Auction. http://www.renewableenergyworld.com/rea/news/article/2011/08/price-of-wind-lower-than-gas-hydro-in-brazil-auction?cmpid=WNL-Friday-August19-2011.

RenewableUK. 2007. Noise from Wind Turbines—The Facts. http://www.bwea.com/ref/noise.html.

Shere, Jeremy. 2011 (March 15). Why People Hate Wind Turbines. *Renewable Energy World*. http://www.renewableenergyworld.com/rea/blog/post/2011/03/why-people-hate-wind-turbines??cmpid=WNL-Friday-March18-2011.

Stant, Jeff. 2010 (August 26). In Harm's Way: Lack of Federal Coal Ash Regulations Endangers Americans and Their Environment. Environmental Integrity Project. http://www.environmentalintegrity.org/news_reports/documents/INHARMSWAY_FINAL.pdf.

SustainableBusiness.com News. 2011 (February 8). Wind Turbine Prices Fall to Lowest Point in Recent Years. http://www.sustainablebusiness.com/index.cfm/go/news.display/id/21839.

Trabish, Herman K. 2011 (February 21). Can Raytheon Keep Wind Farms From Threatening National Security? Greentech Media. http://www.greentechmedia.com/articles/read/can-raytheon-keep-wind-farms-from-threatening-national-security/.

———. 2011 (May 20). How Does U.S. Wind Keep Growing? Greentech Media. http://www.greentechmedia.com/articles/read/how-does-u.s.-wind-keep-growing/.

———. 2011 (July 21). What Do Wind Trends Show? Three Cents per Kilowatt-Hour. Greentech Media. http://www.greentechmedia.com/articles/read/what-do-winds-cost-price-and-performance-trends-show-three-cents-per-kilowa/.

———. 2010 (September 1). Undermining the Critics of Wind Power. Greentech Media. http://www.greentechmedia.com/articles/read/undermining-the-critics-of-wind-power/.

———. 2011 (October 6). Where Is the Best Place to Put a Wind Farm? Greentech Media. http://www.greentechmedia.com/articles/read/where-is-the-best-place-to-put-a-wind-farm/.

———. 2011 (November 17). Wind, National Defense, and Radar Make Peace. Greentech Media. http://www.greentechmedia.com/articles/read/wind-and-national-defense-make-peace/.

Turkel, Tux. 2011 (May 1). Wind Turbine Noise an Issue, Report Finds. *The Portland Press Herald*. http://www.pressherald.com/news/wind-turbine-noise-an-issue-report-finds_2011-05-01.html.

UK Energy Research Centre. 2010 (September). Great Expectations: The Cost of Offshore Wind in UK Waters—Understanding the Past and Projecting the Future. http://www.ukerc.ac.uk/support/tiki-download_file.php?fileId=1164.

U.S. Department of Energy, Energy Efficiency and Renewable Energy, Wind Powering America. 2011 (April). Wind Energy Benefits. DOE/GO/102011-3226. http://www.windpoweringamerica.gov/pdfs/wpa/2011/wind_energy_benefits.pdf.

———. 2011. Wind Turbine Sound.

U.S. Department of Energy, Energy Efficiency and Renewable Energy. 2008 (July). 20% Wind Energy by 2030: Increasing Wind Energy's Contribution to U.S. Electricity Supply. DOE/GP-102008-2567.
http://www1.eere.energy.gov/wind/pdfs/41869.pdf.
http://www.windpoweringamerica.gov/newengland/issues_sound.asp.

———. 2010. Wind Power Today—2010. http://www1.eere.energy.gov/windandhydro/pdfs/47531.pdf.

U.S. Energy Information Administration. 2011 (April). Annual Energy Outlook 2011 with Projections to 2035. DOE/EIA-0383(2011). http://www.eia.gov/forecasts/aeo/pdf/0383(2011).pdf.

U.S. Fish & Wildlife Service. 2011. Links to Worldwide Wind Energy/Wildlife Guidance Documents and Information. http://www.fws.gov/windenergy/docs/Links_Worldwide_Wind_Energy_Wildlife_Guidance_Docs_Info.pdf.

U.S. National Renewable Energy Laboratory, National Wind Technology Center. 2010 (June). Environment and Siting: Supporting Research on Impacts to Birds, Bats, and Habitats From Wind Development. NREL/FS-500-47254. http://www.nrel.gov/wind/pdfs/47254.pdf.

Urgo, Jacqueline L. 2011 (June 12). Atlantic City Wind Turbines Become a Tourist Attraction. *The Philadelphia Inquirer.* http://articles.philly.com/2011-06-12/news/29650149_1_turbines-first-commercial-wind-project-wind-energy.

Utility Wind Integration Group. 2010 (August 20). Frequently Asked Questions about Utility-Scale Wind Power. http://www.uwig.org/UWIG-WINDFAQ.pdf.

Vaira, Douglas. 2011. Win-Win for Wind and Wildlife. The Nature Conservancy. http://www.nature.org/ourinitiatives/urgentissues/conservationlands/conservation-lands-win-win-for-wind-and-wildlife.xml?s_intc=sp2.

Vautard, Robert, et al. 2010 (October 17). Northern Hemisphere Atmospheric Stilling Partly Attributed to an Increase in Surface Roughness. Nature Geoscience. http://www.nature.com/ngeo/journal/v3/n11/full/ngeo979.html.

Weiss, Daniel J., and Caperton, Richard W. 2011 (April 7). Protecting Big Oil at the Expense of Consumers. http://www.greentechmedia.com/articles/read/protecting-big-oil-at-the-expense-of-consumers/.

Wesoff, Eric. 2011 (October 4). Subsidies for Oil, Gas and Nuclear vs. Renewables. Greentech Media. http://www.greentechmedia.com/articles/read/Subsidies-For-Oil-Gas-Nuclear-vs.-Renewables/.

Western EcoSystems Technology, Inc., http://www.west-inc.com/.

Wind Energy Update. 2011 (May 2). Report: Offshore Wind Investment to Surge. http://social.windenergyupdate.com/weekly-brief/weekly-intelligence-brief-april-26-may-02.

WindFacts (Wind Energy—the Facts), http://www.wind-energy-the-facts.org/.

Wind-Wildlife Impacts Literature Database, http://www.nrel.gov/wind/wild.html.

Wiser, Ryan, and Bolinger, Mark. 2011 (August). 2010 Wind Technologies Market Report. http://www1.eere.energy.gov/windandhydro/pdfs/51783.pdf.

A

Appendix A: Wind Turbine Design Equations

The material in Appendix A was excerpted directly from the sources indicated.

Fig. A–1. Air density. *Information courtesy of the University of Oklahoma.*

Air Density

Lesson Number 2 in an Oklahoma Wind Power Tutorial Series

By Tim Hughes, Environmental Verification and Analysis Center,
The University of Oklahoma

Estimating and Calculating Air Density

The wind power density (WPD) term is directly proportional to air density:

$$\text{WPD} = \tfrac{1}{2} \times \rho \times V^3$$

where the Greek letter ρ stands for air density, which is defined by:

$$\rho = \text{M} / \text{Vol}$$

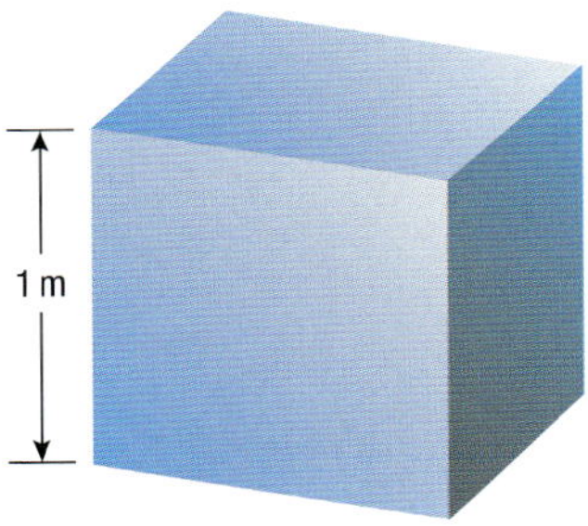

Air density can be defined as the mass of a given sample of air, divided by that sample's volume.

Example: For the cube on the left, measuring 1 meter on each side, the density of the air inside would be determined by its mass divided by 1 cubic meter (or per m^3). At sea level and under standard conditions (temperature of 25°C and pressure = 1 atmosphere), the mass of air in this cube would be 1.225 kg, so the density of the air is **1.225 kg per m^3**.

Air density can be approximated or calculated in the following ways.

Approximating air density

Method 1. If your area of study is close to sea level, and is in a region with moderate climate, using the value of 1.225 kg/m^3 will not introduce a huge error into your estimate of WPD. This is especially so if you are looking at annual average values, so that changes in air density due to seasonal changes in air temperature are averaged over time and a variety of conditions.

Method 2. However, if the area is much above sea level, or if you just want to be more precise, you can use this approximation to account for elevation change:

$$\rho = 1.225 - (1.194 \times 10^{-4}) \times z$$

where z = the location's elevation above sea level in meters.

This approximates the U.S. standard atmosphere profile for air density and will give a good long-term average value of air density in temperate areas.

Calculating air density more exactly

Method 3. If your region of interest has a less-temperate climate (i.e., it is either quite hot or quite cold most of the time), or if you are interested in seasonal estimates of wind power density, the following expression gives a much better value for air density:

$$\rho = P / RT \text{ (kg/m}^3\text{)}$$

where

P = air pressure (in units of Pascals or Newtons/m^2)
R = the specific gas constant (287 J kg^{-1} Kelvin^{-1})
T = air temperature in degrees Kelvin (°C + 273)

Method 4. While air temperature data is not too hard to come by (and is relatively cheap to measure even if you don't have data already), air pressure data can be tougher to find. If air pressure data for your region is unobtainable, you can estimate density as just a function of site elevation and temperature with the following expression:

$$\rho = (P_o / RT) \exp(-gz/RT) \text{ (kg/m}^3\text{)}$$

where

P_o = standard sea level atmospheric pressure (101,325 Pascals) [or you can use a sea-level adjusted pressure reading fronl a nearby weather station];
g = the gravitational constant (9.8 m/s^2); and
z = the region's elevation above sea level (in meters).

Fig. A–2. Energy and power. *Data courtesy of Jack Park,* The Wind Power Book *(page 18).*

Energy and Power

A clear distinction must be made between energy and power—two different but closely related quantities. Briefly, power is the rate at which energy is extracted, harnessed, converted, or consumed. It equals the amount of energy per unit time, or

$$\text{Power} = \frac{\text{Energy}}{\text{Time}} .$$

An equivalent relation between these three entities is:

$$\text{Energy} = \text{Power} \times \text{Time}.$$

The amount of energy extracted or consumed is therefore proportional to the elapsed time. For example, a typical light bulb draws 100 watts of electrical power. One watt (1 W) is the basic unit of power in the metric system. Leave the light bulb on for two hours, and it will consume 200 watt-hours (100 watts times 2 hours equals 200 watt hours, or 200 Wh). Leave it on for 10 hours and it consumes 1,000 watt-hours, or one kilowatt-hour (1 kWh), the more familiar metric unit of energy.

Rarely is it ever a simple task to estimate the wind energy available at a particular site; the windspeed is constantly changing. During one minute, 300 watts of power may be generated by a windmill, or 300 watt-minutes of energy (which equals 5 Wh). During the next minute the wind may die and you get absolutely no energy or power from the machine. The power output is constantly changing with the windspeed, and the accumulated wind energy is increasing with time. The wind energy extracted by the machine is the summation or total of all the minute-by-minute (or whatever other time interval you care to use) energy contributions. For example, if there are 30 minutes during a particular hour when the windmill is generating 5 Wh and another 30 minutes when there is no energy generated, then the machine generates 150 Wh (5 × 30 = 150) of wind energy. If there are 24 such hours a day, then 3,600 Wh or 3.6 kWh are generated that day.

Fig. A–3. Wind energy and wind power. *Data courtesy of Jack Park,* The Wind Power Book *(page 20).*

Wind Energy and Wind Power

Energy and power are derived from the wind by making use of the force it exerts on solid objects, pushing them along. Buildings designed to stand still against this force extract very little energy from the wind. But windmill blades are designed to move in response to this force, and wind machines can extract a substantial portion of the energy and power available.

The wind energy available in a unit volume (one cubic foot or one cubic meter) of air depends on air density ρ (Greek "rho") and the instantaneous windspeed V. This "kinetic energy" of the air in motion is given by the formula:

$$\frac{\text{kinetic energy}}{\text{unit volume}} = ½ \times \rho \times V^2.$$

To find the kinetic energy in a particular volume of air, you just multiply by that volume. The volume of air that passes through an imaginary surface—say the disk swept out by a horizontal-axis windmill—oriented at right angles to the wind direction is equal to:

$$\text{Volume} = A \times V \times t,$$

where t is the elapsed time (in seconds) and A is the area (in square feet or square meters) of the surface in question. Thus, the wind energy that flows through the surface during time t is just:

$$\text{Available Energy} = ½ \times \rho \times V^3 \times A \times t.$$

Wind power is the amount of energy which flows through the surface per unit time, and is calculated by dividing the wind energy by the elapsed time t. Thus, the wind power available under the same conditions as above is given by the formula:

$$\text{Available Power} = ½ \times \rho \times V^3 \times A.$$

Both energy and power are proportional to the cube of the windspeed. If all the available wind power working against a windmill rotor could be harvested by the moving blades, this formula could be used directly to calculate the power extracted.

But getting such an output would require that you stop the wind dead in its tracks and extract every last unit of its kinetic energy. This is an impossible task. Some nonzero windspeed must occur downstream of the blades to carry away the incoming air, which would otherwise pile up. Under ideal conditions, the maximum power that can be extracted from the wind is only 59.3% of the power available, or

$$\text{Maximum Power} = \frac{0.593}{2} \times \rho \times V^3 \times A.$$

In practice, a wind machine extracts substantially less power than this maximum. For example, the windmill rotor itself may capture only 70% of maximum power. Bearings will lose another few percent to friction; generators, gears, and other rotating machinery can lose half of whatever power remains. Pushrods, wires, batteries, and monitoring devices will lose still more. The overall "system" efficiency of the entire wind machine is the fraction of the wind power available that is actually delivered to a load or to a storage device:

$$\text{Efficiency} = \frac{\text{Power Delivered}}{\text{Available Power}} .$$

Thus, the power extracted by a particular wind machine with system efficiency E is given by the formula:

$$\text{Extracted Power} = \tfrac{1}{2} \times \rho \times V^3 \times A \times E .$$

The final output of a wind machine is greatly reduced from the power that is really available in the wind. In practice, values of E commonly range from 0.10 to 0.50, although higher and lower values are possible.

One more factor is needed before the formula above can be used in your calculations—a conversion factor that makes the answer come out in the appropriate power units, whether metric or English. The chart presented here gives the necessary values of this factor, K, which adjusts the calculation so that the result is expressed in horsepower, watts, or kilowatts. The final formula combines everything so far presented:

$$\text{Power} = \tfrac{1}{2} \times \rho \times V^3 \times A \times E \times K \qquad \textbf{(Eq. 1)}.$$

This is a very important formula to help you obtain the various factors needed to use this formula in your design procedure.

Wind Power Conversion Factors

For:	Values of K to get wind power in:		
	Horsepower	**Watts**	**Kilowatts**
Windspeed in mph Rotor area in ft^2	0.00578	4.31	0.00431
Windspeed in ft/sec Rotor area in ft^2	0.697	520.0	0.520
Windspeed in meters/sec Rotor area in (meters)^2	0.00183	1.36	0.00136

Note: Energy conversion losses from generators, gears, and other rotating machinery have been minimized in modern machines.

Fig. A–4. Maximum rotor efficiency. *Data courtesy of Jack Park,* The Wind Power Book *(page 72).*

Maximum Rotor Efficiency

The analysis of maximum possible efficiency for lift-type rotors was originally done by Betz in 1927. Here, the rotor extracts power from the airstream by slowing down the free-stream windspeed V to a lesser speed V_2 far downstream of the rotor blades. The power extracted is just the difference in wind energy upstream and downstream of the rotor, or

$$\text{Power} = \tfrac{1}{2} \times M \times (V^2 - V_2^2),$$

where M is the mass of air that flows through the rotor per second. If V_2 equals zero in the above equation, you might expect that power would be maximized. But no air would flow through the rotor in this case, and the power is zero. The mass flow through the rotor is just the air density times the rotor area times the average wind velocity at the rotor, or:

$$M = \rho \times A \times \frac{V + V_2}{2}.$$

Substituting this formula into the power equation yields:

$$\text{Power} = \tfrac{1}{4} \times \rho \times A \times (V + V_2) \times (V^2 - V_2^2).$$

A graph of the relative power generated versus the ratio of V_2 to V is presented here. Note that maximum power occurs when V_2 equals one-third of V. Under such conditions,

$$\text{Maximum Power} = \tfrac{1}{2} \times \rho \times A \times V^3 \times \frac{16}{27}.$$

Thus, maximum possible (theoretical) efficiency of a lift-type rotor is 16/27, or 59.3%. In reality, swirl in the downwind airstream and other inefficiencies limit the practical efficiency even more.

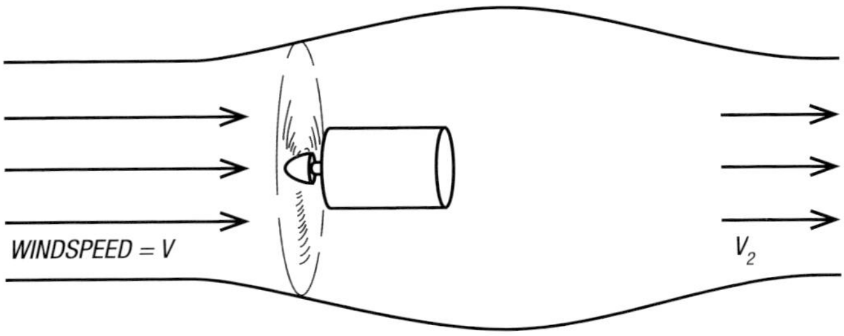

Airflow through lift-type rotor

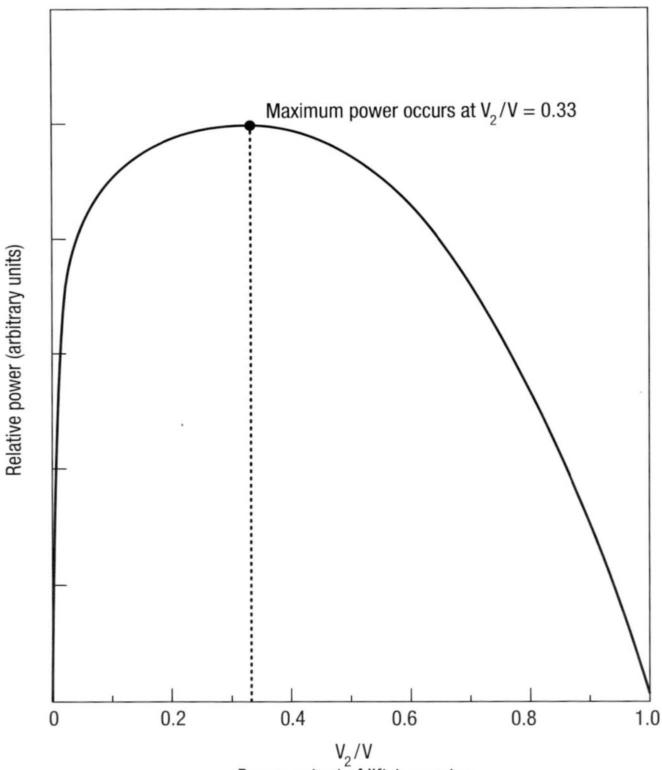

Power output of lift-type rotor

Fig. A–5. Forces on an airfoil. *Data courtesy of Jack Park,* The Wind Power Book *(page 69).*

Forces on an Airfoil

All airfoils, even flat boards tilted into the wind and used as lifting surfaces, have predictable lift and drag characteristics. Lift is the force produced on the airfoil in a direction perpendicular to the "relative wind" approaching the airfoil. This relative wind is the wind that an observer sitting on the airfoil would face. The aerodynamic lift can be calculated from the formula:

$$\text{Lift} = ½ \times \rho \times V_r^2 \times A_b \times C_L ,$$

where

ρ = the air density in lb/ft^3,
V_r = the speed of the relative wind approaching the airfoil, in ft/sec,
A_b = the surface area of the airfoil or blade, in ft^2,
C_L = the lift coefficient of the airfoil.

The drag force on the airfoil occurs in a direction parallel to the relative wind; it acts to retard the forward motion of the airfoil. Its value is calculated by replacing the lift coefficient in the above equation by the airfoil drag coefficient, C_D.

To understand airfoils in more detail, you need to grasp a few other definitions. The "chord line" of an airfoil is a line extending from its leading edge to the trailing edge.

The "angle of attack" is the angle between the chord line and the relative wind approaching the leading edge. The "pitching moment" is a measure of an airfoil's tendency to pitch its leading edge up or down in the face of the wind. It is important to the structural design of the blades and feathering mechanism. Certain airfoils are neutral; they have no pitching moment.

The graph presented on the next page gives values of the lift and drag coefficients for a particular standard airfoil shape—the FX60-126. Similar curves are available for every airfoil tested. The curves shown here give the lift coefficient C_L versus angle of attack and include a "drag polar" that shows how the drag coefficient C_D varies with the lift coefficient. Note that the maximum lift occurs when the angle of attack is 12° and that the minimum drag occurs at $C_D = 0.006$, corresponding to a lift coefficient $C_L = 0.2$.

Example: At an angle of attack equal to 4°, the FX60-126 airfoil has a lift coefficient $C_L = 0.96$. What is the lift force produced if the windspeed at the leading edge equals 40 mph and the blade area is 2 square feet?

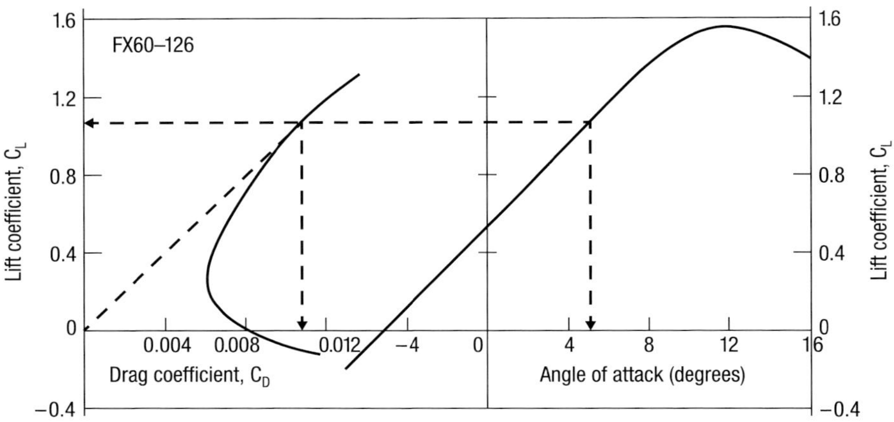

Solution: First convert 40 mph to 58.8 ft/sec by multiplying by 1.47. Then, using the above equation for the lift force,

$$\text{Lift} = 0.5 \times 0.00238 \times (58.8)^2 \times 2.0 \times 0.96 = 7.9 \text{ pounds.}$$

From the graph, $C_D = 0.0098$ when $C_L = 0.96$, so the drag force on the airfoil under the same conditions is:

$$\text{Drag} = 0.5 \times 000238 \times (58.8)^2 \times 2.0 \times 0.0098 = 0.081 \text{ pounds.}$$

By taking the ratio of the lift force to the drag force, you can calculate the lift-to-drag ratio, L/D:

$$L/D = \frac{7.9}{0.081} = 98 .$$

Of course, this is the same result you would obtain if you just took the ratio of the lift coefficient to the drag coefficient. The best airfoil performance occurs at an angle of attack where the lift-to-drag ratio is a maximum. There you get maximum lift for minimum drag, but not necessarily the absolute maximum possible lift. On the FX60-126 airfoil, note that minimum drag occurs when $C_L = 0.2$—a low value compared to the maximum possible ($C_L = 1.6$). To find the angle of attack at which L/D is maximized, simply draw a line from the origin of the drag polar curve to the point where it just touches tangent to this curve. The point of tangency corresponds to maximum L/D for the airfoil. Draw a horizontal line from the point of tangency right to where it intersects the lift coefficient curve, and you get $C/L = 1.08$ in this example. As the drag coefficient here is $C_D = 0.0108$, the lift-to-drag ratio has a maximum value of 100. Note also that the angle of attack for maximum L/D is 5.2°. Setting the blade edge at this angle of attack to the relative wind will allow the airfoil to fly at its optimum performance.

Fig. A–6. Torque. *Data courtesy of Jack Park,* The Wind Power Book *(page 107).*

Powershaft Torque

The powershaft torque, in inch-pounds, on a wind machine can be estimated from the formula:

$$\text{Torque} = 2{,}245 \times \frac{D \times HP}{V \times TSR}$$

where

D = the rotor diameter in feet,
TSR = the tip-speed ratio,
V = the windspeed in mph, and
HP = the horsepower generated by the rotor.

Remember to use the horsepower output of the rotor, not the generator or other device being driven by the powershaft. If your evaluation of power is expressed in watts instead of horsepower, use 3.01 instead of 2,245 in this formula to get a result in inch-pounds.

Problem: Suppose a 12-foot diameter windmill generates 3 horsepower at TSR = 6 in a 25 mph wind. What is the torque on the powershaft?

Solution: From the equation above,

$$\text{Torque} = 2{,}245 \times \frac{12 \times 3}{25 \times 6} = 539 \text{ inch-pounds.}$$

If this is a three-bladed propellor-type rotor, then each blade experiences a bending moment equal to one-third of 539, or 180 inch-pounds. This particular bending moment acts in the plane of rotation—along the thrust direction—and maximum stress occurs at the root of each blade

---

B

Appendix B: Small/Alternative Wind Power Systems

The material in Appendix B was excerpted directly from the sources indicated.

Fig. B–1. Drag-Type Machines. *Data courtesy of Jack Park,* The Wind Power Book *(page 68).*

Drag-Type Machines

A drag-type wind machine harnesses the component of wind force perpendicular to the surfaces of its vanes. Such a machine might be a Savonius rotor or, even more simply, a flat board nailed to the end of a swinging arm. In this case, the drag force on the vane is given by the formula:

$$\text{Drag Force} = ½ \times \rho \times (V - u)^2 \times A_v \times C_D ,$$

where

ρ = the air density in lb/ft^3,
V = the windspeed in ft/sec,
u = the vane speed in ft/sec,
A_v = the area of the vane in ft^2, and
C_D = the drag coefficient of the vane.

Generally, the drag coefficient of a vane has a value between zero and one.

If the rotor is at rest, the vane speed (u in the above equation) is zero, and maximum force occurs when the vane is perpendicular to the wind. If you multiply this maximum drag force by the radius to the center of rotation, you get the starting torque supplied by the vane. Of course, the net torque of the entire machine will be less because the wind is pushing against other vanes on the upwind side of the machine and retarding this rotation.

The power developed by a drag-type machine is just the drag force multiplied by the vane speed:

$$\text{Power} = ½ \times \rho \times (V - u)^2 \times u \times A_v \times C_D .$$

As the vane speed increases, the drag forces drop sharply (see graph), but the power extracted from the wind increases. When the vane speed equals one-third the free-stream windspeed V, maximum power extraction occurs. Of course, you still have to subtract the power wasted in driving other vanes upwind on the other side of the machine.

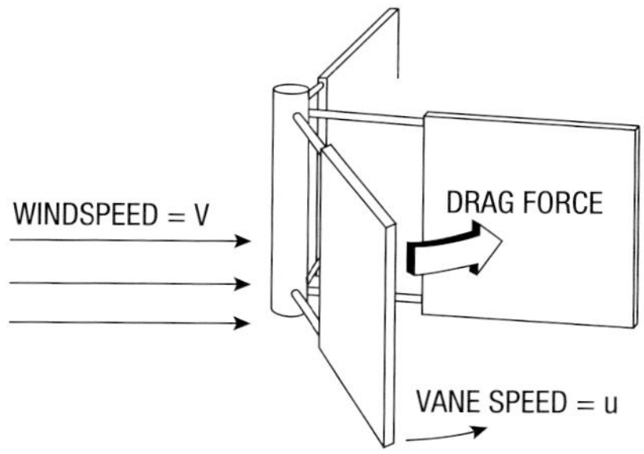

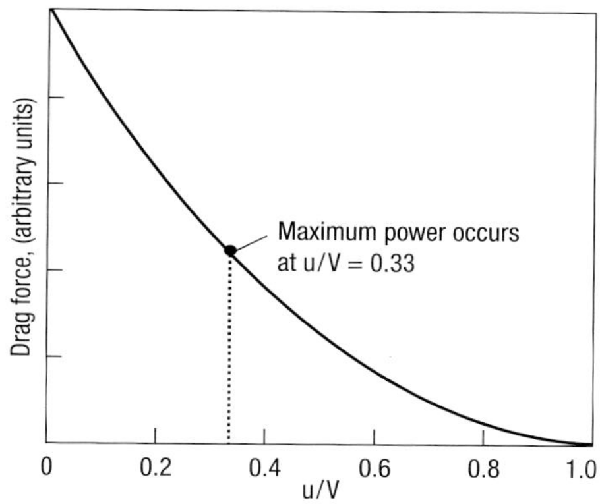

The drag coefficient C_D for a curved, two-vane Savonius rotor is about 1 for the concave, or torque, side and from 0.12 to 0.25 for the opposite, upwind-moving side. With these numbers you can easily calculate the difference in drag force between the two sides and estimate the net torque on the device. But be careful. Note that you should use $V + u$ instead of $V - u$ on the upwind vane. By a similar procedure, you can also estimate the net power developed by a Savonius.

Fig. B–2 (a). Sizing a Small Wind-Power Rotor. *Data courtesy of Jack Park,* The Wind Power Book *(page 88).*

Sizing a Wind Rotor

There are two principal ways to determine the frontal area of a wind machine rotor. You can merely guess how large a machine you want, calculate the power it produces, and stop there. Or you can first determine your average power needs and the wind resources at your site, and then equate the two to determine the rotor area. The first method is the one most often used. The second is more complex but results in a much closer match between your power needs and the wind power available.

Suppose you know in advance your average power needs—denoted here by the letter *P*. Equation 1 (page 370 in fig. A–3) tells you that this power, if supplied by a wind machine, depends on the windspeed *V*, the rotor area *A*, the air density ρ, and the system efficiency *E*:

$$P = \tfrac{1}{2} \times \rho \times V^3 \times A \times E.$$

This formula can be rewritten to express the rotor area *A* in terms of five factors:

$$A = \frac{P}{E \times F \times C_A \times C_T},$$

where *F* is a factor that depends on windspeed and is presented in the first table here, and C_A and C_T are the altitude and temperature correction factors for air density that are listed in the next two tables. The equation above gives you the area in square feet when the power *P* is expressed in watts; if *P* is in horsepower, multiply *A* by 0.737.

Power Factor *F*

V	*F*
6	1.07
7	1.76
8	2.62
9	3.74
10	5.13
11	6.82
12	8.86
13	11.26
14	14.07
15	17.30
16	21.00
17	25.19
18	29.90
19	35.17
20	41.02
21	47.48
22	54.59
23	62.38
24	70.88
25	80.11
26	90.12
27	100.92
28	112.55
29	125.05
30	138.43

Altitude Correction Factor

Altitude, ft	C_A
0	1.000
2,500	0.912
5,000	0.832
7,500	0.756
10,000	0.687

Temperature Correction Factor

Temperature, °F	C_T
0	1.103
20	1.083
40	1.040
60	1.000
80	0.929
100	0.929

If you are purchasing a factory-built machine and know its system efficiency, this formula can tell you whether its frontal area is suited to your power needs. If you intend to design and build your own machine, you need an estimate of the efficiency before you can begin. Use the last table here to get a rapid but rough estimate. Then get values of F, C_A, and C_T for your site from the appropriate tables. The rest is calculation.

Rapid Efficiency Estimator

Wind System	Efficiency, %	
	Simple Construction	Optimum Design
Multibladed farm water pumper	10	30
Sailwing water pumper	10	25
Darrieus water pumper	15	30
Savonius windcharger	10	20
Small prop-type windcharger (up to 2 kW)	20	30
Medium prop-type windcharger (2 to 10 kW)	20	30
Large prop-type wind generator (over 10 kW)	–	30 to 45
Darrieus wind generator 15 35		

Example: You have chosen a three-bladed propellor-type machine to produce 1000 watts. A site survey shows that the energy content of the winds at your site, which is at sea level, peaks at 15 mph. What size rotor is needed?

Solution: Begin by estimating the system efficiency. For small propellor-type systems, you can expect an efficiency from 15% to 30%. You elect to use 25% ($E = 0.25$) for a carefully designed machine. From the first table $F = 17.30$ at 15 mph, and $C_A = C_T = 1$ at sea level, for standard temperature

(60°F). So,

$$A = \frac{1{,}000}{0.25 \times 17.30 \times 1 \times 1} = 231 \text{ ft}^2 .$$

You usually need to know the diameter of the rotor that can do the job. In this case, a diameter of 17.3 feet is needed. Refer to "Rotor Swept Area" on the following page for more detailed information needed to convert rotor frontal area into linear dimensions of the rotor vanes or blades.

Fig. B–2 (b). Sizing a Small Wind-Power Rotor. *Data courtesy of Jack Park,* The Wind Power Book *(page 216).*

Rotor Swept Area

The rotor swept area, or frontal area, is a parameter frequently used in wind power calculations. This area, denoted by the letter A, is the total surface area perpendicular to the wind direction that is swept by the rotor blades. It is measured in square feet or square meters.

The second illustration on the next page indicates the swept area for a propellor-type rotor. This category includes all rotors with a horizontal axis parallel to the windstream; they sweep out an area perpendicular to the wind direction equal to:

$$A = \frac{\pi}{4} \times D^2 = 0.785 \times D^2 ,$$

where $\pi = 3.14159$, and D = diameter is measured from blade tip to blade tip as shown. The first graph on the next page will help you convert diameter measurements into rotor swept area, or vice versa.

For vertical-axis, cross-wind machines similar to the second illustration (i.e., those with a uniform radius about the axis of rotation), the swept area is:

$$A = \text{Height} \times \text{Width}$$

This is the formula to use with Savonius and straight-bladed Darrieus rotors. In an eggbeater-style Darrieus, the blades assume the shape of a *troposkein*—a complex mathematical curve involving elliptic integrals. Fortunately, this shape can be approximated fairly well with a parabola, and the swept area is about equal to:

$$A = 2.67 \times \text{Radius} \times \text{Half-Height}$$

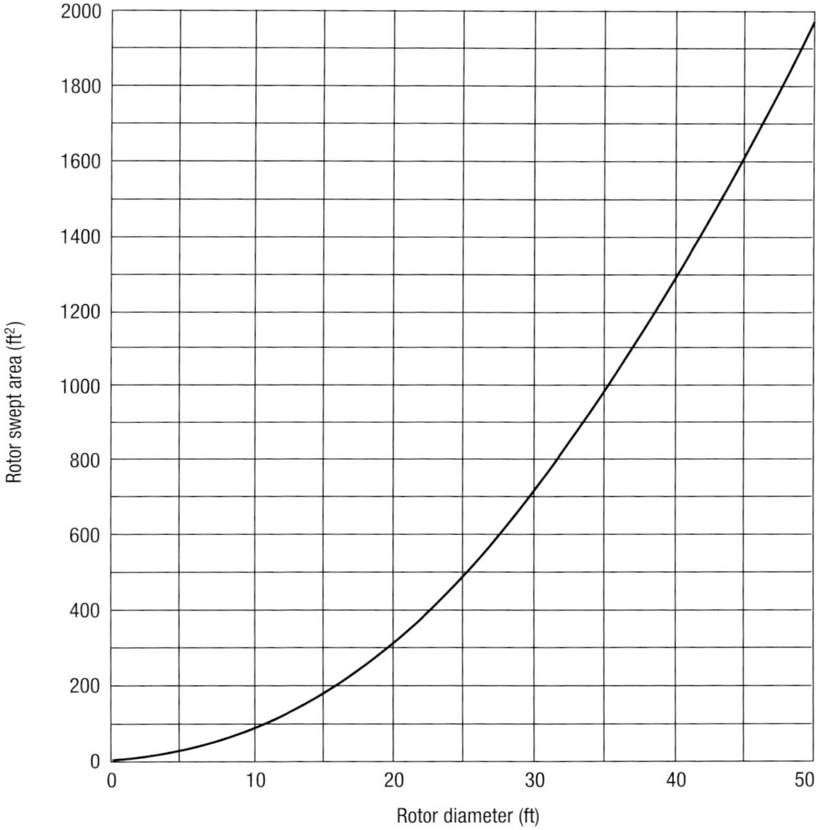

The radius and half-height of a typical eggbeater Darrieus are indicated in the third illustration.

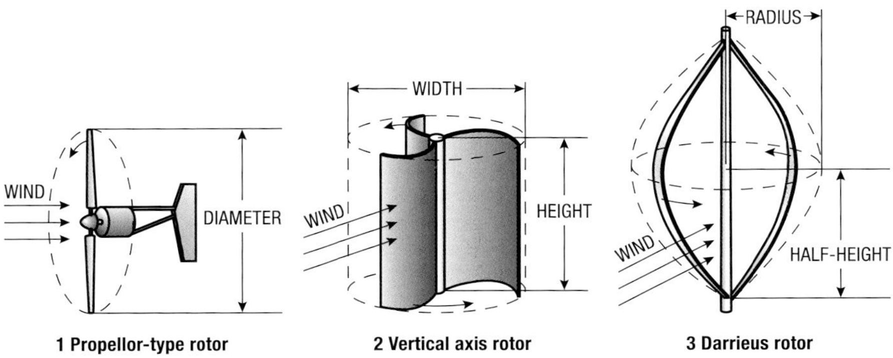

Fig. B–3. Small Wind Power System Costs. *Data courtesy of Jack Park,* The Wind Power Book *(page 150).*

Wind Energy Costs

A complete economic analysis of wind energy versus other energy options requires an estimate of all costs incurred over the life of the wind system. You also need an estimate of the energy it produces per year. The annual cost of that energy can then be compared with the cost of conventional supplies.

To perform such a life-cycle estimate of your wind system costs, you need to know the following:

- Purchase price of equipment, in dollars
- Installation costs, in dollars
- Annual maintenance costs, in dollars
- Annual insurance costs, in dollars
- Other annual costs, in dollars
- Resale value of equipment, in dollars
- Annual interest rate paid, in percent
- Expected system lifetime, in years
- Annual energy yield, in kilowatt-hours

If you don't know all these quantities exactly, try to estimate them as best as possible. System lifetime, for example, is anybody's guess.

Start your analysis by adding up the purchase price and installation costs. Then multiply the sum of all the annual costs by the expected lifetime of the system, and add this product to the total above. Finally, subtract the estimated resale value of the equipment. Exclusive of financing costs, this result is the total cost of your system over its expected lifetime. To get the average annual costs, you divide this result by the expected lifetime and multiply by a factor that includes the costs of borrowing money. An example will help to illustrate.

Problem: Suppose a wind system can be purchased off-the-shelf for $4,000, including wind generator, tower, batteries, wiring, and controls. Suppose that installation of the system costs another $1,000. Based upon discussions with other owners of that model and with the manufacturer, you can expect a system lifetime of 20 years, with annual costs for insurance and maintenance averaging $200 per year. You estimate a resale value for all the equipment to be only $500 at the end of this 20-year period. If the system is projected to yield 3,000 kWh per year at your site, what are you paying per kWh for the energy it produces?

Solution: First add up all the costs and subtract the resale value:

Purchase price	$4,000
Installation cost	1,000
Maintenance & insurance	4,000
	$9,000
Resale value	-500
	$8,500

To get the average annual costs of the system, you should include interest paid as an added expense. Using simple interest at 10% per annum, the average annual costs of this system are:

$$1.10 \times \frac{\$8,500}{20} = \$467.50 \text{ per year.}$$

When divided by the energy produced per year, this number gives you the unit cost of the wind energy delivered by your system:

$$\frac{\$467.50}{3,000 \text{ kWh}} = \$0.16 \text{ per kWh.}$$

This figure may seem high, but wait a minute. If you take the federal tax credit on the capital costs ($5,000) of the installation, you save 40%, or $2,000. This credit reduces your average annual costs to $357.50 and the unit cost to $0.12 per kWh. (State tax credits may reduce this cost even further.)

Note: In the United States, the federal personal tax credit covers 30% of the project cost with no maximum for wind power systems installed by December 31, 2016. Other incentives might also apply. For updates, check:

- DSIREUSA, http://www.dsireusa.org
- U.S. Department of Energy, Energy Star Program, http://www.energystar.gov/index.cfm?c=tax_credits.tx_index
- U.S. DOE-Tax Breaks, http://www.energy.gov/taxbreaks.htm

C

Appendix C: U.S. Offshore Wind Resource

Table C-1 shows the potential U.S. offshore wind resource by wind speed, water depth, and distance from shore for each of the 26 U.S. coastal states (oceans and Great Lakes). For maps and additional details, see the U.S. National Renewable Energy Laboratory (NREL) report, Assessment of Offshore Wind Energy Resources for the United States (http://www.nrel.gov/docs/fy10osti/45889.pdf). *The material in Appendix C was excerpted directly from the sources indicated.*

Table C–1. Detailed U.S. offshore wind resource by state. Data courtesy of NREL.

Distance from Shoreline		0–3 nm[1]			3–12 nm			12–50 nm			
Depth Category (m)		0–30	30–60	> 60	0–30	30–60	> 60	0–30	30–60	> 60	Total
State	Wind Speed at 90m m/s	Area (km^2) *(MW)*	Area (km^2) *(MW)*	Area (km^2) *(MW)*	Area (km^2) *(MW)*	Area (km^2) *(MW)*	Area (km^2) *(MW)*	Area (km^2) *(MW)*	Area (km^2) *(MW)*	Area (km^2) *(MW)*	Area (km^2) *(MW)*
California	7.0–7.5	266.2 *(1,331)*	236.2 *(1,181)*	257.4 *(1,287)*	100.9 *(504)*	456.8 *(2,284)*	4,554.0 *(22,770)*	7.7 *(38)*	22.9 *(115)*	5,536 *(27,684)*	11,438.9 *(57,195)*
	7.5–8.0	239.1 *(1,196)*	256.9 *(1,285)*	189.6 *(948)*	78.8 *(394)*	595.7 *(2,978)*	3,854.6 *(19,273)*	0.0 *(0)*	32.9 *(165)*	19,616.1 *(98,080)*	24.863.7 *(124,318)*
	8.0–8.5	125.2 *(626)*	178.2 *(891)*	281.8 *(1,409)*	7.1 *(36)*	105.8 *(529)*	4,539.1 *(22,695)*	0.0 *(0)*	0.0 *(0)*	17,822.2 *(89,111)*	23,059.3 *(115,296)*
	8.5–9.0	43.2 *(216)*	141.7 *(708)*	176.4 *(882)*	0.6 *(3)*	38.0 *(190)*	4,559.8 *(22,799)*	0.0 *(0)*	0.0 *(0)*	17,892.0 *(89,460)*	22,851.6 *(114,258)*
	9.0–9.5	2.1 *(10)*	18.8 *(94)*	14.9 *(74)*	0.0 *(0)*	0.9 *(4)*	988.0 *(4,940)*	0.0 *(0)*	0.0 *(0)*	12,160.2 *(60,801)*	13,184.8 *(65,924)*
	9.5–10	0.0 *(0)*	6.0 *(30)*	13.9 *(69)*	0.0 *(0)*	0.0 *(0)*	656.1 *(3,280)*	0.0 *(0)*	0.0 *(0)*	14,554.8 *(72,774)*	15,230.7 *(76,153)*
	>10	0.0 *(0)*	0.0 *(0)*	0.1 *(1)*	0.0 *(0)*	0.0 *(0)*	288.1 *(1,441)*	0.0 *(0)*	0.0 *(0)*	6,637.6 *(33,188)*	6,925.8 *(34,629)*
Connecticut	7.0–7.5	500.1 *(2,501)*	30.2 *(151)*	0.0 *(0)*	0.0 *(0)*	0.0 *(0)*	0.0 *(0)*	0.0 *(0)*	0.0 *(0)*	0.0 *(0)*	530.4 *(2,652)*
	7.5–8.0	617.3 *(3,087)*	83.0 *(415)*	1.2 *(6)*	0.0 *(0)*	0.0 *(0)*	0.0 *(0)*	0.0 *(0)*	0.0 *(0)*	0.0 *(0)*	701.6 *(3,508)*
	8.0–8.5	34.6 *(173)*	5.0 *(25)*	0.4 *(2)*	0.0 *(0)*	0.0 *(0)*	0.0 *(0)*	0.0 *(0)*	0.0 *(0)*	0.0 *(0)*	40.1 *(201)*
Delaware	7.0–7.5	223.2 *(1,116)*	0.0 *(0)*	0.0 *(0)*	0.0 *(0)*	0.0 *(0)*	0.0 *(0)*	0.0 *(0)*	0.0 *(0)*	0.0 *(0)*	223.2 *(1,116)*
	7.5–8.0	716.5 *(3,583)*	2.0 *(10)*	0.0 *(0)*	5.2 *(26)*	0.0 *(0)*	0.0 *(0)*	0.0 *(0)*	0.0 *(0)*	0.0 *(0)*	723.7 *(3,618)*

	8.0–8.5	135.4 *(677)*	10.5 *(53)*	0.0 *(0)*	658.0 *(3,290)*	8.6 *(43)*	0.0 *(0)*	240.4 *(1,202)*	8.9 *(44)*	0.0 *(0)*	1,061.9 *(5,310)*
	8.5–9.0	0.0 *(0)*	0.0 *(0)*	0.0 *(0)*	0.0 *(0)*	0.0 *(0)*	0.0 *(0)*	254.0 *(1,270)*	677.4 *(3,387)*	0.0 *(0)*	931.4 *(4,657)*
Georgia	7.0–7.5	547.5 *(2,737)*	0.0 *(0)*	0.0 *(0)*	2,162.3 *(10,811)*	0.0 *(0)*	0.0 *(0)*	1,110.6 *(5,553)*	0.0 *(0)*	0.0 *(0)*	3,820.4 *(19,102)*
	7.5–8.0	85.2 *(426)*	0.0 *(0)*	0.0 *(0)*	529.6 *(2,648)*	0.0 *(0)*	0.0 *(0)*	5,204.3 *(26,021)*	1,922. *(9,610)*	0.0 *(0)*	7,741.2 *(38,706)*
	8.0–8.5	0.0 *(0)*	0.0 *(0)*	0.0 *(0)*	0.0 *(0)*	0.0 *(0)*	0.0 *(0)*	3.9 *(19)*	519.5 *(2,598)*	0.0 *(0)*	523.4 *(2,617)*
Hawaii	7.0–7.5	111.5 *(557)*	97.2 *(486)*	2,631.5 *(13,157)*	0.0 *(0)*	42.6 *(213)*	2,213.3 *(11,067)*	44.4 *(222)*	116.8 *(584)*	13,615.3 *(68,077)*	18,872.6 *(94,363)*
	7.5–8.0	65.5 *(328)*	107.0 *(535)*	2,404.0 *(12,020)*	0.0 *(0)*	145.0 *(725)*	5,051.6 *(25,258)*	6.9 *(34)*	250.7 *(1,254)*	34,267.6 *(171,338)*	42,298.3 *(211,492)*
	8.0–8.5	92.2 *(461)*	114.8 *(574)*	2,364.3 *(11,822)*	0.0 *(0)*	15.0 *(75)*	4,755.7 *(23,778)*	0.2 *(1)*	0.3 *(1)*	25,699.5 *(128,498)*	33,042.0 *(165,210)*
	8.5–9.0	65.6 *(328)*	67.0 *(335)*	2,105.2 *(10,526)*	0.0 *(0)*	0.0 *(0)*	2,625.5 *(13,128)*	0.0 *(0)*	0.0 *(0)*	9,050.0 *(45,250)*	13,913.3 *(69,567)*
	9.0–9.5	25.7 *(129)*	39.0 *(195)*	995.6 *(4,978)*	0.0 *(0)*	0.0 *(0)*	1,853.0 *(9,265)*	0.0 *(0)*	0.0 *(0)*	4,865.5 *(24,328)*	7,778.8 *(38,894)*
	9.5–10	22.0 *(110)*	39.8 *(199)*	666.0 *(3,330)*	0.0 *(0)*	0.0 *(0)*	1,064.9 *(5,324)*	0.0 *(0)*	0.0 *(0)*	4,926.8 *(24,634)*	6,719.6 *(33,598)*
	>10	26.3 *(132)*	70.5 *(353)*	1,344.0 *(6,720)*	0.0 *(0)*	0.0 *(0)*	2,027.6 *(10,138)*	0.0 *(0)*	0.0 *(0)*	1,383.5 *(6,918)*	4,852.0 *(24,260)*
Illinois	7.0–7.5	91.6 *(458)*	0.0 *(0)*	0.0 *(0)*	0.0 *(0)*	0.0 *(0)*	0.0 *(0)*	0.0 *(0)*	0.0 *(0)*	0.0 *(0)*	91.6 *(458)*
	7.5–8.0	164.6 *(823)*	0.0 *(0)*	0.0 *(0)*	0.0 *(0)*	0.1 *(1)*	0.0 *(0)*	0.0 *(0)*	1.3 *(6)*	0.0 *(0)*	166.0 *(830)*
	8.0–8.5	244.2 *(1,221)*	0.1 *(5)*	0.0 *(0)*	830.4 *(4,152)*	512.2 *(2,561)*	140.7 *(704)*	7.1 *(35)*	426.1 *(2,130)*	1,682.8 *(8,414)*	3,844.5 *(19,222)*

Distance from Shoreline		0–3 nm[1]			3–12 nm			12–50 nm			
Depth Category (m)		0–30	30–60	> 60	0–30	30–60	> 60	0–30	30–60	> 60	Total
State	Wind Speed at 90m m/s	Area (km^2) *(MW)*	Area (km^2) *(MW)*	Area (km^2) *(MW)*	Area (km^2) *(MW)*	Area (km^2) *(MW)*	Area (km^2) *(MW)*	Area (km^2) *(MW)*	Area (km^2) *(MW)*	Area (km^2) *(MW)*	Area (km^2) *(MW)*
Illinois (cont.)	8.5–9.0	0.0 *(0)*	0.0 *(0)*	0.0 *(0)*	13.4 *(67)*	76.6 *(383)*	0.0 *(0)*	0.0 *(0)*	0.0 *(0)*	0.0 *(0)*	89.9 *(450)*
Indiana	7.0–7.5	82.1 *(410)*	0.0 *(0)*	0.0 *(0)*	0.0 *(0)*	0.0 *(0)*	0.0 *(0)*	0.0 *(0)*	0.0 *(0)*	0.0 *(0)*	82.1 *(410)*
	7.5–8.0	153.8 *(769)*	0.0 *(0)*	0.0 *(0)*	62.6 *(313)*	0.0 *(0)*	0.0 *(0)*	0.0 *(0)*	0.0 *(0)*	0.0 *(0)*	216.3 *(1,082)*
	8.0–8.5	101.5 *(507)*	0.0 *(0)*	0.0 *(0)*	184.0 *(920)*	0.0 *(0)*	0.0 *(0)*	0.0 *(0)*	0.0 *(0)*	0.0 *(0)*	285.5 *(1,428)*
Louisiana	7.0–7.5	7,759.6 *(38,798)*	28.6 *(143)*	0.0 *(0)*	7,825.2 *(39,126)*	643.4 *(3,217)*	1,460.4 *(7,302)*	11,163.9 *(55,819)*	5,479.1 *(27,396)*	13,682.4 *(68,412)*	48,042.7 *(240,214)*
	7.5–8.0	155.3 *(776)*	0.0 *(0)*	0.0 *(0)*	1,624.5 *(8,123)*	0.0 *(0)*	0.0 *(0)*	8,169.9 *(40,850)*	2,228.7 *(11,143)*	2,854.1 *(14,271)*	15,032.5 *(75,162)*
Maine	7.0–7.5	787.0 *(2,935)*	91.2 *(456)*	11.9 *(59)*	7.8 *(39)*	4.8 *(24)*	3.5 *(18)*	0.0 *(0)*	0.0 *(0)*	0.0 *(0)*	906.2 *(4,531)*
	7.5–8.0	797.2 *(3,986)*	285.4 *(1,427)*	19.4 *(97)*	6.7 *(33)*	19.6 *(96)*	14.1 *(70)*	0.0 *(0)*	0.0 *(0)*	0.0 *(0)*	1,142.3 *(5,711)*
	8.0–8.5	777.0 *(3,885)*	440.8 *(2,204)*	74.2 *(371)*	63.4 *(317)*	385.6 *(1,928)*	234.5 *(1,173)*	0.0 *(0)*	0.0 *(0)*	0.0 *(0)*	1,975.6 *(9,878)*
	8.5–9.0	513.4 *(2,567)*	614.0 *(3,070)*	157.6 *(788)*	18.2 *(91)*	219.1 *(1,095)*	1,401.9 *(7,010)*	0.0 *(0)*	0.0 *(0)*	406.8 *(2,034)*	3,331.1 *(16,655)*
	9.0–9.5	142.2 *(711)*	390.0 *(1,950)*	309.2 *(1,546)*	25.9 *(129)*	469.0 *(2,245)*	3,504.1 *(17,520)*	0.0 *(0)*	57.8 *(289)*	3,530.9 *(17,655)*	8,429.2 *(42,146)*
	9.5–10	5.5 *(28)*	24.9 *(124)*	42.3 *(211)*	1.0 *(5)*	38.3 *(191)*	1,459.8 *(7,299)*	0.0 *(0)*	7.4 *(37)*	13,905.6 *(69,528)*	15,484.7 *(77,424)*
	>10	0.0 *(0)*	0.0 *(0)*	0.0 *(0)*	0.0 *(0)*	0.0 *(0)*	41.6 *(208)*	0.0 *(0)*	0.0 *(0)*	0.0 *(0)*	41.6 *(208)*

Maryland	7.0–7.5	2,175.4 *(10,877)*	16.6 *(83)*	0.0 *(0)*	0.0 *(0)*	0.0 *(0)*	0.0 *(0)*	0.0 *(0)*	0.0 *(0)*	0.0 *(0)*	2,192.0 *(10,960)*
	7.5–8.0	1,922.5 *(9,613)*	14.0 *(70)*	0.0 *(0)*	9.9 *(49)*	0.0 *(0)*	0.0 *(0)*	0.0 *(0)*	0.0 *(0)*	0.0 *(0)*	1,946.4 *(9,732)*
	8.0–8.5	163.3 *(817)*	0.0 *(0)*	0.0 *(0)*	924.0 *(4,620)*	0.0 *(0)*	0.0 *(0)*	435.4 *(2,177)*	17.2 *(16,053)*	0.0 *(0)*	1,539.9 *(7,700)*
	8.5–9.0	0.0 *(0)*	0.0 *(0)*	0.0 *(0)*	0.0 *(0)*	0.0 *(0)*	0.0 *(0)*	531.3 *(2,656)*	3,210.7 *(16,053)*	1,336.1 *(6,681)*	5,078.1 *(25,390)*
Massachusetts	7.0–7.5	201.6 *(1,008)*	0.0 *(0)*	0.0 *(0)*	0.0 *(0)*	0.0 *(0)*	0.0 *(0)*	0.0 *(0)*	0.0 *(0)*	0.0 *(0)*	201.6 *(1,008)*
	7.5–8.0	521.4 *(2,607)*	4.7 *(23)*	0.0 *(0)*	0.0 *(0)*	0.0 *(0)*	0.0 *(0)*	0.0 *(0)*	0.0 *(0)*	0.0 *(0)*	526.1 *(2,631)*
	8.0–8.5	927.4 *(4,637)*	327.3 *(1,636)*	28.6 *(143)*	78.2 *(391)*	152.0 *(760)*	125.5 *(628)*	0.0 *(0)*	0.0 *(0)*	0.0 *(0)*	1,639.1 *(8,195)*
	8.5–9.0	1,508.2 *(7,541)*	378.1 *(1,890)*	12.6 *(63)*	315.0 *(1,575)*	354.5 *(1,773)*	812.2 *(4,061)*	11.4 *(57)*	23.5 *(118)*	190.4 *(952)*	3,606.0 *(18,030)*
	9.0–9.5	1,137.0 *(5,685)*	322.6 *(1,613)*	20.0 *(100)*	2,696.9 *(13,484)*	1,418.6 *(7,093)*	1,006.5 *(5,033)*	1,689.9 *(8,449)*	5,051.8 *(25,259)*	7,007.4 *(35,037)*	20,350.7 *(101,753)*
	9.5–10	2.0 *(10)*	0.0 *(0)*	0.0 *(0)*	8.6 *(43)*	119.2 *(596)*	0.0 *(0)*	472.1 *(2,361)*	3,459.5 *(17,298)*	9,612.5 *(48,063)*	13,674.0 *(68,370)*
Michigan	7.0–7.5	3,411.4 *(17,057)*	330.6 *(1,653)*	243.7 *(1,218)*	142.6 *(713)*	49.8 *(249)*	268.0 *(1,340)*	1.2 *(6)*	0.0 *(0)*	12.1 *(61)*	4,459.3 *(22,297)*
	7.5–8.0	4,632.7 *(23,163)*	1,258.2 *(6,291)*	635.3 *(3,177)*	3,069.0 *(15,345)*	2,916.4 *(13,582)*	2,991.0 *(14,955)*	14.2 *(71)*	108.5 *(542)*	2,448.4 *(12,242)*	18,073.7 *(90,368)*
	8.0–8.5	2,588.1 *(12,940)*	1,253. *(6,267)*	769.0 *(3,845)*	2,700.6 *(13,853)*	5,889.7 *(29,448)*	9,411.9 *(47,059)*	30.9 *(155)*	876.1 *(4,381)*	7,496.4 *(37,482)*	31,085.9 *(155,430)*
	8.5–9.0	138.7 *(693)*	156.2 *(781)*	396.5 *(1,983)*	145.1 *(726)*	550.6 *(2,753)*	8,580.8 *(42,904)*	118.5 *(592)*	1,383.9 *(6,920)*	22,834.5 *(114,172)*	34,304.7 *(171,524)*

Distance from Shoreline		0–3 nm[1]			3–12 nm			12–50 nm			
Depth Category (m)		0–30	30–60	> 60	0–30	30–60	> 60	0–30	30–60	> 60	Total
State	Wind Speed at 90m m/s	Area (km^2) *(MW)*	Area (km^2) *(MW)*	Area (km^2) *(MW)*	Area (km^2) *(MW)*	Area (km^2) *(MW)*	Area (km^2) *(MW)*	Area (km^2) *(MW)*	Area (km^2) *(MW)*	Area (km^2) *(MW)*	Area (km^2) *(MW)*
Michigan (cont.)	9.0–9.5	0.0 *(0)*	0.0 *(0)*	0.0 *(0)*	0.0 *(0)*	0.0 *(0)*	69.9 *(350)*	114.6 *(573)*	249.6 *(1,248)*	8,284.5 *(41,423)*	8,718.7 *(43,593)*
Minnesota	7.0–7.5	5.1 *(25)*	43.6 *(218)*	272.6 *(1,363)*	0.0 *(0)*	31.5 *(158)*	1,718.1 *(8,590)*	0.0 *(0)*	0.0 *(0)*	1,031.1 *(5,155)*	3,101.9 *(15,510)*
	7.5–8.0	0.0 *(0)*	0.1 *(0)*	12.7 *(64)*	0.0 *(0)*	0.0 *(0)*	60.8 *(304)*	0.0 *(0)*	0.0 *(0)*	920.7 *(4,603)*	994.3 *(4,972)*
New Hampshire	7.0–7.5	18.6 *(93)*	0.0 *(0)*	0.0 *(0)*	0.0 *(0)*	0.0 *(0)*	0.0 *(0)*	0.0 *(0)*	0.0 *(0)*	0.0 *(0)*	18.6 *(93)*
	7.5–8.0	45.8 *(229)*	0.0 *(0)*	0.0 *(0)*	0.0 *(0)*	0.0 *(0)*	0.0 *(0)*	0.0 *(0)*	0.0 *(0)*	0.0 *(0)*	45.8 *(229)*
	8.0–8.5	44.6 *(223)*	29.7 *(148)*	0.0 *(0)*	6.9 *(34)*	75.5 *(378)*	14.0 *(70)*	0.0 *(0)*	0.0 *(0)*	0.0 *(0)*	170.6 *(853)*
	8.5–9.0	0.0 *(0)*	8.0 *(40)*	7.2 *(36)*	0.0 *(0)*	12.4 *(62)*	255.7 *(1,279)*	0.0 *(0)*	10.1 *(51)*	42.2 *(211)*	335.7 *(1,678)*
	9.0–9.5	0.0 *(0)*	0.0 *(0)*	0.0 *(0)*	0.0 *(0)*	0.0 *(0)*	0.0 *(0)*	0.0 *(0)*	35.2 *(176)*	66.4 *(332)*	101.6 *(508)*
New Jersey	7.0–7.5	520.2 *(2,601)*	0.0 *(0)*	0.0 *(0)*	8.2 *(41)*	0.0 *(0)*	0.0 *(0)*	0.0 *(0)*	0.0 *(0)*	0.0 *(0)*	528.5 *(2,642)*
	7.5–8.0	960.4 *(4,802)*	0.0 *(0)*	0.0 *(0)*	518.6 *(2,593)*	27.9 *(140)*	0.7 *(4)*	0.0 *(0)*	0.0 *(0)*	0.0 *(0)*	1,507.6 *(7,538)*
	8.0–8.5	776.9 *(3,885)*	0.0 *(0)*	0.0 *(0)*	2,764.8 *(13,824)*	127.1 *(636)*	3.2 *(16)*	899.4 *(4,497)*	372.4 *(1,862)*	21.6 *(108)*	4,965.4 *(24,827)*
	8.5–9.0	13.1 *(66)*	0.0 *(0)*	0.0 *(0)*	214.7 *(1,074)*	0.0 *(0)*	0.0 *(0)*	2,019.5 *(10,098)*	10,382.4 *(51,912)*	303.8 *(1,519)*	12,933.5 *(64,668)*
New York	7.0–7.5	1,047.9 *(5,239)*	54.2 *(271)*	0.0 *(0)*	2.8 *(14)*	0.0 *(0)*	0.0 *(0)*	0.0 *(0)*	0.0 *(0)*	0.0 *(0)*	1,104.8 *(5,524)*

	7.5–8.0	1,983.1 *(9,915)*	354.7 *(1,774)*	11.6 *(58)*	472.7 *(2,363)*	609.0 *(3,045)*	866.8 *(4,334)*	0.0 *(0)*	0.0 *(0)*	59.9 *(300)*	4,357.8 *(21,789)*
	8.0–8.5	1,329.3 *(6,646)*	602.9 *(3,015)*	196.5 *(982)*	375.6 *(1,878)*	535.7 *(2,679)*	3,122.0 *(15,610)*	42.2 *(211)*	12.6 *(63)*	2,107.4 *(10,537)*	8,324.1 *(41,621)*
	8.5–9.0	535.1 *(2,675)*	110.6 *(553)*	0.6 *(3)*	1,042.6 *(5,213)*	93.8 *(469)*	0.0 *(0)*	261.5 *(1,307)*	727.8 *(3,639)*	104.5 *(522)*	2,876.4 *(14,382)*
	9.0–9.5	5.4 *(27)*	0.0 *(0)*	0.0 *(0)*	391.9 *(1,959)*	1,407.6 *(7,038)*	0.0 *(0)*	6.4 *(32)*	4,926.1 *(24,631)*	716.0 *(3,580)*	7,453.3 *(37,267)*
	9.5–10	0.0 *(0)*	0.0 *(0)*	0.0 *(0)*	0.0 *(0)*	0.0 *(0)*	0.0 *(0)*	0.0 *(0)*	1,817.6 *(9,088)*	3,504.8 *(17,524)*	5,322.4 *(26,612)*
North Carolina	7.0–7.5	1,847.4 *(9,237)*	0.0 *(0)*	0.0 *(0)*	0.0 *(0)*	0.0 *(0)*	0.0 *(0)*	0.0 *(0)*	0.0 *(0)*	0.0 *(0)*	1,847.4 *(9,237)*
	7.5–8.0	3,002.1 *(15,010)*	0.0 *(0)*	0.0 *(0)*	1,096.1 *(5,481)*	0.0 *(0)*	0.0 *(0)*	0.0 *(0)*	0.0 *(0)*	0.0 *(0)*	4,098.3 *(20,491)*
	8.0–8.5	3,920.0 *(19,600)*	0.1 *(1)*	0.1 *(1)*	4,653.8 *(23,269)*	0.0 *(0)*	0.0 *(0)*	4,034.9 *(20,174)*	746.5 *(3,733)*	299.2 *(1,496)*	13,654.7 *(68,274)*
	8.5–9.0	105.0 *(525)*	0.0 *(0)*	0.0 *(0)*	2,964.1 *(14,821)*	261.4 *(1,307)*	8.9 *(45)*	6,382.4 *(31,912)*	14,519.6 *(72,598)*	15,633.3 *(78,167)*	39,874.8 *(199,374)*
	9.0–9.5	0.0 *(0)*	0.0 *(0)*	0.0 *(0)*	0.0 *(0)*	0.0 *(0)*	0.0 *(0)*	0.0 *(0)*	16.0 *(80)*	0.0 *(0)*	16.0 *(80)*
Ohio	7.0–7.5	341.0 *(1,705)*	0.0 *(0)*	0.0 *(0)*	0.1 *(1)*	0.0 *(0)*	0.0 *(0)*	0.0 *(0)*	0.0 *(0)*	0.0 *(0)*	341.1 *(1,705)*
	7.5–8.0	1,106.8 *(5,534)*	0.0 *(0)*	0.0 *(0)*	1,959.8 *(9,799)*	0.0 *(0)*	0.0 *(0)*	0.0 *(0)*	0.0 *(0)*	0.0 *(0)*	3,066.7 *(15,333)*
	8.0–8.5	565.2 *(2,826)*	0.0 *(0)*	0.0 *(0)*	2,193.1 *(10,965)*	0.0 *(0)*	0.0 *(0)*	3,071.2 *(15,356)*	0.0 *(0)*	0.0 *(0)*	5,829.5 *(29,147)*
Oregon	7.0–7.5	355.8 *(1,779)*	20.6 *(103)*	0.0 *(0)*	0.9 *(4)*	9.2 *(46)*	1.2 *(6)*	0.0 *(0)*	0.0 *(0)*	0.0 *(0)*	387.6 *(1,938)*

Distance from Shoreline		0–3 nm[1]			3–12 nm			12–50 nm			
Depth Category (m)		0–30	30–60	> 60	0–30	30–60	> 60	0–30	30–60	> 60	Total
State	Wind Speed at 90m m/s	Area (km²) *(MW)*	Area (km²) *(MW)*	Area (km²) *(MW)*	Area (km²) *(MW)*	Area (km²) *(MW)*	Area (km²) *(MW)*	Area (km²) *(MW)*	Area (km²) *(MW)*	Area (km²) *(MW)*	Area (km²) *(MW)*
Oregon (cont.)	7.5–8.0	523.0 *(2,615)*	319.2 *(1,596)*	37.6 *(188)*	46.5 *(232)*	231.8 *(1,159)*	335.0 *(1,675)*	0.0 *(0)*	0.0 *(0)*	0.0 *(0)*	1,493.0 *(7,465)*
	8.0–8.5	198.3 *(991)*	277.0 *(1,385)*	6.6 *(33)*	19.0 *(95)*	595.7 *(2,978)*	2,558.5 *(12,792)*	0.0 *(0)*	0.0 *(0)*	4,989.5 *(24,947)*	8,644.4 *(43,222)*
	8.5–9.0	64.1 *(320)*	98.8 *(494)*	0.6 *(3)*	0.0 *(0)*	108.0 *(540)*	1,967.2 *(9,836)*	0.0 *(0)*	45.6 *(228)*	11,640.3 *(58,201)*	13,924.6 *(69,623)*
	9.0–9.5	64.1 *(321)*	55.4 *(277)*	38.6 *(193)*	0.0 *(0)*	32.6 *(163)*	614.9 *(3,074)*	0.0 *(0)*	0.0 *(0)*	6,588.2 *(32,941)*	7,393.7 *(36,969)*
	9.5–10	47.3 *(237)*	80.4 *(402)*	14.6 *(73)*	0.0 *(0)*	33.8 *(169)*	634.6 *(3,173)*	0.0 *(0)*	0.0 *(0)*	5,254.6 *(26,273)*	6,065.3 *(30,327)*
	>10	0.2 *(1)*	19.4 *(97)*	33.3 *(166)*	0.0 *(0)*	18.2 *(91)*	1,368.5 *(6,843)*	0.0 *(0)*	0.0 *(0)*	4,546.0 *(22,730)*	5,985.5 *(29,928)*
Pennsylvania	7.0–7.5	34.2 *(171)*	0.0 *(0)*	0.0 *(0)*	0.0 *(0)*	0.0 *(0)*	0.0 *(0)*	0.0 *(0)*	0.0 *(0)*	0.0 *(0)*	34.2 *(171)*
	7.5–8.0	113.4 *(567)*	0.0 *(0)*	0.0 *(0)*	53.2 *(266)*	44.2 *(221)*	0.0 *(0)*	0.0 *(0)*	0.0 *(0)*	0.0 *(0)*	210.8 *(1,054)*
	8.0–8.5	276.2 *(1,381)*	0.9 *(4)*	0.0 *(0)*	750.8 *(3,754)*	358.1 *(1,790)*	0.0 *(0)*	222.9 *(1,114)*	70.2 *(351)*	0.0 *(0)*	1,679.1 *(8,395)*
Rhode Island	7.0–7.5	216.3 *(1,082)*	8.0 *(40)*	0.0 *(0)*	0.0 *(0)*	0.0 *(0)*	0.0 *(0)*	0.0 *(0)*	0.0 *(0)*	0.0 *(0)*	224.3 *(1,121)*
	7.5–8.0	123.2 *(616)*	2.8 *(14)*	0.0 *(0)*	0.0 *(0)*	0.0 *(0)*	0.0 *(0)*	0.0 *(0)*	0.0 *(0)*	0.0 *(0)*	126.0 *(630)*
	8.0–8.5	139.9 *(700)*	39.8 *(199)*	0.0 *(0)*	34.5 *(173)*	69.2 *(346)*	0.0 *(0)*	0.0 *(0)*	0.0 *(0)*	0.0 *(0)*	283.5 *(1,417)*

	8.5–9.0	120.3 *(602)*	93.0 *(465)*	0.0 *(0)*	183.6 *(918)*	274.2 *(1,371)*	0.0 *(0)*	0.0 *(0)*	0.0 *(0)*	0.0 *(0)*	671.2 *(3,356)*
	9.0–9.5	53.7 *(269)*	17.6 *(88)*	0.0 *(0)*	176.0 *(880)*	782.7 *(3,914)*	0.0 *(0)*	0.0 *(0)*	430.2 *(2,151)*	1.3 *(6)*	1,461.5 *(7,307)*
	9.5–10	0.0 *(0)*	0.0 *(0)*	0.0 *(0)*	0.0 *(0)*	5.8 *(29)*	0.0 *(0)*	0.0 *(0)*	967.2 *(4,836)*	1,386.8 *(6,934)*	2,359.8 *(11,799)*
South Carolina	7.0–7.5	848.2 *(4,241)*	0.0 *(0)*	0.0 *(0)*	608.4 *(3,042)*	0.0 *(0)*	0.0 *(0)*	0.0 *(0)*	0.0 *(0)*	0.0 *(0)*	1,456.5 *(7,283)*
	7.5–8.0	593.5 *(2,968)*	0.0 *(0)*	0.0 *(0)*	3,053.9 *(15,269)*	0.0 *(0)*	0.0 *(0)*	4,267.6 *(21,338)*	287.0 *(1,435)*	0.0 *(0)*	8,202.1 *(41,010)*
	8.0–8.5	22.9 *(115)*	0.0 *(0)*	0.0 *(0)*	1,609.4 *(8,047)*	0.0 *(0)*	0.0 *(0)*	4,151.4 *(20,757)*	3,925.6 *(19,628)*	674.5 *(3,372)*	10,383.7 *(51,919)*
	8.5–9.0	0.0 *(0)*	0.0 *(0)*	0.0 *(0)*	0.0 *(0)*	0.0 *(0)*	0.0 *(0)*	2,027.0 *(10,135)*	3,109.7 *(15,548)*	869.9 *(4,349)*	6,006.5 *(30,033)*
Texas[1]	7.0–7.5	1,785.5 *(8,9278)*	0.0 *(0)*	0.1 *(0)*	96.2 *(481)*	0.0 *(0)*	0.0 *(0)*	137.2 *(686)*	0.0 *(0)*	0.0 *(0)*	2,019.1 *(10,095)*
	7.5–8.0	9,046.1 *(45,230)*	0.0 *(0)*	0.0 *(0)*	1,165.2 *(5,826)*	0.0 *(0)*	0.1 *(0)*	8,045.2 *(40,226)*	6,174.7 *(30,873)*	391.5 *(1,958)*	24,822.7 *(124,114)*
	8.0–8.5	5,928.5 *(29,643)*	125.4 *(627)*	0.0 *(0)*	742.1 *(3,711)*	136.9 *(684)*	0.0 *(0)*	583.5 *(2,918)*	4,442.8 *(22,214)*	4,597.1 *(22,985)*	16,556.3 *(82,782)*
	8.5–9.0	3,798.0 *(18,990)*	601.0 *(3,005)*	0.0 *(0)*	42.9 *(215)*	945.1 *(4,726)*	0.0 *(0)*	0.0 *(0)*	3,210.9 *(16,055)*	3,674.6 *(18,373)*	12,272.5 *(61,363)*
Virginia	7.0–7.5	889.2 *(4,446)*	0.0 *(0)*	0.0 *(0)*	0.0 *(0)*	0.0 *(0)*	0.0 *(0)*	0.0 *(0)*	0.0 *(0)*	0.0 *(0)*	889.2 *(4,446)*
	7.5–8.0	3,605.5 *(18,028)*	15.3 *(77)*	0.0 *(0)*	37.0 *(185)*	0.0 *(0)*	0.0 *(0)*	0.0 *(0)*	0.0 *(0)*	0.0 *(0)*	3,657.8 *(18,289)*
	8.0–8.5	1,135.8 *(5,679)*	2.0 *(10)*	0.0 *(0)*	2,985.9 *(14,929)*	0.0 *(0)*	0.0 *(0)*	2,355.9 *(11,780)*	69.1 *(345)*	0.0 *(0)*	6,548.7 *(32,743)*

Distance from Shoreline		0-3 nm[1]			3-12 nm			12-50 nm			
Depth Category (m)		0-30	30-60	> 60	0-30	30-60	> 60	0-30	30-60	> 60	Total
State	Wind Speed at 90m m/s	Area (km^2) *(MW)*	Area (km^2) *(MW)*	Area (km^2) *(MW)*	Area (km^2) *(MW)*	Area (km^2) *(MW)*	Area (km^2) *(MW)*	Area (km^2) *(MW)*	Area (km^2) *(MW)*	Area (km^2) *(MW)*	Area (km^2) *(MW)*
Virginia (cont.)	8.5-9.0	0.0 *(0)*	0.0 *(0)*	0.0 *(0)*	23.7 *(119)*	0.0 *(0)*	0.0 *(0)*	2,030.5 *(10,152)*	4,839.8 *(24,199)*	900.0 *(4,500)*	7,794.0 *(38,970)*
Washington	7.0-7.5	622.9 *(3,115)*	133.5 *(667)*	57.3 *(286)*	231.9 *(1,159)*	350.2 *(1,751)*	176.7 *(884)*	0.0 *(0)*	0.0 *(0)*	0.2 *(1)*	1,572.7 *(7,863)*
	7.5-8.0	370.6 *(1,853)*	0.0 *(0)*	0.0 *(0)*	203.0 *(1,015)*	841.6 *(4,208)*	1,173.0 *(5,865)*	0.0 *(0)*	0.0 *(0)*	2,033.1 *(10,166)*	4,621.4 *(23,107)*
	8.0-8.5	61.7 *(308)*	0.0 *(0)*	0.0 *(0)*	211.7 *(1,059)*	977.9 *(4,889)*	475.1 *(2,375)*	0.0 *(0)*	19.6 *(98)*	16,514.8 *(82,574)*	18,260.7 *(91,304)*
Wisconsin	7.0-7.5	1,053.8 *(5,269)*	268.2 *(1,341)*	145.8 *(729)*	349.9 *(1,749)*	616.6 *(3,083)*	1,085.0 *(5,425)*	0.0 *(0)*	34.2 *(171)*	162.0 *(810)*	3,715.4 *(18,577)*
	7.5-8.0	1,417.2 *(7,086)*	235.5 *(1,177)*	23.7 *(119)*	361.7 *(1,808)*	223.0 *(1,115)*	779.3 *(3,896)*	0.0 *(0)*	0.0 *(0)*	364.2 *(1,821)*	3,404.5 *(17,023)*
	8.0-8.5	830.6 *(4,153)*	372.7 *(1,863)*	0.0 *(0)*	451.4 *(2,257)*	1,386.1 *(6,930)*	2,539.0 *(12,695)*	0.0 *(0)*	21.3 *(107)*	2,160.1 *(10,800)*	7,761.1 *(38,806)*
	8.5-9.0	25.5 *(128)*	8.1 *(41)*	0.0 *(0)*	11.7 *(58)*	300.8 *(1,504)*	1,347.3 *(6,737)*	0.0 *(0)*	62.8 *(314)*	6,661.0 *(33,305)*	8,417.3 *(42,087)*

[1] Federal waters begin at 3 nm with the exception of Texas, which begins at 9 nm. For Texas, area reported is 0-9, 9-12, and 12-50 nm.

D

Appendix D: Approximating Wind Speed Distribution

The material in Appendix D was excerpted directly from the sources indicated.

Fig. D–1. The Rayleigh distribution. *Data courtesy of Jack Park,* The Wind Power Book *(page 54).*

The Rayleigh distribution takes the following form:

$$\text{Hours} = 8{,}760 \times \frac{\pi}{2} \times \frac{V}{\overline{V}^2} \times e^{-k},$$

where

V = windspeed
$\overline{V}$ = mean windspeed
$\pi = 3.1416$
$e = 2.718$
$k = \frac{\pi}{4} \times \left(\frac{V}{\overline{V}}\right)^2$

This equation gives you the total number of hours per year you can expect the wind to blow at a windspeed V when the mean windspeed is $\overline{V}$ at that site.

A graph of this complex equation is presented on the next page, with percent time as the vertical scale. To get the result in hours per year, multiply by 8,760. You can read percent of time from this graph. For example, for mean windspeed of 14 mph, you can expect wind to blow at 23 mph for about 2.2% of the time, or 194 hours per year.

The other graph here shows the agreement between a measured windspeed distribution and a calculated Rayleigh distribution for St. Ann's Head, England—with a mean windspeed of 16.2 mph. The Rayleigh distribution is slightly low at high windspeeds and high at low windspeeds (from 10 to 20 mph). As power is proportional to the cube of the windspeed, the higher speed end carries more weight in power calculations, but the greatly reduced duration of time at high windspeed reduces the overall energy impact. Rayleigh calculations are not recommended as a replacement for actually measuring your site's wind characteristics, but they can serve as a reasonable approximation when all you have is the annual average windspeed.

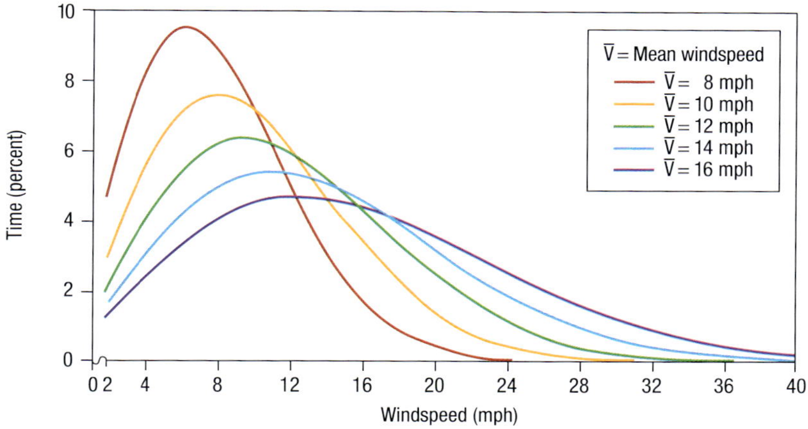

Comparison of Rayleigh and measured windspeed distributions for St. Ann's Head, England.

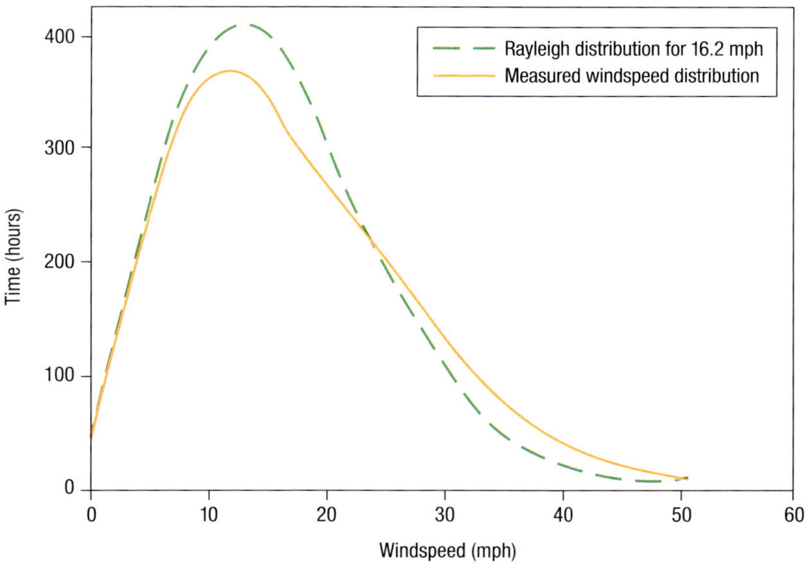

The Rayleigh distribution versus windspeed for sites with mean windspeeds between 8 and 16 mph.

E

Appendix E: Selected Wind Farm Case Studies

The material in Appendix E was excerpted directly from the sources indicated.

Fig. E–1. Case Study 1: Top of Iowa Wind Farm. *Data courtesy of Iowa Department of Natural Resources.*

Top of Iowa Wind Farm	
Project Owner	Northern Iowa Windpower, LLC—Members include Entergy Wholesale Operations, Midwest Renewable Energy Corporation and Zilkha Renewable Energy
Project Designer	Midwest Renewable Energy Corporation
Project Developer	Midwest Renewable Energy Corporation and Zilkha Renewable Energy

Introduction

Northern Iowa Windpower (NIW) built the Top of Iowa 80-megawatt Wind Electric Plant (wind farm) located in western Worth County, IA in 2001. The wind farm is part of a growing effort to meet more of America's energy needs with clean, reliable, and economical electricity. The wind farm consists of 90 NEG-Micon 900 kilowatt wind turbines mounted on 237-ft (72 m) tubular towers. Each turbine blade is 85 feet (26 m) long, making the rotor diameter 171 feet (52 m). The blades turn at 15 or 22 rotations per minute, depending on wind speed. The entire wind turbine, from ground level to the top of the rotor diameter, is 323 feet (98 m) high. The wind farm covers roughly 5,900 acres of cropland on 49 separate land parcels and interconnects to the local electrical grid at a new 161-kilovolt substation installed 6 miles east of the wind farm. The wind farm's energy, equivalent to the electricity consumed by 24,000 typical Midwest homes, is sold to Alliant Energy under a 15-year Power Purchase Agreement.

Project History

Northern Iowa Windpower, LLC (NIW) was created through a joint venture of Midwest Renewable Energy and Zilkha Renewable Energy. Midwest Renewable initiated development of the project in the summer of 2000, and Zilkha provided the expertise and capital required to construct and operate the facility. Construction began in May 2001 and was completed in November 2001. Entergy Corporation then acquired a majority interest in the Top of Iowa Wind Electric Plant in December 2001, retaining Zilkha Renewable Energy as the asset manager.

NIW developed the Top of Iowa Wind Farm largely in response to Wisconsin Act 9, which required Wisconsin utilities to purchase and distribute and annually increasing percentage of renewable energy. Act 9 allows the purchase of energy from out-of-state generating facilities if the electricity is physically delivered to an in-state utility and sold at retail prices. NIW chose to develop the wind farm in Iowa instead of Wisconsin because Iowa generally has better wind resources, making it cheaper to generate wind energy. Equally important, Worth County's proximity to Wisconsin makes it possible to economically transmit the electricity from the wind farm to Wisconsin customers.

The Federal Production Tax Credit (PTC), which was created to stimulate investment in renewable energy, also played a critical role in the wind farm's development. The PTC is adjusted annually for inflation and applies to each kWh of electricity produced from wind energy and sold to unrelated parties during the first 10 years after the facility is placed into service. The PTC has the effect of reducing the cost of wind energy by almost two cents per kWh. Without the PTC, wind energy would have had to be priced higher. The PEC reduced NIW's payments of federal taxes by approximately 1.7 cents for each kWh of energy sold for the first 10 years, lowering the price per kWh paid by Alliant enough to be competitive with traditional renewable energy resources.

The Internal Revenue Code contains a Modified Accelerated Cost Recovery System (MACRS) that allows businesses to recover investments through short term, rapid depreciation deduction of the cost of certain renewable energy systems, such as wind turbines. The current MACRS property class for renewable energy systems in service after 1986 is five years, heavily weighted to the initial years of operation, further lowering the cost of delivered energy.

Western Worth County was selected for the wind farm site due to its excellent wind resource; close proximity to growing power markets such as Chicago, Milwaukee, Minneapolis and Madison; short distance from the project to high voltage transmission lines; transmission capacity; access to energy markets; expansion possibilities, and strong community support. The wind farm is located on a high ridge in Bristol and Gertile townships in Worth County, at an elevation of 1,200 to 1,300 feet above sea level—one of the highest elevations in Iowa. The project site's elevation increases the wind resource dramatically, compared with the surrounding terrain at 1,000 feet or less.

In addition to the greater wind resources, legislation that allows city councils or country boards of supervisors to enact ordinances for the special valuation of "wind energy conservation property" for up to 20 years made Iowa a favorable location for the wind farm. This "Special Assessment of Wind Energy Devices (Iowa Code Chapter 427B.26)" allows for a waiver of or reduction in property taxes for the entire wind plant, including the turbine, tower and electrical equipment, pad mount transformers, power lines, and substations. Also, as noted below, NIW was able to shorten its data collection period, resulting in significant savings, due to the existence of a long-term reference station located in Forest City, IA. This anemometer was installed as part of the Iowa Energy Center's statewide wind energy resource evaluation.

Other renewable energy financial incentives are available in Iowa, and are often influential in other renewable energy projects, but did not apply to this project. In 2001, the Iowa Legislature subsequently passed a law that requires all electric utilities to offer customers the opportunity to purchase power from alternate energy sources by January 1, 2004, which may result in the future development of additional wind farms.

Pre-Construction Activities

Once NIW had tentatively chosen a site for the Top of Iowa wind farm, the first steps in project pre-construction began. Pre-construction activities included meteorological studies, negotiating a power purchase agreement, negotiating land easements, site design, soil studies, project permitting, and engineering of the project interconnection to the electrical grid.

Meteorological Studies

NIW consulted with Ron Nierenberg, a wind resource assessment specialist, to analyze all available data and calculate the wind resource at the Top of Iowa location. Mr. Nierenberg has more than 20 years' experience as a wind energy meteorologist and has managed wind-monitoring programs with an aggregate value of $3 million. He managed two DOE-funded wind research projects, developed a software library for wind energy analysis, and conducted the original wind study of the Altamont Pass in California.

In Fall 2000, NIW installed three 50-meter-tall meteorological stations at several representative points on the site to collect data for a one-year period. Typically, wind data is collected for at least three years, but the existence of a long-term reference station located 15 miles away in Forest City, IA, allowed NIW to correlate the measurements at the site with long-term data, thereby shortening the period of time for which on-site data was required and reducing overall project costs.

Table 1. Electric Energy Output Estimates for Top of Iowa Wind Farm

Average Wind Speed at 72.2-meter Hub Height	Gross Capacity Factor	Net Capacity Factor	Output per Wind Turbine Generator (MWH)	Gross Annual Output for 80.1 MW Wind Farm (MWH)
8.2 meters per second	32.9%	31.6%	2,590	230,510

Nierenberg's analysis compared on-site data from the Top of Iowa to concurrent and historical data from Forest City. He used the data to predict wind speeds at the Top of Iowa. Based on the wind speed distribution, Nierenberg used the wind turbine manufacturer's warranted power curve for the NEG-Micon 900 kilowatt turbine to calculate gross output on an annual, per turbine basis.

This calculation yielded a gross estimate of 2,786 MWH per turbine, per year. Nierenberg then took losses into consideration, including losses for:

- local topography
- wind turbine array
- wind turbine availability
- transformer/line losses
- blade contamination and soiling
- maintenance and control hysteresis

Losses are calculated on a cumulative basis, rather than simply adding them together; the cumulative equation [((1.02 × 1.0175 × 1.01)−1) × 100] equals 4.82% loss, versus the simple sum of 4.75% loss. Net energy production is then calculated at 2,590 MWH per turbine each year. NIW assumes that availability will only reach 97% per year according to the NEG Micon guarantee during the first five years of the project's operation.

The Top of Iowa Wind farm was projected to provide substantial amounts of electricity throughout the year. Turbine production was projected to range from 180 kW to 340 kW per turbine during the year and the projected average turbine production per hour ranged from 224 kWh to 340 kWh.

Chart 1. Projected Average Turbine Production by Month

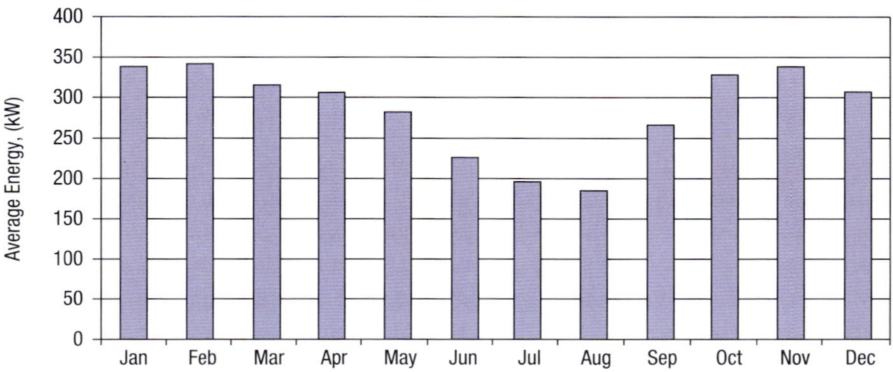

Chart 2. Projected Annual Average Turbine Production by Hour

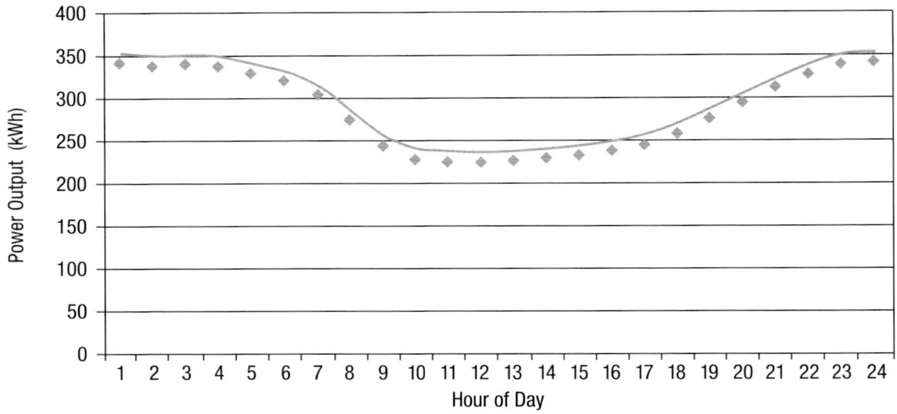

Power Purchase Agreement

Once NIW had a good estimate of the potential output of the Top of Iowa wind farm, they began efforts to secure a purchase agreement for the power. As a first step in the Top of Iowa's marketing efforts, NIW hired Henwood Energy Services, Inc. to conduct a study of the expected long-term energy prices for the MAPP-MAIN-COMED regions. Henwood is the leading energy price forecaster in the United States. The Henwood study included a detailed forecast of electricity market prices in the Midwest region and market revenues for the wind farm project. Henwood created the forecasts using their EMSS/PROSYM modeling software, which simulates actual operation and market interactions of the regional electric transmission system and generating units located throughout the U.S. Eastern Interconnect.

Henwood prepared market prices and revenue forecasts under a Reference Case set of assumptions, which provided a reasonable "base case" forecast of future market conditions driven by various assumptions regarding the future state of power markets. The study's key structural assumptions included:

- Wholesale power markets in the Midwest, including the MAIN, MAPP and ECAR Regions, are already competitive, although they operate primarily on a bilateral basis, with no formal power pool or market clearing exchange in place.
- Midwest electricity markets will continue to operate on a competitive basis, which means that revenues will accrue to owners of non-regulated electric generation based upon market clearing prices established in regional energy markets, and regulatory support for embedded cost generation assets will end over time.

- In forecasting electricity prices, Henwood modeled the regional market as a single, all-in price electricity energy market, wherein participants freely bid prices. During times of relative scarcity of generation, participants are able to bid prices higher than their short run variable costs. The underlying presumption of this market forecast approach is that the market price will converge on an equilibrium state that equates to the long-run cost of operation for new efficient combined cycle entrants, including the return of and on capital. The actual price of electricity in the future will be subject to the actual outcome of key market drivers, such as the price of natural gas and the decisions of new generators to enter the market.

Henwood also considered the seasonal and diurnal variability in the wind resource at the Top of Iowa wind farm in determining its energy value estimates. This is an important calculation, since the project generates more of its energy during off-peak hours and winter months, when energy prices are generally lower.

After discussions and negotiations with several utilities in the Upper Midwest, NIW executed a 15-year Renewable Energy Power Purchase Agreement (PPA) with Alliant Energy's subsidiary, Interstate Power Company, for the sale of the wind farm's output in February 2001. It became clear during negotiations that Alliant was only interested in the output of a 40 MW wind project. However, meeting Alliant's price goals required the economies of scale created by an 80 MW wind project. Thus, NIW created a structure to sell 50 percent of the output as "*Green Energy*" over a 15-year term to meet Alliant's wind energy requirements and to sell the remaining 50 percent of the output as "*Energy Only*," meaning the company sold only the value of the electricity and kept the "green" value that arose from the wind-origins of the electricity, over a 10-year term. These "green" credits (per kWh) are sold to Wisconsin Electric Power Company (WEPCO). NIW negotiated a 10-year term agreement for the *Energy Only* component for several reasons:

- The 10-year term for "Energy Only" creates a contract path into Wisconsin during the time when NIW believes Renewable Resource Credits will have the greatest value.
- Currently, the project site is in a regulated market where bilateral contracts are the norm and there is no "power pool" per se. The project sponsors are confident that a pool, like that developing around ComEd will be in place by the time the "*Energy Only*" term of the PPA expires, thus providing Top of Iowa with a ready market for its output.
- NIW believes the current barriers to economical transmission of wind power will decrease over time, facilitating the sale of the Top of Iowa's output beyond year 10.

The *Green Energy* is sold on an as-delivered basis with an initial fixed price and a small escalator. The *Energy Only* is sold at different prices for peak and off-peak hours. More energy is produced during windier nighttime hours, but there are more peak than off-peak hours. Therefore, peak and off-peak energy output is split approximately 50:50. All capacity value created by the project for the Regional Transmission Operator (RTO) accreditation purposes is assigned to Alliant.

Landowner Agreements

NIW sought to garner public support and landowner cooperation early in the project development by holding six town hall meetings at the Community Center in Joice, IA. At the meetings, citizens were able to ask questions and express concerns on topics such as road construction, electric cables, effects on farming, equipment, tile drainage and fencing.

NIW negotiated easement agreements for the Top of Iowa wind farm. Forty-nine landowners over 5,900 acres have turbines. Each landowner has a non-obstruction clause in his or her easement agreement. The non-obstruction clause obligates the landowner to review with and seek prior approval from NIW for constructing structures greater than 50 feet tall within 650 feet of a turbine. No landowners without turbines have a non-obstruction obligation. The easement agreements also provide unrestricted access to the turbines, a fixed per-turbine payment schedule, reasonable notice for entry and procurement provisions. Neighbors within 1,200 feet of a turbine were offered Neighbor Agreements, which allows the wind farm to cast a shadow caused by the towers and blades across the respective land. The agreements also permit the wind farm to emit audible noise in excess of 50 dBA across the land. Sound levels at the outer walls of existing, occupied homes are kept at or below 50 dBA. The right-of-way easements, encompassing 14.6 acres (24,561 feet long x 25 feet wide), are for additional properties crossed by the transmission line between the Top of Iowa Wind Farm and the substation. The exact terms of the agreements with landowners are confidential, but each landowner receives approximately $2,400 per year per turbine.

Site Design

NIW designed the Top of Iowa Wind Farm to maximize energy capture while minimizing costs of construction and operations. No major difficulties were encountered in siting the wind turbines, except that the developers of the project had to balance maximum energy output with landowner preferences, prudent setbacks, and electrical design to minimize energy losses. To site turbines on the highest topography and to take advantage of the most wind power, NIW configured the wind turbine array in a "Y" shape, spacing the turbines four to five rotor diameters apart. The wind at Top of Iowa blows from a variety of directions. This factor was predicted to result in higher-than-typical array power losses, so NIW designed the turbine spacing to minimize these losses.

To provide a margin of safety and minimize the impact on the local community, the towers were placed at a minimum of 325 feet from public roads, 1,000 feet from residences, and 700 feet from the perimeter of the site. Wind turbines were sited in consideration of agriculture during construction and wind farm operation. Landowner input was the primary factor in the design and placement of new field accesses and turbine access roads. Field access culverts were sized with consultation from the county engineer.

NIW hired Don Keef, an experienced professional who has designed the electrical collection systems for more than half of the wind energy projects built in the United States, to design the electrical layout. The site electrical layout consists of a 33.6 kV underground collection system throughout the project. NIW built an 80 MVA 161kV – 33.6 kV substation approximately five miles east of the project location that connects Top of Iowa Project to Alliant West's 161 kV transmission line. Alliant Energy provided substation design services and construction.

Permitting/Zoning

Permits and government approvals for the Top of Iowa wind farm were relatively straightforward. NIW obtained the following permits:

1. Iowa Utilities Board Construction Approval Waiver
2. Worth County Road Access and Utility Accommodation Approval
3. Army Corps of Engineers Section 404 Wetland
4. Federal Aviation Administration — "Determination of No Hazard to Air Navigation" FAA Lighting was installed because the wind turbine blades extend above 200 feet (61 m), the maximum height limit exclusion under FAA lighting requirements.
5. National Pollutant Discharge Elimination System (NPDES) Storm Water Discharge Permit
6. County Drainage District Permit
7. Iowa DOT Utility Accommodation Permit – this permit was required to cross I–35
8. Iowa DOT Work within Right of Way Permit
9. Iowa DOT Haulage permits — All large truck loads of wind turbine nacelles, rotors, tower sections and transformers were cleared through the Iowa Department of Transportation. Under the construction and supply contracts, the contractors were responsible for obtaining permits.
10. Federal Energy Regulatory Commission — Exempt Wholesale Generator status

11. Worth County Zoning Permit for Substation—NIW obtained zoning permit required to build the substation
12. No Department of Natural Resources permits were required. However, NIW has entered into an agreement with the Iowa Department of Natural Resources to monitor avian mortality caused by collisions with the turbines in the area of the wind farm.

The only land where the townships in Worth County, IA have zoned is that on which the substation is located. A zoning permit was required for the substation and was granted.

Non-obstruction permits from the Federal Aviation Administration (FAA) were more complicated, due to the presence of a FAA repeater station on the project. All FAA permits were obtained, but the development team believed the FAA required more lights than were necessary. The FAA and the American Wind Energy Association are currently working on a joint study of wind turbine lighting requirements.

Interconnection
Under federal law, Alliant Energy's 161 kV line is "open" to any generator who fulfills numerous requirements on a "space available" basis. NIW negotiated an interconnection agreement with Alliant Energy over the space of several months. Negotiations went smoothly, partly because Alliant Energy had agreed to purchase the energy from the project.

Wind Farm Construction
NIW separated the construction of the wind and energy project into two contracts: one with NEG Micon USA as the supplier and erector of the project's equipment; and the second contract to The Mortenson Company of Minneapolis, as the Balance of Plant contractor. Alliant West constructed the electrical substation. Zilkha Renewable Energy's construction manager was Mark Haller. Michael Kelly was the on-site Owner's Representative for NIW to ensure safe construction practices, project success and timely completion. Mr. Kelly handled day-to-day issues throughout the entire construction. All contractors were required to adhere to strict, safe working procedures, weekly safety training, and report any injuries. The tubular, turbine tower doors were locked to restrict access by unauthorized persons. Contractors were held to strict site cleanliness standards to minimize wind-borne debris. The total project cost was approximately $80 million.

The NEG Micon contract contained numerous provisions to ensure the timely and quality construction of the wind project. Important features included the following:

- Strict milestone schedule
- Delay liquidated damages in the event one or more turbines were not commissioned on time, plus damages calculated for Federal Production Tax Credits
- Production Tax Credit liquidated damages in the event one or more turbines did not reach full commercial operations, calculated on an after-tax basis.
- Guarantees for NEG Micon 900's performance at given wind speeds.

The Mortenson contract contained similar provisions, including:

- Delay liquidated damages to compensate the project for losses incurred due to missed schedules.
- Ample retainage on progress.
- Full and clear responsibility for the engineering of roads, foundations and all other required infrastructure.

The contract with Alliant West contained similar provisions to ensure the timely and quality construction of the wind project. The provisions included:

The first phase of construction included improvements to existing roads, construction of new roads, and installation of turbine foundations. It was discovered that low-slung, over-the-road "low-boy" trailers used for parts deliveries, with only a few inches of ground clearance, had difficulty on gravel roads with dips and hills. A special hydraulic dolly was fabricated for transporting trailers with a large, all-wheel-drive farm tractor.

Next, the Mortenson Company built 17 miles of new roads, consisting of 12 inches of crushed rock spread on compacted soil, in corridors for the turbine layout. Total road construction utilized 2,275 truckloads of crushed rock. Road edge elevation was maintained within four inches above the adjacent field. Culverts and tile inlets were installed where new roads could trap water. Temporary silt fencing was installed where needed.

After the corridor roads were completed, the Mortenson Company began excavation and construction of the foundations for the turbines. An octagonal, spread-footing type tower foundation, measuring seven feet deep and forty-two feet in diameter, was constructed. Each foundation was reinforced by 25,713 pounds of reinforcement steel. After the reinforcement steel was in place, 181 cubic yards of concrete was poured into the foundation excavation. The entire project required 2,288,457 pounds of reinforcement steel and 16,752 cubic yards of concrete.

The second phase of construction was installation of the electrical layout. The electrical contractor began trenching and laying 21.8 miles of 33.6 KVA underground cable and 30.7 miles of fiber optic cable. Field tiles that were disturbed during electrical cable installation were repaired and inspected by the landowner before covering. Alliant West then built the 80 MVA 161kV–33.6kV substation approximately five miles east of the project location to connect the Top of Iowa project to Alliant's 161 kV transmission line.

Once the electrical contractors were finished, NEG Micon delivered and erected the wind turbine generators. NEG Micon was responsible for turbine construction from the concrete foundation up and erected the turbines on the foundations. The towers are 237 feet high. Each turbine blade is 85 feet long, making the rotor diameter 171 feet. The entire wind turbine, from ground level to the top of the rotor diameter, is 323 feet high. The transportation of materials for turbine delivery and erection required 511 truckloads with a total weight of 28,312,680 pounds over a total distance of 360,000 miles.

Final Completion under the Turbine Supply and Installation Agreement is pending completion of required performance testing. The project achieved Partial Commercial Operation on October 2, 2001, when grid interconnection was completed, energized, and 25 percent of the turbines were in operation. Full commercial operation was achieved on October 20, 2001, when 75 percent of turbines were fully commissioned and operating. Turbine commissioning was 100% complete on November 14, 2001.

Project Milestones

Date	Milestone
Summer 2000	Wind farm development initiated.
05/02/2001	Construction kick-off meeting. The actual work began two weeks later due to a very wet spring.
08/08/2001	The first turbine was erected.
08/25/2001	kWh first supplied through the temporary substation.
10/03/2001	kWh first supplied through the permanent substation.
10/03/2001	Partial commercial operation was achieved pursuant to the PPA (15 turbines commissioned and producing power).
10/11/2001	The final turbine was erected.
10/17/2001	Balance of Plant Substantial Completion achieved.
10/20/2001	Full commercial operation was achieved pursuant to the PPA (75 percent, or 67 turbines, commissioned and producing power).
11/14/2001	Final turbine commissioned.
11/20/2001	Substantial Completion achieved pursuant to the Turbine Supply and Installation Agreement
05/21/2002	Final completion under the Balance of Plant Agreement achieved.

Operations and Maintenance

NIW maintains property and commercial general liability insurance coverage. All-risk property insurance covers all events, (e.g., flood, fire, lightning, etc.), with some exclusions, such as war, etc. This policy also includes coverage for non-warranty turbine and infrastructure breakdowns. Business interruption coverage addresses potential production and tax credit revenues if the project operation is interrupted or limited. Commercial general liability covers loss or injury to third parties. Fire and theft insurance is also maintained on the maintenance building.

NIW has entered into three contracts for operations & maintenance of the Top of Iowa wind farm. All high voltage preventative maintenance and repair services are subcontracted to Alliant Energy. Wind farm management and administration services are performed by Zilkha Renewable Energy. Operations and Maintenance Services (O&MS) are provided by NEG Micon.

Zilkha provides NIW with accounting, audit, contract management, project performance analysis, regulatory compliance, reporting, warranty administration, risk management and management services. Zilkha's on-site project manager is responsible for management, administration, supervising and maintenance functions. Such activities include comprehensive accounting functions.

Zilkha Renewable Energy's responsibilities include:

- Invoicing of Interstate Power Company
- All accounting, audit and financial reporting requirements
- Marketing and administration of the project's "green credits" or "Renewable Resource Credits"
- Supervising NEG Micon's operations of the wind farm
- Community and landowner relations
- Insurance administration
- Fulfilling all statutory and permitting requirements
- Investor relations
- All warranty claims under the NEG Micon wind turbine supply agreement

NEG Micon provides the Top of Iowa wind farm with O&MS through a five-year operations and maintenance agreement at a fixed cost per year. After the five-year warranty period, it is anticipated that Zilkha Renewable Energy will take over operations and maintenance, or that the contract with NEG Micon would be renewed. NEG Micon's scope of work includes all scheduled and unscheduled inspections and repairs, including all parts and consumables; operation of the turbines; and blade pitching. Furthermore, NEG Micon agreed to use commercially reasonable efforts to provide O&MS services at off-peak times to maximize revenues to the project. NEG Micon guaranteed average availability of 97 percent per year and pays NIW an amount that compensates the wind farm for revenue that would have otherwise been generated if availability drops below this threshold level. In turn, NIW pays

NEG Micon an incentive averaging $0.02 per kWh if the availability exceeds 97 percent.

NEG Micon hired the O&M team, consisting of a site manager and five to six turbine technicians and support staff, while the company's technical staff was on site constructing the wind farm. The O&M staff gained experience during the turbine erection and commissioning process and is currently responsible for day-to-day operations and preventative maintenance of the wind turbines.

The staff will perform daily monitoring of the wind farm operation through the central Supervisory Control and Data Acquisition (SCADA) monitoring and control system and visual inspection of the turbines. Routine maintenance, including lubrication, calibration, and inspection, is performed according to the equipment manufacturer's scheduled plan. Occasional repairs will be performed as required. The interconnection facilities are typically inspected monthly. Routine communications with System Operation and Control is on an "as-required basis." NEG Micon provides remote, manned, 24-hour monitoring and restarting of turbines. On site personnel are on call during non-working hours. NEG Micon will supply turbine spare parts if necessary.

NEG Micon's U.S. headquarters are located about six hours east of the Top of Iowa Wind Farm in Rolling Meadows, IL. The close proximity of these facilities will provide the project quick access to spare parts and to additional NEG Micon technicians if needed. Scheduled maintenance on the turbines is performed semi-annually, according to NEG Micon's recommended O&M procedures. Maintenance outages are scheduled to minimize interruption of delivered energy. Turbine maintenance is accomplished cyclically during the project lifetime. Substation and transformer maintenance is performed annually while pole-line maintenance is accomplished biannually. Maintenance power outages are scheduled during no- or low-wind periods and during off-peak hours.

Unplanned maintenance issues during the first year consisted of:

1. Replacing one gearbox due to the incorrect grinding of gear teeth during the manufacturing process. This caused pitting and ultimate fracture of several teeth.
2. An unusually high rate of bearing failures occurred in the generators. Extensive testing was done to determine the cause of the bearing failures. Preliminary results determined the problem to be ineffectiveness of the lubrication.

Table 3: Unscheduled Maintenance

Date	Wind turbine #	Component	Description
10/13/2001	T88	Generator	Defective winding
11/01/2001	T12	Generator	Lack of grease or faulty bearing
11/01/2001	T72	Generator	Defective winding
12/08/2001	T48	Generator	Front bearing completely damaged
01/18/2002	T29	Generator	Front bearing completely damaged
01/24/2002	T87	Generator	Rear bearing failed
02/01/2002	T75	Generator	Bearing failed
02/27/2002	T76	Generator	Front bearing completely damaged
03/01/2002	T63	Generator	Front bearing completely damaged
03/06/2002	T45	Generator	Both bearings changed
03/23/2002	T10	Generator	Front bearing completely damaged
04/23/2002	T60	Generator	Rear bearing making loud noise
04/15/2002	T11	Gearbox	Intermediate gear has broken teeth
05/14/2002	T73	Generator	Front bearing making loud noise
05/28/2002	T52	Generator	Bearing vibration test failed
05/29/2002	T18	Generator	Bearing vibration test failed
05/31/2002	T15	Generator	Bearing vibration test failed
06/01/2002	T37	Generator	Bearing vibration test failed
07/17/2002	T12	Generator	Rear bearing making noise
11/28/2002	T75	Generator	Rear bearing completely damaged
12/11/2002	T04	Generator	Rear bearing completely damaged
12/13/2002	T73	Generator	Bearing vibration test failed
12/14/2002	T76	Generator	Bearing vibration test failed
12/21/2002	T18	Generator	Rear bearing completely damaged
12/24/2002	T87	Generator	Rear bearing completely damaged

Production data

The Top of Iowa wind farm was projected to produce 240,000 MWH per year, or enough power for approximately 24,000 homes.

The turbines performed as expected for the actual wind experienced. However, the average wind speed experienced for 2002 was below the long-term average, and plant production was at 90 percent of the generation projection, producing 194,316,245 kWh. The annual project output is based upon the long-term average wind speed. Annual variations are anticipated. The first year plant availability was 96.5 percent, or slightly better than the first year forecast of 95.5 percent, and 97.4 percent for the last six months of the year.

Table 1: 2002 Net Electricity Production

Month	Net Electricity Production in kWh
January 2002	17,837,930
February 2002	22,978,315
March 2002	17,028,800
April 2002	21,422,100
May 2002	18,147,000
June 2002	16,173,400
July 2002	10,057,600
August 2002	9,661,300
September 2002	12,177,500
October 2002	12,972,900
November 2002	15,777,700
December 2002	20,081,700

Chart 3. Average Production in 2002

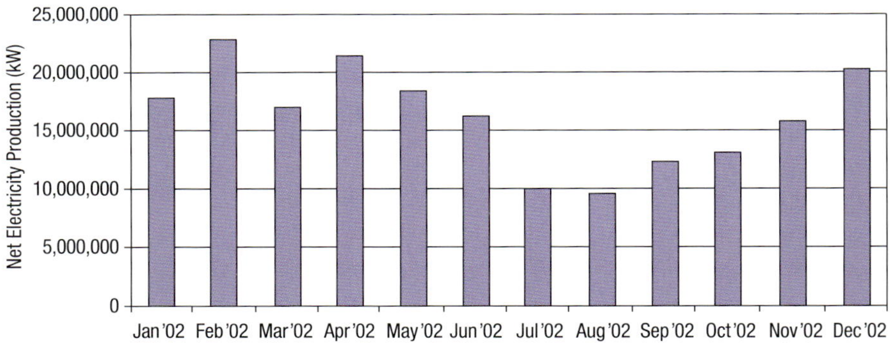

Economic Benefits

The construction process created over 200 jobs, 20 of which went to local residents. Seven permanent operation jobs were created:

- One project administrator employed by Zilkha.
- Six operations and maintenance personnel employed by NEG Micon.

The construction process also resulted in impact payments for soil compaction or damage to crop acres, totaling more than $250,000 to landowners surrounding the turbine sites. The Top of Iowa Wind Farm pays more than $200,000 per year, which ultimately escalates to more than $350,000 by the 25th year for all ongoing payments. Since well under 100 acres of land were taken out of agricultural production for the project, landowners are receiving more than $2,000 per turbine per year for the land they had formerly dedicated to the growing of corn and soybeans.

Even with the special property tax assessment, Top of Iowa will pay Worth County more than $6,000,000 in taxes over the next 20 years.

Federal tax revenue during the first 10 years are off-set by the Production Tax Credit (PTC) and accelerated depreciation deduction for renewable energy property. The total federal Production Tax Credits earned by the project are projected to total more than $4 million per year for the first 10 years of operation. This is calculated by multiplying the PTC by the kWh produced over time. Beginning in approximately the seventh year of the project, the federal tax burden increases to approximately $1.5 million per year and slowly escalates to more than $3 million per year by year fifteen. The increase in projected tax revenue is largely attributable to the end of the accelerated depreciation eligibility. Likewise, state income taxes become effective in the twelfth year of the project and are projected to total several hundred thousand dollars per year in the fifteenth year.

The combination of local job creation, payments to landowners, increased property tax receipts, and utilizing alternative energy resources all produce positive results for the local area.

Environmental Issues

The Top of Iowa wind farm will displace energy ordinarily produced by coal or natural gas. Both of these fuels originate from outside of Iowa. Coal is mined throughout the United States. Domestic gas reserves in the United States are declining and the country is increasingly dependent on foreign sources of natural gas imported through liquefied natural gas (LNG) transport systems. Virtually 100 percent of new electricity generation is fueled by natural gas. Electricity generated from wind, which displaces fossil fuels, is likely to have a net benefit on the environment in terms of avian impacts and air quality benefits.

Expected avian impacts

According to the American Wind Energy Association, studies at wind plants across the country have shown that avian mortality at most potential sites is low, 1 to 2 birds per turbine per year or less. Other studies have shown that birds and bats will collide with wind turbine blades or towers and that, in some instances, it may have a significant impact on local breeding bird populations.

Avian mortality is expected to be minor at the Top of Iowa wind farm site, but NIW has agreed to participate with the Iowa Department of Natural Resources in a two-year avian impact study, starting in the spring of 2003. The Top of Iowa Wind Farm provides a unique opportunity to assess the impacts of a wind farm on resident and migratory birds in a traditionally high bird-use area. The objectives of this study are to determine bird and bat mortality during the spring and fall migration periods, determine bird and bat species abundance, flight patterns and mortality risk at turbine sites versus non-turbine sites, and determine impacts of the turbines on waterfowl use of croplands. This study will be concluded in December 2005.

Air quality benefits

Each turbine offsets more than 11,000 pounds of carbon dioxide, sulfur dioxide, and nitrogen oxides on an average day—pollution that would otherwise be released by fossil fuel use. The wind farm offsets more than 370 million pounds of pollution linked to smog, acid rain, and green house effect annually.

Fig. E–2. Case Study 2: 10-MW European Wind Farm. *Data courtesy of Copper Development Association.*

Wind Farm Case Study

Introduction

Nowadays, most European governments are well aware of the potential of wind energy. From 1980 to 2000, the awareness of wind energy was mainly concentrated in Denmark and Germany where a large number of wind turbines were manufactured and installed. Germany still leads in terms of annually installed capacity with Spain close behind. Over the last 5 to 10 years many European governments have developed policies to promote renewable energy sources, including wind energy. Prominent examples are the United Kingdom, France, Italy, the Netherlands and Eastern European countries where wind energy is already installed or incentives are defined to boost the increase of renewables.

During the last 20 to 25 years the implementation of wind energy has changed dramatically. In the early stages, standalone wind turbines with installed power between 100 and 500 kW were quite common while the development of wind farms was rare. Now, after the development of incentives and planning regimes, most installations are in farms of up to 25 or more turbines. Currently, the installed power per turbine ranges from 750 kW to 3 MW and farms totalling 50 MW are not uncommon. Wind turbines in the range of 4.5 to 6 MW are available as prototypes and test specimens but are not yet commercially exploited.

Offshore wind farms are more expensive that onshore wind farms, requiring about double the initial investment—due to the extra costs of construction, transport to site and interconnection—and double the operational costs. However, the installed power of offshore farms is greater, the wind resource is greater and there is less environmental impact, especially in respect of noise.

The costs of onshore wind energy ranges from 55 to 100 EUR/MWh mainly depending on the wind resource (Figure 1). For most locations, wind energy is not cost-effective and incentives are a prerequisite to make a wind farm profitable.

The time span from first initiative to final commissioning of a wind farm is subdivided into a development period and a construction period. The duration of the construction period is predictable at about 1 year for a small to medium farm (< 15 MW) and at 1 to 2 years for a large wind farm. The duration of the development period is less predictable, especially the time required to obtain planning consent which, depending on the mandatory procedures and the number and kind of objections raised, may vary from approximately 6 months to more than 5 years. The technical life span of a wind turbine is 20 years.

The financing of a wind farm is a crucial issue. Construction of a single turbine was often part of a company's normal capital expenditure and would have been assessed as such. Most wind farms are developed by a dedicated company with capital made up of a limited amount of equity supported by loans, for which guarantees are required. The sole objective is to make a profit, so the financial projections are vitally important.

Generally the development and operation of a wind farm can be subdivided into the following four phases:

- initiation and feasibility (concluded by go/no-go)
- prebuilding (concluded by go/no-go)
- building
- operation and maintenance.

The following case has been selected for this application note:

Installed power	10 MW
Wind turbines of	2 MW
Hub height	80 meters
Rotor diameter	80 meters, 3 bladed rotor
Octangular foundation, gross dimensions	18 × 18 meters
Nacelle weight	100 tons
Tower weight	200 tons
Life span	20 years

Wind Farm Initiation and Feasibility Phase

In this phase, the basic parameters of the project are determined (e.g., the number and type of turbines and total installed power) and potential sites are assessed and compared, taking into account the wind resource available, the proximity of suitable grid connections and the limitations of regional, national and local planning regulations.

At the end of this phase the least suitable sites will have been eliminated and a decision made whether to proceed with further study of the remaining sites or to abandon the project.

Site selection and wind assessment

Wind farms require large sites. Depending on the rotor diameter the required mutual separation is 300 to 500 meters with a further similar separation distance from dwellings and commercial buildings to limit noise nuisance and to provide a safety zone. Even for a medium size wind farm, such as the 5 wind turbines of 2 MW considered here, a substantial land area is required.

Generally speaking, potential wind farm sites are preferably open areas of flat land or on top of hilly areas. Obviously, the sites should be known to be windy, with high and recurrent wind resources. Having selected the site, the next step is to assess the local long-term wind climate by reference to existing data or by long term monitoring. The objective in this phase is to eliminate all sites that may be unsuitable (in other words, unprofitable) in the long term.

If the site screening process does not identify any prohibitive limitations the feasibility study may proceed.

Financial feasibility is obviously a prerequisite for the development of a wind energy project—a nonfeasible project will not attract the required finance. The wind resource assessment and the consequent estimation of the yearly energy yield is of crucial importance since they determine the project yield. The energy available from the wind is proportional to the third power of the wind speed. Based on local wind speed data from meteorological stations a local wind atlas of the planned wind farm can be determined. It is necessary to use at least one full year of wind data to take into account variations in wind speed during the seasons.

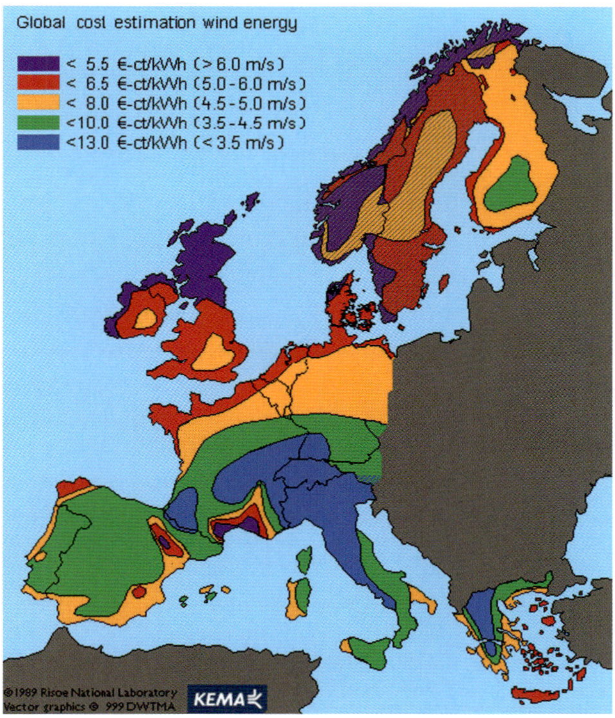

Figure 1. Overview of European Wind Atlas (No data is available for the Balkans and Eastern Europe)

Figure 2. Photograph of 10 MW wind farm on flat land near the Atlantic coastline

The European Wind Atlas (Figure 1) shows that Scandinavia, UK, Ireland and the Atlantic coastline of the European continent have the best wind conditions for the development of wind energy. This wind atlas is combined with a contour map of the area to determine the wind speed at a specific site and height. The estimated wind distribution results in a yearly energy yield representing the gross income of the wind farm.

Figure 3 shows a power/wind speed curve (PV) of a 2 MW wind turbine with optimal efficiency, i.e., without noise reduction measures that usually result in reduced energy generation. Figure 4 shows a typical wind speed distribution based on the statistical Weibull function with shape factor of 2 and average wind speed of 7 m/s. This distribution indicates the number of hours per year that a particular wind speed may be expected. It is site dependent and must be determined.

The annual energy yield is calculated by multiplying the wind turbine power curve with the wind distribution function at site:

$$E_y = \left(\sum_{i=1}^{i=n} f_{w_i} P_{w_i} \right)$$

where:

E_y is annual energy yield in kWh

w is the wind speed in m/s

n is the number of data bins covering the wind speed range of the turbine (0.5 or 1 m/s intervals)

f_{w_i} is the number of hours per year for which wind speed is w m/s (refer to Figure 4)

p_{w_i} is the power resulting from a wind speed of w m/s

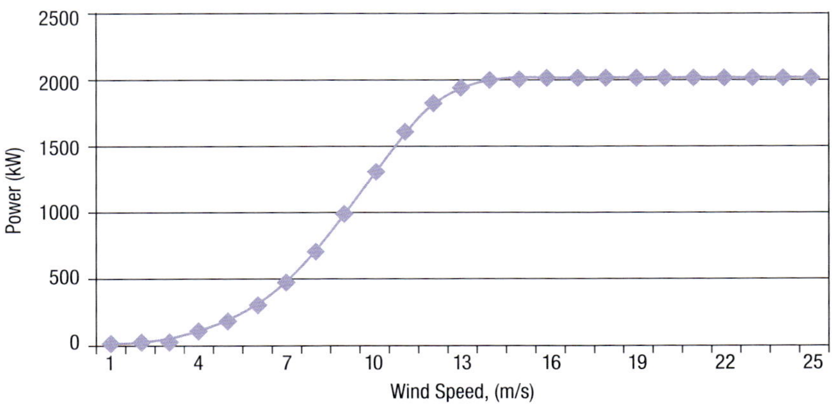

Figure 3. Power/wind speed curve of a 2 MW wind turbine (without noise reduction measures).

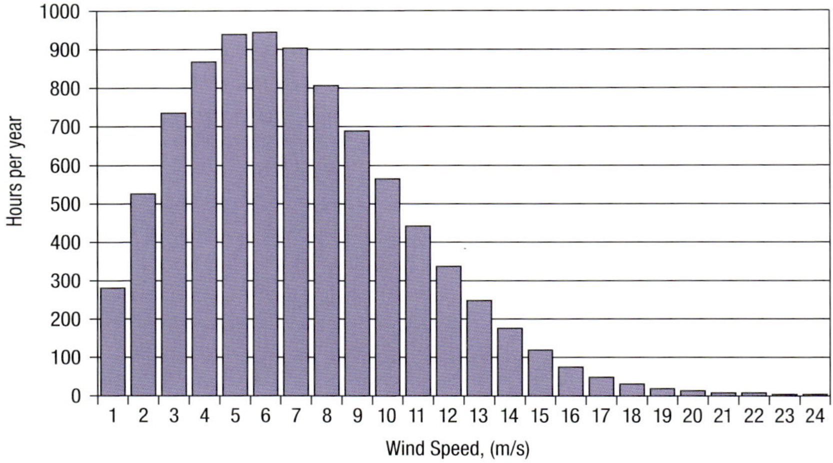

Figure 4. Wind speed distribution, average wind speed 7 m/s and Weibull shape factor of 2.

Based on the PV curve from Figure 3 and Weibull wind speed distribution with a shape factor of 2, the gross energy yield corresponding to 7 to 8.5 m/s is presented in Table 1.

Table 1. Gross yearly energy yield of a 10 MW wind farm as a function of average wind speed.

Average wind speed	7.0 m/s	7.5 m/s	8.0 m/s	8.5 m/s
	Weibull shape factor = 2			
Wind farm power (MW)	10	10	10	10
Gross annual energy yield (MWh)	27,000	31,000	34,000	37,000
Equivalent full load hours	2,700	3,100	3,400	3,700

The yearly gross energy yield of the 10 MW wind farm, assuming:

- 7 m/s average wind speed at hub height
- wind speed distribution according to Weibull distribution function, shape factor 2
- no noise reduction measures required

is 27,000 MWh, equivalent to 2,700 full load hours (utilization 31%).

Technical feasibility

Modern wind turbines are available in the power range from 0.75 to more than 3 MW, having rotor diameters varying from 55 to more than 100 meters. Although two-bladed rotors were also used in the past, nowadays only three-bladed rotors are commercially available. Generally, the hub height varies from 0.9 to 1.25 times the rotor diameter and most manufacturers offer wind turbines with two or three different rotor diameters corresponding to low (large rotor), medium (standard rotor) and high or offshore (small rotor) wind conditions.

At a first screening the outer dimensions of the available terrain are of importance. Wind turbines require a mutual spacing of at least four to five rotor diameters, corresponding to approximately 300 to 500 meters. A flat and undisturbed area is required because buildings, trees and other obstacles lead to a lowering of the wind speed.

It is also important to examine the local grid characteristics at an early stage. The main issue is the distance to the nearest medium or high voltage substation with an adequate feed-in capacity and the cost of installing this interconnection. Construction of this connection may require a lengthy planning cycle and involve a significant cost.

Main risk assessment

In most countries wind turbine rotors may not rotate above roads, railway tracks and waterways and a minimum clearance from public infrastructure must be observed. In northern countries and countries with a continental climate specific attention has to be paid to the possibility of icing. Ice developed on rotating rotor blades can be thrown long distances from the turbine, potentially causing injury and damage.

Planning authorities and regulatory bodies may require an additional risk analysis if the site is subject to icing or close to other infrastructure such as:

- facilities for transport, storage or processing of hazardous goods
- pipelines for transport of hazardous goods (including underground)
- residential, commercial or public buildings
- roads, railway tracks and waterways
- medium and high voltage conductors of transmission lines or cables.

Planning requirements of local authorities

The wind farm site has to meet planning and regulatory requirements. There may be a zoning plan that may prohibit wind turbines or limit the maximum height of structures. Under such circumstances the relevant authorities should be approached to investigate the possibility of obtaining permission at the earliest possible stage.

In most European countries wind turbines must be certified according to the relevant national or international safety standards. Manufacturers have to demonstrate conformance by the production of a valid "type-certificate."

For the proposed area for the 10 MW wind farm the following should be considered:

- check municipality zoning plan on competing activities and maximum building height
- mutual distance wind turbines 400 meters
- wind turbines in line, required length 1,600 meters
- there are to be no dwellings or other buildings and as few obstacles as possible within 300 to 500 meters of the wind turbines
- authorities or concerned parties may request a risk analysis if other activities are to take place within 400 to 500 meters from the wind turbines.

Project financing

Tables 2 and 3 give examples of the investment and operational costs of a 10 MW wind farm consisting of five wind turbines of 2 MW each. The costs are fully proportional to the installed capacity: investment costs per MW installed are about 1.25 M EUR and annual operating cost somewhat more than 40 k EUR per MW.

Table 2. Non-recurring investment costs (at 2006 levels).

	Costs per wind turbine [k EUR]	Wind farm costs [k EUR]
Preparatory costs	100	500
Wind turbines of 2 MW each	2,000	10,000
Wind farm civil and electrical infrastructure	200	1,000
Grid connection	200	1,000
Investment (year 1)	**2,500**	**12,500**
Refurbishment (at year 10)	250	1,250
TOTAL	**2,750**	**13,750**

Table 3. Annual operating and maintenance costs (at 2006 levels).

Annual operating costs	Costs per wind turbine [k EUR]	Wind farm costs [k EUR]
Wind turbine service, maintenance and insurance	50	250
Local taxes and contribution to grid connection	10	50
Land lease	15	75
Daily management	5	25
Own electricity consumption	5	25
TOTAL	**85**	**425**

Investment and operational costs for a 10 MW wind farm (at 2006 levels):

- initial investment costs 12.5 M EUR (1.25 M EUR/MW)
- annual operational costs 0.425 M EUR (42.5 k EUR/MW)
- estimated refurbishment costs halfway through the life span 1.25 M EUR; i.e., 10% of the initial investment.

In most cases the wind farm is financed by a mix of capital (equity) and a bank loan. The amount of equity is usually limited to 20 to 40% of the total investment. It may be attractive for companies, investment groups and individuals to invest in wind energy for possible tax reduction benefits. A number of national governments in Europe have introduced incentives to promote electricity production by renewable sources, for example, by introducing tax allowances for investments in "green" energy sources.

Pre-building phase

Having established the basic installation parameters, the predicted availability of wind, any planning constraints and connection possibilities, the task during this phase is to confirm the details and draw up detailed agreements so that project finance can be secured. To enable this to happen, a power purchase agreement (PPA) must be negotiated and suppliers of equipment and contractors selected. It is during this stage that any assumptions made during the start-up phase are re-examined and justified to avoid expenditure on a non-viable project.

Wind farm design including energy yield predictions

Based on the wind resource assessment the most promising wind farm locations are studied in more detail. "WasP," "Wind farmer," and "WindPRO" are well known computer models to calculate energy yield. The modelling takes into account not only wind speed and direction distributions but also the geography and terrain; for instance, a steep slope in the terrain will cause higher winds at the hill top. Such details in the modelling ensure that wind turbines are optimally sited. Generally the layout is optimized for exposure to the prevailing wind direction.

The mutual distance between the wind turbines has to meet the requirements of the manufacturers. If the wind turbines are too close together output may be reduced. Another, more serious, consequence may be damage to primary structural parts caused by the wake of wind turbines sited upwind. The minimum distance depends on the siting with regard to the prevailing wind direction. For turbines sited perpendicular to the prevailing wind direction, the mutual separation distance has to be at least four and preferably five times the rotor diameter.

The gross energy yield of the wind farm is determined by the local wind distribution and the siting of the wind turbines. To calculate the net energy yield the anticipated losses must be determined. These losses include:

- wake losses
- grid losses
- availability

Wake losses

Downstream of the turbine rotor, in the so-called wake, the wind speed is lower than the undisturbed wind speed, resulting in a somewhat reduced performance of other turbines sited in this area. The wake is characterized by extra turbulence that may lead to premature damage of main structural components.

It is common practice to estimate the wake losses as being in the range of 3 to 4% of the gross energy yield.

Grid losses

In this context, grid losses are defined as the electrical losses between wind turbine switchgear and public grid connection, i.e., the location of the metering position. Depending on the layout the electrical losses are in the range of 2 to 3% of the gross energy yield.

Availability

The availability of a wind turbine is defined as the percentage of time that the wind turbine is either in operation or available for operation should wind conditions permit and the grid connection be available. The technical availability of the turbine is 97% or higher, based on data from modern operational wind farms.

For this case study, we assume:

- 3% wake losses
- 3% grid losses
- 97% availability

The gross energy yield of 27 000 MWh becomes a net yield of 24 500 MWh.

Planning Procedures

Environmental issues

The wind farm must comply with all relevant environmental regulations. This may require a number of studies of, for example, the effects on birds, animals and plant life during the construction and use phases.

Noise

Wind turbines produce noise, mostly caused by the rotor blades and drive train, and the noise impact of wind turbines on the environment is one of the major planning issues. The distance to nearby residential buildings has to be sufficient to ensure that the noise level at the house front is below the statutory limit.

Visual impact

Visual impact of a wind farm is an important planning consideration. Wind farms require open, often elevated, sites and are consequently highly visible from a distance. Many of the potentially most productive sites are in areas of great natural beauty where planning regulation is very restrictive.

Shadow flickering is due to the periodic—about once per second—interruption of the sunlight by the rotating blades. Both flickering and shadow casting on dwellings and offices can be very annoying for the occupants. Shadow flickering is not regulated by law.

Safety

Most planning authorities demand safety and risk assessment studies. Wind turbines are not permitted to rotate above roads or railway-tracks.

Where icing of the blades or nacelle is likely limitations may be placed on the operation of the turbines with the rotor being stopped during icing and only released for restart when the ice has dispersed.

Grid connection

Each of the turbines of the farm is connected to a farm grid, operating at medium voltage (10 to 20 kV) to minimize losses. Usually a ring connection is used to provide a degree of redundancy. Since the generators operate at, typically, less than 1 kV a transformer is required to step the voltage up to the grid voltage. The transformers may be housed in the nacelle or at the base of the tower.

The bulk connection equipment required depends on the operating voltage of the public grid at the connection point. A transformer may be required in addition to switching, metering and protection equipment.

In some locations there may be a suitable connection point or the available connection point may have insufficient capacity. The cost of work to extend or reinforce the grid specifically for the wind farm has to be borne fully by the wind farm developer.

Feed-in contract

In all cases a Power Purchase Agreement (PPA) or feed-in contract is necessary. The feed-in tariff is made up of a price for the production and delivery of electricity and, in most cases, an additional amount for generating renewable energy or corresponding carbon credits. Depending on the wind resources and the investment and operational costs, a feed-in tariff of at least 60 EUR/MWh is required to cover costs and, for an economically viable project, a tariff of 80 to 90 EUR/MWh is needed.

Selection of suppliers

Depending on the working practices of the wind farm developer, the tendering process may be public or restricted to a shortlist of pre-qualified or preferred suppliers. In either case, a Tender Enquiry Document (TED) is required containing all the relevant information.

The TED should include at least the following information:

- basic information such as project time plan and planning status
- information required from bidders, e.g., financial, technical and operational information
- time schedule, e.g., tender closure date, consultation period, site visit opportunities
- contractual issues
- scope of supply
- technical specifications
- maintenance and repair conditions
- insurance and warranty agreements

With regard to the scope of supply it may be decided that a single contractor is responsible for delivering (by sub-contracting) a 'turnkey' wind farm including the wind turbines, foundations, access roads, wind farm grid and grid connection. Alternatively, the wind farm developer may manage the individual subcontracts. The latter may have cost advantages, however the project developer becomes the main contractor and is responsible for the integration commissioning, either internally or by appointing a contractor for this phase of the work. Turnkey delivery by one main contractor is usually the better option since responsibility is clearer.

It is important that the wind turbine is certified by a recognized body and possesses a type-certificate valid for the local wind climate and wind farm layout. In order to avoid any problems on warranties the contractor should be required to state formally that the delivered wind turbines are 'fit for purpose' for the site as described by the developer.

Project financing

As noted earlier, for most sites in Europe wind energy is not yet cost-effective and incentives are essential to encourage investment in wind energy. The most common promotion measures are:

- subsidies (governmental or local) on investments in renewable energy sources
- tax benefits for investing in renewable energy sources
- reduced interest tariffs on loans for renewables
- subsidies on the production of renewable energy (i.e., increased feed-in tariff).

When financing is based for the greater part on loans, the financiers may ask for additional securities to guarantee that the loan can be repaid. The following securities may be requested:

- power purchase agreement with fixed minimum feed-in tariff for the loan period
- guarantee that, even in years with moderate wind resources, the income is sufficient to meet interest and capital repayment
- warranties on supplied components and wind farm performance (availability and power curve) for the loan period
- machine breakdown and business interruption insurance
- service and maintenance contract

Annex A (page 435) gives cash flow calculations for two different feed-in tariffs over an anticipated lifespan of 20 years; the results are summarized in Table 4. Financing costs are not taken into account and all expenses and incomes are based on 2006 price levels so the influence of inflation is implicitly included.

Table 4. Summary of cash flow calculations for a 10 MW wind farm (see Annex A).

Feed-in tariff	60 EUR/MWh	85 EUR/MWh
Gross income	1.47 M EUR/yr	2.08 M EUR/yr
Break even period	> 15 years	> 9 years
Net present value over 20 years (NPV)	6.1 M EUR	17.7 M EUR
Internal rate of return (IRR)	4%	11%

The calculations show that, at a feed-in tariff of 60 EUR/MWh, the rate of return of 4% is just enough to cover the cost of financing the project (assuming the project is fully financed by a bank loan) and no profit is generated. At a tariff of 85 EUR/MWh the project generates an 11% return gross, or 7% after paying the costs of financing the project.

Building Phase

The building phase includes all activities from commencement of the works up to handover of the operational wind farm to the operator.

Overview of the building process

Once all the previous stages have been concluded, and the status reviewed, the manufacturing and build stage can commence.

Wind turbines are large and heavy so suitable site access for transport is required. On-site storage and onsite assembly work will require space at the base of each tower of approximately 80 by 50 meters. In addition, heavy lifting equipment will be required on site—a 2 MW turbine will require a caterpillar crane with a capacity of 600 tonnes for hoisting of the tower parts, nacelle and rotor into position.

Manufacturing and assembly of the main components takes place in the factories of the wind turbine supplier and the following assembled main components are then shipped to site:

- foundation anchor or tube
- three or four tubular tower parts
- ground controller and switchgear
- fully assembled nacelle (including gearbox, generator, yaw mechanism, mechanical break, converter and transformer, if applicable)
- hub and rotor blades
- wind farm Supervisory Control and Data Acquisition system (SCADA)
- transformer (in case of ground based).

For a small and medium size wind farm the time between placing a purchase order and shipment from the factory is 6 to 9 months. In the meantime the civil engineering work, including access roads, turbine foundations and substation, and the electrical infrastructure, is built. The time required for rotor assembly and construction of the main structure takes 2 to 3 working days per wind turbine. Once the turbines are installed, about 7 to 10 working days are needed to complete the installation work, commission the system and make the connection to the grid.

From the time all the material is on site the construction time of a small and medium size wind farm takes only 2 to 3 months.

Quality control during production and construction

It is common practice that the contractor assigns a number of so-called 'hold and witness' points for the client. These hold and witness points are meant for the wind farm owner to audit the progress and quality of work including the verification that the components conform to the specifications. Hold and witness points are mostly planned immediately following a project milestone, for instance a main component ready for transport to site. Often, hold and witness points are linked to staged payments against the contract.

Typical hold and witness points are:

- start of component production including audit of contractors' quality system
- factory acceptance test (FAT) of components ready for shipment
- site acceptance test (SAT) of components delivered to site
- several inspections during building on site, connected to milestones.

Commissioning and handover

Following completion of the building and installation period, and before handover of the wind farm, an overall inspection and commissioning of the works is carried out. Commissioning inspections are performed by representatives of the contractor and the final owner, with participation from the local network operator. The commissioning may involve an elaborate testing and monitoring plan, but the main objective is to verify that the system is complete, correctly installed and functioning properly. Normally a commissioning procedure is formulated in cooperation with all the parties involved.

It is quite common that a number of defects are discovered during the first commissioning inspection, resulting in a defect or 'snag' list. Actions to clear this list are decided between the parties concerned and, if there are serious defects, may result in a second commissioning inspection being required.

Approved commissioning and handover is usually related to final project payment.

Operation and maintenance

Starting from date of handover, the owner is responsible for the daily operation of the wind farm. Also, from that date warranty and maintenance contracts become valid. The technical and economical life span of a wind farm is anticipated as 20 years.

Daily operation

During normal operation it is not necessary to man the site. SCADA equipment allows remote monitoring, via a modem and telephone line or via the internet, of the performance and condition of the turbines and the farm.

The main function of the daily operator is to verify regularly that the wind farm is in the optimum condition and performing according to expectation and to ensure that maintenance and repairs are carried out in accordance with the contract and within a reasonable time.

Warranties and insurance

The following warranties are common for the first five years following handover:

- on delivered goods, including repairs and modifications
- availability of individual wind turbines and wind farm, values of 95% or higher are not uncommon
- performance according to the power curve—warranties of 95% of the certified turbine PV-curve are common. No warranty on actual performance can be given because the exact wind supply cannot be predicted.

If the availability or performance is below the warranted value, the difference between actual and warranted values has to be settled by the supplier. Some suppliers offer warranties of up to 8 to 12 years, or at least for a period comparable to the financing period.

Insurance cover is required for:

- third party liability
- machine breakdown (e.g., material flaws, lightning strikes, fire, vandalism, actions of maintenance engineer and/or operator, etc.)
- business interruption (compensation for unproductive days following a machine breakdown event).

Maintenance and repair

Modern wind turbines require a preventive maintenance service twice a year. For a wind turbine in the MW-segment a planned preventive maintenance overhaul requires 2 to 3 working days for two engineers. The work includes inspection and testing of the control and safety devices, repair of small defects, replacement or replenishment of consumables such as bearing grease and gearbox lubrication. The gearbox is the most vulnerable component and therefore the subject of special interest during maintenance. Oil samples are taken at regular intervals and analyzed for signs of degradation, filters are replaced and gears are inspected for damage.

The number of repairs required varies widely between different wind turbines and wind farms. On average, 3 to 4 corrective actions (i.e. those requiring a visit by a service engineer) are required for each turbine. The mean downtime per failure is 2 to 4 days and the causes are equally divided between mechanical and electrical problems.

Although not formally admitted by the manufacturers, it is common practice that a major overhaul is carried out on wind turbines after 10 to 12 years of operation. The overhaul includes cleaning of, and repair work to the rotor blades and refurbishment of the drive train, i.e., replacement of bearings and, if necessary, replacement of gearbox parts.

Annex A—Examples of two cash flow calculations for a 10 MW wind farm

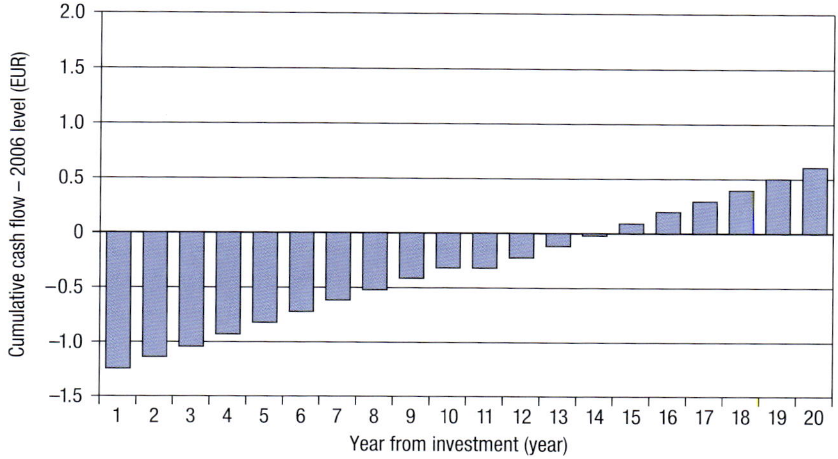

Figure A1. Cash flow for 10 MW wind farm at 60 EUR/MWh.

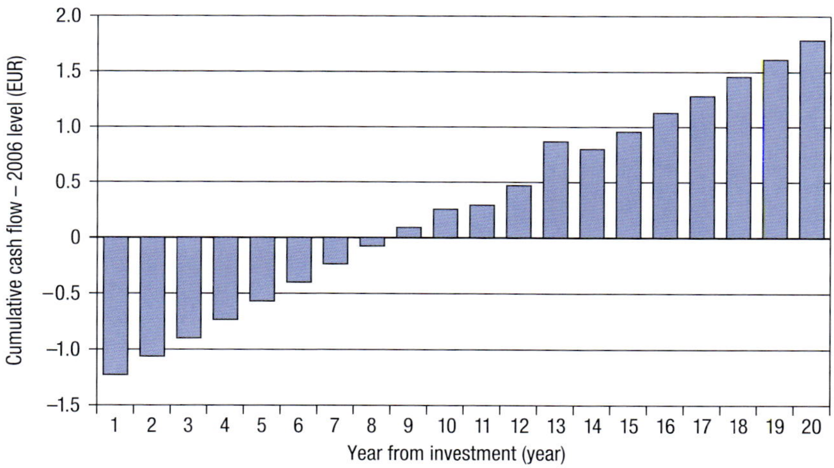

Figure A2. Cash flow for 10 MW wind farm at 85 EUR/MWh.

Table A1. Non-recurring investment costs (at 2006 levels).

Feed-in tariff	0.060	0.085	EUR/kWh
Investment costs (year 1)	12,500,000	12,500,000	EUR
Refurbishment costs (year 10)	1,250,000	1,250,000	EUR
Yearly wind farm production in kWh	24,500,000	24,500,000	kWh
Gross income from energy production	1,470,000	2,082,500	EUR
Yearly recurring costs	425,000	425,000	EUR
Break even period	15	9	years
Net present value (2006 price level, 20 year life)	6,100,000	17,750,000	EUR
Internal rate of return	4	11	%

F

Appendix F: Estimating Wind Energy Costs

The material in Appendix F was excerpted directly from the sources indicated.

Table F–1. Wind Energy Economics Worksheet. *Data courtesy of David M. Eggleston and Forrest S. Stoddard,* Wind Turbine Engineering Design *(pages 109–110).*

CAPITAL COSTS	$/turbine
Preliminary costs	
Site analysis and site control	
Wind resource assessment	
Environmental impact	
Preliminary//detail engineering	
Soil analysis	
Electrical	
Mechanical	
Technical support	
Other	
Permitting/approval, financing, legal, etc.	
Purchasing, shipping, and installation	
Factory price (determine what is included/not)	
Machine	
Tower	
Special electric equipment	
Other	
Shipping	
Export packing	
Factory to dock charges	
Dock receiving charges	
Export declaration papers/brokers' fee	
Wharfage	
Ocean or air freight	
Destination handling charge, dock storage (avoid if possible)	
Inter-island barge shipment (if no direct delivery)	
Trucking to site	
Other	
Site preparation	
Equipment access	
Security fencing	
Other	
Installation	
Footing excavation	
Reinforcement rods	
Anchor rods and plates	
Concrete/delivery to site	
Erection equipment rent/lease/purchase	
Labor	
Electrical hookup	
Special electrical equipment (substation, transformer, telephone poles, etc.)	
Transmission line or energy storage	
Other	

CAPITAL COSTS	$/turbine
GROSS CAPITAL COST	
Tax benefits	
National/federal tax credits	
Regional/local tax credits	
Depreciation allowance (if applicable)	
NET CAPITAL COSTS	
OPERATING COSTS	
Annual costs (based on revenue factors)	
Estimated MWh/yr	
Value of MWh/yr used by owner at utility rate	
Value of MWh/yr sold to utility at PPA or other rate	
GROSS REVENUE/ANNUAL VALUE	
Minus downtime (forced outages, 5–10%)	
NET ANNUAL REVENUE	
Operating expenses	
Debt retirement	
O&M costs	
Routine	
Extraordinary/parts replacement	
Land rent	
Insurance	
Income taxes	
Excise taxes on energy sold to utility	
NET ANNUAL OPERATING EXPENSES	
NET ANNUAL INCOME	

G

Appendix G: World Offshore Wind Power

Table G–1 lists details of Europe's offshore wind farms on 2011, including water depths, distance from shore, turbine manufacturer, type of foundation, and foundation supplier. Table G–2 lists offshore wind farms worldwide in 2011, including operator, number of turbines, turbine model, and foundation type. Table G–3 lists offshore wind capacities by country, including new installations and growth rates. *The material in Appendix G was excerpted directly from the sources indicated.*

Table G–1. Europe's offshore wind farms. *Data courtesy of European Wind Energy Association.*

Country	Project Name	Location	Capacity	No. of Turbines	Water Depth	Distance to Shore	Online	WT Manufacturer	Owner/Operator	Foundation Type	Foundation Supply
Belgium	Thornton Bank phase 1	Off Zeebrugge	30	6	12 to 27	27 to 30	2008	REpower	C-power	Gravity	GeoSea
	Belwind phase 1	Off Zeebrugge	165	55	15 to 37	46 to 51	2010	Vestas	Colruyt, Meevind, PMV, SHV, Rabobank	Monopile	Van Oord
	TOTAL		195	61							
Denmark	Vindeby	NW of Vindeby	4.95	11	2.5 to 5	2.5	1991	Siemens	DONG Energy	Gravity	
	Tunø Knob	Off Aarhus	5	10	0.8 to 4	6	1995	Vestas	DONG Energy	Gravity	
	Middelgrunden	Oresund, E of Copenhagen	40	20	2 to 6	2	2001	Siemens	DONG Energy	Gravity	MT Højgaard
	Horns Rev 1	NW of Esbjerg	160	80	6 to 14	14	2002	Vestas	Vattenfall, DONG, pensionDanmark	Jacket	SIF Smulders, MT Højgaard
	Nysted/Rødsand 1	Off Rødsand	165.6	72	6 to 10	6 to 10	2003	Siemens	DONG Energy (50%), E.On (50%)	Gravity	Per Aarsleft
	Samsø	S of Samsø	23	10	11 to 18	3.5	2003	Siemens	Samsø Kommune	Monopile	Bladt Industries
	Frederikshavn	Frederikshavn Harbour	10.6	4	3	0.8	2003	Vestas, Bonus, Nordex	DONG Energy		
	Horns Rev 2	Blåvandshuk	209.3	91	9 to 17	30	2009	Siemens	DONG Energy	Monopile	Bladt, Per Aarsleft & Bilfinger Berger
	Storebaeld/Sprogø	N of Sprogø	21	7	6 to 16	2	2009	Vestas	Sund & Baelt	Gravity	Per Aarsleft & Bilfinger Berger
	Avedøre Holme	Of Avedøre	7.2	2	0.5	0.001	2009	Siemens	DONG Energy	Gravity	
	Poseidon Wind & Wave	NW of Vindeby	0.033	3	7	3	2010	GAIA Wind	Floating Power plant A/S	Floating	
	Nysted II/Rødsand II	Off Rødsand	207	90	6 to 12	23	2010	Siemens	E.On	Gravity	Per Aarsleft & Bilfinger Berger
	TOTAL		853.7	400							

Country	Project Name	Location	Capacity	No. of Turbines	Water Depth	Distance to Shore	Online	WT Manufacturer	Owner/Operator	Foundation Type	Foundation Supply
Finland	Kemi Ajos	Ajos Harbour	24	8	3	<1	2008	WinWind	Pohjolan Volma	Gravity	
	Pöri offshore 1	Off Pöri	2.3	1	9	2	2010	Siemens	Suomen Hyotytuuli	Gravity	Technip Offshore
	TOTAL		26.3	9							
Germany	Enova Offshore	Ems river in Emden	4.5	1	2	<1	2004	Enercon	Enova, EWE		
	Breitling	Rostock Harbour	2.5	1	2	1	2006	Nordex	Nordex AG		
	Hooksiel	Hooksiel Harbour	5	1	2 to 8	0.4	2008	BARD	BARD Group	Tripile	BARD
	Alpha Ventus	N of Borkum	60	12	30	43	2010	Areva Wind, REpower	DOTI (EWE 48% E.On 26%, Vattenfall 26%)	Tripod, Jacket	Aker Kvaerner, BiFab
	BARD offshore 1 (partially completed)	Western North Sea	20	4	40	100	2010	BARD	BARD Engineering (92.5%), Enovos (7.5%)	Tripile	Cuxhaven Steel Construction
	TOTAL		92	19							
Ireland	Arklow Band	Off Arklow	25.2	7	2.5 to 5	10	2004	GE	GE	Monopile	Sif Smulders
	TOTAL		25.2	7							
Netherlands	Lely	Medembilk, Ijsseimeer	2	4	7.5	0.75	1994	Nedwind	Nuon	Monopile	
	Irene Vorrink	Dronten, Ijsselmeer	16.8	28	2	0.03	1996	NordTank	Nuon	Monopile	
	Offshore Wind Farm Egmond aan Zee	Off Egmond aan Zee	108	36	19 to 22	8 to 12	2007	Vestas	NoordZee Wind	Monopile	Bladt
	Prinses Amalia	Off Ijmuiden	120	60	19 to 24	23	2008	Vestas	Econcern, Eneco	Monopile	Sif Smulders
	TOTAL		246.8	128							
Norway	Floating Hywind	Off Karmoy	2.3	1	220	12	2009	Siemens	Statoil	Floating	
	TOTAL		2.3	1							

Country	Project Name	Location	Capacity	No. of Turbines	Water Depth	Distance to Shore	Online	WT Manufacturer	Owner/Operator	Foundation Type	Foundation Supply
Sweden	Bockstigen	Gotland	2.75	5	6 to 8	3	1998	NEG-Micon		Monopile	
	Utgrunden I	Kalmarsund	10.5	7	4 to 10	7	2001	GE	Vattenfall	Monopile	
	Yttre Stengrund	Kalmarsund	10	5	8 to 12	4	2002	NEG-Micon	Vattenfall	Monopile	
	Lillgrund	Oresund straight	110.4	48	2.5 to 9	10	2007	Siemens	Vattenfall	Gravity	Hocthief
	Gässlingegrund	Värnen	30	10	4 to 10	4	2009	WinWind	Vindpark Värnen Drift	Gravity	PEAB
	TOTAL		163.7	75							
United Kingdom	Blyth Offshore	Blyth Harbour	4	2	6	1	2000	Vestas	E.On	Monopile	
	Horth Hoyle	Prestatyn and Rhyl	60	30	5 to 12	3 to 10	2003	Vestas	Npower (RWE)	Monopile	Sif Smulders
	Scroby Sands	NE of Greater Yarmouth	60	30	2 to 10	2.5	2004	Vestas	E.On	Monopile	Cambrian engineering/Isleburn Mackay Mcleod
	Kentish Flats	Off Whitstable	90	30	5	8.5	2005	Vestas	Vattenfall	Monopile	Sif Smulders, MT Højgaard
	Barrow	Off Walney Island	90	30	21 to 23	7	2006	Vestas	DONG Energy, Centrica	Monopile	Sif Smulders, KBR
	Beatrice	Beatrice oil field	10	2	40	25	2007	REpower	SSE, Talisman	Jacket	BiFab
	Burbo Bank	Off Crosby	90	25	10	5.2	2007	Siemens	DONG Energy	Monopile	MT Højgaard
	Inner Dowsing	Ingoldmells/ Skegness	97.2	27	10	5	2008	Siemens	Centrica	Monopile	MT Højgaard
	Lynn	Ingoldmells/ Skegness	97.2	27	10	5.2	2008	Siemens	Centrica	Monopile	MT Højgaard
	Rhyl flats	Off Rhyl	90	25	4 to 15	8	2009	Siemens	Npower (RWE)	Monopile	MT Højgaard
	Robin Rigg	Off Maryport	180	60	>5	9.5	2010	Vestas	E.On	Monopile	MT Højgaard
	Gunfleet Sands	Off Clacton-on-Sea	172.8	48	2 to 15	7	2010	Siemens	DONG Energy	Monopile	MT Højgaard
	Thanet	Off Margate	300	100	20 to 25	7 to 8.5	2010	Vestas	Warwick Energy (Vattenfall)	Monopile	Sif Smulders
	TOTAL		1,341.2	436							
	TOTAL CAPACITY		**2,946.2**	**1,136**							

Table G–2. Offshore wind farms worldwide. *Data courtesy of Lindø Offshore Renewables Center.*

Wind Farm Name	Country	Operator	Capacity, MW	No. of Turbines	Turbine Models	Foundations
Belwind	Belgium	Belwind	165	55	Vestas V90-3.0 MW	Monopiles
Thornton Bank 1	Belgium	C-Power	30	6	REpower 5M	Gravity-base
Donghai Bridge	China	Shanghai Donghai Wind Power	102	34	Sinovel SL3000/90	Gravity-base
Suizhong 36-1 Oil Field	China	CNOOC	1.5	1	Goldwind GW70/1500	Jackets
Avedore Holme	Denmark	DONG Energy	7.2	2	Siemens SWT-3.6-120	Gravity-base
Frederikshavn	Denmark	DONG Energy	7.6	3	Nordex N90-2300 Vestas V90-3.0 MW Bonus 2.3 MW/82	Monopiles Buckets Monopiles
Great Belt	Denmark	Sund & Bælt	21	7	Vestas V90-3.0 MW	Gravity-base
Horns Rev 1	Denmark	Vattenfall	160	80	Vestas V80-2.0 MW	Monopiles
Horns Rev 2	Denmark	DONG Energy	209.3	91	Siemens SWT-2.3-93	Monopiles
Middelgrunden	Denmark	DONG Energy	40	20	Bonus Energy 2.0 MW/76	Gravity-base
Nysted 1/Rodsand 1	Denmark	DONG Energy	165.6	72	Bonus 2.3 MW/82	Gravity-base
Nysted 2/Rodsand 2	Denmark	E.ON	207	90	Siemens SWT-2.3-93	Gravity-base
Paludans Flak	Denmark	Samsø Havvind	23	10	Bonus 2.3 MW/82	Monopiles
Ronland	Denmark	Thyborøn-Harboøre Vindmøllelaug; Vindenergi/Harboøre Møllelaug	17.2	8	Vestas V80-2.0 MW Bonus 2.3 MW/82	Gravity-base
Tunø Knob	Denmark	DONG Energy	5	10	Vestas V39-500 kW	Gravity-base
Vindeby	Denmark	DONG Energy	4.95	11	Bonus 450 kW	Gravity-base
Kemi Ajos	Finland	Pohjolan Voima	30	10	WinWinD WWD-3	Gravity-base
Pori 1	Finland	Suomen Hyötytuuli	2.3	1	Siemens SWT-2.3-101	Steel Shell Gravity

Wind Farm Name	Country	Operator	Capacity, MW	No. of Turbines	Turbine Models	Foundations
Alpha Ventus	Germany	Stiftung Offshore		12	Areva Multibrid M5000 REpower 5M	Tripods Jackets
BARD 1	Germany	TenneT TSO	400	80	Bard 5.0	Tripiles
Breitling	Germany	WIND-projekt	2.5	1	Nordex N90/2500 Offshore	Gravity-base
Ems Emden	Germany	ENOVA	4.5	1	Enercon E-112	Gravity-base
EnBW Baltic 1	Germany	EnBW	48.3	21	Siemens SWT-2.3-93	Monopiles
Hooksiel	Germany	BARD	5	1	Bard 5.0	Tripiles
Arklow Bank 1	Ireland	GE Energy	25.2	7	GE 3.6 MW Offshore	Monopiles
Kamisu	Japan	Wind Power Ibaraki	14	7	Subaru 80/2.0 MW	Monopiles
Sakata	Japan	Summit Wind Power	16	8	Vestas V80-2.0 MW	Piled Platform
Setana	Japan	Setana Town	1.32	2	Vestas V47-660 kW	n/a
Egmond aan Zee	Netherlands	Nuon Vattenfall	108	36	Vestas V90-3.0 MW	Monopiles
Irene Vorrink	Netherlands	Nuon Vattenfall	16.8	28	Nordtank NKT 600/43	Monopiles
Lely	Netherlands	Nuon Vattenfall	2	4	Nedwind N40/500 kW	Monopiles
Princess Amalia	Netherlands	Offshore Windpark Q7	120	60	Vestas V80-2.0 MW	Monopiles
Hywind	Norway	Statoil	2.3	1	Siemens SWT-2.3-82 VS	Floating
Bockstigen	Sweden		2.5	5	Wind World 500 kW	Monopiles
Lillgrund	Sweden	Vattenfall	110.4	48	Siemens SWT-2.3-93	Gravity-base
Utgrunden Demo.	Sweden	Vattenfall	10.5	7	Enron Wind EW 1.5s	Monopiles
Vänern (Gässlingegrund)	Sweden	Vindpark Vänern Kraft	30	10	WinWinD WWD-3	Gravity-base
Yttre Stengrund	Sweden	Vattenfall	10	5	NEG-Micon NM72/2000	Monopiles
Barrow	UK	Centrica	90	30	Vestas V90-3.0 MW	Monopiles

Beatrice Demo.	UK	SSE Renewables	10	2	REpower 5M	Jackets
Blyth	UK	E.ON	4	2	Vestas V66-2.0 MW (Offshore)	Monopiles
Burbo Bank	UK	Dong Energy	90	25	Siemens SWT-3.6-107	Monopiles
Greater Gabbard	UK	SSE Renewables RWE npower renewables	504	140	Siemens SWT-3.6-107	Monopiles
Gunfleet Sands	UK	DONG Energy	172.8	48	Siemens SWT-3.6-107	Monopiles
Inner Dowsing	UK	Centrica	97.2	27	Siemens SWT-3.6-107	Monopiles
Kentish Flats 1	UK	Vattenfall	90	30	Vestas V90-3.0 MW	Monopiles
Lynn	UK	Centrica	97.2	27	Siemens SWT-3.6-107	Monopiles
North Hoyle	UK	RWE npower renewables	60	30	Vestas V80-2.0 MW	Monopiles
Rhyl Flats	UK	RWE npower renewables	90	25	Siemens SWT-3.6-107	Monopiles
Robin Rigg	UK	E.ON	180	60	Vestas V90-3.0 MW	Monopiles
Scroby Sands	UK	E.ON	60	30	Vestas V80-2.0 MW	Monopiles
Thanet	UK	Vattenfall	300	100	Vestas V90-3.0 MW	Monopiles
Walney 1	UK	DONG Energy	183.6	51	Siemens SWT-3.6-107	Monopiles

Table G–3. Offshore wind power capacity by country. *Data courtesy of World Wind Energy Association.*

Country	End 2008	End 2009	New 2010	End 2010	% Growth 2009–10
		MW			
United Kingdom	574.0	688.0	653.0	1,341.0	94.9
Denmark	426.6	663.6	190.4	854.0	28.7
Netherlands	247.0	247.0	2.0	249.0	0.8
Belgium	30.0	30.0	165.0	195.0	550.0
Sweden	134.0	164.0	0.0	164.0	0.0
China	2.0	23.0	100.0	123.0	434.8
Germany	12.0	72.0	36.3	108.3	50.4
Finland	30.0	30.0	0.0	30.0	0.0
Ireland	25.0	25.0	0.0	25.0	0.0
Japan	1.0	1.0	15.0	16.0	1,500.0
Spain	10.0	10.0	0.0	10.0	0.0
Norway	0.0	2.3	0.0	2.3	0.0
Total	1,491.6	1,955.9	1,161.7	3,117.6	59.4

H

Appendix H: Global Wind Power Capacity

Table H–1 lists capacities by region and country. Table H–2 lists capacity additions and cumulative capacities by country. Table H–3 provides growth rates and previous years' capacities. Table H–4 contains details on Europe's wind power capacities, broken down by European Union members (12, 15, and 27 countries) and by candidate countries that have not yet joined the union. Table H–5 lists wind power capacities by U.S. state. *The material in Appendix H was excerpted directly from the sources indicated.*

Table H–1. Global wind power capacity by region and country. *Data courtesy of Global Wind Energy Council.*

Region	Country	End 2009	New 2010	End 2010
			MW	
Africa & Middle East	Egypt	430	120	550
	Morocco	253	33	286
	Tunisia	54	60	114
	Iran	92	0	92
	Other	37	0	37
	Total	866	213	1,079
Asia	China	25,805	18,928	44,733
	India	10,926	2,139	13,065
	Japan	2,085	221	2,304
	Taiwan	436	83	519
	South Korea	348	31	379
	Philippines	33	0	33
	Other	6	48	54
	Total	39,639	21,450	61,087
Europe	Germany	25,777	1,493	27,214
	Spain	19,160	1,516	20,676
	Italy	4,849	948	5,797
	France	4,574	1,086	5,660
	United Kingdom	4,245	962	5,204
	Portugal	3,535	363	3,898
	Denmark	3,465	327	3,752
	Netherlands	2,215	32	2,237
	Sweden	1,560	604	2,163
	Ireland	1,310	118	1,428
	Turkey	801	528	1,329
	Greece	1,087	123	1,208
	Poland	725	382	1,107
	Austria	995	16	1,011
	Belgium	563	350	911
	Rest of Europe	1,610	1,070	2,684
	Total	76,471	9,918	86,279
	of which EU-27	75,090	9,259	84,278

Region	Country	End 2009	New 2010	End 2010
			MW	
Latin America	Brazil	606	326	931
	Mexico	202	316	519
	Chile	168	4	172
	Costa Rica	123	0	123
	Caribbean	91	8	99
	Argentina	34	27	60
	Other	83	23	106
	Total	1,306	703	2,008
North America	United States	35,086	5,115	40,180
	Canada	3,319	690	4,009
	Total	38,405	5,805	44,189
Pacific	Australia	1,712	167	1,880
	New Zealand	497	9	506
	Pacific Islands	12	0	12
	Total	2,221	176	2,397
Total		158,908	38,265	197,039

Notes: Decommissioning of 151 MW and number-rounding affect totals.
Data for Denmark and the Netherlands are provisional.

Table H–2. Capacity additions and cumulative capacity by country. *Data courtesy of BTM Consult, AWEA, and U.S. DOE.*

Country	New 2010, MW	Country	End 2010, MW
China	18,928	China	44,781
United States	5,113	United States	40,267
India	2,139	Germany	27,364
Germany	1,551	Spain	20,300
United Kingdom	1,522	India	12,966
Spain	1,516	France	5,961
France	1,186	United Kingdom	5,862
Italy	948	Italy	5,793
Canada	690	Canada	4,011
Sweden	604	Portugal	3,837
Rest of world	5,205	Rest of world	28,371
Total	39,402	Total	199,513

Table H–3. Global wind power capacity and growth rates by country. *Data courtesy of World Wind Energy Association.*

Position 2010 Country/Region	Total capacity end 2010 (MW)	Added capacity 2010 (MW)	Growth rate 2010 (%)	Position 2009	Total capacity end 2009 (MW)	Total capacity end 2008 (MW)	Total capacity end 2007 (MW)	Total capacity end 2006 (MW)
1 China	44,733.0	18,928.0	73.3	2	25,810.0	12,210.0	5,912.0	2,599.0
2 USA	40,180.0	5,600.0	15.9	1	35,159.0	25,237.0	16,823.0	11,575.0
3 Germany	27,215.0	1,551.0	6.0.	3	25,777.0	23,897.0	22,247.4	20,622.0
4 Spain	20,676.0	1,527.2	8.0	4	19,149.0	16,689.0	15,145.1	11,630.0
5 India	13,065.8	1,258.8	10.7	5	11,807.0	9,587.0	7,850.0	6,270.0
6 Italy	5,797.0	950.0	19.6	6	4,850.0	3,736.0	2,726.1	2,123.4
7 France	5,660.0	1,086.0	23.7	7	4,574.0	3,404.0	2,455.0	1,567.0
8 United Kingdom	5,203.8	1,111.8	27.2	8	4,092.0	3,195.0	2,389.0	1,962.9
9 Canada	4,008.0	690.0	20.8	11	3,319.0	2,369.0	1,846.0	1,460.0
10 Denmark	3,734.0	309.0	8.9	10	3,465.0	3,163.0	3,125.0	3,136.0
11 Portugal	3,702.0	345.0	10.3	9	3,357.0	2,862.0	2,130.0	1,716.0
12 Japan	2,304.0	211.0	10.1	13	2,083.0	1,880.0	1,528.0	1,309.0
13 The Netherlands	2,237.0	15.0	0.7	12	2,223.0	2,235.0	1,747.0	1,559.0
14 Sweden	2,052.0	603.8	41.7	15	1,448.2	1,066.9	831.0	571.2
15 Australia	1,880.0	3.0	0.2	14	1,877.0	1,494.0	817.3	817.3
16 Ireland	1,428.0	118.0	9.0	16	1,310.0	1,027.0	805.0	746.0
17 Turkey	1,274.0	477.5	59.9	19	796.5	333.4	206.8	64.6
18 Greece	1,208.0	123.0	11.3	17	1,086.0	989.7	873.3	757.6
19 Poland	1,107.0	382.0	52.7	20	725.0	472.0	276.0	153.0
20 Austria	1,010.6	16.0	1.6	18	995.0	994.9	981.5	964.5
21 Brazil	920.0	320.0	53.3	21	600.0	338.5	247.1	236.9

Position 2010 Country/Region	Total capacity end 2010 (MW)	Added capacity 2010 (MW)	Growth rate 2010 (%)	Position 2009	Total capacity end 2009 (MW)	Total capacity end 2008 (MW)	Total capacity end 2007 (MW)	Total capacity end 2006 (MW)
22 Belgium	886.0	340.0	62.0	22	548.0	383.6	286.9	194.3
23 Romania	591.0	577.0	4,121.4	55	14.0	7.0	7.8	2.8
24 Egypt	550.0	120.0	27.6	26	435.0	390.0	310.0	230.0
25 Mexico	521.0	104.5	25.1	27	416.8	85.0	85.0	84.0
26 Chinese Taipeh	518.7	82.6	18.9	24	436.0	358.2	279.9	187.7
27 New Zealand	506.0	8.8	1.8	23	497.0	325.3	321.8	171.0
28 Norway	434.6	18.4	4.3	25	431.0	429.0	333.0	325.0
29 Korea (South)	279.3	48.9	14.0	28	348.4	278.0	192.1	176.3
30 Bulgaria	374.5	198.0	112.2	30	176.5	157.5	56.9	36.0
31 Hungary	295.0	94.0	46.8	31	201.0	127.0	65.0	60.9
32 Morocco	286.0	33.0	13.0	29	253.0	124.0	125.2	64.0
33 Czech Republic	215.0	24.0	12.6	32	191.0	150.0	116.0	56.5
34 Finland	197.0	52.0	35.4	33	147.0	143.0	110.0	86.0
35 Chile	170.0	2.6	1.5	39	167.6	20.1	20.1	2.0
36 Lithuania	154.0	63.0	69.2	36	91.0	54.4	52.3	55.0
37 Estonia	149.0	6.9	4.8	34	142.3	78.3	58.6	33.0
38 Costa Rica	123.0	0.0	0.0	35	123.0	74.0	74.0	74.0
39 Iran	100.0	18.0	22.0	38	82.0	82.0	66.5	47.4
40 Ukraine	87.4	0.6	0.7	37	90.0	90.0	89.0	85.6
41 Cyprus	82.0	82.0	∞	0	0.0	0.0	0.0	0.0
42 Croatia	69.8	43.0	161.0	46	26.7	18.2	17.2	17.2

Position 2010 Country/Region	Total capacity end 2010 (MW)	Added capacity 2010 (MW)	Growth rate 2010 (%)	Position 2009	Total capacity end 2009 (MW)	Total capacity end 2008 (MW)	Total capacity end 2007 (MW)	Total capacity end 2006 (MW)
43 Argentina	54.0	25.3	88.2	43	28.7	29.8	29.8	27.8
44 Tunisia	54.0	0.0	0.0	44	29.7	20.7	20.7	20.7
45 Luxembourg	42.0	7.0	19.8	41	35.3	35.3	35.3	35.3
46 Switzerland	42.0	24.4	138.6	53	17.6	13.8	11.6	11.6
47 Nicaragua	40.0	0.0	0.0	40	40.0	0.0	0.0	0.0
48 Philippines	33.0	0.0	0.0	42	33.0	25.2	25.2	25.2
49 Latvia	31.0	2.0	7.0	45	28.5	26.9	27.4	27.4
50 Vietnam	31.0	22.3	254.3	57	8.8	1.3	0.0	0.0
51 Uruguay	30.5	10.0	48.8	50	20.5	20.5	0.6	0.2
52 Jamaica	29.7	0.0	0.0	52	54.0	20.0	20.0	20.0
53 Netherlands Antilles	24.3	0.0	0.0	47	24.3	12.3	12.3	12.0
54 Guadeloupe	20.5	0.0	0.0	49	20.5	20.5	20.5	20.5
55 Columbia	20.0	0.0	0.0	51	20.0	19.5	19.5	19.5
56 Russia	15.4	1.2	8.6	54	14.0	16.5	16.5	15.5
57 Guyana	13.5	0.0	0.0	56	13.5	13.5	13.5	13.5
58 Cuba	11.7	4.5	62.5	58	7.2	7.2	2.1	0.5
59 South Africa	10.0	2.0	25.0	48	8.0	21.8	16.6	16.6
60 Israel	6.0	0.0	0.0	59	6.0	6.0	6.0	7.0
61 Slovakia	6.0	0.0	0.0	60	6.0	6.0	5.0	5.0
62 Pakistan	6.0	0.0	0.0	61	6.0	6.0	0.0	0.0
63 Faroe Islands	4.0	0.0	0.0	62	4.0	4.1	4.1	4.1

Position 2010 Country/Region	Total capacity end 2010 (MW)	Added capacity 2010 (MW)	Growth rate 2010 (%)	Position 2009	Total capacity end 2009 (MW)	Total capacity end 2008 (MW)	Total capacity end 2007 (MW)	Total capacity end 2006 (MW)
64 Cape Verde	2.8	0.0	0.0	63	2.8	2.8	2.8	2.8
65 Ecuador	2.5	0.0	0.0	64	2.5	4.0	3.1	0.0
66 Nigeria	2.2	0.0	0.0	66	2.2	2.2	2.2	2.2
67 Belarus	1.9	0.0	0.0	67	1.9	1.1	1.1	1.1
68 Antarctica	1.6	0.0	0.0	68	1.6	0.6	0.0	0.0
69 Jordan	1.5	0.0	0.0	69	1.5	1.5	1.5	1.5
70 Indonesia	1.4	0.0	0.0	70	1.4	1.2	1.0	0.8
71 Mongolia	1.3	0.0	0.0	65	1.3	2.4	0.0	0.0
72 Martinique	1.1	0.0	0.0	71	1.1	1.1	1.1	1.1
73 Falkland Islands	1.0	0.0	0.0	72	1.0	1.0	1.0	1.0
74 Eritrea	0.8	0.0	0.0	73	0.8	0.8	0.8	0.8
75 Peru	0.7	0.0	0.0	74	0.7	0.7	0.7	0.7
76 Kazakhstan	0.5	0.0	0.0	75	0.5	0.5	0.5	0.5
77 Syria	0.4	0.0	0.0	77	0.4	0.4	0.3	0.3
78 Namibia	0.2	0.0	0.0	76	0.5	0.5	0.5	0.3
79 Dominican Republic	0.2	0.0	0.0	78	0.2	0.2	0.0	0.0
80 Dominica	0.2	0.0	0.0	79	0.2	0.2	0.0	0.0
81 North Korea	0.2	0.0	0.0	80	0.2	0.2	0.0	0.0
82 Algeria	0.1	0.0	0.0	81	0.1	0.1	0.0	0.0
83 Bolivia	0.01	0.0	0.0	82	0.01	0.01	0.01	0.01
Total World	196,629.7	37,643.0	23.56		159,766.4	120,903.0	93,926.8	74,122.0

Table H–4 (a). European wind power capacity by country. *Data courtesy of European Wind Energy Association.*

	Installed 2009	End 2009	Installed 2010	End 2010
EU Capacity (MW)				
Austria	0	995	16	1,011
Belgium	149	563	350	911
Bulgaria	57	177	198	375
Cyprus	0	0	82	82
Czech Republic	44	192	23	215
Denmark*	334	3,465	327	3,752
Estonia	64	142	7	149
Finland	4	147	52	197
France	1,088	4,574	1,086	5,660
Germany	1,917	25,777	1,493	27,214
Greece	102	1,087	123	1,208
Hungary	74	201	94	295
Ireland	233	1,310	118	1,428
Italy	1,114	4,849	948	5,797
Latvia	2	28	2	31
Lithuania	37	91	63	154
Luxembourg	0	35	7	42
Malta	0	0	0	0
Netherlands	39	2,215	32	2,237
Poland	180	725	382	1,107
Portugal	673	3,535	363	3,898
Slovakia	0	3	0	3
Slovenia	0.02	0.03	0	0.03
Spain	2,459	19,160	1,516	20,676
Sweden	512	1,560	604	2,163
United Kingdom	1,077	4,245	962	5,204
Total EU-27	**10,486**	**75,090**	**9,295**	**84,278**
Total EU-15	**10,025**	**73,516**	**7,997**	**81,406**
Total EU-12	**461**	**1,574**	**1,298**	**2,872**
of which offshore and near shore	582	2,064	883	2,946

Note: Due to a difference in methodology, some figures in this table may differ from figures communicated by national wind energy associations

* 2010 figures are provisional

Table H–4 (b). European wind power capacity by country. *Data courtesy of European Wind Energy Association.*

	Installed 2009	End 2009	Installed 2010	End 2010
Candidate Countries (MW)				
Croatia	10	28	61	89
FYROM**	0	0	0	0
Turkey	343	801	528	1,329
Total	353	829	461	1,290
EFTA (MW)				
Iceland	0	0	0	0
Liechtenstein	0	0	0	0
Norway	2	431	9	441
Switzerland	4	18	25	42
Total	6	449	34	483
Other (MW)				
Faroe Islands	0	4	0	4
Ukraine	4	90	1	87
Russia	0	9	0	9
Total	4	99	1	101
Total Europe	**10,845**	**76,471**	**9,918**	**86,279**

** FYROM = Former Yugoslav Republic of Macedonia

Note: Due to previous-year adjustments, 106.7 MW of project de-commissioning, re-powering, and rounding of figures, the total 2010 end-of-year cumulative capacity is not exactly equivalent to the sum of the 2009 end-of-year total plus the 2010 additions.

Table H–5. U.S. wind power capacity by state. *Data courtesy of U.S. Department of Energy.*

State	Cumulative MW at Year-End											
	1999	2000	2001	2002	2003	2004	2005	2006	2007	2008	2009	2010
Alaska	0.725	0.825	0.825	0.925	0.925	1.189	1.489	1.705	1.705	3.718	8.518	
Arizona	0.000	0.000	0.000	0.000	0.000	0.000	0.000	0.000	0.000	0.000	63.000	
Arkansas	0.000	0.000	0.000	0.000	0.100	0.100	0.100	0.100	0.100	0.100	0.100	
California	1,615.993	1,615.993	1,683.093	1,823.067	2,024.907	2,094.977	2,149.457	2,376.107	2,439.107	2,536.707	2,797.953	3,177
Colorado	21.600	21.600	61.200	61.200	223.200	230.700	230.750	290.750	1,066.750	1,067.650	1,244.250	
Connecticut	0.000	0.000	0.000	0.000	0.000	0.000	0.000	0.000	0.000	0.000	0.000	
Delaware	0.000	0.000	0.000	0.000	0.000	0.000	0.000	0.000	0.000	0.000	0.000	
Florida	0.000	0.000	0.000	0.000	0.000	0.000	0.000	0.000	0.000	0.000	0.000	
Georgia	0.000	0.000	0.000	0.000	0.000	0.000	0.000	0.000	0.000	0.000	0.000	
Hawaii	1.555	1.555	1.555	8.555	8.555	8.555	8.555	42.115	63.115	63.115	63.115	
Idaho	0.000	0.000	0.000	0.000	0.216	0.216	75.216	75.316	75.316	75.640	146.824	
Illinois	0.000	0.000	0.000	0.000	50.400	51.060	107.160	107.160	699.370	915.070	1,547.470	
Indiana	0.000	0.000	0.000	0.000	0.000	0.000	0.000	0.000	0.000	130.500	1,035.950	
Iowa	242.420	242.420	324.170	422.650	471.820	633.993	836.303	932.228	1,272.928	2,791.178	3,603.928	3,675
Kansas	1.500	1.500	113.700	113.700	113.700	113.700	263.700	364.200	364.200	921.000	1,021.000	
Maine	0.100	0.100	0.100	0.100	0.100	0.100	0.100	9.100	42.110	46.610	174.710	
Maryland	0.000	0.000	0.000	0.000	0.000	0.000	0.000	0.000	0.000	0.000	0.000	
Massachusetts	0.300	0.300	0.980	0.980	0.980	0.980	1.080	3.540	5.040	5.740	14.970	
Michigan	0.600	0.600	2.400	2.400	2.400	2.400	2.595	2.595	2.595	144.395	138.495	
Minnesota	273.390	291.210	319.800	337.700	558.250	600.100	745.400	895.850	1,299.790	1,752.845	1,810.045	2,192
Missouri	0.000	0.000	0.000	0.000	0.000	0.000	0.000	0.000	61.700	162.500	308.500	
Montana	0.100	0.100	0.100	0.360	1.145	1.145	136.860	145.860	152.895	271.495	374.995	
Nebraska	2.820	2.820	2.820	13.980	13.980	13.980	73.380	73.380	71.880	116.880	152.880	

State	Cumulative MW at Year-End 1999	2000	2001	2002	2003	2004	2005	2006	2007	2008	2009	2010
Nevada	0.000	0.000	0.000	0.000	0.000	0.000	0.000	0.000	0.000	0.000	0.000	
New Hampshire	0.050	0.050	0.050	0.050	0.050	0.050	0.050	1.100	1.100	25.100	25.200	
New Jersey	0.000	0.000	0.000	0.000	0.000	0.000	7.500	7.500	7.500	7.500	7.600	
New Mexico	0.660	0.660	0.660	0.660	205.980	265.980	405.980	495.980	495.980	497.480	597.480	
New York	0.000	18.200	48.150	48.450	48.450	48.450	185.500	370.300	424.800	831.800	1,274.300	
North Carolina	0.000	0.000	0.000	0.000	0.000	0.000	0.000	0.000	0.000	0.000	0.000	
North Dakota	0.390	0.390	0.390	4.790	66.290	66.290	97.790	178.335	344.835	714.495	1,202.596	
Ohio	0.000	0.000	0.000	0.000	3.600	7.200	7.200	7.425	7.425	7.425	7.425	
Oklahoma	0.000	0.000	0.000	0.000	176.250	176.250	474.500	534.500	689.000	708.050	1,031.150	
Oregon	25.080	25.080	156.840	218.340	259.340	262.640	337.640	438.140	885.390	1,067.190	1,758.140	
Pennsylvania	0.130	10.530	34.530	34.530	129.030	129.030	129.030	179.030	293.530	360.730	748.230	
Rhode Island	0.000	0.000	0.000	0.000	0.000	0.000	0.000	0.660	0.660	0.660	2.360	
South Dakota	0.000	0.000	2.600	3.014	44.264	44.264	44.264	44.264	98.264	186.764	313.164	
Tennessee	0.000	1.980	1.980	1.980	1.980	28.980	28.980	28.980	28.980	28.980	28.980	
Texas	183.520	183.520	1,095.760	1,095.760	1,290.260	1,290.260	1,992.070	2,735.770	4,353.370	7,112.670	9,403.370	10,085
Utah	0.000	0.225	0.225	0.225	0.225	0.225	0.885	0.885	0.885	19.785	223.285	
Vermont	6.050	6.050	6.050	6.050	6.050	6.050	6.050	6.050	6.050	6.050	6.150	
Virginia	0.000	0.000	0.000	0.000	0.000	0.000	0.000	0.000	0.000	0.000	0.000	
Washington	0.000	0.000	180.180	228.280	243.880	240.580	389.980	818.080	1,163.180	1,374.980	1,848.880	2,104
West Virginia	0.000	0.000	0.000	66.000	66.000	66.000	66.000	66.000	146.000	330.000	330.000	
Wisconsin	22.980	22.980	52.980	52.980	52.980	53.045	53.045	53.045	53.045	448.985	449.085	
Wyoming	72.515	90.635	140.635	140.635	284.635	284.635	288.455	288.455	288.455	676.255	1,099.255	
Total	2,472.478	2,539.323	4,231.773	4,687.361	6,349.942	6,723.124	9,147.064	11,574.505	16,907.050	25,410.042	34,863.353	40,180

Glossary

This glossary is divided into three sections:

1. Words and terms
2. Abbreviations and acronyms
3. Units of measurement, conversion formulas, and equation symbols

1. Words and Terms

AC See *alternating current.*

accelerated depreciation A type of government incentive for wind power developers that allows them to deduct capital costs more quickly, for example within five years, rather than over the entire lifetime of a wind farm project.

aerodynamic load Forces on a structure due to wind (air in motion). See also *aerodynamics.*

aerodynamics The motion of air when it interacts with a wind turbine blade or other object.

air density The density of the air in terms of mass per unit volume. Air density is a factor in *wind farm* development.

airfoil The shape of a wind turbine blade, usually shown in cross-section(s). Airfoil can also refer to other blades or wings such as those on airplanes and helicopters.

alternating current A type of electrical current in which the electrons move back and forth. The direction of the flowing electrons alternates, from positive to negative and from negative to positive, in the same conductor.

alternator A device that produces *alternating-current* electricity. Also called a *synchronous* generator or AC generator.

anemometer A device for measuring wind speed at a specific site. The simplest type uses cups mounted on a vertical shaft, but many other types are available. See also *meteorological tower.*

asynchronous generator See *induction generator.*

availability Percent of time that a wind turbine is running or is ready to run.

back-scatter Echoes from waves of sound or light (sonic or laser) that are emitted into the atmosphere by *sodar* and *lidar* instruments. The sonic and laser waves are reflected back to the instrument by atmospheric turbulence and/or particles.

backup power See *spinning reserves* and *non-spinning reserves*.

balance of plant A wind turbine's support system and equipment such as the foundation and the electrical connections. Also called balance of station.

barotrauma Damage to the lungs of bats when they approach spinning turbine blades, which create pockets of low-pressure and high-pressure air. The sudden change in air pressure can severely damage bats' lungs, similar to what happens when a scuba diver comes up from the water too quickly.

baseload Electric power that is available 24/7, usually generated by large nuclear or coal-fired plants that take time to warm up or ramp down. See also *peakshaving*.

battery A device that uses chemical reactions to store and deliver *electricity*.

Betz limit The maximum amount of the wind's kinetic energy that a wind turbine can convert to mechanical power by turning a rotor, defined as about 59%. Named for German physicist Albert Betz.

binning A statistical tool based on categorizing data, used in wind power to develop a *wind distribution curve*. Each small range of speeds (0–1 m/s or mph, 1–2 m/s or mph, etc.) is binned as a single category. Then the number of times that the wind speed falls within each bin is counted up—that is, the frequency of the wind occurring in a particular bin is calculated. This results in a bar graph from which a curve can be derived. See also *wind speed distribution* and *Weibull distribution*.

blade The component of a *wind turbine* that captures the wind's *kinetic energy*. Most turbine blades look like airplane propellers but behave more like airplane wings, capturing aerodynamic lift that causes the blade to rotate.

blown wing A type of turbine blade technology that uses active air circulation control, where compressed air is blown into the turbine, dynamically redirecting the air flow and increasing *lift*.

borescope Optical device used to inspect wind turbine drivetrain components and determine whether they need maintenance.

capacitor An device for storing *energy* and delivering *electricity* on demand. A capacitor consists of two conductors separated by insulation. When voltage is applied across the conductors, the insulator develops an electrostatic field that stores energy. See also *ultracapacitors*.

capacity credit The value of wind power to an electrical system in terms of how much conventional generation capacity is displaced by wind energy. The credit is based on average wind power production, or *capacity factor*.

capacity factor The total amount of *energy* produced by a *wind turbine* during a given period of time divided by how much energy it should have produced during that time as specified by the turbine's *power rating*.

capacity penetration The ratio of wind farm capacity to the power system's peak load. In a system with a 1,000 MW peak load and a 300-MW wind farm, capacity penetration is 30%. See also *energy penetration* and *instantaneous penetration*.

carbon-fiber composite A type of material used to manufacture wind turbine blades, typically fiberglass reinforced with carbon fibers for maximum strength and flexibility.

class A category of wind based on speed and other factors such as *wind power density*.

climate change The warming of the earth's climate over millions of years, which is thought to be caused by carbon dioxide and other greenhouse gases in the atmosphere. Also called global warming.

coefficient of power A measurement of power conversion efficiency. The coefficient of power, designated c_p, is a ratio that expresses the relationship between the total power available in the wind and the amount of power that the *rotor* actually produces or delivers.

collection system Medium-voltage power lines and other electrical equipment that collect the power generated by each turbine in a *wind farm* and connect to one or more utility substations at the farm.

community-scale Refers to a *wind turbine* that generates more than 1 MW, but in small *wind farms* (a dozen turbines or less, and often just one or two turbines).

compact wind-acceleration turbine A type of wind turbine that accelerates the wind speed before the wind enters the turbine generator. Sometimes referred to as CWATs.

compressed-air energy storage A hybrid generation/storage technology for delivering *electricity* on demand. The system uses conventional natural gas–fueled turbines to store energy by compressing air in an underground cavern. When the air is released, its energy is captured by passing it through turbines that power a generator.

condition monitoring Monitoring the condition of wind power system components to prevent unscheduled maintenance and failures. Also called structural health monitoring.

control system A wind power system component that is responsible for monitoring turbine conditions, recording the entire system's performance, alerting operators when maintenance is needed, and slowing down or stopping the rotor when faults or other hazardous conditions arise. The control system uses input from dozens of sensors to control rotor speed, blade pitch, and other operating parameters.

current The number of electrons passing a certain point in a given unit of time, measured in amperes (amps).

curtail Require a wind farm to reduce or shut down its output temporarily when utility transmission lines don't have enough capacity available to accept the wind power.

cyclic stress A stress or load on a wind power system that occurs repeatedly at different times during the rotor's operation. Cyclic stresses can include *gravitational loads*, *flapwise bending* forces, *torque*, and *gyroscopic loads*. Stresses and loads that are constant, rather than cyclic, are more bearable, creating less wear and tear on the machine. See also *fatigue*.

cycloturbine A type of *giromill*.

Darrieus A type of *vertical-axis wind turbine* that uses *aerodynamic lift* as its primary driving force. Darrieus machines are named for their inventor, an aeronautical engineer in France who patented the design in the 1920s. These turbines are often described as looking like an eggbeater.

DC See *direct current*.

decibel A unit of measuring the magnitude of sound or noise, abbreviated dB. A related unit, dBA, is an A-weighted decibel, which approximates the human ear's response to sound.

demand response Reduction in electricity consumption by a large group of customers who have agreed in advance to turn down or turn off certain electrical loads when the power system is running short on supply. Also called load response. See also *negawatt*.

density Mass per unit of volume. Air density varies with temperature and pressure.

diffuser A device for making an air stream more uniform by spreading it out evenly over a wider area.

diffuser-augmented See *compact wind-acceleration turbine*.

direct current A type of electrical current in which the electrons flow in only one direction. See also *alternating current*, *current*.

direct drive A wind turbine system designed without a *gearbox*. Also called gearless wind power systems.

discount rate A measure of the cost of money over time, like an interest rate but more complex.

dispatchable An electric power supply that can be turned on or off, and to some extent up or down, at the command of the power system operator.

distributed energy Electric power (or heat) that is generated at or near the customer's site, usually in relatively small amounts. Distributed energy sources include wind power systems, solar *photovoltaic*, *fuel cells*, and engines. Wind power can also be considered as a distributed resource on a larger scale, with wind farms connected directly to utility grids.

distribution The amount of time that the wind blows at any given speed. See also *wind distribution curve*.

diurnal Daily and nightly variations in wind speed and other characteristics.

doubly fed generator A type of *induction generator* that is equipped with *power electronics* that enable it to operate at variable rotational speeds. Also called a wound-rotor generator.

downwind Refers to the direction in which a wind turbine points relative to its tower. Some turbines are designed to operate with the rotor behind the tower. See also *upwind*.

drag An aerodynamic force that is created in parallel with the wind. Drag can be a penalty in horizontal-axis wind turbines, but some vertical-axis machines are powered by drag. See also *lift*.

driveshaft The rotating shaft that connects the *hub* of a *wind turbine* to the *gearbox*. More generally, a driveshaft transfers rotating motion from one machine to another.

drivetrain The entire assembly of components that transfers the wind's *kinetic energy* into *mechanical power* and then into *electricity*. The drivetrain includes the main *driveshaft*, *gearbox*, and *generator*, all of which are enclosed by the *nacelle*.

ducted turbine See *compact wind-acceleration turbine*.

dynamic reactive power compensation A type of grid voltage support that stabilizes voltage fluctuations. Also called VAR support, because it's measured in volt-amperes reactive, or VAR.

dynamometer A device for measuring the power produced by a gearbox or other rotating prime mover.

easterlies Bands of globally prevailing winds near the North and South Poles of the earth which move from east to west,

edgewise bending Forces that occur through the leading and trailing edges of the blade. Edgewise forces are the aerodynamic lift that causes the turbine's rotor to spin. See also *flapwise bending*.

electromagnetic energy The energy associated with electrical fields that oscillate back and forth with a spectrum of different *frequencies*.

electron Subatomic particle carrying a negative electrical charge.

energy The ability to perform work. Forms of energy include kinetic, mechanical, and electromagnetic.

energy density A term describing the amount of energy that can be stored within a given system per unit of volume.

energy penetration The ratio of the amount of wind energy delivered to the total amount. If a power system supplies 1,000 MWh of energy and 200

MWh of that comes from wind, the energy penetration is 20%. See also *capacity penetration* and *instantaneous penetration*.

external cost The economic cost of damages related to producing and using any form of energy. Also called hidden costs, external costs are not reflected in the market price of the energy and can be difficult to quantify. An example is the damage that air pollution imposes on human health.

fatigue The damage caused by cyclic stress. See also *cyclic loads*.

fault Unplanned events during turbine operation. See also *mean time between faults*.

feather This refers to the practice of angling a wind turbine blade to let some of the wind pass by without capturing its energy. The turbine's pitch control system feathers the blades when wind speeds approach the turbine's maximum capacity. See also *furl*.

feed-in tariff A mechanism for wind farm owners to be paid for the energy they generate. A feed-in tariff is a special rate for electricity that a wind farm (or other producer of renewable power) "feeds" into the grid, based on the cost of generation.

fixed-bottom A type of offshore wind turbine foundation that attaches to the seabed. Fixed-bottom foundations include *monopiles* and *gravity base*.

flapwise bending Forces that are exerted perpendicular to the blade, thrusting it backward. Flapwise forces can cause vibration and fatigue, leading to failures. In some cases, flapwise *thrust* can be a powerfully catastrophic force, resulting in a *tower strike*, a tower collapse, or even an uprooted foundation. Turbine designers account for flapwise thrust by making the blades strong enough to resist it, yet flexible enough not to break. See also *edgewise bending*.

floating wind turbine An offshore wind turbine without a foundation in the seabed. Floating wind turbines are designed for operation in deeper waters.

flywheel A mechanical device for storing *kinetic energy*. A flywheel is a spinning wheel or disc with a fixed axle. Once they are spinning, flywheels resist any change in their rotational speed because of inertia.

footprint The land area required to site a wind turbine or other equipment.

force Any influence that causes an object to accelerate.

fossil fuel Fuels, primarily coal, oil, and natural gas, created from the energy contained in ancient plants.

free stream The wind speed with nothing in its way. In equations, free-stream velocity is designated V_0, or *V-naught*.

frequency The number of cycles per second in a recurring wave of sound, light, vibration, or other physical phenomena. Measured in *hertz*.

frequency converter See *power converter.*

frequency distribution curve See *wind distribution curve.*

fuel cell An electrochemical cell that converts the chemical energy contained in a fuel (such as hydrogen) directly into electricity through a reaction with oxygen. The electrochemical reaction is flameless.

furl This refers to shutting down a turbine when wind speeds are too high for it to operate safely. During furling, the turbine's *pitch control system* turns the blades so that they slice horizontally into the wind (parallel with it), allowing the turbine to stop running. See also *feather.*

gearbox The component of a wind turbine that speeds up the rotation of the *driveshaft* and transfers the rotating energy to the *generator.* The rotor of a large wind turbine rotates at only 10–20 *rpm*, while speeds of 1,200–1,800 rpm are better suited for generators.

gearless See *direct drive.*

generator The component of a wind turbine that converts *mechanical power* from the *driveshaft* into *electricity*. Sometimes the wind turbine itself is called a generator, but it's really this key component that produces the electrical power.

giromill A type of *Darrieus wind turbine* with straight blades, rather than curved.

global warming A gradual process in which greenhouse gases in the earth's atmosphere allow the sun's rays in, but prevent heat from being radiated back into space. See also *climate change.*

gravitational load The force of gravity acting upon different parts of a wind power system. Gravitational loads include *compression* and *tension*. The bigger the wind power system, the greater is the gravity component of its loads and stresses, because every part of the machine is larger and heavier.

gravity base A type of *fixed-bottom* offshore wind turbine foundation that relies on the structure's weight to keep the turbine from overturning.

greenhouse gas Carbon dioxide, methane, and other gases that are thought to cause global warming See also *climate change.*

grid The network of equipment that transmits and distributes electricity.

grid-connected A wind power system or other power generator that interconnects with the utility *grid.*

gyroscopic load A load or stress on a wind power system that occurs when the turbine *yaws*, or pivots, to face into the wind. Gyroscopic loads are cyclic, in that they change with blade position during the yawing motion.

high-temperature superconducting A type of transmission cable that must be kept cold to operate, but can dramatically increase capacity and reduce line losses during transmission. See also *high-voltage direct-current.*

high-voltage direct-current A type of power transmission technology that permits higher electrical current flows with lower power losses compared to *alternating-current* transmission. High-voltage direct-current also lets the system operator control active and *reactive power* independently, which makes the transmission system easier to manage. However, costly equipment is required at each end of the system to convert power from alternating to direct current and back again.

horizontal axis Refers to a wind turbine with a horizontal *driveshaft*, such that the shaft spins in parallel with the wind. Horizontal-axis wind turbines are mounted on towers to capture higher wind speeds at greater heights. See also *vertical axis*.

horizontal load A load created by horizontal forces acting on a vertical structure, such as horizontal winds causing a wind turbine tower to sway.

horizontal wind shear A condition where the wind speed varies horizontally. For example, the wind can be much stronger on the right-hand side of a turbine than on the left, or vice versa. See also *wind shear* and *vertical wind shear*.

hub The center of the rotor where a wind turbine's blade roots are attached.

hybrid system A wind power system that works in tandem with other power generation sources such as solar *photovoltaic* panels. Hybrid systems usually include a battery for storing *electricity*.

hydroelectric power Electricity generated by turbines at river dams. See also *pumped hydropower*.

hydrophobic Water-resistant. *Hydrophobic* coatings are used to prevent ice from accumulating on wind turbine blades in very low-temperature conditions.

independent system operator A regional agency that operates wholesale power markets in the U.S.

induction generator A relatively simple type of generator that runs at a fixed speed, typically 1,200–1,800 rpm. In reality, induction generators have a degree of flexibility, or "slip," in their speed (within about 100 rpm above their fixed speed).

instantaneous penetration The ratio of wind farm output to total power system load at a specific point in time or over a short period. See also *capacity penetration* and *energy penetration*.

inverter A device that converts *direct current* to *alternating current*.

investment tax credit A government incentive that provides credit upfront against the capital expense needed to build a wind farm. See also *production tax credit*.

jacket A type of offshore wind turbine foundation that uses a lattice- or truss-type tubular steel structure.

kinetic energy The energy of motion, which is the primary form of energy provided by the wind. Kinetic energy must be converted to mechanical energy to become useful.

levelize Spread an expense or income that occurs at a particular time over the total life of a project. In wind farm development, costs and income must be levelized to calculate the net cost of the power generated.

levelized cost of energy See *life-cycle cost.*

lidar Light detection and ranging technology, an advanced method of wind resource assessment that uses beams of light, in the form of laser pulses, to define wind speed, direction, and turbulence. See also *back-scatter, sodar.*

life-cycle energy cost A measure of wind farm economics. The life-cycle cost of energy includes not only wind farm installation but also financing, operation and maintenance, and decommissioning, as well as projections of wind power system performance in terms of energy output and long-term reliability.

lift An aerodynamic force that causes a wind turbine blade to rotate. When air blows past a rotor blade, its airfoil shape creates regions of uneven air pressure. The lower pressure air forms on one side of the blade, as the wind begins to flow around the airfoil. This pulls the blade toward the lower pressure region, causing the rotor to turn. The lift force is created perpendicular to the wind, while a small *drag* force acts in opposition to the lift force. See also *drag.*

lift coefficient A dimensionless ratio, c_l, used by design engineers to model the complex interactions of *lift* with *airfoil* shape, blade *pitch* angle, air flow conditions, and other variables.

load-balancing Electric utility strategy for matching power supply exactly to meet demand (load), thus preventing any imbalance between the two. See also *load following.*

load-following Meeting the electrical demand on a power system over a time scale of 10 minutes to several hours.

load response See *demand response.*

low-voltage ride-through A feature of wind turbines that supports the electrical power transmission grid by staying connected during voltage dips or other faults. Modern turbines have power electronics that enable them to remain stable and continue to supply power to the grid.

maglev Magnetic levitation, referring to a type of vertical-axis wind turbine that uses magnetically levitated bearings, reducing friction that impairs turbine efficiency.

mass The weight of an object divided by the accelerating force of gravity, which is equal to 9.8 m/s^2 (32.2 ft/s^2).

mean time between faults The average number of generating hours between *faults*.

mechanical power Mechanical or kinetic energy that is harnessed by machinery to do work. See also *power*.

merchant power Wind farm output that is sold into wholesale markets at spot prices or under short-term contracts. Prices in wholesale markets can be volatile and unpredictable, and the wind farm operator has little or no prior commitment from buyers to take turbines' output. See also *power purchase agreement*.

meteorological tower A structure for measuring wind speed, direction, and other characteristics. The tower, basically a pole on a hinged base, is equipped with *anemometers* to measure wind speed and vanes to measure wind direction. Most towers have additional instrumentation for measuring other variables such as temperature and pressure. A meteorological tower is often called a "met tower" or "met mast."

micro A *wind turbine* generating less than 1 kW (1,000 watts) of power for applications such as battery charging, lighting, and traffic signs.

microsite Pinpoint the location of each turbine on a wind farm during the design process. See also *wake effect*.

mini wind turbine See *micro* wind turbine.

monopile A type of *fixed-bottom* foundation for offshore wind turbines sited in waters up to 30 meters deep (100 feet). The monopile is typically a single column about 4–6 meters (13–20 feet) in diameter.

nacelle A pod-like enclosure that sits on top of a wind turbine tower. The nacelle encloses the gearbox, generator, brakes, and other equipment.

natural frequency A specific point, measured in *hertz*, at which a wind turbine blade, tower, or other component begins to resonate or vibrate. See also *frequency*.

negawatt Nickname for negative watt. Negawatts are supplied by *demand response* providers as a type of electricity generation resource, in that they reduce demand for power rather than increase its supply.

net metering A policy that requires utilities to pay for electricity generated by *grid-connected* wind power systems so that, in effect, the utility customer's electric meter is spinning backward when excess power is being generated (beyond that required by the customer). Many utilities match the actual retail rate when paying homeowners and businesses for excess power, while some reimburse customers at lower rates.

nocebo The opposite of placebo (the beneficial effect of fake medication). Negative effects, such as turbine noise, are felt to increase without any factual basis if a person is predisposed to feel negative.

node A central point of transmission where circuits meet in large grids or power systems.

non-spinning reserves A form of supplemental power that can start up quickly when needed, within 10–30 minutes. Also called supplemental reserves. See also *spinning reserves.*

numerical weather prediction A computer modeling technique used to forecast weather and wind energy output.

off-grid A wind power system or other generator that does not interconnect with the utility *grid*. Often these systems are in remote locations or undeveloped countries and incorporate batteries or other electricity storage components for greater reliability. Also called *stand-alone* systems.

operating reserves See *spinning reserves.*

payback period The time required for investors to recoup their initial costs in a wind power system—that is, the time required for the system to pay for itself in energy cost savings.

peakshaving Power plants designed to deliver electricity when demand peaks seasonally, usually during the summer's heat or winter's cold. See also *baseload.*

permanent-magnet generator A generator that uses permanent magnets, which remain magnetized for a long period of time rather than being magnetized by the flow of electric current (electromagnet) through the *windings*. Permanent magnets in today's machines are made of rare-earth metals for stronger magnetization.

photovoltaic A solar-powered technology that uses panels made up of cells that convert light to electricity.

pitch The angle at which a wind turbine blade encounters the air flow. The ideal pitch angle—that is, the one that captures the most wind energy—varies at different *tip speed ratios.*

pitch control system A wind turbine component that adjusts the angle at which the blades encounter the air flow, turning the blades along their long axis. The pitch control system optimizes wind turbine performance by adjusting the blades to capture the maximum amount of wind energy. In some turbines, the pitch control system also slows down or stops the turbine from running when wind speeds become too high. See also *feather* and *furl.*

planetary gearbox A gearbox consisting of three outer gears, or "planet" gears, revolving around a central "sun" gear. Typically, the planet gears are mounted on a movable arm or carrier, which itself might rotate relative to the sun gear.

power The rate at which work is performed or energy is converted. In mathematical terms, power equals energy divided by time.

power converter A type of power electronics that enables fixed-speed generators to operate at variable speeds and synchronizes the generator's electricity to match conditions required by the utility grid (50–60 Hz AC). Variable-speed operation allows the turbine to capture wind energy over a greater range of wind speeds, thus increasing the amount of power it delivers. Also called *power electronics* and *frequency converter.*

power curve A graph showing how much power a wind turbine rotor produces at different wind speeds. Turbine manufacturers provide power curves as part of their product specifications.

power density See *wind power* density.

power electronics See *power converter.*

power pool Electricity distribution network or regional area encompassing several utility grids.

power purchase agreement A contract between a wind farm developer/owner and a utility or other company to buy wind power at specified rates or prices over a given period of time, typically 20 years. Also called power off-take agreement.

power rating A wind turbine specification indicating its capacity, or maximum output, in kilowatts or megawatts.

predictive maintenance Method of avoiding unplanned maintenance of wind power system equipment. Predictive maintenance includes *condition monitoring*. See also *preventive maintenance* and *reactive maintenance.*

preventive maintenance Method of avoiding unplanned maintenance of wind power system equipment by following the manufacturer's service schedule. See also *condition monitoring*, *predictive maintenance*, and *preventive maintenance.*

production tax credit A government incentive that provides tax credits to wind farm operators based on the amount of energy generated over time. See also *investment tax credit.*

project financing A type of financing where the lender or underwriter focuses on the credit of the project rather than the credit of the borrower. The opportunity for repayment is confined to a wind farm project's performance and its value as collateral.

pumped hydropower A way to store electricity by pumping water from a lower elevation to a higher one, storing it in a reservoir, then releasing it to flow through turbines that generate power when needed. See also *hydroelectric power.*

radius The length of a wind turbine blade, or half of the rotor diameter.

ramp event A sudden increase or decrease in *wind turbine* or *wind farm energy* output due to a corresponding increase or decrease in wind speed.

rated power See *power rating.*

Rayleigh distribution A statistical equation for approximating *wind speed distribution*, rather than measuring it precisely. See also *Weibull distribution* and *wind distribution curve*.

reactive maintenance Unscheduled or unplanned maintenance of wind power system equipment. Reactive maintenance is needed when equipment breaks or fails unexpectedly. See also *predictive maintenance* and *preventive maintenance*.

reactive power A component of electrical power that releases no energy and is associated with losses in *alternating-current* systems.

regional transmission organization An agency that controls the electrical transmission system within a large region or within a country and works with wholesale power market agencies to ensure that electricity flows smoothly where it's needed, without congestion or overloading of vulnerable transmission lines. Also called *transmission system operators*.

regulation service A type of ancillary service that keeps electrical grid voltage and frequency within a narrow range. Excursions in voltage or frequency indicate an imbalance between supply and demand.

reliability A measure of how reliably a system performs, whether it be a wind turbine, wind farm, conventional power plant, or entire power delivery system.

renewable electricity standard See *renewable energy standard*.

renewable energy Sources of energy, such as wind and solar, that are inexhaustible in the world as we know it, and sources that are virtually renewable, such as biomass.

renewable energy credit A type of government incentive that gives wind farm operators and other energy producer a credit for every unit of renewable energy they produce. The credits can be traded on the open market for cash, separately from the electricity itself, or they can be bartered. The credits are also called renewable energy certificates, green tags, etc.

renewable energy standard A government regulation requiring large energy users to obtain certain percentages of their energy or power supply from renewable sources, or requiring utilities to generate a certain percentage of their electricity from renewables. Also called a *renewable portfolio standard*. These should not be confused with renewable energy goals established by many countries to meet their commitments to reducing carbon emissions. Many renewable energy standards include hydropower, biomass, and other renewable sources as eligible generation technologies.

renewable portfolio standard See *renewable energy standard*.

resistance The opposition to a flow of electrical *current*, measured in *ohms*.

rotor The wind turbine's blade assembly, consisting of the blades and their hub, where the blade roots are attached.

rotor diameter The entire diameter of a wind turbine rotor, including two blade lengths plus the diameter of the hub.

Savonius A type of *vertical-axis wind turbine* that produces mechanical power from the wind's *drag* force. When viewed from above, the Savonius rotor is shaped like the letter S.

SCADA system Supervisory control and data acquisition system. SCADA systems, typically located on *wind farms*, collect all of the data from a wind turbine, write a history of the readings to a computer, store and package the data, and send it to a central control site.

shadow flicker A phenomenon that occurs when a turbine's spinning blades pass between the sun and a person who's trying to read, watch TV, or concentrate on some visual task. This moving or flickering shadow can be irritating. Shadow flicker is more of a problem in far northern or southern latitudes where the sun's angle is very low in the sky.

shaftpower The mechanical power of a wind turbine's rotating *driveshaft*.

site suitability assessment A study that determines how site conditions at a wind farm will affect the structural integrity, operation, and performance of the turbines being considered. The assessment covers factors such as wind loads (extreme wind speeds and fatigue loads), seismic loads, temperatures, and other weather conditions (lightning, icing, etc.). It also analyzes the ability of the utility grid to accept wind power without curtailment.

small-scale Refers to a *wind turbine* that generates less than 100 kW. The term includes wind power systems that generate enough electricity for a single home, farm, or small business, typically about 1–10 kW. See also *micro*.

sodar Sonic detection and ranging technology, an advanced method of wind resource assessment that uses acoustic pulses to characterize the wind. The sonic wave is scattered by atmospheric turbulence and echoes back to the sodar system. Changes in the frequency and intensity of this *back-scatter* allow the system's instrumentation to define wind speed, direction, and other turbulent conditions. See also *lidar*.

solidity The total surface area of a turbine's blades divided by the rotor's total swept area. If the blade surface area is large relative to the swept area, the turbine has a high *solidity* and typically would operate at low speeds and high torque. Turbines with low solidity (narrower blades) run at higher speeds and lower torque.

spar A buoy-type float that stabilizes a deepwater *floating wind turbine* with its ballast weight.

spinning reserves A form of contingency power that is online at all times, ready to be ramped up or down nearly instantly to balance power supply and demand. Also called operating reserves. See also *non-spinning reserves.*

square-cube law A mathematical law regarding proportional increases in size, volume, and surface area. In wind power applications, it means that the amount of material needed—and thus the system's weight and cost—increases faster with size than the power output does (by a factor of eight for every doubling of size and power).

stall An *aerodynamic* effect that reduces *lift*. Stall can control power in turbines without an active *pitch control system*.

stand-alone A wind power system or other generator that does not interconnect with the utility *grid*. Often these systems are in remote locations or undeveloped countries and incorporate batteries or other electricity storage components for greater reliability. Also called *off-grid* systems.

structural health monitoring See *condition monitoring*.

supplemental reserves See *non-spinning reserves*.

swept area The circle defined by revolution of a wind turbine's blades. Usually designated as A in equations.

synchronous generator A generator with a power output *frequency* that is perfectly matched (synchronized) to the frequency of the electric utility grid with which it connects.

tension-leg platform A structure for supporting a deepwater *floating wind turbine*. The platform is a tripod with three vertical, anchored mooring lines that stabilize the turbine.

thermographic imaging An infrared scanning technique for finding hot spots in wind turbine drivetrain components, which might indicate the need for *preventive* maintenance.

thermoset resin A type of resin that can be molded or extruded.

thrust A mechanical force resulting from acceleration of a *mass*, such as the wind pushing on a wind turbine structure. In wind power systems, thrust forces affect the whole machine and its individual components. Thrust forces include *flapwise bending*, which pushes a wind turbine blade backward, toward the tower.

thrust coefficient A dimensionless factor calculated using *thrust*, *air density*, *radius*, and the angular speed (a measure of rotation rate).

time-shift Store electricity generated at one time, for example off-peak, and release it at another time, such as on-peak.

tip speed The speed of a wind turbine's blade at the tip. This is typically about four to seven times faster than the *free-stream* wind speed. Tip speed is calculated by multiplying rotor speed by blade length, or *radius*.

tip speed ratio The ratio of *tip speed* to wind speed, a dimensionless factor.

topography The shape and features of the land surrounding a wind turbine.

torque A twisting motion acting upon a wind turbine rotor's shaft. Torque is measured by force and distance, in units such as newton-meters or in foot-pounds or inch-pounds.

tower The structure on which a *horizontal-axis* wind turbine is mounted. Towers are usually made of a huge, tubular steel column.

tower strike This occurs when a wind turbine blade bends backward during rotation, usually because of strong winds, and strikes the tower. Tower strikes are usually catastrophic, damaging the blade or even destroying the whole wind turbine structure.

trade winds Bands of globally prevailing winds near the equator which move from east to west.

transient loads Loads on the wind turbine that result from brief gusts of wind. These include transient torque on the generator and transient thrust on the whole machine.

transmission system operator An agency that controls the electrical transmission system within a large region or within a country and works with wholesale power market agencies to ensure that electricity flows smoothly where it's needed, without congestion or overloading of vulnerable transmission lines. Also called *regional transmission organizations*.

transverse axis A type of *vertical-axis wind turbine* that is tipped on its side rather than mounted vertically.

tripod A type of offshore wind turbine foundation that uses three legs made of steel tubes welded together.

turbine A rotary engine that extracts energy from a flow of fluid such as air, other gases, or water.

turbine design class A specific standard for turbine performance, as defined by the International Electrotechnical Commission. The standard specifies wind speeds and turbulence for various classes and subclasses of turbines.

turbulence Sudden, frequent variations in the speed and/or direction of the wind, including unpredictable gusts. Turbulence is the opposite of a steady wind. Typically, turbulence weakens at higher elevations above the ground. Turbulent conditions can arise anywhere depending on the land's topography, but turbulence is prevalent in areas such as cities with a lot of obstacles.

twist coupling An experimental way to reduce loads on wind turbine blades by designing a blade that twists as it bends, thus passively adapts itself to changing loads. Twist coupling lets the blade adjust its effective pitch angle as aerodynamic loads fluctuate.

ultracapacitor An electrical device that can store an unusually great amount of energy compared to a common capacitor. Ultracapacitors are used in many wind turbine *pitch control systems*.

unit commitment A commitment to operate a large power generating unit at a particular time scheduled by the power system operator. Unit commitments are needed for large generators that require hours or even days to come up to speed.

upwind Refers to the direction in which a wind turbine points relative to its tower. Most horizontal-axis turbines face into the wind, or upwind, with the tower behind the rotor. See also *downwind*.

utility-scale Refers to wind turbines generating more than 1 MW of power, usually in large *wind farms* with hundreds of turbines.

velocity duration curve See *wind distribution curve*.

vertical axis Refers to a wind turbine with a vertical *driveshaft*, such that the shaft spins perpendicular to the ground. Vertical-axis wind turbines are typically mounted near the ground like a flagpole, rather than on a tower, although sometimes the driveshaft is called a tower. See also *horizontal axis*.

vertical load The dead weight of a structure such as a wind turbine system. See also *horizontal load*.

vertical wind shear Sudden, sharp changes in wind speed with height above the ground. Typically, wind speed increases with height, so that the turbine's rotor sees much higher wind speeds than the tower base. See also *wind shear*, *horizontal wind shear*, and *vertical wind speed*.

vertical wind speed The speed at which the wind blows vertically, upward or downward. Vertical wind speed should not be confused with *vertical wind shear*.

V-naught The *free-stream* velocity of the wind, designated V_0.

voltage The force that drives electrical *current* between two points, measured in *volts*.

wake effect The loss of available energy in the wind downstream from a wind turbine. The turbine extracts wind energy, casting a "wind shadow" called a wake in the downwind direction. The wake is a long trail of wind that is not only slowed down, but also quite turbulent. Proper spacing of wind turbines to avoid the wake effect is critical to obtaining the maximum energy output of a wind farm.

Weibull distribution A statistical equation that estimates how frequently the wind blows at any particular range of speeds. See also *wind distribution curve* and *wind speed distribution*.

westerlies Bands of globally prevailing winds moving from west to east at about 30 to 60 degrees in latitude within the Northern and Southern hemispheres.

wind distribution curve A graph that shows the amount of time (in hours or as a percentage of total time measured) that the wind blows at any given speed. Also called a velocity duration curve or frequency distribution curve. See also *wind speed distribution.*

wind energy The amount of energy available from the motion of air.

wind farm Large area of land where dozens or even hundreds of wind turbines have been built to provide power to electric utilities. Also called wind park, especially in Europe.

windings A coil of wire inside a *generator.* The windings create a magnetic field when an electric current flows through them, or alternatively, the windings produce an electric current when rotated in a magnetic field.

wind park See *wind farm.*

wind power density A measure of the strength of a wind resource, in units of watts per square meter (W/m^2). Power density indicates how much potential wind energy is available over a given area of land for a turbine to convert to power.

wind rose A diagram used in analyzing the wind resource which shows not only the direction of the wind, but also the amount of wind power that comes from that direction. A wind rose can also display the average wind speed by direction.

wind shear Sudden, frequent changes in the speed and direction of the wind, such as strong gusts or downdrafts. See also *horizontal wind shear* and *vertical wind shear.*

wind speed distribution The amount of time that the wind blows at any given speed. See also *wind distribution curve.*

wind turbine A rotary engine that extracts energy from the air flow of the wind.

wound-rotor generator See *doubly fed generator.*

yaw The direction in which the wind turbine is pointing. The turbine's yaw system uses a wind-direction sensor and a motor to pivot the turbine to face into the wind. See also *gyroscopic load.*

2. Abbreviations and Acronyms

AC *alternating current*
ASTAE Asia Alternative Energy Program
AWEA American Wind Energy Association
AWWI American Wind and Wildlife Institute (U.S.)
BNDES Brazilian Development Bank
BOEM Bureau of Ocean Energy Management (U.S.)
BOEMRE Bureau of Ocean Energy Management, Regulation, and Enforcement (U.S.)
BPA Bonneville Power Administration (U.S.)
CAES compressed-air energy storage
CalISO California Independent System Operator (U.S.)
CanWEA Canadian Wind Energy Association
CDM clean development mechanism (international)
CENER Centro Nacional de Energías Renovables (National Renewable Energy Center, Spain)
CEQA California Environmental Quality Act (U.S.)
CFD computational fluid dynamics
CREIA Chinese Renewable Energy Industry Association
CRP China Resources Power
CSIRO Commonwealth Scientific and Industrial Research Organization (Australia)
Danak Danish Accreditation and Metrology Fund
DC *direct current*
DNV Det Norske Veritas (Norway)
DOE Department of Energy (U.S.)
DOI Department of the Interior (U.S.)
DSAT Distributed Wind Site Analysis Tool (U.S.)
DWEA Distributed Wind Energy Association (U.S.)
EC European Commission
ECN Energy Research Centre of the Netherlands
EERE Energy Efficiency and Renewable Energy, an office of the U.S. Department of Energy
EIA Energy Information Administration (U.S.)
EnBW EnBW Energie Baden-Württemberg AG (Germany)
Entso-e European Network of Transmission System Operators for Electricity
EPC engineering, procurement, and construction
EPFL Ecole Polytechnique Fédérale de Lausanne (Switzerland)
Ercot Electric Reliability Council of Texas (U.S.)
ETI Energy Technologies Institute (UK)

EU European Union
EWEA European Wind Energy Association
FEED front-end engineering and design
FERC Federal Energy Regulatory Commission (U.S.)
FIT feed-in tariff
FLOWE Field Laboratory for Optimized Wind Energy (U.S.)
GDP gross domestic product
GWEC Global Wind Energy Council
Heden Haveneiland duurzame energie North Sea (Offshore port sustainable energy North Sea, Netherlands)
HVDC high-voltage direct-current
IEA International Energy Agency
IEC International Electrotechnical Commission
IOU investor-owned utility (U.S.)
IPO initial public offering
IREC Catalonia Institute for Energy Research (Institut de Recerca en Energia de Catalunya, Spain)
ISO independent system operator (U.S.)
ISO-NE Independent System Operator-New England (U.S.)
LCOE levelized cost of energy
lidar light detection and ranging
LORC Lindø Offshore Renewables Center (Denmark)
Measnet Measuring Network of Wind Energy Institutes
MISO Midwest Independent System Operator (U.S.)
MIT Massachusetts Institute of Technology (U.S.)
MTBF mean time between faults
MVP multi-value project (U.S.)
NaREC National Renewable Energy Center (UK)
NCAR National Center for Atmospheric Research (U.S.)
NEPA National Environmental Policy Act (U.S.)
NOAA National Oceanic and Atmospheric Administration (U.S.)
NREL National Renewable Energy Laboratory (U.S.)
NWCC National Wind Coordinating Collaborative (U.S.)
NWTC National Wind Technology Center (U.S.)
NYISO New York Independent System Operator (U.S.)
OECD Organisation for Economic Co-operation and Development
OREC offshore wind renewable energy certificate (U.S.)
OSWInD Offshore Wind Innovation and Demonstration Initiative (U.S.)
PILOT payment in lieu of taxes (U.S.)
PJM Pennsylvania-[New] Jersey-Maryland Interconnection (U.S.)
PPA *power purchase agreement*

REC *renewable energy credit*; renewable energy certificate
REPP Renewable Energy Policy Project (U.S.)
RES *renewable energy standard* or renewable electricity standard (U.S.)
Risø DTU National Laboratory for Sustainable Energy (Denmark)
ROC renewable obligation certificate (UK)
RPM reliability pricing model
RPO renewable purchase obligation
RPS renewable portfolio standard
RTO *regional transmission organization* (U.S.)
SCADA Supervisory Control and Data Acquisition
SIROCCO silent rotors by acoustic optimization (Europe)
sodar sonic detection and ranging
SSE Scottish and Southern Energy (UK)
SWCC Small Wind Certification Council (U.S.)
SWERA Solar and Wind Energy Resource Assessment
SWIIS Small Wind Industry Strategy (Europe)
TIF tax increment financing
TIV turbine installation vessel
TSO *transmission system operator*
TSR *tip speed ratio*
U.N. United Nations (international)
UWIG Utility Wind Integration Group (U.S.)
WAsP Wind Atlas Analysis and Application Program (Denmark)
WECC Western Electricity Coordinating Council (U.S.)
WEICan Wind Energy Institute of Canada
WILD Wind-Wildlife Impacts Literature Database (international)
WISE World Institute for Sustainable Energy (India)
WTO World Trade Organization
WWEA World Wind Energy Association (Germany)
Y2K year 2000

3. Units of Measurement, Conversion Formulas, and Equation Symbols

dB decibel
dBA decibel (A-weighted)
decibel unit of measuring the magnitude of sound
foot unit of measuring distance
foot-pound unit of measuring *torque* (force and distance)
ft-lb foot-pound
GB gigabyte

gigabyte unit of measuring data, equal to 109 (1 billion) bytes
gigawatt unit of measuring energy conversion, equal to 1,000 MW
gigawatt-hour metric unit of measuring energy, equal to 1 GW for a period of 1 hour
GW gigawatt
GWh gigawatt-hour
hertz unit of measuring frequency, which is the number of cycles per second in a recurring wave of sound, light, vibration, or other physical phenomena
Hz Hertz
inch-pound unit of measuring *torque* (force and distance)
in.-lb inch-pound
J Joule
Joule metric unit of measuring the amount of work done or energy expended by a force of 1 newton acting through a distance of 1 meter (a newton-meter)
kilovolt unit of measuring voltage, equal to 1,000 volts
kilowatt unit of measuring energy conversion, equal to 1,000 watts
kilowatt-hour metric unit of measuring energy, equal to 1 kW for a period of 1 hour
kV kilovolt
kW kilowatt
kWh kilowatt-hour
megajoule metric unit of measuring energy, equal to 1,000 joules
megawatt unit of measuring energy conversion, equal to 1 million W or 1,000 kW
megawatt-hour metric unit of measuring energy, equal to 1 MW for a period of 1 hour
meter metric unit of measuring distance
mJ megajoule
mph miles per hour
m/s meters per second
MW megawatt
MWh megawatt-hour
newton metric unit of measuring *force*, named after English scientist Sir Isaac Newton
newton-meters metric unit of measuring *torque* (force and distance)
N-m newton-meters
ohm unit of measuring *resistance* to flow through an electrical *current*
rpm revolutions per minute
terawatt unit of measuring energy conversion, equal to 1,000 GW

terawatt-hour metric unit of measuring energy, equal to 1 TW for a period of 1 hour
TLP tension-leg platform
TW terawatt
TWh terawatt-hour
V volt
VAR volt-amperes reactive
volt unit of measuring the force that drives electrical *current*
W watt
watt unit of measuring energy conversion (the rate at which energy is converted to power). A watt, named after Scottish engineer James Watt, equals 1 Joule per second.
watts/square meter metric unit of measuring *wind power density*
W/m² watts per square meter

Conversion formulas

Distance

1 meter	=	3.28 foot
1 foot	=	0.3 meter
1 mile	=	1.61 km
1 km	=	0.62 mile

Energy/Work

kW	=	1.34 hp
	=	1,000 W
MW	=	1,000,000 W
kJ	=	0.948 Btu
kWh	=	3,600,000 J
	=	3,600 kJ
	=	3.6 MJ
hp	=	746 W
	=	0.746 kW

Mass

1 kg	=	2.2046 lb
1 lb	=	0.4536 kg
1 ton	=	0.978 metric ton
1 metric ton	=	1.023 tons
	=	2,204.6 lb
	=	1,000 kg

Temperature

°F	=	1.8°C + 32
°C	=	(°F—32)0.555

Pressure

1 psi	=	0.06895 bar
	=	6.895 kPa
1 bar	=	14.5 psi
	=	100 kPa
	=	0.1 MPa
1 MPa	=	145 psi

Equation symbols

A	*swept area*
beta (β)	blade *pitch* angle
c_l	*lift coefficient*
c_p	*coefficient of power*
c_t	thrust coefficient
omega (Ω)	rotor speed, *rpm*
n	number of blades
R	*radius*
rho (ρ)	*air density*
t	time, seconds
TSR	*tip speed ratio*
V	wind speed (instantaneous), m/s (mph)
V_0	*free-stream* velocity (*V-naught*)
β	*beta*
ρ	*rho*
Ω	*omega*
ΩR	*tip speed*
$\Omega R / V_0$	*tip speed ratio*

Bibliography

Web sites

Alpha Ventus, http://www.alpha-ventus.de/index.php?id=80.
Alstom, http://www.alstom.com/power/renewables/wind/.
Altaeros Energies, http://www.altaerosenergies.com/.
American Wind Energy Association, http://www.awea.org.
American Wind Wildlife Institute, http://www.awwi.org/.
Ampyx Power, http://www.ampyxpower.com.
Answers.com, http://wiki.answers.com/Q/What_is_reactive_power.
Areva, http://www.areva.com/.
Asia Sustainable and Alternative Energy Program, http://www.worldbank.org/astae.
Atlantic Wind Connection, http://atlanticwindconnection.com/.
Atlas CTB, http://www.atlasctb.com/.
Avent Lidar Technology, http://www.aventlidartechnology.com/.
AWEA Seal of Approval Program, http://www.awea.org/learnabout/education/awea_soa.cfm.
AWS Truepower, http://www.awstruepower.com/.
Bard, http://www.bard-offshore.de/en.
Basic principles of wind resource evaluation, http://www.awea.org/faq/basicwr.html.
Bat Conservation International, http://www.batcon.org/.
Bats and Wind Energy Cooperative, http://www.batsandwind.org/.
Bergey WindPower, http://www.bergey.com/.
Better Generation, http://www.bettergeneration.com.
Boulder Wind Power, http://www.boulderwindpower.com/.
BTM Consult, http://www.btm.dk/.
Burke Porter Machinery, http://www.bepco.com/.
Caitin, http://www.caitin.com.
Canadian Wind Energy Association, http://www.canwea.ca.
Carbon Trust, http://www.carbontrust.co.uk/.
Catalonia Institute for Energy Research (IREC), http://www.irec.cat/.
Catch the Wind, Inc., http://www.catchthewindinc.com.
Centurion Energy LLC, http://centurionenergy.net.
Coalition for Fair Transmission Policy, http://www.thecftp.org/.
Commonwealth Scientific and Industrial Research Organization, http://www.csiro.au/.
Community Windpower Ltd., http://www.communitywindpower.co.uk/index.asp.

CREZ Transmission Program Information Center, http://www.texascrezprojects.com/.
DeTect, Inc., http://www.detect-inc.com/index.html.
Det Norske Veritas, http://www.dnv.com/.
Distributed Wind Energy Association, http://www.distributedwind.org/.
Distributed Wind Policy Comparison Tool, http://www.eformativeoptions.com/dwpolicytool/.
Distributed Wind Site Analysis Tool, http://dsat.cadmusgroup.com/.
EnBW, http://www.enbw.com/content/en/index.jsp.
Enercon, http://www.enercon.de/en-en/Windenergieanlagen.htm.
Energy Basics, Wind Energy Technologies. U.S.DOE EERE, http://www.eere.energy.gov/basics/renewable_energy/wind.html.
Energy Research Centre of the Netherlands, http://www.ecn.nl/units/wind/.
Enflo Windtec, http://www.enflo-windtec.ch/english/turbine.html.
EOLE Wind Energy Project, http://collaboration.cmc.ec.gc.ca/science/rpn/modcom/eole/.
European Master in Renewable Energy, http://www.master.eurec.be/en/.
European Network of Transmission System Operators for Electricity, https://www.entsoe.eu.
FloDesign, http://www.flodesign.org.
Fraunhofer IWES, http://www.iwes.fraunhofer.de/en/.
Frequently Asked Questions on Small Wind Systems, www1.eere.energy.gov/windandhydro/small_wind_system_faqs.html.
Gamesa, http://www.gamesa.es/en/.
GE, http://www.ge-energy.com/products_and_services/products/wind_turbines/index.jsp.
Generator, http://en.wikipedia.org/wiki/Generator.
Global Wind Energy Council (GWEC), www.gwec.net.
Glosten Associates, http://www.glosten.com/.
Goldwind, http://www.goldwindglobal.com/web/index.do.
Greentech Media, http://www.greentechmedia.com.
Guided Tour (English). Danish Wind Energy Association. http://guidedtour.windpower.org/en/tour/wres/index.htm.
Hawaiian Electric Co., http://www.heco.com.
Heden, http://www.haveneilandopzee.nl/.
Home Energy International, http://www.home-energy.com/int/windpower.htm.
Iberdrola, http://www.iberdrolarenewables.us/.
IEA Wind, International Energy Agency, http://www.ieawind.org.
Illustrated History of Wind Power Development, Dodge, Darrell M., http://www.telosnet.com/wind.
Infinite Wind Energy, http://www.infinitewindenergyllc.com/index.html.
International Energy Agency, http://www.iea.org.
International Wind Resource Maps. U.S. National Renewable Energy Laboratory. http://www.nrel.gov/wind/international_wind_resources.html.
IREC Zèfir Test Station, http://www.irec.cat/index.php/en/zefir-test-station.
Irish Wind Energy Association, http://www.iwea.com/.
ISO/RTO Council, http://www.isorto.org.

Joby Energy, http://www.jobyenergy.com.
KidWind, http://www.kidwind.org.
KiteGen, http://www.kitegen.com/en/.
KITEnergy, http://www.kitenergy.net/.
Knowledge Centre WMC, http://www.wmc.eu/.
Landscape Assessment Tool, http://wind.tnc.org/awwi/.
Lindø Offshore Renewables Center, http://www.lorc.dk/Home.
LM Wind Power, http://www.lmwindpower.com/.
London Array, http://www.londonarray.com/.
Magenn Power, http://www.magenn.com.
Makani Power, http://www.makanipower.com.
Mass Megawatts Wind Power, http://www.massmegawatts.com/index.php.
Master Flo Technology, http://www.masterflowind.com/Home_Page.html.
Maxwell Technologies, Inc., http://www.maxwell.com.
Meteodyn, http://www.meteodyn.com/en/town-planning.html.
MIT Energy Initiative, http://web.mit.edu/mitei/.
National Center for Atmospheric Research, http://ncar.ucar.edu/
National Weather Service-National Centers for Environmental Prediction, http://www.ncep.noaa.gov/
National Wind Coordinating Collaborative, http://www.nationalwind.org.
National Wind Technology Center, http://www.nrel.gov/wind/nwtc.html.
Natural Power, http://www.naturalpower.com.
Nordex, http://www.nordex-online.com/en.
Norsewind, http://www.norsewind.eu/public/index.html.
Northern Power Systems, http://www.northernpower.com.
Northwest Community Energy, http://nwcommunityenergy.org/wind/resource-assessment.
Northwest Power and Conservation Council, http://www.nwcouncil.org/.
NRG Systems, http://www.nrgsystems.com.
Oak Ridge Carbon Fiber Composites Consortium, http://www.cfcomposites.org/index.shtml.
Offshore Wind Farms, Lindø Offshore Renewables Center, http://www.lorc.dk/Knowledge/Offshore-renewables-map/Offshore-wind-farms.
OptiWind, http://www.optiwind.com.
Poseidon. http://www.floatingpowerplant.com.
Power Predictor, http://www2.powerpredictor.com.
Principle Power, http://www.principlepowerinc.com/.
Proven Energy, http://www.provenenergy.co.uk/.
REN21 Renewable Energy Policy Network for the 21st Century, http://ren21.net/.
Renewable Energy Jobs, http://www.renewableenergyjobs.com/content/global_guide_to_education_and_training_within_renewable_energy.
Renewable Energy Policy Project, http://www.repp.org/.
REpower, http://www.repower.de/.
Risø DTU, http://www.dtu.dk/English/Public_Sector_Consultancy/Departments/Ris%C3%B8_DTU.aspx.
Rohn Products, http://www.rohnnet.com.
Russian Wind Atlas, http://www.wasp.dk/Download/DownloadFiles/General/Russian%20Wind%20Atlas.pdf.

Sandia National Laboratories Wind Power Technologies, http://windpower.sandia.gov/index.htm.
Sany Group, http://www.sanygroup.com/products/en-us/turbine/Offshore.htm.
Second Wind, Inc., http://www.secondwind.com/index.html.
Siemens, http://www.energy.siemens.com/hq/en/power-generation/renewables/wind-power.
Sinovel, http://www.sinovel.com/en/index.aspx.
Sinovel National Offshore Wind Power Technology and Equipment R&D Center, http://www.sinovel.com/en/RandDcenter.aspx.
SkySails, http://www.skysails.info/index.php?id=472&L=2.
Small Wind Certification Council, http://www.smallwindcertification.org/.
Small Wind Implementation Strategy (SWIIS), http://www.smallwindindustry.org/index.php?id=114.
Solar and Wind Energy Resource Assessment, http://swera.unep.net.
Southwest Windpower, http://www.windenergy.com/.
Statistical Review of World Energy 2010, http://www.bp.com/productlanding.do?categoryId=6929&contentId=7044622.
Subsidy FAQs, http://www.earthtrack.net/subsidy-faqs.
Sway, http://www.sway.no/.
3Tier, Inc., http://www.3tier.com/en.
UpWind, http://www.upwind.eu/default.aspx.
Urban Green Energy, http://www.urbangreenenergy.com.
U.S. Energy Information Administration, http://www.eia.gov/.
Vertical Axis Wind Turbine. Wikipedia, http://en.wikipedia.org/wiki.
Vestas, http://www.vestas.com/.
V Squared Wind, http://www.vsquaredwind.com.
WAsP—Wind Atlas Analysis and Application Program. http://en.openei.org/wiki/Wind_Atlas_Analysis_and_Application_Program_(WAsP).
WEICan, http://www.weican.ca/.
Western EcoSystems Technology, Inc., http://www.west-inc.com/.
Western Electricity Coordinating Council, http://www.wecc.biz/.
Wind and Water Program, U.S. DOE EERE, http://www1.eere.energy.gov/windandhydro.
Wind Atlases of the World, http://www.windatlas.dk/index.htm.
WindData.com, http://www.winddata.com.
Wind Energy Solutions, http://www.windenergysolutions.nl/.
WindFacts (Wind Energy—the Facts), http://www.wind-energy-the-facts.org/.
Wind Maps and Wind Resource Potential Estimates. Wind Powering America. http://www.windpoweringamerica.gov/wind_maps.asp.
Windlift, http://www.windlift.com/.
WindMade, http://www.windmade.org/.
Wind Maps. EcoBusinessLinks, http://www.ecobusinesslinks.com/wind_maps.htm.
Wind power, http://en.wikipedia.org/wiki/Wind_power.
Wind Tower Systems, http://www.windtowersystems.com/.
Wind turbine, http://en.wikipedia.org/wiki/Wind_turbine.
Wind-Wildlife Impacts Literature Database, http://www.nrel.gov/wind/wild.html.
WindTamer, http://www.windtamerturbines.com/.

WindTronics, http://www.windtronics.com/home.
Winergy, http://www.winergy-group.com/cms/website.php.
WinFlex, http://winflex.co.il/.
WinWind, http://www.winwind.com/.
World Institute for Sustainable Energy, http://www.wisein.org/.
World Wind Energy Association, http://www.wwindea.org/home/index.php.
Xtreme Power, http://www.xtremepower.com/.
ZK Energy, http://www.zkenergy.com/en/index.html.

Reports and Presentations

Ackerman, Frank, and Stanton, Elizabeth A. 2011 (July 13). Climate Risks and Carbon Prices: Revising the Social Cost of Carbon. E3 Network (Economics for Equity & Environment. http://www.e3network.org/papers/Climate_Risks_and_Carbon_Prices_executive_summary+full_report+comments.pdf.

Acoustic Ecology Institute. 2011. Wind Farm Noise 2011: Science and Policy Overview. http://acousticecology.org/wind/winddocs/AEI_WindFarmNoise2011.pdf

———. 2010. Wind Farm Noise: 2009 in Review. http://acousticecology.org/docs/AEI_WindFarmNoise_2009inReview.pdf.

American Wind Energy Association. 2011 (April 7). U.S. Wind Industry Annual Market Report. http://www.awea.org/learnabout/publications/reports/AWEA-US-Wind-Industry-Market-Reports.cfm.

———. 2010. AWEA Small Wind Turbine Global Market Study. http://www.awea.org/documents/2010_AWEA_Small_Wind_Turbine_Global_Market_Study.pdf.

———. 2010. U.S. Wind Industry Annual Market Report. http://www.awea.org/documents/reports/Annual_Market_Report_Press_Release_Teaser.pdf.

———. Basic Principles of Wind Resource Evaluation. http://www.awea.org/faq/basicwr.html.

———. Building a Green Power Superhighway. http://www.awea.org/documents/Green_Power_Superhighways_Factsheet.pdf.

———. Green Power Transmission and Consumer Savings. http://www.awea.org/documents/factsheets/Transmission_and_Consumer_Savings.pdf.

———. Offshore Wind Energy. http://www.awea.org/documents/factsheets/Offshore_Factsheet.pdf.

———. Wind Power—Clean AND Reliable. http://www.awea.org/documents/factsheets/Reliability_Factsheet.pdf.

———. Wind Power and Energy Storage. http://www.awea.org/documents/factsheets/Energy_Storage_Factsheet.pdf.

———. 20% Wind Energy by 2030: Wind, Backup Power, and Emissions. http://www.awea.org/documents/factsheets/Backup_Power.pdf.

Ancona, Dan, and McVeigh, Jim. 2001 (August 29). Wind Turbine—Materials and Manufacturing Fact Sheet. Princeton Energy Resources International, LLC. http://www.perihq.com/documents/WindTurbine-MaterialsandManufacturing_FactSheet.pdf.

Assessment of Offshore Wind Energy Resources for the United States. U.S. National Renewable Energy Laboratory. http://www.nrel.gov/docs/fy10osti/45889.pdf.

Bajdek, Christopher. 2007 (October 22-24). Communicating the Noise Effects of Wind Farms to Stakeholders. Harris Miller Miller & Hanson Inc. http://www.hmmh.com/cmsdocuments/Bajdek_NC07.pdf.

Bell, Benjamin. 2009 (September 2). Operating Efficiently in a New Age of Renewable Power Generation. Presentation at the GreenPower Conferences: Optimizing Wind Power Performance. Houston.

———. 2010 (January 12). Wind Project Development Value Chain. Maine Wind Seminar Series 2010.

Bentek Energy LLC. 2011 (July 19). The Wind Power Paradox. http://www.bentekenergy.com/InTheNews.aspx#Article4701.

———. 2010 (April 16). How Less Became More: Wind, Power and Unintended Consequences in the Colorado Energy Market. http://docs.wind-watch.org/BENTEK-How-Less-Became-More.pdf.

Better Generation. Comparing Wind Turbine Power Curves. http://www.bettergeneration.com/wind-turbines/comparing-wind-turbine-power-curves.html.

BlueGreen Alliance. 2010 (June). Winds of Change: A Manufacturing Blueprint for the Wind Industry. http://www.bluegreenalliance.org/admin/private_publications/files/BGA-Report-062510_FINAL.pdf.

Bolinger, Mark and Ryan Wiser. Understanding Trends in Wind Turbine Prices Over the Past Decade. Lawrence Berkeley National Laboratory. LBLN-5119E. http://eetd.lbl.gov/ea/ems/reports/lbnl-5119e.pdf.

Bonnon, Ian D. 2010 (September 21). Project Planning—Concept to Finance to Build. Presentation at the Offshore Wind Best Practice Summit. Boston.

Brattle Group. 2011 (January 25). Brattle Estimates that Approximately 1,000 MW of Additional Generator Regulation Service Will Be Needed to Integrate 50,000 MW of Wind Generation in Large Power Systems. http://www.brattle.com/NewsEvents/NewsDetail.asp?RecordID=902.

Butterfield, S., et al. 2007 (September). Engineering Challenges for Floating Offshore Wind Turbines. NREL/CP-500-38776. http://gisceu.net/PDF/U444.pdf.

California Independent System Operator. 2010 (August 31). Integration of Renewable Resources: Operational Requirements and Generation Fleet Capability at 20% RPS. http://www.caiso.com/2804/2804d036401f0.pdf.

Canadian Wind Energy Association. 2006 (August). The Sights and Sounds of Wind. http://www.canwea.ca/images/uploads/File/NRCan_-_Fact_Sheets/7_visual_sound.pdf.

———. WindVision 2025. http://www.canwea.ca/images/uploads/File/Windvision_summary_e.pdf.

Center for Energy Efficiency & Renewable Energy, University of Massachusetts at Amherst. Wind Power: Resource Assessment. http://www.ceere.org/rerl/about_wind/RERL_Fact_Sheet_5_Resource_Assessment.pdf.

Coppin, P. A., Ayotte, K. A., and Steggel, N. 2003 (October). Wind Resource Assessment in Australia—A Planner's Guide. Australian Commonwealth Scientific and Industrial Research Organization (CSIRO), Version 1.1. http://www.csiro.au/files/files/pis7.pdf.

Corbus, Dave, et al. 2010 (November). Oahu Wind Integration and Transmission Study. NREL/TP-5500-48632. http://www.nrel.gov/wind/systemsintegration/pdfs/2010/owits_summary_report.pdf.

Couture, Toby D., et al. 2010 (July). A Policymaker's Guide to Feed-in Tariff Policy Design. NREL/TP-6A2-44849. http://www.nrel.gov/docs/fy10osti/44849.pdf.

Crotty, Marty. 2010 (September). What Gets Measured Gets Done. Upwind Solutions, Engineering White Papers. http://www.engineeringwhitepapers.com/wp-content/uploads/2011/01/Day-1_Upwind-Sol_Marty_Crotty.pdf.

Davis, Brian, and Wheelock, Clint. 2011 (March). Energy & Environment Consumer Survey: Consumer Attitudes and Awareness about 12 Energy & Environmental Concepts. Pike Research. https://www.pikeresearch.com/wordpress/wp-content/uploads/2011/02/EECS-11-Pike-Research.pdf.

Denholm, Paul, et al. 2009 (August). Land-Use Requirements of Modern Wind Power Plants in the United States. NREL/TP-6A2-45834. http://www.nrel.gov/docs/fy09osti/45834.pdf.

Dmitriev, Grigori. 2001 (June). Wind Energy in Russia. VetrEnergo for Gaia Apatity and Inforse-Europe. http://www.inforse.org/europe/word_docs/ruswind2.doc.

Doris, Elizabeth, and Gelman, Rachel. 2011 (January). State of the States 2010: The Role of Policy in Clean Energy Market Transformation. NREL/TP-6A20-49193. http://www.nrel.gov/applying_technologies/state_local_activities/pdfs/49193.pdf.

EcoBusinessLinks. Wind Maps. http://www.ecobusinesslinks.com/wind_maps.htm.

EnerNex Corp. 2011 (February). Eastern Wind Integration and Transmission Study. NREL/SR-5500-47078, revised edition. http://www.nrel.gov/wind/systemsintegration/pdfs/2010/ewits_final_report.pdf.

Environment Canada. Wind Energy Atlas. http://www.windatlas.ca/en/index.php.

Erickson, Wallace P., et al. 2002 (December). Synthesis and Comparison of Baseline Avian and Bat Use, Raptor Nesting and Mortality Information from Proposed and Existing Wind Developments. Western EcoSystems Technology, Inc http://www.bpa.gov/Power/pgc/wind/Avian_and_Bat_Study_12-2002.pdf.

Erickson, Wallace P., Johnson, Gregory D., and Young, David P., Jr. 2005. A Summary and Comparison of Bird Mortality from Anthropogenic Causes with an Emphasis on Collisions. USDA Forest Service Gen. Tech. Rep. PSW-GTR-191. http://www.fs.fed.us/psw/publications/documents/psw_gtr191/Asilomar/pdfs/1029-1042.pdf.

Ernst & Young. 2011 (February). Renewable Energy Country Attractiveness Indices. http://www.ey.com/Publication/vwLUAssets/Renewable_energy_country_attractiveness_indices_-_Issue_28/$FILE/EY_RECAI_issue_28.pdf

European Commission. 2010 (October). Wind Energy Developments and Natura 2000. http://ec.europa.eu/environment/nature/natura2000/management/docs/Wind_farms.pdf.

———. 2003. External Costs: Research Results on Socio-Environmental Damages Due to Electricity and Transport. ExternE. EUR 20198. http://www.externe.info/externpr.pdf.

European Union. 2010 (April). Roadmap 2050: A Practical Guide to a Prosperous, Low-Carbon Europe. Volume 1, Technical Analysis. http://www.roadmap2050.eu/attachments/files/Volume1_fullreport_PressPack.pdf.

European Wind Energy Association 2010 (April). Roadmap 2050: A Practical Guide to a Prosperous, Low-Carbon Europe. Volume 2, Policy Recommendations. http://www.roadmap2050.eu/attachments/files/Volume2_Policy.pdf.

———. 2009 (January). Wind at Work: Wind Energy and Job Creation in the EU. http://www.ewea.org/fileadmin/ewea_documents/documents/publications/Wind_at_work_FINAL.pdf.
———. 2009 (February 13). Wind Energy—the Facts, Executive Summary. http://www.ewea.org/fileadmin/ewea_documents/documents/publications/WETF/1565_ExSum_ENG.pdf.
———. 2009 (February). Integrating Wind: Developing Europe's Power Market for the Large-Scale Integration of Wind Power. http://www.trade-wind.eu/fileadmin/documents/publications/Final_Report.pdf.
———. 2009 (March). The Economics of Wind Energy. http://www.ewea.org/fileadmin/ewea_documents/documents/publications/reports/Economics_of_Wind_Main_Report_FINAL-lr.pdf.
———. 2010 (June). The European Wind Initiative: Wind Power Research and Development for the Next Ten Years. http://www.ewea.org/fileadmin/ewea_documents/documents/publications/EWI/EWI_2010_final.pdf.
———. 2010 (November). Powering Europe: Wind Energy and the Electricity Grid. http://www.ewea.org/fileadmin/ewea_documents/documents/publications/reports/Grids_Report_2010.pdf.
———. 2011 (January). The European Offshore Wind Industry Key Trends and Statistics 2010. http://www.ewea.org/fileadmin/ewea_documents/documents/00_POLICY_document/Offshore_Statistics/110120_Offshore_stats_Exec_Sum.pdf.
———. 2011 (February). Wind in Power. http://www.ewea.org/fileadmin/ewea_documents/documents/statistics/EWEA_Annual_Statistics_2010.pdf.
———. 2011 (March). New Research Shows 20 MW Wind Turbines Are Feasible. http://www.ewea.org/index.php?id=60&no_cache=1&tx_ttnews%5Btt_news%5D=1901&tx_ttnews%5BbackPid%5D=1&cHash=b4462d717ccc0ee61e55922ce236f70c.
———. 2011. Fact Sheets: Climate. http://www.ewea.org/index.php?id=1611.
———. 2011. Fact Sheets: Electricity Markets. http://www.ewea.org/index.php?id=1611.
———. 2011. Fact Sheets: Employment. http://www.ewea.org/index.php?id=1611.
———. 2011. Fact Sheets: Offshore. http://www.ewea.org/index.php?id=1611.
———. 2011. Fact Sheets: The Power Grid. http://www.ewea.org/index.php?id=1611.
———. 2011. Operational Offshore Wind Farms in Europe, End 2010. http://www.ewea.org/fileadmin/ewea_documents/documents/statistics/110214__public_offshore_wind_farms_in_Europe_2010.pdf.
———. Wind Energy—the Facts. http://www.wind-energy-the-facts.org.
Eyer, Jim, and Corey, Garth. 2010 (February). Energy Storage for the Electricity Grid: Benefits and Market Potential Assessment Guide. Sandia National Laboratories. SAND2010-0815. http://www.goodcompanyassociates.com/files/manager/SAND_Sandia_BenefitsAndMarketGuide_FINAL_Feb.pdf.
Fingersh, L., Hand, M., and Laxson, A. 2006 (December). Wind Turbine Design Cost and Scaling Model. NREL/TP-500-40566. http://www.nrel.gov/wind/pdfs/40566.pdf.

GlobalData. 2009 (September). The US Small Wind Market Set to Become BIG. http://www.reportsandreports.com/reports/15111-the-us-small-wind-market-set-to-become-big.html.

———. 2009 (December). Global Wind Energy Market Analysis and Forecasts to 2020.

Global Security. 1993. Chapter 3. Affected Environment. Figure 3-8, Wind Rose for SRS, 1987 through 1991. http://www.globalsecurity.org/wmd/library/report/enviro/eis-0217/eis0217_3.html.

Global Wind Energy Council. 2011 (August 24). 1,928.8MW of Wind Power Contracted at Brazilian Auctions. http://www.gwec.net/index.php?id=30&no_cache=1&tx_ttnews[tt_news]=298&tx_ttnews[backPid]=97&cHash=a56c316695.

———. 2011 (March). Global Wind Report: Annual Market Update 2010. http://www.gwec.net/fileadmin/documents/Publications/Global_Wind_2007_report/GWEC%20Global%20Wind%20Report%202010%20low%20res.pdf.

———. 2012 (February). Global Wind Statistics 2011. http://www.gwec.net/uploads/media/GWEC-PRstats-2011_20120206_06__1_.pdf.

Global Wind Energy Council, Greenpeace International. 2010 (October). Global Wind Energy Outlook 2010. http://www.gwec.net/fileadmin/documents/Publications/GWEO%202010%20final.pdf.

Hastings, Rob. 2010 (September 21). Wind Energy Development in the UK. Presentation at the Offshore Wind Best Practice Summit. Boston.

Hoen, Ben, et al. 2009 (December). The Impact of Wind Power Projects on Residential Property Values in the United States: A Multi-Site Hedonic Analysis. Lawrence Berkeley National Laboratory. LBNL-2829E. http://eetd.lbl.gov/ea/ems/reports/lbnl-2829e.pdf.

Holttinen, Hannele, et al. 2006 (September 18-21). "Design and Operation of Power Systems with Large Amounts of Wind Power." Global Wind Power Conference, Adelaide, Australia. http://www.ieawind.org/AnnexXXV/Meetings/Oklahoma/IEA%20SysOp%20GWPC2006%20paper_final.pdf.

International Energy Agency. 2011. Clean Energy Progress Report. http://www.iea.org/papers/2011/CEM_Progress_Report.pdf

——— 2010 (July). IEA Wind Energy 2009 Annual Report. http://www.ieawind.org/AnnualReports_PDF/2009/2009AR_92210.pdf (ISBN-0-9786383-4-4).

Johnson, David. 2010 (May 11). NCAR's Xcel Energy Project Overview. http://ral.ucar.edu/projects/wind_energy_workshop/presentations/Xcel_Energy_Project_Overview_Johnson_5.pdf.

Jones, Gareth, et al. 2009 (May). Scoping and Method Development Report: Determining the Potential Ecological Impact of Wind Turbines on Bat Populations in Britain. University of Bristol. http://www.bats.org.uk/data/files/determining_the_impact_of_wind_turbines_on_british___bats_final_report_29.5.09_website.pdf.

Jones, Lawrence E. 2011 (December 19). Strategies and Decision Support Systems for Integrating Variable Energy Resources in Control Centers for Reliable Grid Operations: Global Best Practices, Examples of Excellence and Lessons Learned. Alstom Grid Inc. and U.S. Department of Energy. http://www1.eere.energy.gov/wind/pdfs/doe_wind_integration_report.pdf.

Kassakian, John G., et al. 2011 (December 5). The Future of the Electric Grid: An Interdisciplinary MIT Study. Massachusetts Institute of Technology Energy Initiative. http://web.mit.edu/mitei/research/studies/documents/electric-grid-2011/Electric_Grid_Full_Report.pdf.

Kirby, Brendan, and Milligan, Michael. 2009 (July). Capacity Requirements to Support Inter-Balancing Area Wind Delivery. NREL/TP-550-46274. http://www.nrel.gov/docs/fy09osti/46274.pdf.

Kitasei, Saya. 2011 (February 17). Powering the Low Carbon Economy: The Role of Natural Gas in a Shifting Power Industry. Worldwatch Institute. Saya Kitasei 2011-02-17_Flexible_Power_Symposium_(SSK)[1].pdf.

Knight, Matthew. 2010 (September 21). Offshore Wind Electrical Infrastructure. Presentation at the Offshore Wind Best Practice Summit, Boston.

Komanoff, Charles. 2009 (April 21). Wind Power's Displacement of Fossil Fuels. Komanoff Energy Associates. http://www.komanoff.net/wind_power/Wind_Power's_Displacement_of_Fossil_Fuels.pdf.

Lantz, E., and Tegen, S. 2009 (April). Economic Development Impacts of Community Wind Projects: A Review and Empirical Evaluation. NREL/CP-500-45555. http://www.nrel.gov/docs/fy09osti/45555.pdf.

Li, Junfeng, et al. 2010 (October). China Wind Power Outlook 2010. http://www.greenpeace.org/raw/content/eastasia/press/reports/wind-power-report-english-2010.pdf.

Massachusetts Department of Environmental Protection and Massachusetts Department of Public Health. 2012 (January). Wind Turbine Health Impact Study: Report of Independent Expert Panel. http://www.mass.gov/dep/energy/wind/turbine_impact_study.pdf.

Meroney, Robert N. 1977 (Oct. 11–12). Jousting with the Wind—a General Review of the Wind Energy Problem, New Zealand Meteorological Service Symposium on Meteorology and Energy.

Milligan, Michael, and Kirby, Brendan. 2010 (July). Utilizing Load Response for Wind and Solar Integration and Power System Reliability. NREL/CP-550-48247. http://www.nrel.gov/docs/fy10osti/48247.pdf.

Monteiro, C., et al. 2009 (November 6). Wind Power Forecasting: State-of-the-Art 2009. INESC Porto and Argonne National Laboratory. ANL/DIS-10-1. http://www.dis.anl.gov/pubs/65613.pdf.

Musial, Walt. 2010 (January 13). Trends in U.S. Wind Industry Technology. Presented at the Maine Wind Seminar, January 12–14, 2010, Portland, Maine.

Musial, Walt, and Ram, Bonnie. Large-Scale Offshore Wind Power in the United States. NREL/TP-500-40745. http://www.nrel.gov/wind/pdfs/40745.pdf.

National Conference of State Legislatures. 2009. Integrating Wind Power into the Electric Grid. http://www.ncsl.org/documents/energy/Wind_Integration.pdf.

National Health and Medical Research Council. 2010 (July). Wind Turbines and Health. http://www.nhmrc.gov.au/_files_nhmrc/file/publications/synopses/public_statement_wind_turbines_and_health.pdf.

National Research Council. 2010. Hidden Costs of Energy: Unpriced Consequences of Energy Production and Use. http://www.nap.edu/openbook.php?record_id=12794&page=1.

National Wind Coordinating Collaborative. 2002 (August). Permitting of Wind Energy Facilities. http://www.nationalwind.org/assets/publications/permitting2002.pdf.

National Wind Coordinating Committee. 2010 (Spring). Wind Turbine Interactions with Birds, Bats, and Their Habitats: A Summary of Research Results and Priority Questions. https://www.nationalwind.org/assets/publications/Birds_and_Bats_Fact_Sheet_.pdf.

Nelson, Guy, et al. 2011 (December). Establishing an In-House Wind Maintenance Program, 2nd Ed. U.S. Department of Energy, Energy Efficiency and Renewable Energy, Wind Powering America. DOE/EE 0644. http://www.repartners.org/wind/WindO&Mreport.pdf.

North American Electric Reliability Corp. 2009 (April). Accommodating High Levels of Variable Generation. http://www.nerc.com/files/IVGTF_Report_041609.pdf.

———. 2010 (October). 2010 Long-Term Reliability Assessment. http://www.nerc.com/files/2010%20LTRA.pdf.

Northwest Power and Conservation Council. 2011 (August 17). The Effects of an Increasing Surplus of Energy Generating Capability in the Pacific Northwest. http://www.nwcouncil.org/library/2011-09.pdf.

Parsons, Brian, et al. 2006 (July). Grid Impacts of Wind Power Variability Recent Assessments from a Variety of Utilities in the United States. NREL/CP-500-39955. http://www.nrel.gov/docs/fy06osti/39955.pdf.

Pernick, Ron, Wilder, Clint, and Winnie, Trevor. 2010 (October). Clean Tech Job Trends 2010. http://www.cleanedge.com/reports/pdf/JobTrends2010.pdf.

The Pew Charitable Trusts. 2011. Who's Winning the Clean Energy Race? http://www.pewenvironment.org/uploadedFiles/PEG/Publications/Report/G-20Report-LOWRes-FINAL.pdf.

Pfund, Nancy, and Healey, Ben. 2011 (September). What Would Jefferson Do? DBL Investors. http://www.dblinvestors.com/documents/DBL_energy_subsidies_paper.pdf.

Pike Research. 2011 (April). Offshore Wind Power: Market Opportunities and Challenges, Technology Issues, Key Industry Players, and Global Capacity and Production Revenue Forecasts. http://www.pikeresearch.com/research/offshore-wind-power.

———. 2011 (July 12). Global Wind Power Capacity to Triple by 2017. http://www.pikeresearch.com/newsroom/global-wind-power-capacity-to-triple-by-2017.

———. 2011 (July 21). Submarine Electricity Cable Demand to Place a Significant Strain on Supply during the Next 10 Years. http://www.pikeresearch.com/newsroom/submarine-electricity-cable-demand-to-place-a-significant-strain-on-supply-during-the-next-10-years.

Porter, K., and Rogers, J. 2010 (April). Status of Centralized Wind Power Forecasting in North America. NREL/SR-550-47853. http://www.nrel.gov/wind/pdfs/47853.pdf.

Potomac Economics. 2010. 2009 State of the Market Report for the Midwest ISO. http://www.potomaceconomics.com/uploads/midwest_documents/2009_State_of_the_Market_Report.pdf.

Rand, Joseph. Wind Turbine Blade Design. Kidwind. http://www.kidwind.org/Presentations/BladeDesign%208.13.09.pdf.

REN21 Renewable Energy Policy Network for the 21st Century. 2010 (July 15). *Renewables 2010 Global Status Report*. http://www.ren21.net/globalstatusreport/g2010.asp.

REN21 Renewable Energy Policy Network for the 21st Century. 2011 (July 12). *Renewables 2011 Global Status Report*. http://www.ren21.net/Portals/97/documents/GSR/REN21_GSR2011.pdf.

Renewable Energy Atlas. Government of Australia. http://climatechange.gov.au/en/what-you-need-to-know/renewable-energy/~/media/publications/renewable-energy/atlas/atlas-help.ashx.

Rowan, Tony. 2011 (August 24). July 2011 Wind Curtailments. Midwest Independent System Operator. https://www.midwestiso.org/Library/Repository/Meeting%20Material/Stakeholder/RSC/2011/20110823-24/20110823-24%20RSC%20Item%2014b%20Wind%20Curtailments.pdf.

Salerno, Liz. 2011 (April). Opportunities in Wind Manufacturing. http://www.americanbearings.org/resource/resmgr/2011_annual/salerno.pdf.

Schmidt, Michael. Wind Turbine Design Optimization. Strategic Energy Institute, Georgia Institute of Technology. http://www.clemson.edu/scies/wind/Poster-Schmidt.pdf.

Schwartz, M., Heimiller, D., Haymes, S., and Musial, W. 2010 (June). Assessment of Offshore Wind Energy Resources for the United States. NREL/TP-500-45889. http://www.nrel.gov/docs/fy10osti/45889.pdf.

Sterzinger, George, and Svrcek, Matt. 2004 (September). Wind Turbine Development: Location of Manufacturing Activity. Renewable Energy Policy Project. http://www.repp.org/articles/static/1/binaries/WindLocator.pdf.

Thresher, R., Robinson, M., and Veers, P. 2008 (August). Wind Energy Technology: Current Status and R&D Future. NREL/CP-500-43374. http://www.nrel.gov/wind/pdfs/43374.pdf

———. "Wind Energy Technology: Current Status and R&D Future, NREL/CP-500-43374, August 2008.

Tuohy, Aidan. 2011. Integrating Wind Generation in the Bulk Electricity System. Electric Power Research Institute. http://mydocs.epri.com/docs/CorporateDocuments/Newsroom/WashingtonUpdate/Tuohy.pdf.

UK Energy Research Centre. 2010 (September). Great Expectations: The Cost of Offshore Wind in UK Waters—Understanding the Past and Projecting the Future. http://www.ukerc.ac.uk/support/tiki-download_file.php?fileId=1164.

U.S. Department of Energy. 2009 (December). National Electric Transmission Congestion Study. http://congestion09.anl.gov/documents/docs/Congestion_Study_2009.pdf.

U.S. Department of Energy, Energy Efficiency and Renewable Energy. 2011. U.S. Installed Wind Capacity and Wind Project Locations. http://www.windpoweringamerica.gov/wind_installed_capacity.asp.

——— 2010. Wind Power Today—2010. http://www1.eere.energy.gov/windandhydro/pdfs/47531.pdf.

——— 2008 (July). 20% Wind Energy by 2030: Increasing Wind Energy's Contribution to U.S. Electricity Supply. DOE/GP-102008-2567. http://www1.eere.energy.gov/wind/pdfs/41869.pdf.

———. 2005 (May). Wind Energy Myths. DOE/GO-102005-2137. http://www.nrel.gov/docs/fy05osti/37657.pdf.

———. Energy Basics, Wind Energy Technologies. http://www.eere.energy.gov/basics/renewable_energy/wind.html.

——— Wind and Water Program. http://www1.eere.energy.gov/windandhydro.

U.S. Department of Energy, Energy Efficiency and Renewable Energy, Wind and Water Program. Frequently Asked Questions on Small Wind Systems. www1.eere.energy.gov/windandhydro/small_wind_system_faqs.html.

U.S. Department of Energy, Energy Efficiency and Renewable Energy, Wind Powering America. 2011 (April). Wind Energy Benefits. DOE/GO/102011-3226. http://www.windpoweringamerica.gov/pdfs/wpa/2011/wind_energy_benefits.pdf.

———. 2011. Wind Turbine Sound. http://www.windpoweringamerica.gov/newengland/issues_sound.asp.

———. 2005 (May). Wind Energy Myths. DOE/GO-102005-2137. http://www.nrel.gov/docs/fy05osti/37657.pdf.

———. Wind Maps and Wind Resource Potential Estimates. http://www.windpoweringamerica.gov/wind_maps.asp.

U.S. Energy Information Administration. 2011 (April). Annual Energy Outlook 2011 with Projections to 2035. DOE/EIA-0383(2011). http://www.eia.gov/forecasts/aeo/pdf/0383(2011).pdf.

U.S. Fish & Wildlife Service. 2011. Links to Worldwide Wind Energy/Wildlife Guidance Documents and Information. http://www.fws.gov/windenergy/docs/Links_Worldwide_Wind_Energy_Wildlife_Guidance_Docs_Info.pdf.

U.S. National Renewable Energy Laboratory, National Wind Technology Center. 2010 (June). Environment and Siting: Supporting Research on Impacts to Birds, Bats, and Habitats from Wind Development. NREL/FS-500-47254. http://www.nrel.gov/wind/pdfs/47254.pdf.

Utility Wind Integration Group. 2010 (August). Frequently Asked Questions About Utility-Scale Wind Power. http://www.uwig.org/UWIG-WINDFAQ.pdf.

Wind Energy Update. 2010 (September). The European Offshore Wind Energy Standards, Permitting and Markets Report: A Country-By-Country Comparison and Practical Guide. http://www.windenergyupdate.com/standards/report-summary.shtml.

Wiser, Ryan, and Bolinger, Mark. 2011 (August). 2010 Wind Technologies Market Report. http://www1.eere.energy.gov/windandhydro/pdfs/51783.pdf.

Wiser, Ryan, and Bolinger, Mark. 2010 (August). 2009 Wind Technologies Market Report. http://www1.eere.energy.gov/windandhydro/pdfs/2009_wind_technologies_market_report.pdf.

World Wind Energy Association. 2011 (April). World Wind Energy Report 2010. http://www.wwindea.org/home/images/stories/pdfs/worldwindenergyreport2010_s.pdf.

Magazines and Journals

North American Clean Energy-Wind
Wind Engineering
Wind Systems
Wind Today
Windpower Engineering
Windpower Monthly

Articles

Airlie, Catherine, and Lundgren, Kari. 2011 (March 2). U.K. Government Ready to Pay Power Users to Switch off in 'Negawatt' Plan. Bloomberg. http://www.bloomberg.com/news/2011-03-02/u-k-ready-to-pay-power-users-to-switch-off-in-negawatt-plan.html.

Alliance for Renewable Energy. 2011. Status of FIT Legislation in Each State. http://www.allianceforrenewableenergy.org/us-reps.html.

Angelou, Nikolas. 2010 (October 14). Predicting the Future Using Lidar. *Renewable Energy World*. http://www.renewableenergyworld.com/rea/news/article/2010/10/lidar-forecasts-boost-wind-harvests.

Appleyard, David. 2011 (June 8). Spring Growth for Wind Turbines. *Renewable Energy World*. http://www.renewableenergyworld.com/rea/news/article/2011/06/spring-growth-for-wind-turbines.

———. 2011 (February 16). Record Growth for EU Offshore Wind in 2010. *Renewable Energy World*. http://www.renewableenergyworld.com/rea/news/article/2011/02/wind-power-market-statistics??cmpid=WNL-Wednesday-February16-2011.

———. 2011 (February 6). Where Next for the Supergrid? *Renewable Energy World*. http://www.renewableenergyworld.com/rea/blog/post/2011/02/where-next-for-the-supergrid?cmpid=WNL-Wednesday-February9-2011.

———. 2010 (September 30). Latin America Warms to Wind. *Renewable Energy World*. http://www.renewableenergyworld.com/rea/news/article/2010/09/latin-america-warms-to-wind.

Archer, Cristina L., and Jacobson, Mark Z. 2005. Evaluation of Global Wind Power. *Journal of Geophysical Research—Atmospheres*, 110, D12110, doi:10.1029/2004JD005462. http://www.stanford.edu/group/efmh/winds/2004jd005462.pdf.

Arnett, Edward B., et al. 2010 (November). Altering Turbine Speed Reduces Bat Mortality at Wind-Energy Facilities. Bats and Wind Energy Cooperative. http://www.batsandwind.org/pdf/Arnett%20et%20al.%202011.pdf.

———. 2009 (April). Effectiveness of Changing Wind Turbine Cut-in Speed to Reduce Bat Fatalities at Wind Facilities. Bats and Wind Energy Cooperative. http://www.iberdrolarenewables.us/pdf/curtailment-final-report-05-15-10-v2.pdf.

Beattie, David. 2010 (December 22). Key Players in the Wind Energy: Pausing for Thought. *Renewable Energy World*. http://www.renewableenergyworld.com/rea/news/article/2010/12/key-players.

Black, Richard. 2011 (May 8). Nuclear "Cheapest Low-Carbon Option" for UK Energy. BBC News. http://www.bbc.co.uk/news/mobile/science-environment-13300595.

Blau, John. 2011 (March 17). Spanish Wind Power Manufacturer Expanding into India. *Renewable Energy World*. http://www.renewableenergyworld.com/rea/news/article/2011/03/spanish-wind-power-manufacturer-expanding-into-india??cmpid=WNL-Wednesday-March23-2011.

———. 2011 (March 29). Germans Push for More Renewable Energy; Offshore Wind to Benefit. *Renewable Energy World*. http://www.renewableenergyworld.com/rea/news/article/2011/03/germans-push-for-more-renewable-energy-offshore-wind-to-benefit.

———. 2011 (May 11). Baltic Sea Could Be Home to 1 GW of Wind Power Capacity. *Renewable Energy World*. http://www.renewableenergyworld.com/rea/news/article/2011/05/baltic-sea-could-be-home-to-1-gw-of-wind-power-capacity?cmpid=WNL-Wednesday-May11-2011.

———. 2011 (May 18). Europe Leads Offshore Wind Market. *Renewable Energy World*. http://www.renewableenergyworld.com/rea/news/article/2011/05/wind?cmpid=WNL-Friday-May20-2011.

Bloomberg News. 2011 (May 12). Huaneng Renewables Said to Revive $1 Billion Hong Kong Initial Share Sale. http://www.bloomberg.com/news/2011-05-12/huaneng-renewables-said-to-revive-1-billion-hong-kong-initial-share-sale.html.

Bonafoux, John. 2011 (March 10). Getting into Ship Shape for Far Offshore. *Renewable Energy World*. http://www.renewableenergyworld.com/rea/news/article/2011/03/getting-into-ship-shape-for-far-offshore??cmpid=WNL-Friday-March11-2011.

Boyle, Rebecca. 2010 (October 30). Increasing Wind Turbine Turn-On Speeds Could Help Reduce Bat Deaths, New Study Says. *Popular Science*. http://www.popsci.com/science/article/2010-10/increasing-wind-turbine-turn-speeds-could-help-reduce-bat-deaths-new-study-says.

Bradsher, Keith. 2010 (December 14). To Conquer Wind Power, China Writes the Rules. *New York Times*. http://www.nytimes.com/2010/12/15/business/global/15chinawind.html.

Bryce, Robert. 2011 (July 28). T. Boone's Windy Misadventure and the Global Backlash against Wind Energy. *Energy Tribune*. http://www.energytribune.com/articles.cfm/8120/T-Boones-Windy-Misadventure-And-the-Global-Backlash-Against-Wind-Energy.

———. 2011 (August 15). The Party's Over for Big Wind. *Energy Tribune*. http://www.energytribune.com/articles.cfm/8255/The-Partys-Over-for-Big-Wind.

———. 2010 (August 27). Wind Energy Gets Huge Subsidies. So Where Are the CO_2 Reductions? *Energy Tribune*. http://www.energytribune.com/articles.cfm/5103/Wind-Energy-Gets-Huge-Subsidies-So-Where-Are-The-CO2-Reductions?

———. 2010 (September 14). Wind Energy's Real Problems: (Hint: It Has Nothing to Do with *The Wall Street Journal*). *Energy Tribune*. http://www.energytribune.com/articles.cfm/5261/Wind-Energys-Real-Problems--Hint-It-Has-Nothing-to-Do-With-The-Wall-Street-Journal.

BTM Consult. 2011 (April). BTM Consult Releases New Wind Report: World Market Update 2010. *Wind Today*. http://www.windtoday.net/info/articles.html?ID=106940.

———. 2011 (April). Strong Year Predicted as BTM Reveals Its Top Ten Leading Turbine Markets in 2010 by Market Share. http://www.btm.dk/news/strong+year+predicted+by+btm-+a+part+of+navigant+consulting/?s=9&p=1&n=44&p_id=1.

———. 2010 (November 22). Offshore Wind Power 2010. http://www.btm.dk/news/offshore+wind+power+2010/?s=9&p=&n=39.

Buffett, Warren. 2011. Annual Letter to Shareholders. Berkshire Hathaway. http://www.berkshirehathaway.com/letters/2010ltr.pdf.

Cameron, Alasdair. 2011 (June 30). Offshore Wind Costs to Fall 15 Percent by 2022. *Renewable Energy World*. http://www.renewableenergyworld.com/rea/news/article/2011/06/offshore-wind-costs-to-fall-15-percent-by-2022.

Castano, Ivan. 2011 (April 7). China Installing Wind-power Capacity as Fast As It Can. *Renewable Energy World*. http://www.renewableenergyworld.com/rea/news/article/2011/04/china-installing-wind-power-capacity-as-fast-as-it-can?cmpid=WindNL-Thursday-April21-2011.

Chelsea Green. A Case Study in Community Wind: Denmark. http://www.chelseagreen.com/content/a-case-study-in-community-wind-denmark.

Coates, Christine. 2011 (April-June). Using Radar to Protect Birds and Bats. *Wind Today*. http://www.bluetoad.com/publication/?i=68530&p=26.

Comfort, Nicholas. 2011 (June 1). Merkel Faces Achilles Heel in Grids to Unplug German Nuclear. Bloomberg. http://www.bloomberg.com/news/2011-05-31/merkel-faces-achilles-heel-in-grids-to-unplug-german-nuclear.html.

Crowe, Robert. 2011 (August 17). Energy Storage Industry Grows to Integrate Wind, Solar. *Renewable Energy World*. http://www.renewableenergyworld.com/rea/news/article/2011/08/energy-storage-industry-grows-to-integrate-wind-solar.

Cummings, Jim. 2011 (February 22). Looking for Wind Industry Leadership in Reducing Noise Impacts. Acoustic Ecology Institute. *Renewable Energy World*. http://www.renewableenergyworld.com/rea/news/article/2011/02/looking-for-wind-industry-leadership-in-reducing-noise-impacts?cmpid=WNL-Wednesday-February23-2011.

Curry, Dick, and Kerlinger, Paul. What Kills Birds. Curry & Kerlinger, LLC. http://www.currykerlinger.com/birds.htm.

Daley, Beth. 2011 (February 16). New Harvard Study Examines Full Cost of Coal. Boston.com. http://www.boston.com/lifestyle/green/greenblog/2011/02/new_harvard_study_looks_at_ful.html.

Datamonitor. 2011 (January 19). Brazil: Policy Changes Aim to Drive the Domestic Wind Market Forward. http://about.datamonitor.com/cop15/archives/552.

Dawei, Yu, and Yue, Yang. 2010 (June 18). Cool Start for Offshore Wind Energy in China. http://english.caing.com/2010-06-18/100153345.html.

Deese, Heather, and Schmitt, Catherine. 2010 (January 29). Fathoming: How Do Offshore Wind Turbines Affect Marine Life (USA). The Working Waterfront. http://www.workingwaterfront.com/articles/Fathoming-What-are-the-marine-impacts-of-offshore-wind-turbines/13667/.

Deign, Jason. 2011 (June 13). Will Wind Energy Cut CO_2? Wind Energy Update. http://social.windenergyupdate.com/uncategorised/will-wind-energy-cut-co2.

———. 2011 (March 11). Wind Industry Needs to Put Safety First. Wind Energy Update. http://social.windenergyupdate.com/industry-insight/wind-industry-needs-put-safety-first?utm_source=newsletter&utm_medium=email&utm_campaign=WEU.

Del Franco, Mark. 2010. Permanent Magnet Generators Join Turbine Conversation. *North American Windpower*. 7(8): 48–50. http://www.nawindpower.com/new_home.php.

Diaz, Chris. 2010 (November). Community Wind Projects Still Have Financing Options. http://www.seminolefinancialservices.com/wp-content/themes/seminole/images/NA_Windpower_Article.pdf

Dodge, Darrell M. Illustrated History of Wind Power Development. http://www.telosnet.com/wind.

Doty, F. David, 2011 (February 2). Kicking Oil Addiction with Windfuels. Greentech Media. http://www.greentechmedia.com/articles/read/guest-post-kicking-oil-addiction-permanently-with-windfuels.

Drewitt, Allan L., and Langston, Rowena H.W. 2006. Assessing the Impact of Wind Farms on Birds. Ibis. http://www.gflrpc.org/programareas/wind/TechnicalDocuments/AvianImpactsSocietyforProtectionofBirds.pdf.

Duggan, Ed. 2011. Interviewed by Rebecca Busby. E-mail message to author. July 12.

Earth Track. 2010. Subsidy FAQs. http://www.earthtrack.net/subsidy-faqs.

Economist. 2010 (August 19). Wind Energy and Politics: Not on My Beach Please. http://www.economist.com/node/16846774?story_id=16846774.

———. 2010 (March 4). Smarting from the Wind. http://www.economist.com/node/15582251.

———. 2011 (June 2). Japan's Winds of Change. http://www.economist.com/node/18750586?story_id=18750586.

———. 2011 (March 12). Europe's Climate Policy: Being Ambitious. http://www.economist.com/node/18333149.

Eggleston, Eric. What Are Vertical-Axis Wind Turbines (VAWTs)? American Wind Energy Association. http://www.awea.org/faq/vawt.html.

Electrical Power Systems Technology. 2010 (November 25). HVDC Transmission Market to Grow 44% by 2015. http://eps-technology.blogspot.com/2010/11/hvdc-transmission-market-to-grow-44-by.html.

Ernst & Young. 2011 (February). Renewable Energy Country Attractiveness Indices. http://www.ey.com/Publication/vwLUAssets/Renewable_energy_country_attractiveness_indices_-_Issue_28/$FILE/EY_RECAI_issue_28.pdf

European Wind Energy Association. 2011 (April). 2011 Wind Power Meeting Kyoto Commitments While Climate Negotiations Stall. http://www.ewea.org/index.php?id=60&no_cache=1&tx_ttnews[tt_news]=1904&tx_ttnews[backPid]=1&cHash=d3188cd02e6ec4ad617f42b1000fc17e.

Eurotunnel Group. 2011 (May 28). Groupe Eurotunnel and Star Capital Partners Establish Joint Venture to Develop an Electricity Interconnector Between Great Britain and France. http://www.eurotunnelgroup.com/uploadedFiles/assets-uk/Media/Press-Releases/2011-Press-Releases/110526ELECLINK-UK.pdf.

Ewing, Jack, and Bajaj, Vikas. 2011 (March 23). German Energy Company Hits Headwinds in India. *New York Times*. http://www.nytimes.com/2011/03/24/business/global/24enercon.html?ref=todayspaper.

Fairley, Peter. 2008 (April 17). Scheduling Wind Power: Better Wind Forecasts Could Prevent Blackouts and Reduce Pollution. *MIT Technology Review*. http://www.technologyreview.com/energy/20646.

———. 2011 (January). A Less Mighty Wind. IEEE Spectrum. http://spectrum.ieee.org/green-tech/wind/a-less-mighty-wind/1.

Farrand, Tim. 2011 (September 27). Scotland to Invest 35 Million Pounds in Wind Turbine Production. Bloomberg. http://www.bloomberg.com/news/2011-09-27/scotland-to-invest-35-million-pounds-in-wind-turbine-production.html.

Farrell, John. 2011 (May 23). Michigan to Grid Operator: We Prefer to Generate Our Own Renewable Energy. *Renewable Energy World*. http://www.renewableenergyworld.com/rea/blog/post/2011/05/michigan-to-grid-operator-we-prefer-to-generate-our-own-renewable-energy??cmpid=WNL-Wednesday-May25-2011.

———. 2010 (October 27). Hermann Scheer's Feed-in Tariff Legacy: 75% of World's Solar PV. http://energyselfreliantstates.org/content/hermann-scheers-feed-tariff-legacy-75-worlds-solar-pv.

Fisher, Curtis, et al. 2010. Offshore Wind in the Atlantic: Growing Momentum for Jobs, Energy Independence, Clean Air, and Wildlife Protection. National Wildlife Federation. http://www.nwf.org/News-and-Magazines/Media-Center/News-by-Topic/Global-Warming/2010/~/media/PDFs/Global%20Warming/Reports/NWF-Offshore-Wind-in-the-Atlantic.ashx.

Fraunhofer Institute IWU and Physik Instrumente. 2010 (July). Piezoelectric Vibration Absorbers to Reduce Wind Turbine Noise Emissions. *Solar & Wind Power*. http://ntbpdf.techbriefs.net/2010/GDM_SW_0710.pdf

Galbraith, Kate. 2010 (September 9). Fighting the Power Lines to Protect Hill Country Vistas. *The Texas Tribune*. http://www.texastribune.org/texas-energy/wind-energy/fighting-power-lines-protect-hill-country-vistas.

Gassert, Jochen, and Heckmann, Wilhelm. 2010 (January). Too Many Question Marks? GL Garrad Hassan. Energize Renewables. http://www.gl-group.com/pdf/energize-renewables_KOMPLETT.pdf.

Gay, Paul L. Why Two Bladed Rotors Make Sense. Aerostar Wind. http://www.aerostarwind.com/Why%20Two%20Bladed%20Rotors%20Make%20Sense.html.

Gipe, Paul. 2009. Wind Energy—the Breath of Life or the Kiss of Death: Contemporary Wind Mortality Rates. http://www.wind-works.org/articles/BreathLife.html.

———. Community Wind: The Third Way. Wind-Works.org. http://www.wind-works.org/articles/communitywindthethirdway.html.

———. 2011 (August 11). How to Design Feed-in Tariffs in the U.S. without Fear of Federal Preemption. *Renewable Energy World*. http://www.renewableenergyworld.com/rea/news/article/2011/08/how-to-design-feed-in-tariffs-in-the-u-s-without-fear-of-federal-preemption?cmpid=WNL-Friday-August12-2011.

———. 2011 (September 13). Nova Scotia Community Feed-in Tariffs to Launch September 19th. *Renewable Energy World*. http://www.renewableenergyworld.com/rea/news/article/2011/09/nova-scotia-community-feed-in-tariffs-to-launch-september-19th?cmpid=WNL-Wednesday-September14-2011.

———. 2011 (February 3). Taiwan Increases Offshore Wind Tariff 30 Percent. *Renewable Energy World*. http://www.renewableenergyworld.com/rea/news/article/2011/02/taiwan-increases-offshore-wind-tariff-30-percent.

———. 2011. Debunking the Myths about Wind Energy. Wind-Works.org. http://www.wind-works.org/articles/large_turbines.html#Debunking the Myths about Wind Energy.

———. 2010 (August 10). NREL Releases Feed-in Tariff Guide. *Renewable Energy World*. http://www.renewableenergyworld.com/rea/news/article/2010/08/nrel-releases-feed-in-tariff-guide.

Goggin, Michael. 2010 (September 1). The Facts about Wind Energy and Emissions. *Renewable Energy World*. http://www.renewableenergyworld.com/rea/news/article/2010/09/the-facts-about-wind-energy-and-emissions.

Golding, E.W. (1977). Jousting with the Wind, a General Review of the Wind Industry Problem.

Gopal, Prashant. 2009 (July 23). Now Hiring: Green-Collar Workers. *Businessweek*. http://www.businessweek.com/managing/content/jul2009/ca20090721_381476.htm.

Gosman, Taylor. 2010 (May 4). Cutting Transport Costs Down to Size. Windpower Monthly. http://www.windpowermonthly.com/news/EmailThisArticle/1000918/Cutting-transport-costs-down-size.

Green, John. 2011 (April 6.) Change in the Wind. *The Hutchinson News*. http://www.hutchnews.com/Todaystop/dillon-lecture-speaker--1.

Gross, Martin. 2011 (January). The Transmission Imperative for Renewables. *Utility Automation Engineering*. http://www.elp.com/index/display/article-display/1065111962/articles/utility-automation-engineering-td/volume-16/issue-1/feature/the-transmission-imperative-for-renewables.html.

Hagens, Nate. 2010 (September 18). The Fake Fire Brigade—How Can Renewable Sources Support Our Current Energy Delivery Expectations? *Oil Drum*. http://www.theoildrum.com/node/6957.

Hamm, Steve. Wind: The Power, The Promise, The Business. *Business Week*, July 3, 2008 (businessweek.com).

Harst, David, "Wind Turbine Design—The Keys to a High Efficiency Generator," ezine.

Hazlett, Maril. 2011 (January 4). Climate Change Could Have Major Impacts on Wind Resources. http://www.nawindpower.com/e107_plugins/content/content.php?content.7130.

Heller, Arnie. 2010 (April). Extracting More Power from the Wind. Lawrence Livermore National Laboratory, Science and Technology Review. https://str.llnl.gov/AprMay10/mirocha.html.

———. 2010 (April). Extracting More Power from the Wind. Lawrence Livermore National Laboratory, Science and Technology Review. https://str.llnl.gov/AprMay10/mirocha.html

Hsu, Jeremy. 2010 (January). Wind Turbines Leave Clouds and Energy Inefficiency in Their Wake. *Popular Science*. http://www.popsci.com/technology/article/2010-01/wind-turbines-leave-clouds-and-energy-inefficiency-their-wake.

Hughes, Tim. Lesson Number 2 in an Oklahoma Wind Power Tutorial Series. University of Oklahoma, Environmental Verification and Analysis Center. http://www.ocgi.okstate.edu/owpi/EducOutreach/Library/Lesson2_airdensity.pdf.

Hull, Dana. 2011 (August 29). Altamont Pass Wind Farm Gets Major Upgrade. *San Jose Mercury News*. http://www.energycentral.com/generationstorage/wind/news/en/21338730/Altamont-Pass-wind-farm-gets-major-upgrade.

IHS Emerging Energy Research. 2010 (May 4). Brazil Leads Latin America Wind Energy Markets to 46 GW by 2025. http://www.emerging-energy.com/content/press-details/Brazil-Leads-Latin-America-Wind-Energy-Markets-to-46-GW-by-2025/29.aspx.

———. 2010 (November 9). Recent Policy Moves a Start, but Much Stronger Action Is Needed to Accelerate the Transformation of the Global Energy System, Says the IEA's Latest *World Energy Outlook*. http://www.iea.org/press/pressdetail.asp?PRESS_REL_ID=402.

International Energy Agency. 2011 (October 4). World Energy Outlook 2011: IEA analysis of fossil-fuel subsidies. http://www.iea.org/weo/Files/ff_subsidies_slides.pdf.

Irish Wind Energy Association. 2011 (March 1). Report Launched on Impact of Wind on Pricing within the Single Electricity Market. http://www.iwea.com/index.cfm?page=viewnews&id=73.

Jones, Jackie. 2011 (March 1). Europe's Policymakers Sign up for a New Offshore Transmission Grid. *Renewable Energy World*. http://www.renewableenergyworld.com/rea/news/article/2011/03/along-the-right-lines??cmpid=WNL-Wednesday-March2-2011.

Jordan, Kim. 2011 (June 20). Developing a Wind Farm Support Campaign. *Renewable Energy World*. http://www.renewableenergyworld.com/rea/news/article/2011/06/developing-a-wind-farm-support-campaign?cmpid=WNL-Wednesday-June22-2011.

Kalinoski, Gail. 2010 (November 10). Underperformance Issues Deserve Fresh Examination. *North American Windpower*. 7(10): 1, 41–43.

Kanellos, Michael. 2010 (September 8). Power from the Atmosphere: Is It Real? Greentech Media. http://www.greentechmedia.com/articles/read/power-from-the-atmosphere-is-it-real.

———. 2010 (September 28). Will the Netherlands Rule High-Altitude Wind? Greentech Media. http://www.greentechmedia.com/articles/read/will-the-netherlands-rule-high-altitude-wind.

———. 2010 (September 29). Honeywell's Take on Wind Turbines and the Flying Donut from MIT. Greentech Media. http://www.greentechmedia.com/articles/read/honeywells-take-on-wind-turbines-and-the-flying-donut-from-mit.

———. 2011 (January 12). Energy Storage Originators: A New Type of Company? Greentech Media. http://www.greentechmedia.com/articles/read/energy-storage-originators-a-new-type-of-company.

———. 2011 (February 11). New Storage Technology: A 100 Million Ton Stone. http://www.greentechmedia.com/articles/read/new-storage-technology-a-100-million-ton-stone.

———. 2011 (April 21). Alstom Gets Contract for High Voltage DC at Tres Amigas. Greentech Media. http://www.greentechmedia.com/articles/read/alstom-gets-contract-for-high-voltage-dc-at-tres-amigas/.

———. 2011 (May 19). Tornadoes, Otter Pops, and Other Unlikely Energy Technologies. Greentech Media. http://www.greentechmedia.com/articles/read/tornadoes-pot-pies-and-other-unlikely-energy-technologies/.

———. 2011 (June 15). Water or Rocks: What's the Secret Ingredient for Energy Storage? Greentech Media. http://www.greentechmedia.com/articles/read/water-or-rocks-whats-the-secret-ingredient-for-energy-storage1/.

Keane, Brian. 2011 (February 9). Polls: Anti-Wind NIMBYs Are Small Minority. Huffington Post. http://www.huffingtonpost.com/brian-keane/polls-antiwind-nimbys-are_b_819905.html.

KEMA. 2010 (June 23). KEMA Sees Opportunities for Storage Applications in High-Penetration Renewables. http://www.kema.com/news/pressroom/press-releases/2010/KEMA-sees-opportunities-for-storage-applications-in-high-penetration-renewables.aspx.

Kirsner, Scott. 2011 (April 17). A Generator That's Lighter Than Air—and Relatively Light on the Wallet. *Boston Globe*. http://www.boston.com/business/technology/articles/2011/04/17/a_generator_thats_lighter_than_air__and_relatively_light_on_the_wallet/?page=full.

Kjaer, Christian. 2010 (November 9). More Than $5 Fossil Fuel Subsidies for Every $1 of Support for Renewables. http://www.renewableenergyworld.com/rea/blog/post/2010/11/more-than-5-fossil-fuel-subsidies-for-every-1-of-support-for-renewables.

Kline, Jack. 2011. Interviewed by Rebecca Busby. E-mail message to author. June 23.

Kremer, Josiane. 2011 (March 17). Norway Says Offshore Wind Too Expensive to Develop, Aftenbladet Reports. Bloomberg. http://www.bloomberg.com/news/2011-03-17/norway-says-offshore-wind-too-expensive-to-develop-aftenbladet-reports.html.

Kropper, Steve. 2011 (January 26). Wind Measurement Towers: How High Should They Be? *Renewable Energy World*. http://www.renewableenergyworld.com/rea/news/article/2011/01/wind-measurement-towers-how-high-should-they-be.

Lacey, Stephen. 2010 (July 21). Will Vertical Axis Wind Ever Scale Up? *Renewable Energy World*. http://www.renewableenergyworld.com/rea/news/podcast/2010/07/will-vertical-axis-wind-ever-scale-up.

———. 2010 (August 26). When Is Wind Energy Noise Pollution? *Renewable Energy World*. http://www.renewableenergyworld.com/rea/news/print/article/2011/03/when-is-wind-energy-noise-pollution.

———. 2011 (December 22). Wind Makes Up 80% of Contracts in Brazil's Latest Power Auction. Climate Progress. http://thinkprogress.org/romm/2011/12/22/394455/wind-contracts-in-brazil-power-auction/?utm_source=feedburner&utm_medium=email&utm_campaign=Feed%3A+climateprogress%2FlCrX+%28Climate+Progress%29.

Lacy, Virginia. 2010. The Answer Is Growing in the Wind. Rocky Mountain Institute. http://www.rmi.org/rmi/AnswerGrowinginWind.

LaMonica, Martin. 2010 (November 17). Start-Up Squeezes Flowing Wind to Boost Turbine Power. Green Tech. http://news.cnet.com/8301-11128_3-20023117-54.html#ixzz15ePYQwG2.

Laughlin, James. Drive Train Innovation Raises Wind Turbine Efficiency, Offers Gearbox Improvements. Power-Gen Worldwide. http://www.powergenworldwide.com/index/display/articledisplay/258472/articles/power-engineering/volume-110/issue-6/field-notes/drive-train-innovation-raises-wind-turbine-efficiency-offers-gearbox-improvements.html.

Lawson, James. 2011 (August 22). What Are the Challenges and Innovations in Offshore Wind Design and Development? *Renewable Energy World*. http://www.renewableenergyworld.com/rea/news/article/2011/08/what-are-the-challenges-and-innovations-in-offshore-wind-design-and-development?cmpid=WNL-Wednesday-August24-2011.

Leirgulen, Svein Inge. 2011 (September 27). DNV Releases Updated Standard for Offshore Wind Turbine Structures. Det Norske Veritas. http://www.dnv.com/press_area/press_releases/2011/DNV_releases_updated_standard_for_offshore_wind_turbine_structures.asp.

Levesque, Carl. 2011 (October 4). Wildlife Groups Team with AWS on Multi-Layered Wind Resource and Wildlife Map. *Renewable Energy World*. http://www.renewableenergyworld.com/rea/news/article/2011/10/wildlife-groups-team-with-aws-on-multi-layered-wind-resource-and-wildlife-map?cmpid=WindNL-Thursday-October6-2011.

Liebreich, Michael, and Young, William. 2005 (July 19). Offshore Wind: Europe's EUR 90 Billion Funding Requirement. White Paper, New Energy Finance Ltd. http://bnef.com/Download/NEF_HTML_Press/Offshore-wind-funding.pdf.

Link, David. 2010 (September 8). Pumped Hydro: Part of the Solution to Our Energy Storage Needs? Pike Research. http://www.pikeresearch.com/blog/articles/pumped-hydro-part-of-the-solution-to-our-energy-storage-needs.

Loos, Marcio R., et al. 2011 (August). World's First Carbon Nanotube Reinforced Polyurethane Wind Blades. Case Western Reserve University. http://polymers.case.edu/stories/World's%20First%20PU%20CNT%20Blades.html.

MacCleery, Brian. 2011 (January). The Advent of Airborne Wind Power. Wind Systems. http://windsystemsmag.com/article/detail/187/the-advent-of-airborne-wind-power.

Martin, Christopher. 2011 (April 6). Google, Warren Buffett May Be Next Wind-Turbine Buyers, Suzlon, Vestas Say. Bloomberg. http://www.bloomberg.com/news/2011-04-06/google-buffett-may-be-wind-turbine-buyers-say-suzlon-vestas.html.

Melnyk, Markian. 2011 (Spring). Offshore Wind Could Get Its Backbone. *Offshore Wind: A Supplement to North American Windpower*. 8(2): 14–15.

Milton, Joseph. 2010 (October 17). Why Winds Are Slowing. Nature Geoscience. http://www.nature.com/news/2010/101017/full/news.2010.543.html.

Moore, Mike. 2010 (August). Maintaining the Wind Fleet. *Power Engineering*. http://www.powergenworldwide.com/index/display/articledisplay/2077898726/articles/power-engineering/volume-114/Issue-8/Features/Maintaining-the-Wind-Fleet.html.

Morales, Alex. 2011 (February 7). Wind Turbine Prices Fall Below 1 Million Euros Per Megawatt. Bloomberg. http://www.bloomberg.com/news/2011-02-07/wind-turbine-prices-fall-below-1-million-euros-per-megawatt-bnef-says.html.

———. 2011 (May 9). U.K. Should Build More Nuclear, Less Offshore Wind, Panel Says. *Renewable Energy World*. http://www.bloomberg.com/news/2011-05-08/britain-should-rely-on-more-nuclear-power-less-offshore-wind-panel-says.html.

———. 2011 (April 15). Clean Energy Investment Fell 34% as Incentives Cut in Europe, BNEF Says. Bloomberg New Energy Finance. http://www.bloomberg.com/news/2011-04-15/clean-energy-investment-fell-34-as-incentives-cut-in-europe-bnef-says.html.

Muckosy, Pamela. 2010 (June 9). True Cost of Wind Turbine Operations & Maintenance. *Renewable Energy World*. http://www.renewableenergyworld.com/rea/partner/first-conferences/news/article/2010/06/true-cost-of-wind-turbine-operations-maintenance.

NASA. 2011 (February 1). NASA Engineer Finds Answer to Green Energy in the Air. NASA Tech Briefs. http://www.techbriefs.com/component/content/article/9168.

National Audubon Society. Audubon's Position on Wind Power. http://policy.audubon.org/wind-power-overview-0.

National Renewable Energy Laboratory and GE Energy. 2010 (May). Western Wind and Solar Integration Study. NREL/SR 550-47434. http://www.nrel.gov/wind/systemsintegration/pdfs/2010/wwsis_final_report.pdf.

O'Brien, George. 2008 (April 29). FloDesign Has Innovation Down to a Science. Economic Development Council of Western Massachusetts. http://www.westernmassedc.com/news/flodesignhasinnovationdowntoascience_469.

Orlowski, Aaron. 2011 (February 7). Largest Wind Turbine Testing Facility Nears Opening in Charlestown. *Boston Globe*. http://www.boston.com/yourtown/news/charlestown/2011/02/largest_wind_turbine_testing_f.html.

Pacella, Rena Marie. 2010 (March 26). How It Works: The Next-Gen Wind Turbine. *Popular Science*. http://www.popsci.com/technology/article/2010-03/next-gen-wind-turbine.

Palmer, Adrian. Delivering Power to Meet Britain's 2020 Targets. Utility Automation Engineering. http://www.elp.com/index/display/article-display/4283246046/articles/utility-automation-engineering-td/volume-16/issue-2/features/eye-on-europe-feature-delivering-power-to-meet-britains-2020-targets.html.

Patel, Sonal. 2010 (February 1). The Age of the 800-kV HVDC. Power. http://www.powermag.com/issues/departments/global_monitor/The-Age-of-the-800-kV-HVDC_2450.html.

———. 2011 (April 1). Changing Winds: The Evolving Wind Turbine. Power. http://www.powermag.com/renewables/wind/Changing-Winds-The-Evolving-Wind-Turbine_3557_p4.html.

Pearson, Natalie Obiko. 2011 (April 28). India's Rusting Wind Farms Help Gamesa, Vestas Expand Foothold. Bloomberg. http://www.bloomberg.com/news/2011-04-28/india-s-rusting-wind-farms-help-gamesa-vestas-expand-foothold.html.

Pernick, Ron. 2011 (May 3). Reassessing the Chinese Clean-Tech Juggernaut (or Why the U.S. Shouldn't Be Written Off Quite Yet). *Clean Edge*. http://www.cleanedge.com/views/.

Petersen, John. 2011 (March 8). Two Stocks for Grid Storage: ZBB Energy and Axion Power. *Seeking Alpha*. http://seekingalpha.com/article/257113-two-stocks-for-grid-storage-zbb-energy-and-axion-power.

———. 2011 (March 23). Grid-Based Energy Storage: Widely Misunderstood Challenges and Opportunities. *Seeking Alpha*. http://seekingalpha.com/article/259651-grid-based-energy-storage-widely-misunderstood-challenges-and-opportunities?source=email_alternative_energy_investing.

———. 2011 (April 10). Wind Power Investors Get Another Reality Check. *Seeking Alpha*. http://seekingalpha.com/article/262713-wind-power-investors-get-another-reality-check?source=email_alternative_energy_investing.

———. 2011 (December 15). Hype Busters From Lux Research Explain Grid Based Energy Storage. Renewable Energy World. http://www.renewableenergyworld.com/rea/news/article/2011/12/hype-busters-from-lux-research-explain-grid-based-energy-storage?cmpid=WindNL-Thursday-December15-2011.

Pfahl, Greg. 2010 (July 10). ITC or PTC for Your Renewable Energy Project? *Renewable Energy World*. http://www.renewableenergyworld.com/rea/news/article/2010/07/itc-or-ptc-for-your-renewable-energy-project

Pfeifenberger, Johannes. 2010 (October 12). Renewable Energy Development and Transmission Expansion—Who Benefits and Who Pays. http://www.brattle.com/_documents/UploadLibrary/Upload887.pdf

Pöyry. 2011 (March 30). Groundbreaking Study into the Impact of Wind and Solar Generation on Electricity Markets in North-West Europe. http://www.poyry.com/media/media_2.html?Id=1301471113.html&C=0&Y=2011.

Relich, Natalie. 2010 (December 29). A Profile of China's Involvement in the U.S Wind Energy Industry. *In News*. http://leadenergy.org/2010/12/a-profile-of-chinese-involvement-in-the-u-s-wind-energy-industry/.

Remington, K. 2010 (August 12). Laser-Based Wind Sensor Lets Turbines Grab More Power. *Windpower Engineering*. http://www.windpowerengineering.com/design/electrical/sensors/laser-based-lets-turbines-grab-more-power/.

———. 2010 (August 9). Ultrasonic Wind Sensors or Cup Anemometers? That is the Question. *Windpower Engineering*. http://www.windpowerengineering.com/construction/simulation/ultrasonic-wind-sensors-or-cup-anemometers/.

———. 2010 (August 7). 4-MW Direct Drive Headed Offshore. *Windpower Engineering*. http://www.windpowerengineering.com/featured/business-news-projects/4-mw-direct-drive-headed-offshore/.

Renewable Energy World. 2010 (November 22). Measuring Wind Turbine Noise. http://www.renewableenergyworld.com/rea/news/article/2010/11/measuring-wind-turbine-noise.

———. 2010 (December 1). Spanish Companies Plan a 15-MW Wind Turbine. http://www.renewableenergyworld.com/rea/news/article/2010/12/spanish-companies-plan-a-15-mw-wind-turbine.

———. 2011 (August 18). Price of Wind Lower Than Gas, Hydro in Brazil Auction. http://www.renewableenergyworld.com/rea/news/article/2011/08/price-of-wind-lower-than-gas-hydro-in-brazil-auction?cmpid=WNL-Friday-August19-2011.

RenewableUK. 2007. Noise from Wind Turbines—the Facts. http://www.bwea.com/ref/noise.html.

Rensselaer Polytechnic Institute. 2011 (January 28). Rensselaer Researcher Working To Make Smarter Wind Turbine Blades. http://news.rpi.edu/update.do?artcenterkey=2815.

Risø DTU. 2010 (August 17). A New Wind Turbine Test Centre at Østerild in North-Western Jutland to Ensure Denmark's Leading Position Within Wind Energy. http://www.risoe.dtu.dk/News_archives/News/2010/0817_Oesterild.aspx?sc_lang=en.

——— 2010 (February 17). Controllable Rubber Trailing Edge Flap May Ease Stress on Wind Turbine Blades. *Renewable Energy World*. http://www.renewableenergyworld.com/rea/news/article/2010/02/controllable-rubber-trailing-edge-flap-may-ease-stress-on-wind-turbine-blades.

Ritchie, Ed. 2010 (September-October). Tipping Point. *Distributed Energy*. 8(5): 38-42. http://www.distributedenergy.com/september--october-2010/tipping-point-windturbines.aspx.

Robb, Drew. 2010 (December 1). Finding Fault: Improving Wind Farm Availability. *Power*. http://www.powermag.com/business/3218.html.

Rock, Mary, and Parsons, Laura. 2010 (October). Offshore Wind Energy. Environmental and Energy Study Institute. http://www.eesi.org/files/offshore_wind_101310.pdf.

Rouda, Jennifer, and Yanez, Silvia. 2010. Connecting Renewable Energy to the Electric Transmission Grid: Trending Toward Innovative HVDC

Technology. *North American Clean Energy*. http://www.nacleanenergy.com/index.php?option=com_content&view=article&id=3676:connecting-renewable-energy-to-the-electric-transmission-grid-&catid=45:solar-feature-articles&Itemid=112.

Saugrain, Jean-Maxime. 2010 (October). Successfully Operating 200 kV Superconducting HVDC Transmission Highlights Reliability, Efficiency. *Utility Automation Engineering*. http://www.elp.com/index/display/article-display/3784736836/articles/utility-automation-engineering-td/volume-15/issue-10/features/successfully-operating-200-kv-superconducting-hvdc-transmission-highlights-reliability-efficiency.html.

SBI Energy. 2011 (February 3). Offshore Wind Farm Manufacturing Worldwide. SBWire. http://www.sbwire.com/press-releases/sbwire-76729.htm.

Schlez, Wolfgang, and Tindal, Andrew. Wind Farm Siting and Layout Design. World Wind Energy Association. http://www.wwindea.org/technology/ch02/en/2_4_2.html

Shankleman, Jessica. 2011 (May 11). Japan Ditches Nuclear Plans Raising Hopes for Wind Industry. *Business Green*. http://www.businessgreen.com/bg/interview/2069773/japan-ditches-nuclear-plans-raising-hopes-wind-industry.

Sharp, Duane. 2010 (December 20). In Depth: Profiling Engineering and Manufacturing in Canada's Wind Energy Sector. World.edu. http://world.edu/worldedu_posts/depth-profiling-engineering-manufacturing-canadas-wind-energy-sector/.

Shere, Jeremy. 2011 (January 6). Knowing Which Way the Wind Blows. *Renewable Energy World*. http://www.renewableenergyworld.com/rea/blog/post/2011/01/knowing-which-way-the-wind-blows?cmpid=WindNL-Thursday-January13-2011.

———. 2011 (March 15). Why People Hate Wind Turbines. *Renewable Energy World*. http://www.renewableenergyworld.com/rea/blog/post/2011/03/why-people-hate-wind-turbines??cmpid=WNL-Friday-March18-2011.

Song, Wayne. 2011 (April). Wind Turbine Market Outlook. *North American Clean Energy*. http://www.morganlewis.com/pubs/WindMarketOutlook.pdf.

Stant, Jeff. 2010 (August 26). In Harm's Way: Lack of Federal Coal Ash Regulations Endangers Americans and Their Environment. Environmental Integrity Project. http://www.environmentalintegrity.org/news_reports/documents/INHARMSWAY_FINAL.pdf.

Starr, Scott. 2011 (February). Newer Safety Standards Include Precautions for Turbine Fires. *North American Windpower*. 8 (1): 64, 68, 70.

St. John, Jeff. 2011 (November 23). Viridity, Enbala Try Negawatts to Balance Pennsylvania's Grid. Greentech Media. http://www.greentechmedia.com/articles/read/viridity-enbala-try-negawatts-to-balance-pennsylvanias-grid/.

Susman, Jacob. The "Coming of Age" of Community Wind. Electric Energy Online. http://www.electricenergyonline.com/?page=show_article&mag=61&article=470.

SustainableBusiness.com News. 2011 (February 8). Wind Turbine Prices Fall to Lowest Point in Recent Years. http://www.sustainablebusiness.com/index.cfm/go/news.display/id/21839.

Sykes, Benj. 2010. Supporting Innovation. *Carbon Trust*. http://www.taplondon.co.uk/bwea31/files/benjaminsykes.pdf.

Titchen, John. 2010 (May). Direct Drive Permanent Magnet Wind Turbine Experience. Goldwind. http://www.cleanenergyweek.com.au/mediaObject/events/2010-conference/presentations/1500-Titchen/original/1500%20Titchen.pdf.

Totaro, Philip. 2011 (February 18). Five Trends to Watch in Wind. Greentech Media. http://www.greentechmedia.com/articles/read/guest-post-five-trends-to-watch-in-wind.

Trabish, Herman K. 2010 (May 4). Wind Turbines, the Next Generation. Greentech Media. http://www.greentechmedia.com/articles/read/wind-tubines-the-next-generation.

———. 2010 (June 10). Deepwater Drilling for U.S.? Try Deepwater Wind. Greentech Media. http://www.greentechmedia.com/articles/read/deep-water-drilling-no-deep-water-wind.

———. 2010 (July 8). Measuring the Wind and the Sun. Greentech Media. http://www.greentechmedia.com/articles/read/measuring-the-wind-and-the-sun.

———. 2010 (August 13). The Birth of a U.S. Wind Power Manufacturing Industry. Greentech Media. http://www.greentechmedia.com/articles/print/the-birth-of-a-u.s.-wind-power-manufacturing-industry/.

———. 2010 (August 31). Advanced Wind Measurement Technologies Take Hold in the Industry. Greentech Media. http://www.greentechmedia.com/articles/read/advanced-wind-measurement-technologies-take-hold-in-the-industry.

———. 2010 (September 1). Undermining the Critics of Wind Power. Greentech Media. http://www.greentechmedia.com/articles/read/undermining-the-critics-of-wind-power/.

———. 2010 (September 10). Wind's Future Ally: Better Transmission. http://www.greentechmedia.com/articles/read/winds-future-ally-better-transmission.

———. 2010 (October 1). Operations and Maintenance: Keys to Wind Farm Profitability. Greentech Media. http://www.greentechmedia.com/articles/read/operations-and-maintenance-the-key-wind-farm-profitability.

———. 2010 (October 7). Energy Storage for the Grid: Better Than the Holy Grail. Greentech Media. http://www.greentechmedia.com/articles/read/energy-storage-for-the-grid-better-than-the-holy-grail/.

———. 2010 (November 18). Will Goldwind Rule the Wind World? Greentech Media. http://www.greentechmedia.com/articles/read/will-goldwind-rule-the-wind-world/.

———. 2010 (November 24). Offshore Wind Moves to Full Speed Ahead. Greentech Media. http://www.greentechmedia.com/articles/read/offshore-wind-moves-to-full-speed-ahead.

———. 2010 (December 9). The Texas CREZ Plan to Get Wires for Wind. http://www.greentechmedia.com/articles/read/the-texas-crez-plan-to-get-wires-for-wind.

———. 2010 (December 24). Federal Regulators Approve 100 Percent Cost Sharing for New Transmission. https://www.greentechmedia.com/articles/read/federal-regulators-approve-100-percent-cost-sharing-for-new-transmission.

———. 2010 (December 29). Wind's 2010 Top Ten: Bust and Building, Coming and Going. Greentech Media. http://www.greentechmedia.com/articles/print/winds-2010-top-ten-bust-and-building-coming-and-going/.

———. 2010 (December 30). The Emerging Opportunity in Offshore Wind Vessels. Greentech Media, http://www.greentechmedia.com/articles/read/the-emerging-opportunity-in-offshore-wind-vessels.

———. 2011 (January 11). The Money in the Wind. Greentech Media. http://www.greentechmedia.com/articles/read/the-money-in-the-wind/.

———. 2011 (January 14). Wind Turbines Get Sensitive. Greentech Media. http://www.greentechmedia.com/articles/print/wind-turbines-get-sensitive/.

———. 2011 (January 17). Higher Education Gets Wind's Approval. Greentech Media. http://www.greentechmedia.com/articles/read/higher-education-gets-winds-approval/.

———. 2011 (January 21). Flow, Good Vibrations and Other Wind Machine Profit Opportunities. Greentech Media. http://www.greentechmedia.com/articles/read/flow-good-vibrations-and-other-wind-machine-profit-opportunities.

———. 2011 (January 27). When Solar Power Plants and Wind Farms Talk to the Grid. Greentech Media. http://www.greentechmedia.com/articles/read/when-solar-power-plants-and-wind-farms-talk-to-the-grid.

———. 2011 (January 28). The Transmission Quandary of California. http://www.greentechmedia.com/articles/read/the-transmission-quandary-of-california.

———. 2011 (February 3). The Truth about the Cape Wind Turbines. Greentech Media. http://www.greentechmedia.com/articles/read/the-truth-about-the-cape-wind-turbines.

———. 2011 (February 15). U.S. Wind Tech Moving into China? Greentech Media. http://www.greentechmedia.com/articles/read/a-u.s.-wind-company-moving-into-china/.

———. 2011 (February 21). Can Raytheon Keep Wind Farms From Threatening National Security? Greentech Media. http://www.greentechmedia.com/articles/read/can-raytheon-keep-wind-farms-from-threatening-national-security/.

———. 2011 (February 24). Will Google's Offshore Wind Transmission Be Slowed by Regulatory Red Tape? Greentech Media. http://www.greentechmedia.com/articles/read/will-googles-offshore-wind-transmission-be-slowed-by-regulatory-red-tape.

———. 2011 (March 8). Will China Take Over the U.S. Wind Market? Greentech Media. http://www.greentechmedia.com/articles/read/why-is-china-coming-for-u.s.-wind/.

———. 2011 (March 15). Have Small Wind Manufacturers Exploited Loopholes in California Rebates? Greentech Media. http://www.greentechmedia.com/articles/read/have-small-wind-manufacturers-exploited-loopholes-in-california-rebates.

———. 2011 (March 24). Why Aren't Those Wind Turbines Turning? Greentech Media. http://www.greentechmedia.com/articles/read/why-arent-those-wind-turbines-turning/.

———. 2011 (April 11). At 35 Percent of New Capacity, Wind Moves from Alternative to Mainstream. Greentech Media. http://www.greentechmedia.com/articles/read/is-wind-still-an-alternative-energy/.

———. 2011 (April 19). Goldwind: Will Wind Be China's Next Big Export to the US? Greentech Media. http://www.greentechmedia.com/articles/read/goldwind-will-wind-be-chinas-next-big-export-to-u.s/.
———. 2011 (April 27). Psst—Hey Buddy, Wanna Build a Wind Farm? Greentech Media. http://www.greentechmedia.com/articles/read/psst-hey-buddy-wanna-build-a-wind-farm/.
———. 2011 (May 18). How Will the California System Operator Cope with 33% Renewables? Greentech Media. http://www.greentechmedia.com/articles/read/how-will-the-california-system-operator-cope-with-33-renewables/.
———. 2011 (May 20). How Does U.S. Wind Keep Growing? Greentech Media. http://www.greentechmedia.com/articles/read/how-does-u.s.-wind-keep-growing/.
———. 2011 (May 23). Financial Tools and Tourism: Some Ideas for US Wind. Greentech Media. http://www.greentechmedia.com/articles/read/three-solutions-for-wind/.
———. 2011 (May 24). Smackdown: Wind vs. Washington State Grid Operator Over Renewable Integration. Greentech Media. http://www.greentechmedia.com/articles/read/smackdown-wind-vs-washington-state-grid-operator-over-renewable-integration/.
———. 2011 (June 10). From Oil and Gas to Offshore Renewables. Greentech Media. http://www.greentechmedia.com/articles/read/from-oil-and-gas-to-offshore-renewables/.
———. 2011 (June 15). Will German Offshore Wind RAVE On? Greentech Media. http://www.greentechmedia.com/articles/read/will-german-offshore-wind-rave-on/.
———. 2011 (July 21). What Do Wind Trends Show? Three Cents per Kilowatt-Hour. Greentech Media. http://www.greentechmedia.com/articles/read/what-do-winds-cost-price-and-performance-trends-show-three-cents-per-kilowa/.
———. 2011 (August 8). Can Germany's North Sea Winds Blow Away Nuclear? Greentech Media. http://www.greentechmedia.com/articles/read/can-germanys-north-sea-winds-blow-away-nuclear/.
———. 2011 (August 18). How Wind Helps Coal Plants. Greentech Media. http://www.greentechmedia.com/articles/read/how-wind-helps-coal-plants/.
———. 2011 (September 15). Can Wind Generate Electricity at $0.04 per Kilowatt-Hour? Greentech Media. http://www.greentechmedia.com/articles/read/how-can-wind-generate-electricity-at-four-cents-per-kilowatt-hour/.
———. 2011 (September 16). A Tale of Intrigue and IP Theft in the Wind Market. Greentech Media. http://www.greentechmedia.com/articles/read/american-superconductor-files-suit-against-china-wind-giant-sinovel/.
———. 2011 (September 19). Is Distributed Wind the Answer? (Part Three). Greentech Media. http://www.greentechmedia.com/articles/read/is-distributed-wind-the-answer-part-three/.
———. 2011 (September 26). Wind Power: AMSC Employee Pleads Guilty to Passing Code to Sinovel. Greentech Media. http://www.greentechmedia.com/articles/read/amsc-employee-pleads-guilty-to-passing-code-to-sinovel/.

———. 2011 (October 6). Where Is the Best Place to Put a Wind Farm? Greentech Media. http://www.greentechmedia.com/articles/read/where-is-the-best-place-to-put-a-wind-farm/.

———. 2011 (November 17). Wind, National Defense, and Radar Make Peace. Greentech Media. http://www.greentechmedia.com/articles/read/wind-and-national-defense-make-peace/.

———. 2011 (December 13). Grid Integration, Part 2: The Rubik's Cube of Renewable Energy. Greentech Media. http://www.greentechmedia.com/articles/read/grid-integration-part-2-the-rubiks-cube-of-renewable-energy/.

———. 2012 (January 6). Floating the Future of Offshore Wind. Greentech Media. http://www.greentechmedia.com/articles/read/floating-the-future-of-offshore-wind/solar-wind-hybrid-goes-down-another-goes-up/.

———. 2012 (January 25). Wind Tower Reliability Means Being Proactive in O&M. Greentech Media. http://www.greentechmedia.com/articles/read/Wind-Tower-Reliability-Means-Being-Proactive-in-OM/.

———. 2012 (January 31). One Solar-Wind Hybrid Goes Down, Another Goes Up. Greentech Media. http://www.greentechmedia.com/articles/read/one-

———. 2012 (February 3). Multinationals Invade Scotland for Offshore Wind Wealth. Greentech Media. http://www.greentechmedia.com/articles/read/multinationals-invade-scotland-for-offshore-wind-wealth/.

Turkel, Tux. 2011 (May 1). Wind Turbine Noise An Issue, Report Finds. The *Portland Press Herald*. http://www.pressherald.com/news/wind-turbine-noise-an-issue-report-finds_2011-05-01.html.

Tweed, Katherine. 2011 (March 14). Demand Response and Renewables: Too Good to Be True? Greentech Media. http://www.greentechmedia.com/articles/read/demand-response-and-renewables-too-good-to-be-true.

———. 2011 (March 17). Demand Response, Rejoice! Negawatts Put on Equal Footing with Generation. Greentech Media. http://www.greentechmedia.com/articles/read/demand-response-rejoice-negawatts-put-on-equal-footing-with-generation.

———. 2010 (May 5). Tehachapi Renewable Transmission Project Completes Phase One. http://www.greentechmedia.com/articles/read/phase-one-completed-in-tehachapi-renewable-transmission-project.

Types of Wind Turbines. Centurion Energy LLC. http://centurionenergy.net/types-of-wind-turbines.

Urgo, Jacqueline L. 2011 (June 12). Atlantic City Wind Turbines Become a Tourist Attraction. *The Philadelphia Inquirer*. http://articles.philly.com/2011-06-12/news/29650149_1_turbines-first-commercial-wind-project-wind-energy.

Vaira, Douglas. 2011. Win-Win for Wind and Wildlife. The Nature Conservancy. http://www.nature.org/ourinitiatives/urgentissues/conservationlands/conservation-lands-win-win-for-wind-and-wildlife.xml?s_intc=sp2.

Varma, K.J.M. 2011 (February 17). Wind Power Wasted in China Due to Poor Grid Connectivity. http://www.powergenworldwide.com/index/display/wire-news-display/1361827140.html.

Vautard, Robert, et al. 2010 (October 17). Northern Hemisphere Atmospheric Stilling Partly Attributed to an Increase in Surface Roughness. *Nature Geoscience*. http://www.nature.com/ngeo/journal/v3/n11/full/ngeo979.html.

Webb, Chris. 2011 (March 22). Small Wind: Market Update. *Renewable Energy World*. http://www.renewableenergyworld.com/rea/news/article/2011/03/small-wind-market-update?cmpid=WNL-Wednesday-March23-2011.
Webb, Jim. 2011. Interviewed by Rebecca Busby. E-mail message to author. June 28.
Weiner, Jon. 2010 (May 17). Schooling Fish Offer New Ideas for Wind Farming. Caltech University. http://media.caltech.edu/press_releases/13347.
Weiss, Daniel J., and Caperton, Richard W. 2011 (April 7). Protecting Big Oil at the Expense of Consumers. http://www.greentechmedia.com/articles/read/protecting-big-oil-at-the-expense-of-consumers/.
Wesoff, Eric. 2011 (March 9). Update: Gigawatt-Scale Energy Storage by Rubenius. Greentech Media. http://www.greentechmedia.com/articles/read/update-gigawatt-scale-energy-storage.
———. 2011 (March 10). MIT's Star Prof. Don Sadoway on Innovations in Energy Storage. Greentech Media. http://www.greentechmedia.com/articles/read/mits-don-sadoway-on-energy-storage-and-his-liquid-battery.
———. 2011 (March 24). Xtreme Blends 15MW of Energy Storage With Wind. Greentech Media. http://www.greentechmedia.com/articles/read/xtreme-power-on-hawaii/.
———. 2011 (April 18). Ionex Has a 1 MW Battery Ready for the Grid. Greentech Media. http://www.greentechmedia.com/articles/read/ionex-has-a-1-mw-battery-ready-for-the-grid/.
———. 2011 (August 17). Beacon Power's Spin on Energy Storage: It's a Service. Greentech Media. http://www.greentechmedia.com/articles/read/beacon-power-a-new-spin-on-energy-storage/.
———. 2011 (September 13). AES Brings Real Utility-Scale Energy Storage to the Grid. Greentech Media. http://www.greentechmedia.com/articles/read/aes-bringing-real-utility-scale-energy-storage-to-the-grid/.
———. 2011 (October 4). Subsidies for Oil, Gas and Nuclear vs. Renewables. Greentech Media. http://www.greentechmedia.com/articles/read/Subsidies-For-Oil-Gas-Nuclear-vs.-Renewables/.
———. 2010 (November 9). Gravity Power's New Take on Pumped-Hydro Energy Storage. Greentech Media. http://www.greentechmedia.com/articles/read/Gravity-Power.
Wharton, S., and Lundquist, J. K. 2010 (February 23). Atmospheric Stability Impacts on Power Curves of Tall Wind Turbines—an Analysis of a West Coast North American Wind Farm. LLNL-TR-424425. http://atoc.colorado.edu/~jlundqui/wharton_lundquist2010LLNLTR424425.pdf.
Wickless, Andy. 2011 (July 27). Financing Offshore Wind Farms in the U.S. Navigant Consulting. http://www.renewableenergyworld.com/rea/news/article/2011/07/financing-offshore-wind-farms-in-the-u-s?cmpid=WNL-Wednesday-July27-2011.
Widman, Miriam. 2012 (February 6). Germany Invests in Hydrogen Technology for Renewables Storage, Vehicles. Renewable Energy World. http://www.renewableenergyworld.com/rea/news/article/2012/02/germany-invests-in-hydrogen-technology-for-renewable-storage-vehicles?cmpid=WNL-Wednesday-February8-2012.

Williams, Andrew, 2011 (March 17). Seeking Wind Power Innovation: Europe's Largest Wind Turbine Test Facility Now Open. *Renewable Energy World*. http://www.renewableenergyworld.com/rea/news/article/2011/03/seeking-wind-power-innovation-europes-largest-wind-turbine-test-facility-now-open.

———. 2011 (April 15). Direct Drive Turbines: Developers Square Up to Large-Scale Offshore Turbine Challenge. Wind Energy Update. http://social.windenergyupdate.com/industry-insight/direct-drive-turbines-developers-square-large-scale-offshore-turbine-challenge?utm_source=newsletter&utm_medium=email&utm_campaign=weunewsletter1904.

——— 2011 (May 3). A Buoyant Future for Floating Wind Turbines? *Renewable Energy World*. http://www.renewableenergyworld.com/rea/news/article/2011/05/a-buoyant-future-for-floating-wind-turbines?cmpid=WNL-Wednesday-May4-2011.

———. 2011 (July 15). Hedging Against Wind Variability: A Risky Business? *Renewable Energy World*. http://www.renewableenergyworld.com/rea/news/article/2011/07/hedging-against-wind-variability-a-risky-business??cmpid=WNL-Wednesday-July20-2011.

Wilson, Ryan D. 2011 (January 20). Wind Farms Built from the Ground Up. *Clay Center Dispatch*. http://www.zwire.com/site/news.cfm?newsid=20454645&BRD=1160&PAG=461&dept_id=190958&rfi=6.

Wind Energy Math Calculations: Calculating the Tip Speed Ratio of Your Wind Turbine. KidWind. http://www.kidwind.org.

Wind Energy Update. 2011 (May 2). Report: Offshore Wind Investment to Surge. http://social.windenergyupdate.com/weekly-brief/weekly-intelligence-brief-april-26-may-02.

Wind Today. 2010 (July). GE Energy Services Increases Wind Farm Profitability by Reducing Gearbox Maintenance Costs with Condition-Based Maintenance. http://www.windtoday.net/info/articles.html?ID=96752.

Yuanyuan, Liu. 2011 (April 26). China's Wind Power Sector Expected to See Tremendous Growth. *Renewable Energy World*. http://www.renewableenergyworld.com/rea/news/article/2011/04/chinas-wind-power-sector-expected-to-see-tremendous-growth??cmpid=WNL-Friday-April29-2011.

———. 2011 (May 19). China Seeks Big Wind in a Push to Standardize Project Size. *Renewable Energy World*. http://www.renewableenergyworld.com/rea/news/article/2011/05/china-to-standardize-wind-power-projects.

———. 2011 (September 22). China Releases Technical Standards for Wind Power Industry. *Renewable Energy World*. http://www.renewableenergyworld.com/rea/news/article/2011/09/china-releases-technical-standards-for-wind-power-industry?cmpid=WNL-Friday-September23-2011.

———. 2012 (January 12). China Increases Target for Wind Power Capacity to 1,000 GW by 2050. Renewable Energy World. http://www.renewableenergyworld.com/rea/news/article/2012/01/china-increases-target-for-wind-power-capacity-to-1000-gw-by-2050?cmpid=WNL-Friday-January6-2012.

Zervos, Arthouros. 2009 (March 19). Large Scale Integration of Wind Energy into Electricity Grids. Global World Energy Council. http://www.gwec.net/index.php?id=146.

Zipp, Kathleen. 2010 (October 1). Wind Power—More Efficient Than You Thought. *Windpower Engineering*. http://www.windpowerengineering.com/policy/wind-power-more-efficient-than-you-thought/.

———. 2010 (October 3). A Safer Way to Reach High Places. *Windpower Engineering*. http://www.windpowerengineering.com/maintenance/safety/a-safer-way-to-reach-high-places.

Books

Children's Books

Hansen, Amy S. 2010. *Wind Energy: Blown Away!* New York: The Rosen Publishing Group, Inc..

Morris, Neil. 2007. *Wind Power*. Smart Apple Media: North Mankato, MN.

Parks, Peggy J. 2010. *Wind Power*. San Diego: Reference Point Books.

Technical books

Asmus, Peter. 2000 (December). *Reaping the Wind: How Mechanical Wizards, Visionaries, and Profiteers Helped Shape Our Energy Future*. Washington, DC: Island Press. (ISBN 978-1559637077).

Baxter, Richard. 2005. *Energy Storage: A Nontechnical Guide*. Tulsa: PennWell. (ISBN 978-1-59370-027-0).

Chambers, Ann. 2003. *Renewable Energy in Nontechnical Language*. Tulsa: PennWell. (ISBN 978-1-59370-005-8).

Eggleston, David M., and Stoddard, Forrest S. 1987. *Wind Turbine Engineering Design*. New York: Van Nostrand Reinhold Co. Inc. (ISBN 0-442-22195-9).

Gipe, Paul. 2009 (April). *Wind Energy Basics Revised: A Guide to Home- and Community-Scale Wind Energy Systems*. White River Junction, VT: Chelsea Green Publishing. (ISBN 978-1-06358-030-4). http://www.chelseagreen.com/bookstore/item/wind_energy_basics_second_edition:paperback.

Golding, E. W. 1976. *The Generation of Electricity by Wind Power*. New York: Halsted Press. (ISBN 978-0470149867).

Jamieson, Peter. 2011 (July 11). *Innovation in Wind Turbine Design*. Hoboken, NJ: Wiley. (Print ISBN 9780470699812; Online ISBN 9781119975441).

Manwell, James F., McGowan, Jon G., and Rogers, Anthony L. 2010 (February 23). *Wind Energy Explained: Theory, Design and Application*. 2nd edition. Hoboken, NJ: Wiley. (ISBN-10: 0470015004; ISBN-13: 978-0470015001). http://www.wiley.com/WileyCDA/WileyTitle/productCd-0470015004.html

Melnyk, Markian M.W., and Andersen, Robert M. 2009. *Offshore Power: Building Renewable Energy Projects in U.S. Waters*. Tulsa: PennWell. (ISBN 978-1-59370-211-3).

Park, Jack. 1981. *The Wind Power Book*. Palo Alto, CA: Cheshire Books. (ISBN 0-917352-05-X).

Righter, Robert W. 2011 (December 3). *Windfall: Wind Energy in America Today*. Norman, OK: University of Oklahoma Press. (ISBN 9780806141923).

Robinson, Abraham. 1974. *Non-Standard Analysis*. Princeton, NJ: Princeton University Press. (ISBN: 9780691044903). http://press.princeton.edu/titles/5812.html.

Shields, Craig. 2010. *Renewable Energy: Facts and Fantasies*. 2GreenEnergy.com. http://2greenenergy.com/renewable-energy-facts-fantasies-ebook.

Thomsen, Kurt E. 2011 (November). *Offshore Wind: A Comprehensive Guide to Successful Wind Farm Installation*. London: Elsevier. (ISBN-13: 9780123859365; ISBN-10: 0123859360).

Index

A

B

C

D

E

F

I

J

K

L

M

N

O

R

S

T

V

W

X

Y

Z